INTRODUCTION TO MAINTENANCE ENGINEERING

INTRODUCTION TO MAINTENANCE ENGINEERING

MODELING, OPTIMIZATION, AND MANAGEMENT

Mohammed Ben-Daya
King Fahd University of Petroleum & Minerals, Dhahran, Saudi Arabia

Uday Kumar
Luleå University of Technology, Sweden

D.N. Prabhakar Murthy
The University of Queensland, Brisbane, Australia

Registered Office
John Wiley & Sons, Ltd, The Atrium, Southern Gate, Chichester, West Sussex, PO19 8SQ, United Kingdom

For details of our global editorial offices, for customer services and for information about how to apply for permission to reuse the copyright material in this book please see our website at www.wiley.com.

Library of Congress Cataloging-in-Publication Data

Names: Ben-Daya, Mohammed, author. | Kumar, Uday, author. | Murthy, D. N. P., author.
Title: Introduction to maintenance engineering : modeling, optimization, and management / Mohammed Ben-Daya, Dhahran, Saudi Arabia, Uday Kumar, Lulea, Sweden, D. N. Prabhakar Murthy, Brisbane, Australia.
Description: Hoboken : John Wiley & Sons, Inc., 2016. | Includes bibliographical references and index.
Identifiers: LCCN 2015036759 | ISBN 9781118487198 (cloth)
Subjects: LCSH: Maintenance.
Classification: LCC TS174 .B455 2016 | DDC 620/.0046–dc23 LC record available at http://lccn.loc.gov/2015036759

A catalogue record for this book is available from the British Library.

Set in 10/12pt Times by SPi Global, Pondicherry, India

1   2016

To our wives *Faouzia*, *Renu*, and *Jayashree* for their patience and understanding.

Contents

Preface

The Metal Age, which started around 3000 BC, saw the appearance of metal tools and the evolution of civilizations in different regions of the world. This led to the development of tools for warfare and farming, and the building of roads, boats, houses, and so on. The Industrial Revolution created new mechanical devices and machines. This, in turn, led to the development of the electrical, hydraulic, and other devices and equipment that are used nowadays in nearly all sectors – farming, processing, mining, manufacturing, transport, communication, and so on, all with specific needs for maintenance. The construction of infrastructures (such as electricity, water, gas and sewage networks, dams, roads, railways, bridges, etc.) resulted in new maintenance challenges in order to keep them operational.

The reason why engineered objects (be they products, plants, or infrastructures) need maintenance is that every object is unreliable, in the sense that it degrades with age and/or usage and ultimately fails when it is no longer capable of discharging its function. Maintenance actions compensate for the inherent unreliability of an object and may be grouped broadly into two categories: (i) preventive maintenance (PM) to control the degradation process and (ii) corrective maintenance (CM) to restore a failed object to the operational state.

Maintenance actions were mainly of the corrective type until the middle of the last century – the adage being, "don't touch if it ain't broke." Also, preventive actions were viewed as money wasted. Maintenance was done by trained technicians who were very good at fixing failures (often fondly referred to as "grease-pit monkeys" in the popular literature). Maintenance was an afterthought in the design of new objects and was simply viewed as an unavoidable cost to be incurred after these objects were built and put into operation.

There was a dramatic change after the Second World War. Reliability evolved as a new discipline, and the theory of reliability dealt with various aspects, such as (i) the science of degradation, (ii) the use of statistical methods to assess reliability, and (iii) mathematical models to predict item failures and the importance of preventive maintenance. Investing in preventive maintenance lowers the cost of corrective maintenance but results in additional costs. Operational Research (the application of scientific methods to solve industrial problems) focused on models to decide on the optimal level of preventive maintenance to achieve a proper trade-off between the costs of corrective and preventive maintenance.

The next stage of evolution was the emergence of alternative approaches to the maintenance of objects in different industry sectors. Two methods that have been used extensively across the world are: (i) reliability-centered maintenance (RCM), which had its origins in the airline industry, and (ii) total productive maintenance (TPM), which had its origins in manufacturing. Advances in technology (sensors, data collection, computers, and communication) have resulted in the evolution of condition-based maintenance (CBM) and e-maintenance.

Maintenance in the twenty-first century has moved from the trial and error approach of the technicians of the early twentieth century to a multi-disciplinary subject with science, engineering, and technology as its foundations. A maintenance engineer is a professional engineer with this background, and so is different from a maintenance technician, who is skilled in carrying out specified maintenance tasks. An understanding of the basic principles of management is also an important element of modern maintenance practice. Furthermore, maintenance engineers/managers need advanced techniques for maintenance data analysis and also need to build models to assist effective maintenance decision making. The need to interact with other disciplines (such as law, accounting, etc.) is also needed by senior-level maintenance managers. The figure below shows this in a schematic format.

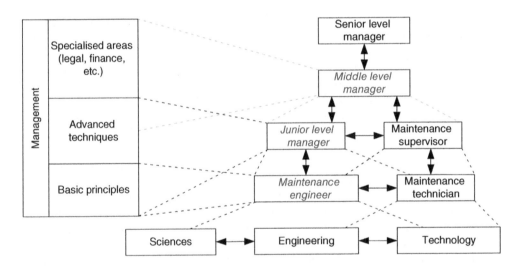

Over the last few decades, hundreds of books on maintenance have appeared in print. The authors are not aware of any book for use in a first course on maintenance that takes the comprehensive view needed for the twenty-first century. This book aims to fill this gap and is meant for use as a textbook on maintenance at the senior undergraduate or graduate level in engineering programs. The unique features of the book are as follows:

- It provides a unified approach linking science, engineering, technology, mathematics and statistics, and management.
- It focuses on concepts, tools, and techniques.
- It links theory and practice using real, illustrative cases involving products, plants, and infrastructures (many chapters have three sections dealing with specific issues for these different types of items).

The book provides a good foundation for a new graduate to work as a maintenance engineer and to build a career by moving through the ranks of junior- and middle-level management responsible for maintaining the various types of engineered objects. It can also be used as a reference book by practicing maintenance engineers/managers to understand the modern knowledge-based approach to maintenance. The book can also be used as a starting point for researchers in maintenance.

The book is flexible enough to be used as a textbook in various undergraduate and graduate programs. A suggested sequence for four programs is as follows:

- Undergraduate level
  - Industrial engineering programs: Chapters 1–7, 8–9, 17, 19–22
  - Other engineering programs: Chapters 1–4, 6–7, 17, 19, 22
- Graduate level
  - Maintenance engineering programs: Chapters 1–4, 6, 8–12, 13–20
  - Engineering management programs: Chapters 1–7, 17–22

Each chapter deals with several topics. The book is suitable for one or two full courses or part of one or more courses depending on the topics selected.

The background needed to understand and fully appreciate the contents of the book is an understanding of the basic concepts from the following disciplines:

- Mathematics;
- Physics and Chemistry;
- Engineering (covering design, manufacturing, construction, and operations);
- Probability and Statistics.

This book evolved through a joint partnership between three researchers/educators from three different continents and is based on the experiences of the authors in teaching and research in maintenance over the last three decades.

Mohammed Ben-Daya, *Dhahran, Saudi Arabia*
Uday Kumar, *Lulea, Sweden*
D.N. Prabhakar Murthy, *Brisbane, Australia*

Acknowledgments

Many colleagues and ex-doctoral students from several universities have helped in different ways. A special thanks to the following people:

- Dr Nat Jack for discussions and proof reading;
- Dr Mohamed Rezaul Karim from the University of Rajasahi in Bangladesh for carrying out the data analysis and modeling (using Minitab) for the examples in Part B of the book;
- Dr Iman Arasteh Khouy from Lulea University of Technology in Sweden for writing the second case study of Chapter 23;
- Professor Alireza Ahmadi from Lulea Technical University in Sweden for assistance with the first case study of Chapter 23.

Others who have been very helpful include Professor Renyan Jiang from Changsha University in China and Dr Sami Elferik from King Fahd University of Petroleum & Minerals. The authors are grateful for their help and contributions.

The authors wish to acknowledge the support from King Fahd University of Petroleum & Minerals (project # IN 121004), Luleå University of Technology, and the University of Queensland. Finally, a special thanks to Anne Hunt, Tom Carter, and Clive Lawson from the editorial staff of Wiley for their support and encouragement.

Abbreviations

A–D	Anderson and Darling
AE	Acoustic emission
AI	Artificial Intelligence
AIC	Akaike information criterion
ASCE	American Society of Civil Engineers
BIT	Built-in-testing
BOT	Build, operate, transfer
CAPEX	Capital expenditure
CBM	Condition-based maintenance
CC	Cycle cost
CEN	Comité Européenne de Normalisation (French) European Committee for Standardization
CEO	Chief Executive Officer
CL	Cycle length
CM	Corrective maintenance
CMMS	Computerized maintenance management system
CPM	Critical path method
DBFO	Design, build, finance, and operate
DIKW	Data, information, knowledge, and wisdom
DoD	Department of Defense
DOM	Design out maintenance
DTC	Diagnostic trouble code
EAC	Equivalent annual cost
ECC	Expected cycle cost
ECL	Expected cycle length
EDF	Empirical distribution function
EMMS	e-maintenance management system
EN	Europäische Norm (German): European Standard
EPP	Exponential probability plot

ET	Electromagnetic testing
FF	Failure finding
FHWA	Federal Highway Authority
FM	Facilities management
FMEA	Failure mode and effects analysis
FMECA	Failure mode, effects, and criticality analysis
FMMEA	Failure modes, mechanisms, and effects analysis
FRW	Free repair/replacement warranty
FT	Fault tree
FTA	Fault tree analysis
GPR	Ground-penetrating radar
GPRS	General packet radio service
HPP	Homogeneous Poisson process
HSE	Health, safety, and environmental
ICT	Information and communication technology
IEC	International Electrotechnical Commission
IEEE	Institute of Electrical and Electronics Engineers
IEV	International electrotechnical vocabulary
IMU	Intelligent monitoring unit
IQR	Inter-quartile range
ISO	International Standards Organization
JIT	Just-in-time
KPI	Key performance indicator
K–S	Kolmogorov and Smirnoff
LC	Lease contract
LCC	Life cycle cost
LCC_C	Life cycle cost (customer perspective)
LCC_M	Life cycle cost (manufacturer perspective)
LCCA	Life cycle cost analysis
LIDAR	Laser imaging detection and ranging
LORA	Level of repair analysis
LT	Leak testing
LTM	Laser testing method
M&R	Maintenance and rehabilitation
MCF	Mean cumulative function
MEMS	Micro-electromechanical sensor
MFL	Magnetic flux linkage
MGT	Million gross tons
MIL-HDBK	Military handbook
ML	Maximum likelihood
MLE	Maximum likelihood estimate
MMS	Maintenance management system
MPI	Maintenance performance indicator
MPM	Maintenance performance metric
	Maintenance performance management
MPMS	Maintenance performance management system

MPT	Magnetic particle testing
MTBF	Mean time between failures
MTTF	Mean time to failure
MTTR	Mean time to repair
NASA	National Aeronautic and Space Administration
NDT	Non-destructive testing
NFF	No fault found
NHPP	Non-homogeneous Poisson process
NN	Neural network
NPD	New product development
NPV	Net present value
NRT	Neutron radiographic testing
NTC	Negative temperature coefficient
O&M	Operation and maintenance
OBD	On-board diagnostics
OEE	Overall equipment effectiveness
OEM	Original equipment manufacturer
OOR	Out-of-round
OPEX	Operating expenditure
OPG	One-pass grinding
OR	Operations Research
PDCA	Plan, do, Check, Act
PFI	Private financing initiative
PI	Performance indicator
PLC	Product life cycle
PM	Preventive maintenance
PMS	Performance management system
	Production management system
PPE	Property, plant, and equipment
PPP	Public–private partnership
PT	Penetrant testing
PTC	Positive temperature coefficient
R&D	Research and development
R&M	Reliability and maintainability
RAIB	Rail Accident Investigation Branch
RAM	Reliability, availability, and maintainability
RBD	Reliability block diagram
RBM	Risk-based maintenance
RCD	Residual current device
RCF	Rolling contact fatigue
RCM	Reliability-centered maintenance
RFID	Radio frequency identification
RIW	Reliability improvement warranty
RLC	Regional logistic center
ROCOF	Rate of occurrence of failures
RP	Renewal process

RT	Radiographic testing
RTD	Resistance temperature detector
RTF	Run to failure
SAE	Society of Automotive Engineers
SCC	Stress corrosion cracking
SOLE	The International Society of Logistics
SRB	Solid rocket booster
TAM	Turn around maintenance
TLC	Technology life cycle
TPM	Total productive maintenance
TQM	Total quality management
TR	Thermal/infrared testing
UHF	Ultra high frequency
UT	Ultrasonic testing
VA	Vibration analysis
VHF	Very high frequency
VT	Visual testing
WLAN	Wireless local area network
WPAN	Wireless personal area network
WPP	Weibull probability plot
WSN	Wireless sensor network
WT	Wind turbine
WWAN	Wireless wide area network

1

An Overview

Learning Outcomes

After reading this chapter, you should be able to:

- Define maintenance and explain its importance from a strategic business perspective;
- List the three main aspects of maintenance;
- Provide a classification of engineered objects;
- Describe reliability and non-reliability performance measures of engineered objects;
- Describe the factors that affect performance degradation;
- Recognize the consequences of poor maintenance;
- Describe the main categories of maintenance costs;
- Explain that there is a trade-off between preventive maintenance effort and maintenance costs;
- Explain that there are maintenance decision-making problems at the strategic, tactical, and operational levels;
- Describe the evolution of maintenance over time and the new trends;
- Understand the structure of the book.

Introduction to Maintenance Engineering: Modeling, Optimization, and Management, First Edition.
Mohammed Ben-Daya, Uday Kumar, and D.N. Prabhakar Murthy.
© 2016 John Wiley & Sons, Ltd. Published 2016 by John Wiley & Sons, Ltd.

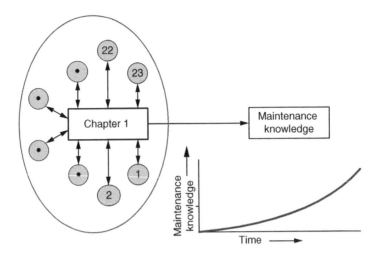

1.1   Introduction

Modern societies use a range of engineered objects for many different purposes. The objects are designed and built for specific functions. These include a variety of products (used by households, businesses, and government in their daily operations), plants, and facilities (used by businesses to deliver goods and services) and a range of infrastructures (networks such as rail, road, water, gas, electricity; dams, buildings, etc.) to ensure the smooth functioning of a society.

Every engineered object is unreliable in the sense that it degrades with age and/or usage and ultimately fails. A dictionary definition of failure is "falling short in something expected, attempted, desired, or in some way deficient or lacking." From an engineering point of view, an engineered object is said to have failed when it is no longer able to carry out its intended function for which it was designed and built. Failures occur in an uncertain manner and are influenced by several factors such as design, manufacture (or construction), maintenance, and operation. In addition, the human factor is also important in this context.

The consequence of a product failure may vary from mere inconvenience (for example, a dishwasher failure) to something serious (for example, an automobile brake failure leading to economic and possibly human loss). The failure of an industrial plant or commercial facility may have major economic consequences for a business as it affects the delivery of goods and services (outputs of the business) and the revenue generation. The daily loss in revenue as a result of the product being out of action due to failure may be very high. Rough estimates (circa 2000) for the revenue lost due to engineered objects being out of action are as follows:

- Large aircraft (A340 or Boeing 747) ~ $500 000/day;
- Dragline (used in open cut mining) ~ $1 million/day;
- A large manufacturer (for example, Toyota) ~ $1–2 millions/hour.

> **Definition 1.1**
>
> Maintenance is the combination of all technical, administrative, and managerial actions during the life cycle of an item intended to retain it in, or restore it to, a state in which it may perform the required function (CEN, 2001).

In a sense, maintenance may be viewed as actions to compensate for the unreliability of an engineered object. Building in reliability is costly and is constrained by technical limits and economic considerations. However, not having adequate reliability is costlier due to the consequence of failures. Thus, maintenance becomes an important issue in this context. Table 1.1 shows the maintenance costs (as a fraction of the operating costs) in different industry sectors, as reported in Campbell (1995).

There are several aspects to maintenance and they may be grouped broadly into the following three categories:

- Technical (engineering, science, technology, etc.);
- Commercial (economics, legal, marketing, etc.);
- Management (from several different perspectives – manufacturer, customer and maintenance service provider when maintenance is outsourced).

This implies that maintenance decisions need to be made in a framework that takes into account these issues from an overall business perspective. Figure 1.1 shows the link between maintenance (strategic and operational) and production from a business perspective.[1]

In this book we discuss all of these aspects and this chapter gives a broad overview of the book.

The outline of the chapter is as follows. Section 1.2 deals with the classification of engineered objects and presents some examples that are used in later chapters to illustrate different concepts and issues. The performance of an engineered object degrades with age and/or usage and this is the focus of Section 1.3, where we look at both reliability and non-reliability performance measures. Maintenance consists of actions to ensure the desired performance and this is discussed in Section 1.4, where we look at a range of such types of maintenance, the consequence of poor maintenance, maintenance costs, and so on. Although

Table 1.1 Maintenance as a percentage of operating cost.

Industry sector	Maintenance cost (%)
Mining (highly mechanized)	20–50
Primary metals	15–20
Electric utilities	5–15
Manufacturing processing	3–15
Fabrication/assembly	3–5

---

[1] The state of an object is discussed briefly in Section 1.3.3 and in more detail later in Chapter 2.

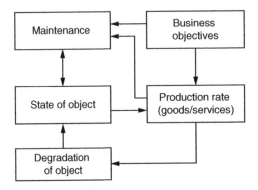

Figure 1.1   Maintenance from a business perspective.

maintenance has been practiced since the dawn of civilization (maintaining shelters to live, stone tools, etc.), the theory of maintenance evolved only recently (in the early part of the twentieth century). Since then it has been growing at an ever-increasing pace and this issue is discussed in Section 1.5, where we look at both the past and future trends. These sections provide the background to highlight the focus of the book, which is discussed in Section 1.6. We conclude the chapter with a brief outline of the various chapters of the book in Section 1.7.

1.2   Classification of Engineered Objects

Engineered objects may be grouped into three broad categories, as indicated in Table 1.2.
    Each of these categories may be subdivided, and this is discussed in subsequent sections.

1.2.1   Products

Products may be classified into three groups, as indicated in Table 1.2. Each group may be divided into two subgroups: (i) standard (or off-the-shelf) and (ii) custom-built.

- *Consumer products:* These are mostly standard products (for example, television sets, appliances, automobiles, and personal computers) that are consumed by society at large. (These products are also consumed by businesses and government agencies.) As such, the number of customers is large, with a small to medium number of manufacturers. The complexity of the product may vary considerably, and the typical small consumer is often not sufficiently well informed to evaluate product performance, especially in cases involving complex products (computers, cars, etc.).
- *Commercial and industrial products:* These may be either standard or custom-built (for example, mainframe computers, CNC machines, pumps, X-ray machines, and aircraft), with a small number of customers and manufacturers. The technical complexity of such products and their mode of usage may vary considerably. The products may be either complete units, such as cars, trucks, pumps, and so forth, or product components needed by another manufacturer, such as batteries, drill bits, electronic modules, turbines, and so on.[2]

---

[2] Industrial products include machines, equipment, tools, and so on.

Table 1.2 Classification of engineered objects.

Products	*Consumer:* Household appliances, automobiles, and so on
	Commercial and Industrial: Also referred to as equipment, machinery, and so on
	Defense: Ships, tanks, planes, and so on
Plants	Collection of several elements: Power plant composed of boiler turbine, generators, and so on
Infrastructures	*Discrete:* Buildings, dams, and so on
	Distributed networks: Rail, road, gas, water, and so on

Table 1.3 Classification of plants.

Industry sector	Operations and outputs
Mining	Extracting and enriching raw materials (for example, ore, fuels)
Processing	Converting ore to metal, crude oil to gasoline, and so on
Manufacturing	Converting processing plant outputs to goods
Power	Producing electricity from coal, oil, nuclear fuel, and so on

Table 1.4 Classification of facilities based on sectors.

Health	Provide healthcare in hospitals, nursing homes, and so on
Transport	Help move goods and people by road, rail, sea, or air
Maintenance	Provide maintenance services for a variety of industrial and commercial products (such as elevators, buses, etc.)
Educational	Provide education (such as in schools and universities)

- *Defense products:* These are specialized products (for example, military aircraft, ships, rockets) with a single customer and a relatively small number of manufacturers. The products are usually complex and expensive and involve state-of-the-art technology with considerable research and development effort required from the manufacturers. These products are usually designed and built to customers' requirements.

1.2.2 Plants and Facilities

Plants are used to produce a variety of goods. They may be classified into several categories, as indicated in Table 1.3.

Facilities (such as hospitals, schools, sport centers, entertainment centers, etc.) also use a range of products to deliver different services. We include these under plants.

A facility is a collection of products used to produce different types of services. They may be classified into several categories, as indicated in Table 1.4.

Table 1.5 Classification of infrastructures.

Industry sector	Examples of infrastructures
Transport	Road and rail networks, ferries, airports, pavements, bridges, and so on
Energy	Electricity networks, gas and petroleum pipelines, and so on
Water management	Water network, sewerage network, dams, and so on
Communication	Telephone and mobile phone networks, cable television, Internet, and so on
Others	Public buildings such as school and hospital buildings, and so on

1.2.3 Infrastructures

Infrastructures are physical structures and facilities that provide services essential for plant operation and also to enable, sustain, or enhance societal living conditions. In other words, they are needed for the smooth operation of a society and the effective functioning of the economy. Infrastructures facilitate the production of goods and services and their delivery to customers; they may be classified into several groups, as indicated in Table 1.5.

Infrastructures may be further classified as being (i) distributed (involving a spatial dimension, such as networks) or (ii) discrete or lumped (where the spatial dimension is not significant, such as buildings, dams, terminals, etc.).

1.2.4 Assets and Systems

The term *asset* is often used in the context of maintenance. In financial accounting, assets are economic resources – tangible or intangible with a positive economic value. A business balance sheet records the monetary value of the assets owned by the business. Tangible assets contain various subclasses, including current assets and fixed assets. Current assets include inventory (such as spares and material needed for carrying out maintenance), whilst fixed physical assets (such as buildings, plants, and equipment) are purchased for continued and long-term use to earn profit for a business. This group includes buildings, machinery, furniture, tools, equipment, and so on. They are written off against profits over their anticipated lives by charging depreciation expenses. Accumulated depreciation is shown in the balance sheet.

The term *system* is used to denote a collection of interconnected elements. Thus, a product, a plant, and an infrastructure may all be viewed as a system.

1.2.5 Illustrative Examples

The following examples will be used in later chapters to illustrate the different concepts, tools, and techniques needed for effective maintenance.

Example 1.1   Automobile (Consumer Product)

The automobile is a self-propelled passenger vehicle designed to operate on ordinary roads. Automobiles may be classified into several types based on (i) structure and usage – passenger cars, light trucks, heavy trucks, vans, buses, and so on, and (ii) the primary energy source – gasoline, diesel, electric, hybrid (combination of gasoline and electric) and others such as hydrogen, solar, and so on, which are still in the experimental stages. Individuals buy one automobile at a time whereas a business might buy a fleet, either for use by its staff or for renting out.[3]   ∎

Example 1.2   Photocopier (Commercial Product)

Photocopying (also referred to as *xerography* – a word derived from two Greek words – *xeros* meaning dry and *graphy* meaning writing) is a dry process for making paper copies of documents and was invented by Chester F. Carlson (an American Physicist) in 1938. The process of xerography involves the following steps:[4]

1. The clean surface of a "photoreceptor" drum (or belt) is coated with a light-sensitive (photo-conductive) material that acts as an insulator in the dark and as a conductor when exposed to light.
2. The photoreceptor material is electrically charged positively through a "corona wire."
3. Light is reflected from the original through a lens on to the drum.
4. The light dissipates the charge on the drum in the areas of the image that are blank. A positively charged image forms on the light-sensitive surface.
5. The negatively charged *toner* (also referred to as *dry ink*) is dusted on the drum and sticks to the positively charged image on the drum. This leaves a "toner image" of the original on the drum.
6. A paper charged positively with the corona wire is pressed against the drum so that the toner image is transferred.
7. The "fuser" heats the positively charged paper for a short period so that the toner is permanently attached to the paper.

The drum surface is cleaned by a "cleaning blade" to remove the remainder of the toner and transferred into a waste bin so that the process may be repeated.   ∎

---

[3] See http://auto.howstuffworks.com for a discussion of the principles of how the different subsystems work.
[4] For more details, see Bruce and Hunt (1984). See also, http://www.howstuffworks.com/photocopier.htm.

Example 1.3   Diesel Engines (Commercial/Industrial Product)

A diesel engine (also known as a compression-ignition engine) is an internal combustion engine (ICE) that uses the heat of compression to initiate ignition to burn the fuel, which is injected into the combustion chamber. This is in contrast to spark-ignition engines such as a gasoline engine (petrol engine) or gas engine (using a gaseous fuel as opposed to gasoline), which uses a spark plug to ignite an air–fuel mixture. The engine was developed by Rudolf Diesel in 1893.

Diesel engines are manufactured in two- or four-stroke versions. Since the 1910s they have been used in ships, and their use in locomotives, trucks, and electricity-generating plants followed later. Since the 1970s, diesel engines have been used in on- and off-road vehicles.[5]   ■

Example 1.4   Thermal Power Stations (Plant)

A power plant generates electrical power. At the center is a generator, a rotating machine that converts mechanical power into electrical power by creating relative motion between a magnetic field and a conductor. The energy source used to turn the generator varies widely and sources include: (i) burning fossil fuel such as coal, oil, and natural gas, (ii) fission in a nuclear reactor, and (iii) cleaner renewable sources such as solar, wind, and hydroelectric power.

A thermal power station is a power plant in which the prime mover is steam-driven. Water is heated, turns into steam, and spins a steam turbine which drives an electrical generator. After it passes through the turbine, the steam is condensed in a condenser and recycled to where it was heated. Some thermal power plants also deliver heat energy for industrial purposes, for district heating, or for desalination of water as well as delivering electrical power.   ■

Example 1.5   Oil Refineries (Plant)

An oil refinery is an industrial process plant where crude oil is processed or refined to produce useable products such as gasoline, kerosene, diesel fuel, heating oil, and asphalt base. The process is very complex and involves both chemical reactions and physical separations, as the crude oil is composed of thousands of different molecules. Mixtures of molecules are isolated according to the mixture's boiling point range (gasoline molecules boil in the range 90–400 °F and kerosene in the range 380–520 °F) through a separation process called *distillation*. These fractions are mixed or blended to satisfy specific properties that are important in allowing the refined product to perform as desired in an engine. To have an effective and efficient operation, process optimization and advanced process control are used to run a refinery.   ■

---

[5] Currently the world's largest diesel engine is a Wartsila Sulzer RT96-C with 108 920 hp (81 220 kW) output.

Example 1.6   Dragline (Industrial Plant)

A dragline is a moving crane with a bucket at the end of a boom. It is used primarily in coal mining for removing the dirt to expose the coal. The bucket volume varies around 90–120 cubic meters and the dragline is operated continuously (24 hours/day and 365 days/year) except when it is down undergoing either corrective or preventive maintenance actions. A performance indicator of great importance to a mining business is the yield (annual output) of a dragline. This is a function of the dragline (bucket) load, speed of operation, and availability. Availability depends on two factors – (i) degradation of the components over time and (ii) maintenance (corrective and preventive) actions used. Degradation depends on the stresses on different components and these, in turn, are functions of the dragline load. As a result, availability is a function of the dragline load and the maintenance effort. Availability decreases as the dragline load increases and increases as the maintenance effort increases. This implies that the annual total output is a complex function of dragline load.   ∎

Example 1.7   Rail Transport (Infrastructure)

Rail transport involves wheeled vehicles running on rail tracks. Tracks usually consist of steel rails installed on sleepers and ballast on which the rolling stock (wagons and carriages), fitted with metal wheels, moves. Rolling stock in railway transport systems has a lower frictional resistance than vehicles on highways and roads and is coupled to form longer trains. In some countries the rail transport is public (owned by the government) and in others it is private (owned by private businesses) or jointly private and public. The two major subsystems are (i) infrastructure and (ii) rolling stock. The infrastructure is managed by the track operator (often a publicly owned company or agency) and the rolling stock is managed by rolling stock operators (may be either public or private). Together, they provide transport between train stations (for passenger and freight transport) and between two terminals (for freight) – such as a mine or manufacturing/processing plant and a port. Power is provided by locomotives which either draw electrical power from an electrical network or produce their own power (usually using diesel engines). Most tracks are accompanied by a signaling system to ensure smooth and safe operation of trains.   ∎

Example 1.8   Road Transport (Infrastructure)

Road transport involves wheeled vehicles (automobiles, buses, trucks, etc.) moving on roads. Road infrastructure consists of pavements (or roads) and other items such as traffic signals, signs, and so on. There are two types of pavement – rigid and flexible. Rigid pavements consist of a thick concrete top surface. Flexible pavements have a flexible layer on top of the surface. The infrastructure may be owned and managed by a public entity (Federal, State, or Local government) or by a private agency under some form of private–public partnership. In contrast to rail transport, many types of vehicles use roads and they provide transport between different points of the road network. Most roads are accompanied by a signaling system to ensure smooth and safe movement of vehicles.   ∎

Example 1.9   Pipe Networks (Infrastructure)

Pipeline infrastructures have been employed as one of the most practical and low-cost methods for large oil and gas transport for decades. Pipeline networks for water distribution and sewage systems are everywhere and they pose critical management and maintenance problems. Aging infrastructure and replacement costs are major challenges for municipal water utilities. With populations increasing and available freshwater resources decreasing, water and wastewater distribution pipelines need to be maintained to prevent water loss and damage to the surrounding environment. Since replacements are certain to become more costly over time, the burdens, if shifted to the future, are bound to get heavier.

America's drinking water systems face an annual shortfall of at least $11 billion to replace aging facilities that are near the end of their useful life and to comply with existing and future federal water regulations. This does not account for growth in the demand for drinking water over the next 20 years. Leaking pipes lose an estimated 7 billion gallons of clean drinking water a day.   ■

Example 1.10   Concrete Structures (Infrastructure)

Reinforced concrete is the world's most important structural material due to its versatility and relatively low cost. A large part of its worldwide appeal is that the basic constituent materials – cement, sand, aggregate, water, and reinforcing bars – are widely available and that it is possible to construct a structure using local sources of labor and materials. Massive concrete structures include multi-story buildings, dams, bridges, and so on. Although most of these structures have high durability, they are susceptible to deterioration due, for instance, to corrosion of bars, which may lead to a reduction in the strength, serviceability, and esthetics of the structure. As such, proper inspection, monitoring, and timely maintenance interventions may reduce the massive investment sometimes needed to restore the deteriorated structure. For example, chloride-induced corrosion is a progressive problem and if it is identified before or just after initiation, treatment is far simpler and cheaper than if the structure is permitted to degrade further.

According to the ASCE 2009 Report Card for America's Infrastructure, $2.2 trillion needs to be invested over five years to "bring the nation's infrastructure to a good condition."   ■

1.3   Performance of Engineered Objects

The performance of an engineered object is a complex entity involving many dimensions and it depends on the perspective – manufacturer or customer – and is best characterized through a vector of variables, where each variable is a measurable property of the object. These measures may be divided broadly into two categories: –non-reliability performance and reliability performance measures.

1.3.1   Non-Reliability Performance Measures

The non-reliability performance measures include technical, operational, economic, environmental impact, and so on. They are specific to the engineered object. Table 1.6 lists a few of the non-reliability performance measures for four of the illustrative examples discussed in the previous section.

1.3.2   Reliability Performance Measures

Some of the reliability measures used in the designing of products and plants are as follows:

- *Interval reliability:* The probability of no failure over a specified interval.
- *Interval availability:* The fraction of the time in which the product or system is in the operational (non-failed) state over a specified interval.
- The number of failures over a specified interval.

1.3.3   Degradation of Performance

The *desired performance* is the starting point for the designing and building/manufacturing of every engineered object. The design process is complicated, starting with components and materials and then building/manufacturing to produce the object. Since performance depends on usage (mode, intensity, etc.) and operating environment, the design process involves selecting components and materials to ensure the desired performance for some nominal values (or ranges) for usage and operating environment.

The performance of the object degrades due to the degradation of the material and components of the object. These are functions of age and/or usage and are influenced by operating environment. The degradation phenomenon is discussed in more detail in Chapter 3.

Table 1.6   Non-reliability performance measures for some engineered objects.

Engineered object	Type	Performance measures
Automobile	Consumer product	Fuel efficiency (km/l)
		Economic efficiency (cost/km/kg)
		Quality of ride
		Emissions (PPM)
Photocopier	Commercial product	Quality of image
		Throughput (copies/min)
Oil refinery	Plant	Efficiency
		Downtimes
		Emissions
Rail network	Infrastructure	Train delays (min/wk)
		Ride quality (noise, vibration, etc.)

1.4   Maintenance

Maintenance involves actions to (i) control or prevent the deterioration process leading to failure of an engineered object and (ii) restore the object to its operational state through corrective actions after a failure. The former is called *preventive maintenance* (PM) and the latter *corrective maintenance* (CM).

Maintenance is the combination of all technical and associated administrative actions intended to retain an item in, or restore it to, a state in which it may perform its required function.

1.4.1   Consequences of Poor Maintenance

We illustrate the consequences of poor maintenance through two examples.

Example 1.11   Vehicle Maintenance

Head gaskets: The head gasket in an automobile engine seals the cylinder head of the engine to the engine block. There are coolant and oil passages that transfer the oil and coolant from the engine to the head and back. The reason for these passages is for the oil to lubricate the valve train and the coolant to remove heat from the cylinder head. The other job of the head gasket is to seal the top of the cylinder to keep the compression contained. Head gasket problems arise generally due to poor maintenance of the cooling system. Acidic coolant may begin to eat away or erode the sealing area of the coolant passages in the gasket. This may cause a weak area and a leak may start to form. The head gasket leakage may travel either internally or externally. An external leak is visible outside the engine; an internal leak means that coolant may seep into oil passages or erode the compression sealing ring in the head gasket, allowing coolant to enter the cylinder or compression to enter the cooling system. This is what is called a *blown* head gasket.

Brakes: A coach driver and his business partner were jailed for the manslaughter of a couple who died in a road crash in the UK. An investigation by police concluded that the cause of the crash was acute brake failure due to poor maintenance. At the Crown Court, the coach driver was sentenced to five years and three months in prison after he admitted charges of causing death by dangerous driving and gross negligence manslaughter. His business partner, who pleaded guilty to gross negligence manslaughter, was jailed for three years. This resulted in a warning to all drivers who ignore vehicle maintenance warning signs – particularly in relation to tire and brake wear – that could result in a fatal road crash and land themselves and their bosses in court.   ∎

Example 1.12   Rail System Operations

An investigation by the Rail Accident Investigation Branch (RAIB) into a passenger train that overshot a station in East Sussex, UK by almost two-and-a-half miles revealed that it was because of poor maintenance. The report said that the train did not deposit sand (needed to assist the braking process) when the driver braked because the leading sand-hoppers were almost empty. Maintenance procedures did not ensure the sand-hoppers were refilled despite there being information that the sand was low.   ∎

1.4.2 Maintenance Costs

The costs of maintenance may be divided into two major categories:

- *Direct costs:* These costs are incurred due to maintenance and repair actions, broadly represented by the cost of labor, the cost of material and spare parts, the cost of contractors, and the costs of infrastructures used and related tax (service tax, etc.). Often, these are the costs which may be tracked down easily in account books.
- *Indirect costs:* These are costs resulting from the consequences associated with failure or unplanned maintenance actions and include loss of revenue due to the production stops owing to maintenance and repair actions, cost of accidents, demurrages, insurance policies, and so on.

The maintenance costs shown in Table 1.1 are the direct costs. The indirect costs are, in general, higher and depend on the engineered object. These indirect costs are difficult to measure but, in general, they are roughly equal to or greater than the direct costs The maintenance costs increase with time due to the aging effect and increasing labor costs. This implies that maintenance is a significant issue for businesses and government agencies.

1.4.3 Preventive versus Corrective Maintenance

As mentioned earlier, there are two types of maintenance: preventive (PM) and corrective (CM). Carrying out maintenance involves additional costs to the owners (individuals, businesses, and government agencies). As the level of PM effort increases, the PM costs increase and the CM costs decrease, as shown in Figure 1.2.

The total cost (PM and CM costs) has a convex shape, indicating that there is an optimal level of PM effort.

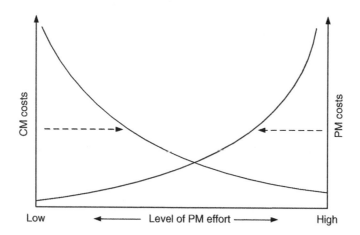

Figure 1.2 PM and CM costs versus PM effort.

1.4.4 Maintenance Management

Maintenance management deals with maintenance-related decision making (for example, recruiting of skilled labor, resource allocation, and scheduling of resources, etc.) at the strategic, tactical, and operational levels, and then initiating actions to implement the decisions.

Businesses and government agencies need to make decisions relating to maintenance of engineered objects at three different levels: strategic, tactical, and operational. Figure 1.3 lists some of the decision problems at each of these three levels.

Proper maintenance with periodic in-service inspections of an engineered object has a positive influence on the technical state of the object and may extend its lifetime considerably. A proper framework is required for planning and executing the decisions, with data playing an important role. Figure 1.4 shows the sequence of activities for implementing the decisions at the operational level.

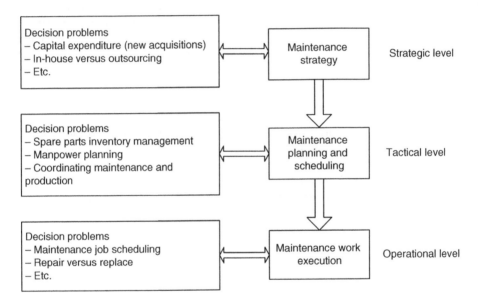

Figure 1.3 Decision problems in maintenance.

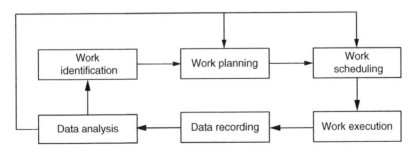

Figure 1.4 Implementation of decisions at the operational level.

1.4.5   Role of Science and Technology

Every object consists of several elements and each of these is comprised of one or more components. An understanding of the degradation process at the component level is critical to understanding the degradation of the object. Reliability science (discussed in the next chapter) deals with this topic. Technology has played an important role in assessing the degradation through the use of sensors. Both of these have played an important role in more effective management of maintenance.

1.5   Evolution of Maintenance

The approach to maintenance has changed significantly over the last century. In this section, we give a very brief historical overview of the evolution[6] as well as current trends in maintenance.

1.5.1   Historical Perspective

Until about 1940, maintenance was considered an unavoidable cost and the only maintenance used was corrective maintenance. When equipment failed it was the task of a specialized maintenance workforce to return the failed item to its operational state. Maintenance was not addressed during the design of the system, nor was the impact of maintenance on system and business performance recognized.

The evolution of Operations Research (OR) from its origin and applications during the Second World War to its subsequent use in industry led to the widespread use of preventive maintenance at component and higher levels. Since 1950, OR models for maintenance have appeared at an ever-increasing pace. The models examine many different maintenance policies and the optimal selection of the parameters of these policies. The impact of maintenance actions on the overall business performance is not addressed.

Starting in 1970, a more integrated approach to maintenance evolved in both the government and private sectors. New, costly defense acquisitions by the US government required a life cycle costing approach, with maintenance cost being a significant component. The close link between reliability and maintainability formed the basis for this change. The term "R&M" began to be used more widely in defense acquisitions to denote reliability and maintainability. This concept was also adopted by manufacturers and operators of civilian aircraft and formed the basis for Reliability Centered Maintenance (RCM) in the USA.

In the RCM approach, maintenance is carried out at the component level and the maintenance effort for an item (component or higher level) is a function of the reliability of the item and the consequence of its failure under normal operation. The core of the RCM philosophy is that maintenance will be performed only after evaluating the consequences of failures (safety, economic, operational, and environmental) at component level. In other words, it deals with optimization of preventive maintenance activities considering failure consequences. The RCM approach is system-oriented and may be implemented free of a company's organizational culture.

---

[6] For a more detailed discussion of the evolution of maintenance, see Pintelon and Parodi-Herz (2008).

At the same time, the Japanese evolved the concept of Total Productive Maintenance (TPM) in the context of manufacturing. Here, maintenance is viewed in terms of its impact on the manufacturing (or production process) through its effect on equipment availability, production rate, and output quality. In TPM the focus is on autonomous maintenance through involvement of all employees and is a human- and employee-centered maintenance approach.

Both RCM and TPM are now widely used in various industrial sectors and many variants have been developed to extend their original functions and/or facilitate their application. Many businesses use elements of both as part of their maintenance strategies.

Since the late 1970s and early 1980s there has been a trend toward Condition Based Maintenance (CBM). This became possible with developments in sensor technologies which enabled PM actions to be based on the condition (or level of degradation) as opposed to age and/or usage.

Maintenance needs to be viewed from a long-term perspective. It needs to take into account the commercial aspects (which determine the load on components), the science aspect (to model the effect of load on equipment degradation), the socio-political aspect, demographic trends, and the capital needed. It needs to address issues such as in-house versus outsourcing of maintenance and their impact on the overall costs of maintenance and the associated risks. This requires an approach where maintenance decisions are made from a strategic perspective using a framework that integrates both technical and commercial issues in an effective manner from an overall business perspective.

1.5.2 Trends in Maintenance

Engineered objects are becoming more complex to meet the ever-increasing demand of customers. Detecting failures and faults is becoming harder and more time-consuming. The cost of labor to carry out maintenance has also been increasing. As a result, maintenance will continue to evolve and the two main drivers for this are (i) technology and (ii) management.

1.5.2.1 Technology Trends

Many different types of technologies are beginning to impact on maintenance. These include:

- *Sensor technologies:* These are used to monitor the condition of an object and to decide on maintenance based on the condition.
- *Information and communication technologies (ICTs):* These technologies are used to access, store, transmit, and manipulate relevant information for maintenance decision making.

1.5.2.2 Management Trends

Maintenance is no longer viewed as a cost but as a function which creates additional value in the business process. The focus has shifted from fail-and-fix to root cause elimination, and from functional thinking to a process-oriented approach with the end customer being the focus. Trends include:

- *A risk-based approach to maintenance:* The focus is to reduce the business risk.
- *Maintenance outsourcing:* Here, a business outsources some or all of the maintenance actions to an external agent under a maintenance service contract.

1.6 Focus of the Book

Maintenance of engineered objects requires finding and implementing the solutions to a wide range of decision problems. The starting point is the list of business objectives. These determine the production rates and they, in turn, impact on the state of the asset which degrades with age and usage. Maintenance strategies need to take these issues into account. Formulating effective maintenance strategies requires (i) proper data collection and analysis and (ii) models to assist the decision-making process. This, in turn, requires a proper understanding of many different concepts, tools, and techniques. Figure 1.5 (from Murthy, Atrens, and Eccleston, 2002) shows the key elements and the linking between them.

A proper understanding of maintenance requires a comprehensive framework. There are many different definitions of a framework and the one that is appropriate in the context of the book is the following:

Definition 1.2

A framework is a logical structure that identifies key concepts, the relationships among the concepts to provide a focus, a rationale, and a tool for the integration and interpretation of information relevant to a decision problem. The structure serves as a starting point for developing models for solving the decision problem.

The framework needs to deal with one or several of the following issues depending on the maintenance problem under consideration:

- Use of scientific methods to understand the degradation processes;
- Proper collection and analysis of relevant data;
- Use of models for decision making;

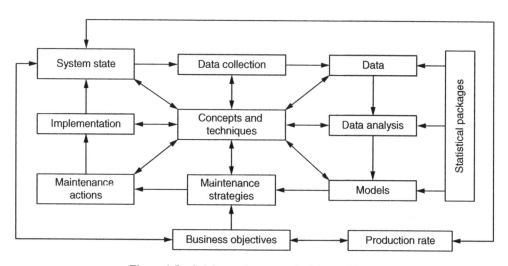

Figure 1.5 Solving maintenance decision problems.

- Use of appropriate technologies;
- Effective maintenance management.

The focus of the book is on introducing students to a comprehensive framework that looks at maintenance from a "big picture" perspective, combining the above elements in a unified manner including the latest trends in maintenance.

1.7 Structure and Outline of the Book

The book is structured in five parts (Parts A–E), each containing one or more chapters, and Part F with five appendices. For Parts A–E, the logical linking between chapters and appendices is given below.

Part A: Maintenance Engineering and Technology
This part consists of the following six chapters:

- Chapter 2: Basics of Reliability Theory
- Chapter 3: System Degradation and Failure[7]
- Chapter 4: Maintenance – Basic Concepts
- Chapter 5: Life Cycle of Engineered Objects
- Chapter 6: Technologies for Maintenance
- Chapter 7: Maintainability and Availability

Chapter 2 looks at basic concepts from reliability theory, since they are needed for a proper understanding of maintenance. Chapter 3 deals with system degradation and failure. A proper understanding of failure mechanisms is crucial to effective maintenance planning. Basic maintenance concepts are covered in Chapter 4, including types of maintenance actions, maintenance requirements of different engineered objects as complexity increases from product to plant to infrastructure, and the important elements of effective maintenance. To be effective, maintenance needs to be viewed from a life cycle perspective, and this is discussed in Chapter 5. Chapter 6 deals with maintenance technology, as technology plays a crucial role in maintenance in terms of data collection, transmission, and processing. Technology also plays a key role in CBM. The final chapter (Chapter 7) in this part of the book is devoted to maintainability, since designing engineered objects for ease of maintenance and efficient use of resources may reduce a good proportion of maintenance costs and enhance the performance of the engineered object in terms of its reliability and availability.

Part B: Reliability and Maintenance Modeling
This part consists of the following five chapters:

- Chapter 8: Models and the Modeling Process
- Chapter 9: Collection and Analysis of Maintenance Data

---

[7] A system is a collection of interconnected elements. As such, products, plants, and infrastructures can be viewed as systems.

- Chapter 10: Modeling First Failure
- Chapter 11: Modeling CM and PM Actions
- Chapter 12: Modeling Subsequent Failures

Models play an important role in understanding and solving maintenance problems. Models and modeling issues, including the steps of the mathematical model-building process, are discussed in Chapter 8. Chapter 9 deals with maintenance data and information. Types and sources of maintenance data, data collection, and preliminary analysis of data (which requires concepts from the theory of statistics) are among the issues presented in this chapter. Chapter 10 discusses probability models for time to first failure. Important issues discussed include different probability distributions, their properties, parameter estimation methods, and model validation. Chapter 11 deals with modeling maintenance actions which affect subsequent failures. The modeling of subsequent failures is the focus of Chapter 12 and this requires an understanding of point processes because failures occur as random points along the time axis.

Part C: Maintenance Decision Models and Optimization
This part consists of the following four chapters:

- Chapter 13: Optimal Maintenance
- Chapter 14: Maintenance Optimization for Non-Repairable Items
- Chapter 15: Maintenance Optimization for Repairable Items
- Chapter 16: Condition-Based Maintenance

Building on the knowledge gained from Part B, Part C deals with various replacement, preventive maintenance, and condition maintenance models. Chapter 13 looks at the process needed to build models for the optimal maintenance of an item. Chapter 14 deals with models for the optimal maintenance of non-repairable items, while Chapter 15 deals with similar issues for repairable items. Chapter 16 deals with condition-based maintenance, building on Chapter 6.

Part D: Maintenance Management
This part consists of the following six chapters:

- Chapter 17: Maintenance Management
- Chapter 18: Maintenance Outsourcing and Leasing
- Chapter 19: Maintenance Planning, Scheduling, and Control
- Chapter 20: Maintenance Logistics
- Chapter 21: Maintenance Economics
- Chapter 22: Computerized Maintenance Management Systems and e-Maintenance

Chapter 17 deals with maintenance management issues ranging from maintenance strategic planning to maintenance control and the two well-known and commonly used methodologies – RCM and TPM. Maintenance outsourcing is a strategic maintenance issue discussed in Chapter 18. Chapter 19 is devoted to the important issue of maintenance planning and scheduling. Maintenance logistics, the supply chain, and spare parts management issues are presented in Chapter 20. Chapter 21 deals with maintenance economics and includes life cycle costing methods and capital replacement models. Maintenance performance measurement is a key to continuous improvement and computerized maintenance management systems (CMMSs) are vital for storing, processing, and producing timely reports and information required for informed decision making. These issues are the subject of Chapter 22 along with e-maintenance, an area where technology plays a key role in gathering and delivering information where it is needed.

Part E: Case Studies
This part consists of the following chapter:

• Chapter 23: Case Studies

The chapter looks at two real cases that illustrate the linking of concepts provided in the various chapters of the book.

Part F: Appendices
Part F consists of the following five appendices:

• Appendix A: Introduction to Probability Theory
• Appendix B: Introduction to Stochastic Processes
• Appendix C: Introduction to the Theory of Statistics
• Appendix D: Introduction to Optimization
• Appendix E: Data Sets

Review Questions

1.1   What is maintenance?

1.2   What are the consequences of product failure? Give examples.

1.3   Explain the following statement: "Building in reliability is costly. However, not having adequate reliability is costlier."

1.4   What are the three main aspects of maintenance?

1.5   What is an adequate classification of engineered objects?

1.6   How is the performance of an engineered object measured?

1.7   What are the factors that affect performance degradation?

1.8   What are the different types of maintenance costs?

1.9   What is the effect of PM effort level on CM and PM cost and what are the implications?

1.10 What are some of the maintenance management decisions at the strategic, tactical, and operational levels?

1.11 What are the main historical developments in maintenance up to the present day?

1.12 What are the new trends in maintenance?

1.13 What is the main focus of this book?

Exercises

1.1 Describe the operations and list some non-reliability performance measures for the following engineered objects:
(a) Room air-conditioner.
(b) Commercial refrigerator in a restaurant.
(c) Back-up generator in a large hospital.
(d) Elevators in an underground mine.
(e) Food-processing plant.
(f) Pipe networks in an urban area.
(g) Railway infrastructure (tracks, bridges, power supply, communication system).

1.2 Comment on the following statement: "Building in reliability is costly but the consequences of not having adequate reliability are costlier" in the context of the following engineered objects:
(a) Aircraft.
(b) Nuclear reactor.
(c) Implant in a human (for example, a heart pacemaker).

1.3 Discuss how production (output in the case of plants and throughput in the case of infrastructures) and maintenance affect each other in the context of the following:
(a) Road network in a region.
(b) Manufacturing plant.
(c) Power station.
(d) Pipe network distributing water to an urban area.
(e) Railway bridges and tunnels.

1.4 Many consumer products need regular maintenance. Make a list of a few products and discuss the maintenance (preventive and/or corrective) requirements for each of these items.

1.5 Explain in a paragraph the relevance of each of the topics listed below from a maintenance perspective:
(a) Sustainability.
(b) Risk.
(c) Economic impact.
(d) Societal impact.
(e) Product quality.
(f) Product reliability.

1.6 Individuals may lease automobiles and other household objects (such as refrigerators, televisions, etc.) as opposed to buying. List the reasons and the advantages and disadvantages of leasing.

1.7 Buildings (large complexes and apartments) need regular maintenance. List the different types of maintenance activities that are needed.

1.8 Explain the link between maintenance and production from a business perspective using Figure 1.1.

1.9 Explain, using Figure 1.5, why a proper understanding of maintenance requires a comprehensive framework that requires making use of several disciplines.

References

Bruce, N. and Hunt, J. (1984) How the photocopier works. *The Science Corner.*

Campbell, J. (1995) *Uptime: Strategies for Excellence in Maintenance Management*, Portland, Productivity Press.

CEN (European Committee for Standardization) EN 13306:2001 (2001) Maintenance Terminology, European Standard, Brussels.

Murthy, D.N.P., Atrens, A. and Eccleston, J.E. (2002) Strategic maintenance management, *Journal of Quality in Maintenance Engineering*, **8**: 287–305.

Pintelon, L. and Parodi-Herz, A. (2008) Maintenance: An evolutionary perspective, in *The Complex System Maintenance Handbook*, Kobbacy, K.A.H. and Murthy, D.N.P. (eds), Springer-Verlag, London.

Part A

Maintenance Engineering and Technology

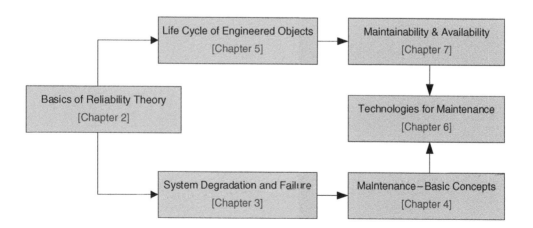

Part A

Maintenance Engineering and Technology

2

Basics of Reliability Theory

Learning Outcomes

After reading this chapter, you should be able to:

- Describe an engineered object as a multi-level system that facilitates the description of the physical structure of the system and the relationship between its components;
- Define system functions and describe their classification;
- Define system failure and faults and provide a proper classification of failure;
- Define system failure causes and describe their classification;
- Describe system degradation leading to failure through two state and multi-state characterizations involving finite and infinite numbers of states;
- Define system reliability;
- Define the failure rate function and provide its interpretation;
- Describe different methods for the linking of component failures to system failure such as failure modes and effects analysis (FMEA), fault tree analysis (FTA), and structure functions;
- Conduct FMEA, construct FTA, and derive structure functions for simple systems.

Introduction to Maintenance Engineering: Modeling, Optimization, and Management, First Edition.
Mohammed Ben-Daya, Uday Kumar, and D.N. Prabhakar Murthy.
© 2016 John Wiley & Sons, Ltd. Published 2016 by John Wiley & Sons, Ltd.

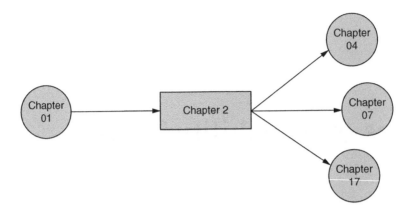

2.1   Introduction

Every engineered object (product, plant, or infrastructure) is unreliable in the sense that it degrades and eventually fails. The reliability of the object is determined by decisions made during the design and building (manufacturing) of the object and is affected by factors such as operating environment, usage mode, and intensity. Maintenance actions are needed to counteract the unreliability of the object. Effective maintenance decision making needs to take into account the reliability of the object. This requires a proper understanding of some basic concepts from reliability theory, and this is the focus of this chapter. In addition, this chapter is the starting point for an understanding of the engineering and technology aspects that are relevant in the context of maintaining an engineered object discussed in the later chapters of Part A, and for the chapters of Part B, which deal with modeling issues.

The outline of the chapter is as follows. Section 2.2 deals with the decomposition of an engineered object. This is important, as the degradation and failure of an object are related to the degradation and failure of its components. Section 2.3 deals with three key aspects of an engineered object: functions, failures, and faults. Section 2.4 looks at three different approaches to characterizing the degradation of an object. In Section 2.5 we define reliability and its mathematical characterization. Section 2.6 deals with linking component failures and system failure and we discuss an alternative approach to describe the link. Section 2.7 provides a brief discussion of reliability theory and the main issues involved, before we conclude with a summary of the chapter in Section 2.8.

2.2   Decomposition of an Engineered Object

Even the simplest engineered product (for example, a small consumer product such as a toaster) is comprised of several interacting elements and can be viewed as a multi-level system. The number of levels needed depends on the system under consideration. A seven-level characterization that is appropriate for products and plants is shown in Figure 2.1.

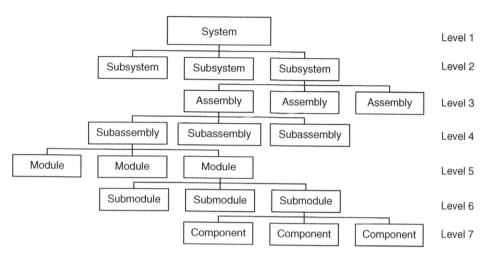

Figure 2.1   Decomposition of a product or plant.

Example 2.1   Rail System

Level	Label
1	Rail system
2	Track (spatially distributed), rolling stock, stations, and so on
3	For rolling stock: Wagons, coaches, and so on
	For track: Rail, bridges
4	For wagons: Axles, body, and so on
5	For axles: Wheels
6	For wheels: Brakes, control mechanisms
7	For brakes: Brake pads, bolts

In the case of some infrastructure, one or more of the elements can be distributed with a spatial dimension.   ∎

2.3   Functions, Failures, and Faults

2.3.1   Functions

Table 1.1 listed the primary (essential) functions of several engineered objects. There are many other functions for an object, and the elements are as indicated below in the context of a power plant:

- *Essential function:* This defines the intended or primary function. In the case of a thermal power plant, it is to provide electrical power on demand to the consumers who are part of the network.

- *Auxiliary functions:* These are required to support the primary function. They are usually less clear than essential functions. For example, "preserving fluid integrity" is an auxiliary function of a pump and its failure may cause a critical safety hazard if the fluid is toxic or corrosive.
- *Protective functions:* The two-fold goal here is to protect people from injury and protect against damage to the environment. Examples of these are relays that offer protection against current surges and scrubbers on smokestacks that remove particulate matter to protect the environment.
- *Information functions:* These comprise condition monitoring, gauges, alarms, and so on. In a power plant, the main control panel displays various bits of information about the different subsystems – for example, voltage and current output of generators, pressure and temperature of steam in the various parts of the plant, and so on.

2.3.2   Failures

Definition 2.1

Failure is the termination of the ability of an item to perform a required function (IEC 50(191), 1990).

A system failure occurs due to the failure of one or more of its components. Henley and Kumamoto (1981) propose the following classification for failures:

- *Primary failure:* A primary failure of a component occurs when the component fails due to natural causes (for example, aging). An action (for example, repair or replacement by a working unit) is needed to make the component operational.
- *Secondary failure:* A secondary failure is the failure of a component due to one or more of the following causes: (i) the (primary) failure of some other component(s) in the system, (ii) environmental factors, and/or (iii) actions of the user.
- *Command failure:* A command failure occurs when a component is in the non-working (rather than a failed) state because of improper control signals or noise (for example, a faulty action of a logic controller switching off a pump). Often, no corrective action is needed to restore the component to its working state in this case.

Example 2.2   Stereo System

Consider the stereo system shown in Figure 2.2, which is made up of a CD player, tuner, amplifier, and two speakers for stereophonic sound.

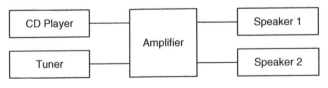

Figure 2.2   Stereo system.

The system is working if one can listen to either a CD or the radio (tuner) in stereophonic sound. There are several types of failures:

- *Partial failures:*
  - One can listen to either a CD or the radio in monophonic sound (as one speaker has failed).
  - One can listen to only one source in stereophonic sound (either the CD or the radio has failed).
  - One can listen to only one source in monophonic sound (either the CD or the radio and one of the speakers has failed).
- *Failure:* No sound (both the CD and the radio or the amplifier or both speakers have failed). ∎

2.3.3 Faults

Definition 2.2

A *fault* is the state of an item characterized by its inability to perform its required function (IEC 50(191), 1990).

Note that this excludes situations arising from preventive maintenance or any other intentional shutdown period during which the system is unable to perform its required function. A fault is, hence, a state resulting from a failure.

2.3.4 Failure Modes

Definition 2.3

A *failure mode* is a description of a fault (IEC 50(191), 1990).

It is sometimes referred to as a fault mode (for example, IEC 50(191); 1990). Failure modes are identified by studying the (performance) function of the item. Blache and Shrivastava (1994) suggest a classification scheme for failure modes:

- *Intermittent failures:* Failures that last for only a short time. A good example of this is a software fault which occurs only under certain conditions that occur intermittently.
- *Extended failures:* Failures that continue until some corrective action rectifies the failure. They can be divided into the following two categories:
  - *Complete failures:* Failures which result in a total loss of function.
  - *Partial failures:* Failures which result in a partial loss of function.

Each of these can be further subdivided into the following:

- Sudden failures: Failures that occur without any warning.
- Gradual failures: Failures that occur with signals to warn of the occurrence of a failure.

A complete and sudden failure is called a *catastrophic* failure and a gradual and partial failure is designated a *degraded* failure.

Example 2.3   Hydraulic Valves

Hydraulic valves are used in refineries to control the flow of liquids. If a valve does not shut properly, the flow is not reduced to zero and this can be viewed as a partial failure. If a valve fails to operate (due, for example, to the spring not functioning properly), then the failure is a complete failure. A valve usually wears out with usage and this corresponds to a gradual failure.   ■

2.3.5   Failure Causes and Severity

Definition 2.4

Failure cause is the circumstances during design, manufacture, or use which have led to a failure (IEC 50(191), 1990).

The failure cause is useful information in the prevention of failures or their reoccurrence. Failure causes may be classified (in relation to the life cycle of the object[1]), as indicated below:

- *Design failure:* Due to inadequate design;
- *Weakness failure:* Due to weakness (inherent or induced) in the system so that the system cannot stand the stress it encounters in its normal environment;
- *Manufacturing failure:* Due to non-conformity during manufacturing;
- *Aging failure:* Due to the effects of age and/or usage;
- *Misuse failure:* Due to misuse of the system (operating in environments for which it was not designed);
- *Mishandling failure:* Due to incorrect handling and/or lack of care and maintenance.

The *severity* of a failure mode signifies the impact of the failure mode on the system as a whole and on the outside environment. A severity ranking classification scheme (MIL-STD-882D, 2000) is as follows:

- *Catastrophic:* Failures that result in death or total system loss;
- *Critical:* Failures that result in severe injury or major system damage;
- *Marginal:* Failures that result in minor injury or minor system damage;
- *Negligible:* Failures that result in less than minor injury or system damage.

---

[1] Life cycles are discussed in Chapter 5.

2.4   Characterization of Degradation

In Section 1.3.3, the degradation of an object was described in terms of the degradation in its performance, and this can be characterized by a variable, $X(t)$. This variable indicates the state or condition of the item (system, component, or something in between) as a function of age. We now look at three different characterizations.

2.4.1   Two-State Characterization

The state $X(t)$ is binary valued with:

- $X(t) = 1$ corresponding to the object being in the *working state* (performance satisfactory or acceptable) and
- $X(t) = 0$ corresponding to the object being in the *failed state* (performance is unsatisfactory or unacceptable).

The item starts in the working state and changes to the failed state after a period T, as shown in Figure 2.3. T is the time to failure (or lifetime of the component). This is a random variable,[2] as the time instant of change from working to failed is uncertain.

Some items for which this characterization is appropriate are (i) an electric bulb and (ii) the heating element in a kettle.

2.4.2   Multi-State Characterization (Finite Number of States)

Here, $X(t)$ can assume values from the set $\{1, 2, \ldots, K\}$ with:

- $X(t) = 1$ corresponding to item performance being fully acceptable (item is in a good *working state*);
- $X(t) = i, 1 < i < K$, corresponding to item performance being partially acceptable (item is in a working state, with a higher value of i implying a higher level of degradation); and
- $X(t) = K$ corresponding to item performance being unacceptable (item is in a *failed state*).[3]

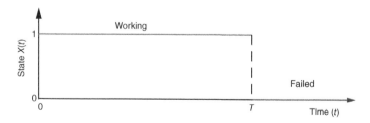

Figure 2.3   Time to failure. (binary-state characterization.)

[2] See Appendix A for a definition of a random variable.

[3] The numbering of states is arbitrary. One can easily reverse the order so that the lower the state, the greater the degradation.

The time to failure of the item is given by $T = \inf\{t : X(t) = K\}$, as shown in Figure 2.4.[4] Let T_i denote the duration of the time for which the component state is $i, 1 \leq i < K - 1$. This is a random variable and, as a result, the time to failure is the sum of $(K-1)$ random variables.

An object for which this characterization is appropriate is a pump, where state 1 corresponds to the pump functioning normally, state 2 to a minor leak, and state 3 to a major leak.

2.4.3 Multi-State Characterization (Infinite Number of States)

This is an extension of the above case, with $K = \infty$. $X(t)$ is now a non-decreasing, continuous-time stochastic process, as shown in Figure 2.5. Here, a higher value of $X(t)$ implies greater degradation and the item failure time is given by $T = \inf \{t : X(t) = x^*\}$.[5]

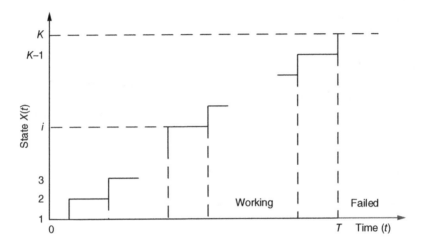

Figure 2.4 Time to failure. (multi-state characterization with a finite number of states.)

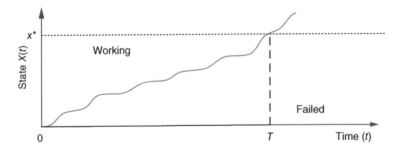

Figure 2.5 Time to failure. (multi-state characterization with an infinite number of states.)

---

[4] Inf{ } (infimum of a set) is the greatest lower bound of the set.
[5] In some cases, $X(t)$ could be non-increasing, with lower values corresponding to greater degradation. In this situation, the curve in Figure 2.5 would be downward sloping.

Some objects for which this characterization is appropriate are (i) a turbine, where $X(t)$ is the diameter of the shaft which is decreasing due to wear and (ii) a pipe, where $X(t)$ is the crack growth which is increasing with age.

2.5   Reliability Concept and Characterization

The reliability of an item conveys the concept of dependability, successful operation or performance, and the absence of failures. It is an external property of great interest to both manufacturer and consumer. Unreliability (or lack of reliability) conveys the opposite.

Definition 2.5

The reliability of an item is the ability of the item to perform a required function, under given environmental and operational conditions and for a stated period of time (ISO 8402, 1986).

Here, we adopt the following definition:

Definition 2.6

The reliability of an item is the probability that the item will perform its intended function for a specified time period when operating under normal (or stated) environmental conditions.

If the operational conditions are the same as the nominal conditions assumed in the designing of the object, then we are referring to the *design reliability*. However, when put into operation, the operational condition will differ from the nominal design conditions and, as such, the reliability (called the *field reliability*) will differ from the design reliability. There are several other notions of reliability, and these are discussed in Chapter 5.

The reliability of an object depends on a complex interaction of the laws of physics, engineering design, manufacturing processes, management decisions, random events, and usage. Usage can be continuous (for example, refineries) or intermittent (for example, an aircraft with flying and non-flying periods) and the item can be used for short periods followed by large idle periods (for example, a coffee grinder or aircraft landing gear).

2.5.1   Time to First Failure

Definition 2.7

Time to first failure is the time elapsed between when a new item is put into operation (in continuous mode) and when it fails for the first time.[6]

---

[6] Here, we assume continuous usage and an age-based clock, where $t = 0$ corresponds to the object being put into operation. There are other types of clocks such as the calendar clock, the usage clock (hours flown by an engine, number of landings, etc.). This is discussed further in Chapter 4.

The time to failure is a random variable, T, with cumulative distribution function, $F(t)$.[7] Note that the item is in the working state at time t if $T > t$ and in the failed state if $T \le t$ and no action is being initiated to rectify the failure. As a result, we have:

$$F(t) = \Pr\{T \le t\} \tag{2.1}$$

The connection between the state variable $X(t)$ and the random variable T has been illustrated in Figures 2.1–2.3. If $F(t)$ is differentiable, then its derivative is given by the probability density function for time to failure:

$$f(t) = \frac{dF(t)}{dt} \tag{2.2}$$

This function is useful as it characterizes the probability of failure in $[t, t+\delta t)$, which is given by $f(t)\delta t + O(\delta t^2)$ (one can ignore the second term if δt is small).

2.5.2 Reliability Function

The reliability of an item is the probability that the item does not fail before t and is therefore given by:

$$R(t) = 1 - F(t) = P(T > t), \quad t > 0. \tag{2.3}$$

$\bar{F}(t)$ is often used instead of $R(t)$, and this has the following properties:

- $R(t)$ is a non-increasing function of t, $0 \le t < \infty$.
- $R(0) = 1$ and $R(\infty) = 0$.

Typical plots of $R(t)$ are shown in Figure 2.6.

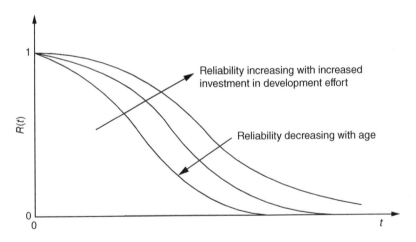

Figure 2.6 Plots of reliability functions.

---

[7] See Appendix A for a definition of a random variable.

Note that the reliability improves with better design (through greater investment in development effort) and decreases with age.

2.5.3   Mean Time to First Failure

The mean time to first failure for an item is the expected value of the random variable T; that is:

$$\text{MTTF} = E[T] = \int_0^\infty t f(t) dt \qquad (2.4)$$

Integrating by parts, we have the following result:

$$\text{MTTF} = \int_0^\infty R(t) dt \qquad (2.5)$$

2.5.4   Failure Rate Function

Consider an item that has survived an interval of time $[0, t]$. The conditional probability that the item will fail in the time interval $(t, t + \Delta t]$, given that it is functioning at time t is given by:

$$P\big(t < T \le t + \delta t \,|\, T > t\big) = \frac{P\big(t < T \le t + \delta t\big)}{P\big(T > t\big)} = \frac{F\big(t + \delta t\big) - F(t)}{R(t)} \qquad (2.6)$$

By dividing this conditional probability by δt and letting $\delta t \to 0$, we have:

$$\lim_{\delta t \to 0} \frac{F\big(t + \delta t\big) - F(t)}{\delta t R(t)} = \frac{1}{R(t)} \lim_{\delta t \to 0} \frac{F\big(t + \delta t\big) - F(t)}{\delta t} = \frac{f(t)}{R(t)}$$

This is called the *failure rate* (or hazard) function and is denoted by $h(t)$, so that:

$$h(t) = \frac{f(t)}{R(t)} \qquad (2.7)$$

Equations (2.6) and (2.7) imply that when δt is small,

$$P\big(t < T \le t + \delta t \,|\, T > t\big) \approx h(t) \delta t \qquad (2.8)$$

The significance of this is that $h(t)\delta t$ is the probability that the item will fail in an interval $(t, t + \delta t]$ given that it has not failed before t. The failure rate function can have many different shapes. One is the bathtub shape shown in Figure 2.7, where, in region A (decreasing failure rate), the failure is due to manufacturing and/or assembly errors (often referred to as *teething problems*), in region B (constant failure rate), the failure is purely due to chance (and is not

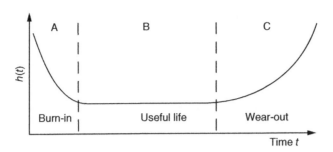

Figure 2.7   Bathtub curve.

affected by age), and in the final region C (increasing failure rate), failure is due to the aging effect. This has implications for the appropriate type of maintenance and is discussed further in Chapter 4 and in Part B.

2.6   Linking System and Component Failures

A system is a collection of interconnected components. The failure of a system is due to the failure of one or more of the components of the system. The linking of component failures to system failures can be done using two different approaches. The first is referred to as the forward (or bottom-up) approach and the second as the backward (top-down) approach.

In the forward approach, one starts with failure events at the component level and then proceeds forward to the system level to evaluate the consequences of such failures on system performance. *Failure modes and effects analysis* (*FMEA*) uses this approach. In the backward approach, one starts at the system level and then proceeds downward to the part level to link system performance to failures at the part level. *Fault tree analysis* (*FTA*) uses this approach.

2.6.1   Failure Modes and Effects Analysis

FMEA is a structured, logical, and systematic approach. FMEA involves reviewing a system in terms of its subsystems, assemblies, and so on, down to the component level, to identify failure modes and causes and the effects of such failures on a system's function. The exercise of identifying component failure modes and determining their effects on the system function assists the design engineer/analyst in developing a deeper understanding of the relationships among the system components. The analyst can then use this knowledge to suggest changes to the system that can eliminate or mitigate the undesirable consequences of a failure. FMEA is used to assess system safety and to identify design modifications and corrective actions needed to mitigate the effects of a failure on the system. It is also used in planning system maintenance activities, as we shall see later in Chapter 17.

According to IEEE Standard 352, some of the objectives of FMEA are as follows:

- To ensure that all conceivable failure modes and their effects on operational success of the system have been considered;
- To list potential failures and identify the magnitude of their effects;

- To provide historical documentation for future reference to aid in the analysis of field failures and consideration of design changes;
- To provide a basis for establishing corrective action priorities;
- To assist in the objective evaluation of design requirements related to redundancy, failure detection systems, fail-safe characteristics, and automatic and manual override.

2.6.1.1   FMEA Procedure

The FMEA methodology is based on a hierarchical, inductive approach to analysis; the analyst must determine how every possible failure mode of every system component affects the system operation. The basic procedure consists of:

1. Determining the item functions;
2. Identifying all item failure modes – usually these are the ways in which the item fails to perform its functions;
3. Determining the effect of the failure for each failure mode, both on the component and on the overall system being analyzed;
4. Classifying the failure by its effects on the system operation and mission;
5. Determining the failure's probability of occurrence;
6. Identifying how the failure mode can be detected (this is especially important for fault-tolerant configurations);
7. Identifying any design changes to eliminate the failure mode, or if that is not possible, mitigate or compensate for its effects.

The details of the FMEA analysis are documented on special worksheets. Figure 2.8 is an example which indicates the format to describe the failure modes and their consequences.

Component	Function	Failure mode	Failure cause	Failure effects	
				Component	System

Figure 2.8   Illustrative FMEA worksheet.

Example 2.4   Photocopier

The paper feeder is an important element of a photocopier. The feeder uses a vacuum belt to hold and feed paper. The normal operation requires feeding one sheet at a time at set times. Maintaining appropriate timing, speed, vacuum pressure, and gap between the paper and the platform is essential to avoid failure of the paper feeder. The following are some of the failure modes for the paper feeder:

- No paper being fed;
- More than one sheet of paper being fed at the same time;

- Paper jam resulting in paper flow blockage;
- Paper fed skewed so that the image is misaligned;
- Paper curling so that the quality is affected.

The root causes of the failure due to a paper jam are as follows:

- Condition of the paper;
- Paper path;
- Measurement control;
- Environment;
- Operator.

An online control system is used to reduce the occurrence of this failure mode. The control system consists of three modules: sensor, controller, and actuator. The sensors (such as photo, pressure, and motor current sensors) feed the data to the controller. These are processed and the output is used to adjust the settings of various factors (such as pressure, gap, timing, etc.) to improve the reliability of the feeder. One of the above failure modes is used in Table 2.1 to illustrate the use of the FMEA worksheet in Figure 2.8.   ■

Table 2.1   Completion of an FMEA worksheet for a photocopier failure mode.

Component	Function	Failure mode	Failure cause	Failure effects	
				Components	System
Paper feeder	Feed papers	Paper jam	Condition of paper	√	√
			Paper path	√	√
			Measurement control	√	√
			Environment	√	√
			Operator	√	√

The terms FMEA and FMECA (failure modes, effects, and criticality analysis) are often used interchangeably. Some authors consider that FMECA extends FMEA by including a ranking of the failure modes.

2.6.2   Fault Tree Analysis

FTA is concerned with the identification and analysis of conditions and factors that cause, or may potentially cause or contribute to, the occurrence of a defined *top event* (such as failure of a system). A fault tree is an organized graphical representation of the conditions or other factors causing or contributing to the occurrence of the top event. FTA can be used for analysis of systems with complex interactions between the components, including software–hardware interactions.

FTA is a deductive (top-down) method of analysis aimed at pinpointing the causes or combinations of causes that can lead to the defined top event. The analysis can be qualitative or quantitative, depending on the scope of the analysis. FTA may be undertaken independently of, or in conjunction with, other reliability analyses and its objectives include:

- Identification of the causes or combinations of causes leading to the top event;
- Determination of whether a particular system reliability measure meets a stated requirement;
- Determination of which potential failure mode(s) or factor(s) would be the highest contributor to the system probability of failure (unreliability) or unavailability, when a system is repairable, for identifying possible system reliability improvements; and
- Analysis and comparison of various design alternatives to improve system reliability.

It is important to understand that a fault tree is tailored to its top event. Therefore, the fault tree includes only those faults that contribute to this particular top event. Moreover, the generated faults are not exhaustive – they contain only the faults that are deemed to be realistic by the design engineer/analyst.

2.6.2.1 Construction of a Fault Tree

There are many symbols commonly used for constructing a fault tree. Table 2.2 contains the most basic ones and their descriptions.

The construction of a fault tree begins with the top event and proceeds downward to link to basic events through the use of different gates. Completion of the fault tree requires specification of the output of each gate, as determined by the input events to the gate. Example 2.5 illustrates this.

Table 2.2 Basic fault tree symbols.

Symbol	Name	Description
	Resultant event	An event which results from the combination of other events
	Basic event	An event that cannot be developed any further
	OR gate	The output event occurs if any of the input events occur
	AND gate	The output event occurs only if all of the input events occur
	Undeveloped event	A primary event that represents a part of the system that is not yet developed

Example 2.5   Electric Motor

Consider the simple electric motor circuit shown in Figure 2.9. The steps described above can be used to construct a fault tree for it.

The top event is "motor fails to operate." One primary failure is the failure of the motor itself. This event is a basic event because no details of the motor are given and the event cannot be developed further. The other possibility is the event that no current is supplied to the motor. The event "no current in motor" is the result of other events and is therefore developed further. The complete fault tree is shown in Figure 2.10.   ■

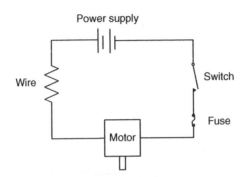

Figure 2.9   Electric motor circuit.

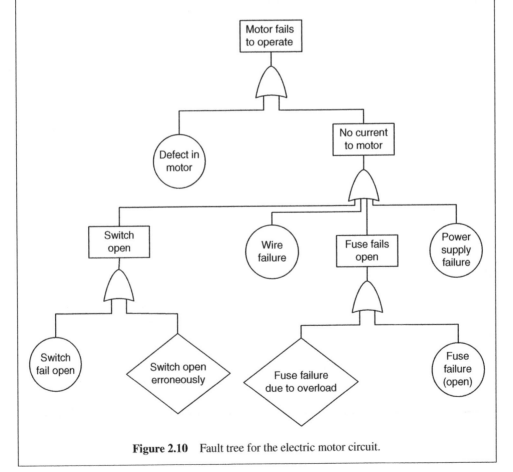

Figure 2.10   Fault tree for the electric motor circuit.

2.6.3   Reliability Block Diagram (RBD)

A reliability block diagram (RBD) is an alternative way of describing a system that is useful for reliability analysis. Each component is represented by a block with two end points. When the component is in its working state, there is a connection between the two end points and this connection is broken when the component is in a failed state. A multi-component system can be represented as a network of such blocks with two end points. The system is in a working state if there is a connected path between the two end points. If no such path exists, then the system is in a failed state. Two well-known network structures are the series and parallel configurations shown in Figure 2.11 for a two-component system. In a series config-uration (with two or more components), the system is functioning only when all of its components are functioning. In contrast, in a parallel structure (with two or more components), the system is not working (i.e., is in a failed state) only when all of its components are not working.

The RBD for most systems has a general network structure with series and parallel sub-structures. Also, note that a block diagram representation is not a physical layout of the system but a representation that indicates the link between system performance and component performance.

2.6.3.1   Link between RBD and FTA

A fault tree representation of a system can be converted easily into a block diagram represen-tation and vice versa. Figure 2.11 shows the equivalence for the series and parallel configura-tions with two components.

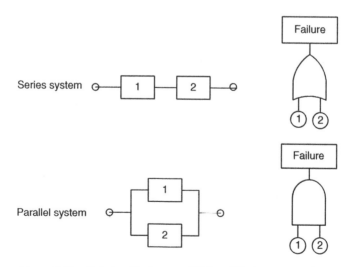

Figure 2.11   Relationship between RBD and FTA representations.

2.6.4   Structure Function

We confine our attention to the case where each component is characterized as being in either a working or a failed state (the two-state characterization of Section 2.4). Let $X_i(t)$, $1 \leq i \leq n$, denote the state of component i (a random variable) at time t, with:

$$X_i(t) = \begin{cases} 1 & \text{if component } i \text{ is in the working state} \\ 0 & \text{if component } i \text{ is in a failed state} \end{cases} \tag{2.9}$$

Let $X(t) = (X_1(t), X_2(t), \ldots, X_n(t))$, with this n-dimensional vector denoting the state of the n components. $X(t)$ can assume any one of 2^n values, corresponding to the possible different combinations of the states (working or failed) for the n components.

The state of the system $X_s(t)$ is also characterized by a binary characterization and is a function of the states of the components. Let $\phi(X)$ denote this function so that:

$$X_s(t) = \phi(X(t)) = \begin{cases} 1 & \text{if the system is in the working state} \\ 0 & \text{if the system is in a failed state} \end{cases} \tag{2.10}$$

$\phi(X)$ is called the *structure function* of the system. One can derive the structure function using either a fault tree or the RBD, as illustrated below.

2.6.4.1   Series Configuration

The system is in a working state if and only if (*iff*) all the components are in the working state. This implies the following logical statement:

$$X_s(t) = 1 \quad \textit{iff} \quad \{X_i(t) = 1,\ 1 \leq i \leq n\}$$

As a result, the structure function is given by:

$$\phi(X(t)) = X_1(t) X_2(t) \ldots X_n(t) = \prod_{i=1}^{n} X_i(t) \tag{2.11}$$

2.6.4.2   Parallel Configuration

The system is in the failed state if and only if all the components are in the failed state. This implies the following logical statement:

$$X_s(t) = 0 \quad \textit{iff} \quad \{X_i(t) = 0,\ 1 \leq i \leq n\}$$

As a result, we have:

$$1 - \phi(X(t)) = (1 - X_1(t))(1 - X_2(t)) \ldots (1 - X_n(t)) = \prod_{i=1}^{n} (1 - X_i(t)).$$

or

$$\phi\big(X(t)\big)=1-\big(1-X_1(t)\big)\big(1-X_2(t)\big)\ldots\big(1-X_n(t)\big)=1-\prod_{i=1}^{n}\big(1-X_i(t)\big) \qquad (2.12)$$

2.6.4.3 General Configuration

For many RBDs, the results for series and parallel configurations can be used repeatedly to obtain the structure function, as illustrated in Example 2.6.

Example 2.6   Stereo System

Consider the stereo system of Example 2.2. Let $X(t)=\{X_1(t),X_2(t),X_3(t),X_4(t),X_5(t)\}$ denote the state of the tuner, CD player, amplifier, speaker 1, and speaker 2 at a particular time, respectively. Failure is defined as not being able to listen to any music at all. The RBD for this case is as shown in Figure 2.12.

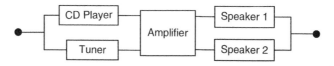

Figure 2.12   Reliability block diagram of a stereo system

The tuner and CD player can be viewed as a module and the structure function (using Equation (2.12)) is given by:

$$\phi\big(X_{12}(t)\big)=1-\big(1-X_1(t)\big)\big(1-X_2(t)\big)$$

Similarly, for the two speakers:

$$\phi\big(X_{45}(t)\big)=1-\big(1-X_4(t)\big)\big(1-X_5(t)\big)$$

These two modules and the amplifier can be viewed as a series system, so that the structure function (using Equation (2.11)) yields:

$$\phi\big(X(t)\big)=\phi\big(X_{12}(t)\big)X_3(t)\phi\big(X_{45}(t)\big)=\Big[1-\big(1-X_1(t)\big)\big(1-X_2(t)\big)\Big]X_3(t)$$
$$\Big[1-\big(1-X_4(t)\big)\big(1-X_5(t)\big)\Big].\quad\blacksquare$$

2.6.5 System Reliability

Since $X_s(t)$ is a binary-valued random variable, its expected value is given by:

$$E\big[X_s(t)\big]=1\times P\{X_s(t)=1\}+0\times P\{X_s(t)=0\}=P\{X_s=1\}.$$

Note that $P\{X_s = 1\}$ is the reliability $R_s(t)$. As a result, we have:

$$R_S(t) = E\big[\phi\big(X(t)\big)\big]. \tag{2.13}$$

This is the basic formula for linking component reliability to system reliability. If component failures are statistically independent, then:

$$R_S(t) = E\big[\phi\big(X(t)\big)\big] = \phi\big(E\big(X(t)\big)\big) = \phi\big(R(t)\big) \tag{2.14}$$

where $R(t) = \{R_1(t), R_2(t), \ldots, R_n(t)\}$.

2.6.5.1 Series Configuration

The reliability of a system with a series configuration (and statistically independent component failures) obtained using Equation (2.13) is given by:

$$R_S(t) = \phi\big(R(t)\big) = \prod_{i=1}^{n} R_i(t) \tag{2.15}$$

From Equation (2.15) we can see that the reliability of a series system decreases rapidly as the number of components increases, and the system reliability will always be less than or equal to that of the least reliable component.

2.6.5.2 Parallel Configuration

The reliability of a system with a parallel configuration (and statistically independent component failures) obtained using Equation (2.13) is given by:

$$R_S(t) = 1 - \prod_{i=1}^{n} \big(1 - R_i(t)\big) \tag{2.16}$$

From Equation (2.16) we see that the system reliability increases as n increases.

2.6.5.3 General Parallel Configuration

For some RBDs, the results for series and parallel configurations can be used repeatedly to obtain the overall system reliability, as illustrated by Example 2.7.

Example 2.7   Stereo System

Consider the stereo system of Example 2.6, where failure is defined as "one cannot listen to any music at all." We consider the case where failed components are not replaced and we are interested in the reliability of a new system over three years. The component reliabilities for three years are as follows:

$$R_1(3) = 0.90, R_1(3) = 0.90, R_2(3) = 0.95, R_3(3) = 0.99, \text{and } R_4(3) = 0.75.$$

The equivalent series component reliability for the tuner and CD player is:

$$R_{12}(3) = 1 - (1 - 0.90)(1 - 0.95) = 0.995,$$

and the equivalent series component reliability for the two speakers is:

$$R_{45}(3) = 1 - (1 - 0.85)(1 - 0.75) = 0.9625$$

So the system reliability of the stereo system is that of the resulting series arrangement; that is:

$$R_S(3) = R_{12}(3) R_3(3) R_{34}(3) = 0.995 \times 0.99 \times 0.9625 = 0.948. \qquad \blacksquare$$

2.7   Reliability Theory

Reliability theory deals with the interdisciplinary use of probability, statistics, and stochastic modeling, combined with engineering insights into the design and the scientific understanding of the failure mechanisms, to study the various aspects of reliability. As such, it encompasses issues such as: (i) reliability modeling, (ii) reliability analysis and optimization, (iii) reliability engineering, (iv) reliability science, (v) reliability technology, and (vi) reliability management.

* *Reliability modeling* deals with model building to obtain solutions to problems in predicting, estimating, and optimizing the survival or performance of an unreliable system, the impact of the unreliability, and actions to mitigate this impact.
* *Reliability analysis* can be divided into two broad categories: (i) qualitative and (ii) quantitative. The former is intended to verify the various failure modes and causes that contribute to the unreliability of a product or system. The latter uses real failure data in conjunction with suitable mathematical models to produce quantitative estimates of product or system reliability.
* *Reliability engineering* deals with the design and construction of systems and products, taking into account the unreliability of their parts and components. It also includes testing and programs to improve reliability. Good engineering results in a more reliable end product.
* *Reliability science* is concerned with the properties of materials and the causes for deterioration leading to part and component failures. It also deals with the effect of manufacturing processes (for example, casting, annealing) on the reliability of the part or component produced.
* *Reliability management* deals with the various management issues in the context of managing the design, manufacture, and/or operation of reliable products and systems. Here, the emphasis is on the business viewpoint, since unreliability has consequences in terms of cost, time wasted and, in certain cases, the welfare of an individual or even the security of a nation.

We shall discuss some of these topics in more detail in later chapters.

2.8   Summary

This chapter has dealt with basic reliability concepts. A seven-level hierarchical decomposition of an engineered object has been presented to facilitate better understanding of system function and reliability analysis. Definitions and classifications of failures, faults, failure modes, and failure causes have been presented.

The degradation of a system can be described in terms of the degradation in its performance, and that is characterized by a variable, $X(t)$. This variable indicates the system state or condition of the object as a function of age. Three different characterizations were presented: (i) a two-state characterization where the system can be either in a working or a failed state; (ii) a multi-state characterization involving a finite number of states; and (iii) a multi-state characterization with an infinite number of states.

The reliability of a system was defined as the probability that the system will perform its intended function for a specified time period when operating under normal (or stated) environmental conditions, and the related mean time to failure and failure rate function were also introduced.

The linking of component failures to system failures can be done using two different approaches:

- *FMEA* is a method used to identify the potential failure modes of the components of a system and determine the local and system effects of these failures.
- *FTA* is a deductive approach that starts with a specified system failure and then considers the possible causes of this failure. FTA uses logic gates to combine potential failure and events.

Formulae for the structure function and reliability function for various system configurations were developed. System RBDs are useful in developing these functions. Finally, a brief definition was provided for various components of reliability theory.

Review Questions

2.1 Explain the difference between essential, auxiliary, and protective functions of a system and provide examples.

2.2 Explain the difference between primary and secondary failures.

2.3 What is a command fault? Provide an example.

2.4 What is the difference between a fault and a failure?

2.5 Provide a classification of failure causes.

2.6 Provide three characterizations of the state of a system.

2.7 Define the reliability of a system.

2.8 Explain the relationship between the failure and reliability distributions.

2.9 Describe briefly two approaches for linking component failures to system failures.

2.10 List three objectives of failure modes and effects analysis (FMEA).

2.11 Describe, using a flow chart, the steps involved in FMEA.

2.12 Explain briefly the objective of fault tree analysis (FTA).

2.13 Discuss one limitation of FMEA.

2.14 Define reliability engineering, reliability science, and reliability management.

Exercises

2.1 A bicycle is used for transport of people and goods based on pedal power. It can be viewed as a system comprised of several components.
   (a) Describe the operation of the bicycle.
   (b) Carry out a multi-level decomposition of the system.
   (c) Make a list of the different failure modes of a bicycle.
   (d) Construct a fault tree for one of the failure modes identified in (c).
   (e) Construct an RBD for the fault tree constructed in (d).
   (f) Derive the reliability function of the bicycle in terms of the reliability function of the components based on the RBD in (e).

2.2 Diesel engines produce mechanical power by combusting diesel oil, as described in Example 1.3 of Chapter 1.
   (a) Carry out a multi-level decomposition of a diesel engine.
   (b) Make a list of the different failure modes of a diesel engine.
   (c) Construct a fault tree for one of the failure modes identified in (b).
   (d) Construct an RBD for the fault tree constructed in (c).
   (e) Derive the reliability function of the bicycle in terms of the reliability function of the components based on the RBD in (e).

2.3 A refinery plant converts crude oil into various products such as naphtha, gasoline, jet fuel, diesel, and so on. Among the subsystems of the plant is the pump subsystem (composed of several pumps and electrical motor units) that moves the liquid throughout the plant. Consider a pump motor unit.
   (a) Describe the operation of such a unit.
   (b) Carry out a decomposition of the unit.
   (c) Make a list of the different failure modes of the unit.
   (d) Construct a fault tree for one of the failure modes identified in (c).
   (e) Construct an RBD for the fault tree constructed in (d).

2.4 A wind power plant is an engineered object that produces electrical power from wind energy.
   (a) Describe the operation of such a plant.
   (b) List the different subsystems of such a plant.
   (c) A turbine is one of the subsystems. Carry out a multi-level decomposition of the turbine.
   (d) Identify some of the failure modes of a wind turbine.
   (e) Construct a fault tree for one of the failure modes identified in (d).

2.5 Rail transport (described briefly in Example 1.7 of Chapter 1) involves an infrastructure, and one of the main elements of it is the rail track.
   (a) Describe the main components of rail track.
   (b) Make a list of the different failure modes of rail track.
   (c) Construct a fault tree for one of the failure modes identified in (b).
   (d) Construct an RBD for the fault tree constructed in (c).

2.6 Road transport (described briefly in Example 1.8 of Chapter 1) involves an infrastructure consisting of highways (connecting cities and towns) and urban roads

(within a city). Consider a highway that links two big cities with several small towns
in between.

(a) Describe the various elements of the highway.

(b) One of the elements is a steel suspension bridge. Carry out a detailed decomposition
    of the bridge.

(c) Define partial and total bridge failures.

(d) Make a list of the different failure modes that can lead to total failure.

(e) Construct a fault tree for one of the failure modes identified in (d).

2.7 The failure rate function may have many complex shapes. Two of these are:

(a) Inverted bathtub.

(b) Roller coaster – increasing – decreasing – increasing.

Sketch the shapes of these two functions. Can you think of a real item whose failure rate
will exhibit these shapes?

2.8 Conduct an FMEA for the simple lighting circuit shown below.

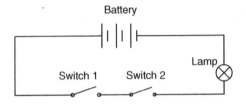

2.9 Match each block diagram shown below with the corresponding fault tree.

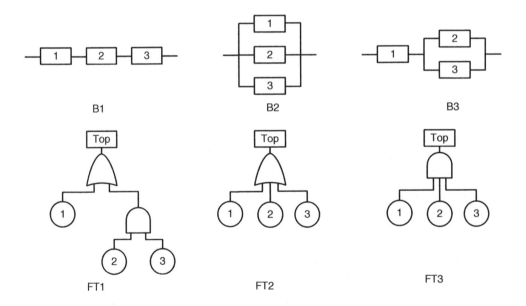

2.10 Construct a fault tree for the redundant fire pump system depicted below. Consider the top event "No water from the fire water system."

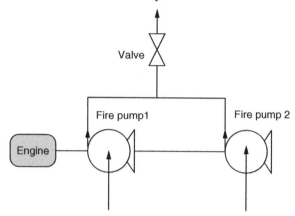

2.11 Consider the stereo system of Example 2.6. Now define system failure as corresponding to the system not being capable of producing stereophonic sound.
(a) Determine the corresponding RBD.
(b) Construct the corresponding fault tree.

2.12 Consider the domestic water heater shown below.

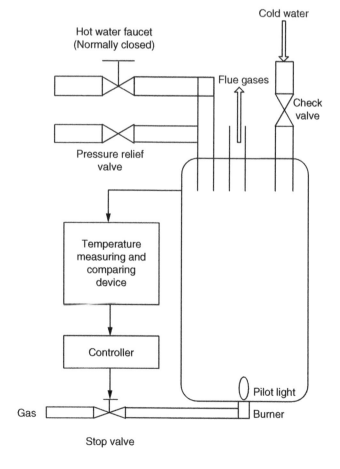

The heater takes cold water and gas as inputs and produces hot water as output. The water temperature is regulated by a controller that closes and opens the gas valve depending on pre-set temperature limits of the water in the tank. A device that measures and compares the temperature operates the controller. Reverse flow due to excessive pressure is prevented by the check valve in the water inlet pipe. The relief valve opens if the pressure exceeds a certain threshold.

(a) Carry out a three-level decomposition of the water heater system.
(b) Identify the function of the various components of the heater.
(c) Conduct the FMEA of the stop valve component.
(d) Construct a fault tree for the top event "No hot water."

References

Blache, K.M. and Shrivastava, A.B. (1994) Defining failure of manufacturing machinery and equipment. *Proceedings of the Annual Reliability and Maintainability Symposium*, pp. 69–75.

Henley, E.J. and Kumamoto, H. (1981) *Reliability Engineering and Risk Assessment*, Prentice-Hall, Englewood Cliffs, NJ.

IEC (International Electro-Technical Commission) IEC 50 (191) (1990) *International Vocabulary, Chapter 191: Dependability and Quality of Service*.

ISO 8402 (1986) *Quality Vocabulary*, International Standards Organization, Geneva.

US Department of Defense MIL-STD-882D (2000) *Standard Practice for System Safety*, US Department of Defense, Washington, DC.

3

System Degradation and Failure

Learning Outcomes

After reading this chapter, you should be able to:

- Define failure mechanisms;
- Explain the importance of understanding failure mechanisms for effective maintenance decision making;
- Classify failure mechanisms according to two broad categories: overstress and wear-out;
- Classify failure mechanisms by type of stress (mechanical, electrical, chemical, etc.);
- Describe briefly some common failure mechanisms;
- Describe the factors that affect the rate of degradation on components' material leading to failure;
- Identify failure mechanisms for different types of engineered objects.

Introduction to Maintenance Engineering: Modeling, Optimization, and Management, First Edition.
Mohammed Ben-Daya, Uday Kumar, and D.N. Prabhakar Murthy.
© 2016 John Wiley & Sons, Ltd. Published 2016 by John Wiley & Sons, Ltd.

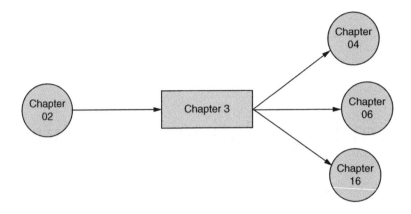

3.1 Introduction

Engineered objects can be viewed as systems of interconnected elements having multi-level decomposition, with components at the lowest level. The performance of the system degrades due to the degradation of the material and components of the object. Understanding the ways in which degradation leads to failure requires understanding of the underlying failure mechanisms.

The focus of this chapter is on an understanding of the underlying failure mechanisms involved in various maintained systems. Examples will be provided in different business areas. Knowledge of the failure mechanisms that cause engineering object failure is essential, not only to design and qualify reliable products but also to determine their remaining useful life once they are in operation. Estimation of remaining useful life (the operating time between detection and an unacceptable level of degradation) of an object is essential for condition-based maintenance.

This chapter is organized as follows. In Section 3.2 we define stress and strength of components, and these are used later to characterize failure of components. Section 3.3 provides various classifications of failure mechanisms and brief descriptions of some common mechanisms. The dynamic nature of stress and strength and the factors affecting them are discussed in Section 3.4. Section 3.5 illustrates common failure mechanisms causing degradation of products and plants and Section 3.6 deals with infrastructures. The importance of failure mechanism understanding in maintenance is discussed in Section 3.7. A summary of this chapter is provided in Section 3.8.

3.2 Failure Mechanisms

Failure is usually a result of the effect of deterioration. The deterioration process leading to a failure (also called the *failure mechanism*) is a complex process.

Definition 3.1

Failure mechanisms are physical, chemical, or other processes which lead or have led to failure (EN, 2010).

Failure mechanisms depend on the type of component (electrical, mechanical, pneumatic, etc.), material (wood, metal, composite, plastic, glass, etc.), manufacturing processes (annealing, casting, machining, etc.) and the operating environment – load (electrical, mechanical, thermal, etc.), chemical properties (pH level of gas or fluid in a pipe network), and so on. There are different classifications of failure mechanisms, as discussed in Section 3.3.

3.2.1   Effect of Materials and Manufacturing

Different materials have different mechanical, physical, chemical, and fabrication properties. Mechanical properties describe their response to applied loads or forces in terms of elastic deformation, plastic deformation, and fracture. Physical properties such as densities and thermal properties are important in many engineering applications. For example, in aerospace applications, the strength-to-weight ratio for a commercial aircraft is an important parameter to control weight and meet design requirements. Chemical properties describe the interaction between materials and various substances in the environment, such as oxygen, sulfur, and chlorine, to form other substances, leading, for example, to corrosion that degrades the performance of the material. Fabrication properties determine the response of materials to various fabrication methods. For example, castability is the ability of the material to be formed into useful shapes. Manufacturing processes may introduce some flaws in a component such as voids, pores, and small cracks that may be the starting points of some failure mechanism. Flaws can also be introduced by improper heat treatment. Heat can harden (annealing) or soften material.

3.2.2   Stress and Strength of a Component

Components are subjected to load when put into operation and the load induces stress. Stress and strength concepts can provide a useful explanation of the mechanisms leading to component failure.

Definition 3.2

Stress is the intensity of the internally distributed forces or components of forces that resist a change in the volume or shape of a material that is or has been subjected to external forces. Stress can be either direct (tension or compression) or shear.

Definition 3.3

Strength is the property of a metal part that resists the stresses imposed upon the part.

There are many forms of stress – mechanical (force on component), electrical (voltage across or current flowing through), hydraulic (pressure – fluid, gas, etc.), and thermal (heat). Many of these stresses may be acting simultaneously on a particular component. Components fail because their atomic structures can no longer take the stress due to the imposed load.

Atomic structures fail for two reasons – stress causes the atomic bonds to separate or the atomic bonds are attacked and removed. The stress on the component is determined by environmental and operating conditions when put into use. This is also an uncertain variable that needs to be characterized by a probability density function.

The strength of a new component depends on the decisions made during design and manufacture. Design issues will be discussed in Chapter 5. Due to variability in manufacturing, the strength of a new component is uncertain and needs to be characterized by a probability density function. The two possible scenarios are as follows.

In scenario 1 the two density functions do not overlap (see Figure 3.1a) and the stress is always less than the strength. In this case, the component can withstand the load and is 100% reliable when put into operation. In scenario 2 the two density functions overlap (see Figure 3.1b). In this case, there is a non-zero probability of the component failing immediately when put into use. This occurs when the strength and stress are in the overlapping region (indicated by the thick line in Figure 3.1b) and stress exceeds strength. In this case the reliability of the component is less than one.

Static reliability is the probability that the new component has strength greater than the stress to which it is subjected.

The strength of a component decreases with age and/or usage. In this case, the component will operate satisfactorily until the instant when strength falls below stress. This issue is discussed later in the chapter.

3.3 Classification of Failure Mechanisms[1]

Failure mechanisms can be grouped into two broad categories: (i) overstress mechanisms and (ii) wear-out mechanisms. In the former case, an item fails only if the stress to which the item is subjected exceeds the strength of the item. If the stress is below the strength, the stress has no permanent effect on the item. In the latter case, however, the stress causes damage that usually accumulates irreversibly. The accumulated damage does not disappear when the stress is removed, although sometimes annealing is possible. The cumulative damage does not cause any performance degradation as long as it is below the endurance limit. Once this limit is reached, the item fails. The effects of stresses are influenced by several factors – geometry of

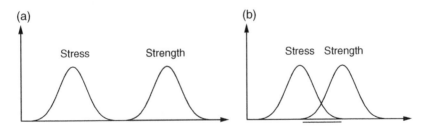

Figure 3.1   Stress and strength – two scenarios. (a) No overlap and (b) overlap.

---

the part, constitutive and damage properties of the materials, manufacturing, and operational environment, as discussed earlier.

There are many different failure mechanisms in each of the above two groups and some examples are listed in Table 3.1.

We will briefly discuss some of these mechanisms later in the chapter.

The stresses that trigger the failure mechanism can be mechanical, electrical, thermal, radiation and/or chemical. This forms the basis for the alternative categorization summarized in Table 3.2.

Often an item failure can be the result of interactions among these various types of stresses. For example, a mechanical failure can be due to a thermal expansion mismatch or stress-assisted corrosion.

A proper understanding of the response of materials to the stresses that may be encountered, as well as the effect of other factors (geometry of the part, raw materials, manufacturing, etc.) is essential for the modeling of component reliability based on the mechanism of failure. In the next two sections, we give brief descriptions of the various failure mechanisms leading to overstress and wear-out failures.

3.3.1 Overstress Failure Mechanisms

In this section we give a qualitative characterization of the various mechanisms that may lead to overstress failures.

Table 3.1 Examples of overstress and wear-out failure mechanisms (Dasgupta and Pecht, 1991).

Overstress failures		Wear-out failures	
Brittle fracture	Large elastic deformation	Wear	Diffusion
Ductile fracture	Interfacial de-adhesion	Corrosion	Radiation
Yield		Dendritic growth	Fatigue crack initiation
Buckling		Inter-diffusion	Creep
		Fatigue crack propagation	

Table 3.2 Failure mechanism classification by type of stress.

Types of failure	Failure mechanisms
Mechanical failures	Elastic and plastic deformation, buckling, brittle and ductile fracture, fatigue crack initiation and crack growth, and creep and creep rupture
Electrical failures	Electrostatic discharge, dielectric breakdown, junction breakdown in semiconductor devices, hot electron injection, surface and bulk trapping, and surface breakdown
Thermal failures	Heating beyond critical temperatures (for example, melting point), and thermal expansions and contractions
Radiation failures	Radioactive containment, and secondary cosmic rays
Chemical failures	Corrosion, oxidation, and surface dendritic growth

3.3.1.1 Large Elastic Deformation

Elastic deformation typically occurs in slender items. Failure is due to excessive deformation under overstress. Being elastic, the deformations are reversible and therefore do not cause any permanent change in the material. These types of failures occur in structures such as long antennas and solar panels, where large deformations can trigger unstable vibration modes and thereby affect the performance. The overstress is due to external factors and failure occurs when the stress exceeds some safe limit.

3.3.1.2 Yield

A component that is stressed past its yield strength results in an irreversible plastic strain. This strain causes a permanent deformation due to a permanent change in the material. This phenomenon most often occurs in components made of metals which are ductile and are crystalline materials (as opposed to brittle materials which fracture with no plastic deformation). Plastic deformation is an instantaneous deformation occurring due to a slip motion (called *dislocation*) of planar crystal defects in response to an applied stress. In contrast, *creep* (to be discussed later) is a time-dependent deformation. The two main types of dislocations are edge and screw dislocations. Permanent deformation may or may not constitute a failure, depending on the context. In the case of a spring, a yield failure occurs when the item fails to operate as a spring after being subjected to an overstress. Another example is a connector losing its clamp pressure.

3.3.1.3 Buckling

Buckling is a phenomenon that occurs in slender structures under compressive overstress. Deformation in the direction of the compressive load can suddenly change at a critical point, resulting in an instantaneous and catastrophic deformation in a direction perpendicular to the loading direction. The load at which buckling occurs is called the *critical load* and the new deformation mode is termed *post buckling*. Buckling is a structural rather than a material failure mechanism. Post-buckling deformation often involves large deformation and/or finite rotation of the structure.

3.3.1.4 Brittle Fracture

In brittle materials (such as glass and ceramics), high stress concentration can occur at local microscopic flaws under overstress. This excessive stress can cause a failure by sudden catastrophic propagation of the dominant micro-flaw. The failure is not only related to the applied stress on the component but also depends on the flaw size. Failure resulting from brittle fracture is also referred to as *cracking*.

Brittle fracture typically occurs at pre-existing microscopic flaws as a result of nucleation and sudden propagation of cracks. Cleavage fracture is the most common brittle type of fracture and it occurs by direct separation along crystallographic planes and is due to tensile breaking of molecular bonds.

3.3.1.5 Ductile Fracture

In ductile fracture, the failure is due to sudden propagation of a pre-existing crack in the material under external stress. It differs from brittle fracture in the sense that there is large-scale yielding at the tip of the crack which preceded the crack propagation.

Ductile fracture is dominated by shear deformation, and occurs by nucleation and coalescence of micro voids due to pile-ups of dislocation at defects such as impurities and grain boundaries.

Example 3.1   Gas Pipeline[2]

A natural gas transmission pipeline ruptured in San Bruno, CA in 2010. Eight people were killed, many injured, and 38 homes were destroyed. The examination of the seam by optical and electron microscopy concluded that the rupture was initiated at a pre-existing crack in the weld that had grown first by ductile fracture and then by fatigue.   ∎

3.3.2   Wear-Out Failure Mechanisms

In this section we give a qualitative characterization of the different failure mechanisms leading to wear-out failures.

3.3.2.1   Fatigue Crack Initiation and Propagation

When a component is subjected to a cyclic stress, failure usually occurs at stresses significantly below the ultimate tensile strength and is due to the accumulation of damage. Such failures are a result of incremental damage that occurs during each load cycle and accumulates with the number of cycles. Failures of this type are termed *fatigue failures*. Fatigue causes more failures than any other mechanism.

Fatigue failure is comprised of two stages: crack initiation and crack propagation. A crack typically develops at a point of discontinuity (such as a bolt hole or a defect in the material grain structure) because of local stress concentration. Once initiated, a crack can propagate stably under cyclic stress until it becomes unstable and leads to overstress failure. Fatigue is the leading cause of wear-out failures in engineering hardware. Typical examples are fatigue cracking in airframes (the cyclic loading being the result of landings and take-offs), rotating shafts, reciprocating components, and large structures such as buildings and bridges.

3.3.2.2   Corrosion and Stress Corrosion Cracking

Corrosion is the process of chemical or electrochemical degradation of materials. The three common forms of corrosion are uniform, galvanic, and pitting corrosion.

As the name suggests, in uniform corrosion, the reactions occurring at the metal–electrolyte interface are uniform over the surface of the item. Continuation of the process depends on the

---

[2] For more details see Richards (2013).

nature of the corrosion product and the environment. If the corrosion product is washed off or otherwise removed, fresh metal is exposed for further corrosion.

Galvanic corrosion occurs when two different metals are in contact. In this case, one acts as a cathode (where a reduction reaction occurs) and the other as an anode (where corrosion occurs as a result of oxidation).

Pitting corrosion occurs at localized areas and results in the formation of pits. The corrosive conditions inside the pit accelerate the corrosion process.

Stress corrosion cracking (SCC) is an interaction between the mechanisms of fracture (for example, resulting from fatigue) and corrosion that occurs because of the simultaneous action of mechanical stress and corrosion phenomena. The corrosion reduces the fracture strength of the material. The process is synergistic – one process assisting the other in leading to item failure.

Corrosion is a very pervasive problem in some engineering environments – for example, off-shore rigs operating in a salty environment and chemical plants, to name a couple.

Example 3.2   Chemical Plant[3]

A Methyl Methacrylate (MMA) process plant was shut down after four months of operation. Several types of stainless steel are utilized in process plants for their high resistance to corrosion, good weldability, and superior material properties at high temperatures. For this plant, failure analysis revealed that the failure was due to SCC caused by the chloride that remained in the pipe. Corrosion pitting occurred on the inside surface of the pipe. The stress corrosion cracking started from the pits and grew out through the thickness of the pipe. Concentrated chloride was found in the deposit stuck to the pipe in addition to the pre-process MMA materials. Many work-hardened grains were observed in the area of the SCC, providing evidence of high residual stress due to welding, which could have served as the driving force for SCC.   ■

3.3.2.3   Wear

Wear is the erosion of material resulting from the sliding motion of two surfaces under the action of a contact force. Erosion can be due to physical and chemical interactions between the two surfaces. The various microscopic physical processes, by which the particles are removed as wear debris, are called *wear mechanisms* and the engineering science dealing with the study of such contacts is called *tribology*.

Wear mechanisms can be classified broadly into five categories:

- *Adhesive wear:* The molecular attractions existing between two relatively moving surfaces create adhesion between the touching asperities. If the adhesive strength is greater than the internal cohesive strength of the material, there is a tendency to create a wear particle after several cycles of contact.
- *Abrasive wear:* When a hard material is sliding against a soft material.
- *Surface-fatigue wear.*

---

[3] For more details see Yang, Yoon, and Moon (2013).

- *Corrosive (chemical) wear:* Sliding surfaces may wear by chemically reacting with the partner surface or the environment or both. The oxide layers resulting from reactions with the environment may have a protective role unless the thickness tends to grow during the cycle contact process. If the oxide layer grows, it becomes liable to break in brittle fraction, producing wear particles. Hard, broken-off oxide particles may then profoundly affect subsequent wear life as abrasive agents. If soft, ductile debris results, however, it may assume a protective role for the contact.
- *Thermal wear (mainly in polymers):* In polymers, temperature greatly affects mechanical behavior.

Wear erosion may be uniform (for example, wearing-away of piston rings in an internal combustion engine) or non-uniform (for example, pitting in gear teeth and cam surfaces).

Example 3.3   Aviation Gas Turbine[4]

Modern aviation gas turbine engines are considered to be very reliable since failures in service are rare. However, the fact that in-service failures are rare is due to the stringent standards imposed during turbine inspections. Most failures are detected at the incipient stage, so that appropriate action is taken to prevent in-service failure. The following are the common failure mechanisms found in gas turbine blades:

- *Mechanical damage:* Gas turbines ingest large amounts of air that may contain solid materials which can cause damage through erosion or impact. In particular, abrasive material such as sand causes damage to the blades by abrasive wear.
- *High temperature damage:* Ingested solid materials usually affect compressor blades. Turbine blades operate at high temperatures. The most probable failure mechanisms affecting turbine blades include creep, fatigue, and high-temperature corrosion.
- *Creep failures:* All turbine blades and often the high stages of compressor blades are subject to creep, since they operate at high temperatures and stresses. In fact, creep is the life-limiting mechanism for all blades operating under such conditions.
- *Fatigue failures:* Gas turbine blades are designed to avoid high cycle fatigue unless there is some initiating damage due to ingested debris or the presence of a manufacturing defect.
- *Corrosion failures:* Compressor and turbine blades are exposed to corrosive conditions due to the presence of sodium and chlorine in the form of sea air, for example, or due to the presence of harmful elements in the fuel such as sulfur, vanadium, or lead.   ∎

The following example shows the typical failure mechanisms for pumps.

---

[4] For more details, see Carter (2005).

Example 3.4   Pump[5]

In the case of a pump, there are many failure modes and several causes for each. A short list is given below.

1. *Seal leakage:* Several failure mechanisms can cause this type of failure:
   - seal surfaces wear when no water is present in the pump during operation;
   - thermal shock due to sudden cooling of heated seals causes fracture;
   - vibration of the pump (for example, by cavitation in the impeller, misalignment, or wrong assembly) yields high loads on the seals, producing overload damage.
2. *Insufficient yield:* The impeller gets damaged due to one of several mechanisms:
   - corrosion due to prolonged exposure to sea water;
   - crevice corrosion due to elevated temperature in the pump;
   - erosion due to (sand) particles in sea water flow;
   - fatigue damage due to cavitation in the impeller.
3. *Bearing damage:* This damage is caused by the following mechanisms:
   - local fatigue and wear damage due to vibration of non-operative pumps (caused by vibration of other machines in the same room);
   - wear damage due to misalignment (high loads), bad lubrication, or bad assembly of the bearings.   ∎

3.3.3   Other Failure Mechanisms

The following three mechanisms are common for electronic products.

3.3.3.1   Electromigration

Electromigration refers to the migration or displacement of metal atoms due to the impact of moving electrons. This atom displacement causes voids in some areas of the lattice of the metal and hillocks (accumulation of atoms) in other areas. Voids decrease the cross-sectional area of the conductor, which increases the current density, which, in turn, accelerates the phenomenon. Voids accumulate, resulting in an open failure. Hillocks lead to a short circuit failure.

3.3.3.2   Dielectric Breakdown

A dielectric is an insulator that can be polarized by an applied electric field. The insulating material between the metallic plates of a capacitor is a common example of a dielectric. Each dielectric material has a maximal electric field that it can sustain beyond which the insulator is destroyed, allowing current to flow and causing a short or leakage at the point of breakdown.

There are also failure mechanisms due to radiation.

---

[5] For more details, see Tinga (2012).

3.3.3.3 Radiation Mechanisms

Materials used to construct nuclear power facilities are exposed to high-energy radiation. Neutron flux is the most damaging mode of radiation to a material as it alters its microstructure. The higher the radiation level, the higher the damage to the material microstructure by creating voids and dislocations. These voids function as sites of stress concentration for premature fracture of the material. Since a defective crystalline structure has lower density, swelling of metals also occurs. This causes metal components to exceed dimensional tolerances, with drastic consequences for shafts and other moving components.

3.4 Dynamic Nature of Stress and Strength

At the design stage, engineered objects are designed for some nominal level of performance. Components and materials are selected to ensure the desired level of performance under some nominal ranges for usage and operating environment. However, the performance of the object degrades due to the degradation of the materials and components of the object.

The rate of change in strength (component degradation) depends on the material selected, the manufacturing method, the operating environment, and the usage mode and intensity.

The usage of engineering objects may vary in terms of mode and intensity. Usage may be intermittent or continuous under various conditions of load and speed. Higher loads imply a higher rate of degradation. Other usage conditions that may have an effect include inadequate maintenance, improper start-up or shutdown procedure, and unanticipated service conditions. For example, during start-up or shutdown, certain components may experience severe conditions different from those during normal operation.

All these factors can act jointly or independently to increase the stress and/or reduce the strength of the component, in time leading to its degradation and failure, as shown in Figure 3.2. Note that the rate of deterioration is a random process that depends on time and usage intensity, as illustrated by Example 3.5.

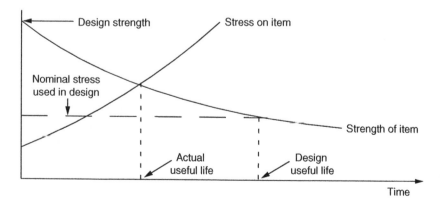

Figure 3.2 Change of strength and stress with time, leading to failure.

Example 3.5   Aircraft – Body and Engine[6]

An aircraft is a complex system made up of many subsystems. Table 3.3 looks at the different failure mechanisms leading to failure for the body and engine of an aircraft and the timescales and usage factors that affect the rate of deterioration involved.

Table 3.3   Failure mechanisms for the body and engine of an aircraft.

Subsystem	Timescale/usage intensity	Failure mechanism
Body	Calendar year	Corrosion
	Time in air	Wear
		Accumulation of fatigue
	Number of landings (take offs) or flights	Accumulation of fatigue
		High-amplitude loading
Engine	Calendar year	Corrosion
	Operation time	Wear
		Accumulation of fatigue
	Number of operation cycles	Temperature cycling
	Time in take-off regime	High temperature

Failure is often the cumulative and combined effect of several of these failure mechanisms.   ■

3.5   Degradation of Products and Plants

Plants are a collection of elements and the degradation of a plant is due to degradation of the different elements. The degradation of elements is due to degradation of their components. As such, the focus is on degradation at the component level. In this section, we look at examples of degradation and failure mechanisms in the context of various engineered objects.

Example 3.6   Automobile

We consider some components of a combustion engine to illustrate two of the failure mechanisms and their impacts:

- Mechanical losses resulting from friction in piston assemblies, bearings, and the valve train account for nearly 15% of the total energy loss in an internal combustion engine (Taylor, 1998).
- Wear between pistons and cylinders causes the leakage of combustion gases from inside the cylinders, leading to a degradation of performance. The engine becomes less efficient as the vehicle consumes more fuel than would otherwise be the case. This degradation can lead to failure if the clearance between the piston ring and cylinder increases to the point where there is no compression of combustion gases.

---

[6] Adapted from Kordonsky and Gertsbakh (1995).

Example 3.7   Boiler Feed-water Pumps

Boiler feed-water pumps are used in chemical and power plants. There are three forms of mechanical failure mechanisms that typically exist in the working environment of a pump:

- *Corrosion:* Water is not very corrosive by itself. However, boiler feed-water is usually treated using processes that involve acid and caustic substances. If the treatment process is not controlled properly, it is possible for the feed-water to become corrosive.
- *Erosion:* Boiler feed-water is typically very hot and when the suction pressure gets too low, cavitation will occur, generating small bubbles in the pump suction. These bubbles will have the same effect as solid debris and will result in erosion of pump components in the flow path. If cavitation is not managed properly, the pump impeller can be damaged to the extent where the pump can no longer produce the required discharge pressure.
- *Fatigue:* Fatigue is one of the most common failure mechanisms affecting rotating equipment. The speed of rotating components will result in a significant number of fatigue cycles being quickly accumulated if there is any problem that places the rotating element under stress. Common situations that can place the rotating element under stress include:
  - Misalignment of the pump shaft with the shaft of the driver;
  - Improper balance of the coupling or some other rotating element;
  - Improper installation of the inlet and outlet piping that places stress on those connections transmitting the stress to the pump case.

Example 3.8   Artificial Heart Valve[7]

Artificial heart valves must be designed to survive greater than 10^9 cycles[8] over 40 years of operation. Thus, fatigue represents one of the primary driving forces for safe operation of these devices. More specifically, fatigue cracking from initial weld defects is the main failure mechanism in this device. The inability to maintain the long-term performance of such a critical device may lead to catastrophic failure and patient loss of life.   ■

Example 3.9   Space Shuttle O-ring

The Space Shuttle *Challenger* disaster took place on 28 January 1986, when the vehicle burst into flames 73 seconds after lift-off, leading to the death of all seven crew members. The vehicle broke apart after an O-ring seal in its right solid rocket booster (SRB) failed. This failure was due to a design problem. A warm compressed O-ring returns to its original shape more quickly than a cold O-ring when compression is relieved. So the behavior of the ring material at different temperatures is the source of the problem. In the orbiter's configuration, an O-ring is designed to follow the opening in the SRB joint; however, a cold one may not. At launch, the ambient air temperature was too cold. The O-ring was very slow in returning to its normal rounded shape and did not follow the gap, thus allowing hot gases to escape, causing the disaster.   ■

---

[7] Koppenhoefer, Crompton, and Dydo (2006).
[8] The cardiac cycle refers to a complete heartbeat from its generation to the beginning of the next beat.

3.6   Degradation of Infrastructures

As discussed in the previous section, the degradation of an infrastructure is due to the degradation of the material used in the construction of the object.

3.6.1   Rail Infrastructures

Rail track is an important component of rail infrastructure. The elements of a rail track are depicted in Figure 3.3. In this section we focus on the degradation of rail tracks.

The ballast deteriorates over time due to crushing and movement by the weight of trains. This leads to the tracks becoming uneven, causing swaying, rough riding, and possibly derailments. The deterioration of the rails is mainly due to wear and fatigue. They occur due to a number of causes such as rail manufacture (for example, a *tache ovale*[9]), damage caused by inappropriate handling, installation, and use (for example, a wheel burn defect[10]), or the exhaustion of the material of the rail's (steel) inherent resistance to fatigue damage.[11] Many forms of RCF-initiated defects are within this group (for example, squats[12]).

There are several major contributors to rail deterioration and these include:

- Operational factors (such as train speed, axle load, million gross tons (MGT) moved, etc.);
- Design factors (rail-wheel material type, size, and profile, track construction, curvature, traffic type, etc.);
- Environmental factors (temperature, rain, snow, etc., leading to, for example, the track bed condition being affected by water blockage, soil subsidence, etc.).

The rail head is worn away by wheels running on its surface and deteriorates due to abrasive contact with the base plate or sleeper on its underside. Corrosion leads to loss of a rail section

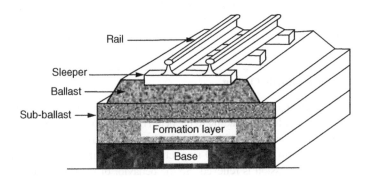

Figure 3.3   Rail track elements.

---

[9] *Tache ovale*: A subsurface defect formed around 10–15 mm below the rail head surface and caused by hydrogen accumulation during manufacture of the rail or due to poor welding.
[10] A *wheel burn defect* is caused by spinning wheels.
[11] This process is called *rolling contact fatigue* (RCF).
[12] Squats are surface-initiated crack defects formed by RCF that appear in the crown area of straight rail sections.

and the surface crack itself reduces the fatigue resistance of the rail. Rail wear, rolling contact fatigue (RCF), and plastic flow are major problems for railway operators.

Rail break is the last phase of the crack development process. As the crack continues to increase in length as well as depth, stress concentration also increases and finally a rail break occurs. Some of the cracks disappear early on in their development as a result of wear and tear, whilst most of them are removed by grinding operations.

3.6.2 Road Infrastructures[13]

The deterioration of a road depends on the materials used in the construction of the road and several other factors. In the case of asphalt pavement, the deterioration is because the materials that make up asphalt begin to break down over time and are affected by elements such as rain, sunlight, and chemicals that come into contact with the pavement surface. The liquid asphalt binder that is the "glue" of the pavement begins to lose its natural resistance to water, allowing it to penetrate into and underneath the pavement. Once this happens, the surface can quickly fall prey to a number of different types of deterioration. The premature deterioration of asphalt pavement is usually due to failures in construction and/or human error, including the following factors:

- Insufficient or improperly compacted base below the asphalt;
- Over or under compaction of the asphalt;
- Improper temperature of the asphalt when applied;
- Poor drainage.

Cracks, potholes, edge defects, depressions, and corrugations are the significant road defects observed in the field. Traffic, age, road geometry, weather, drainage, and construction quality as well as construction material and maintenance policy are the major factors that affect the deterioration of a road.

The damage to a road surface can be classified into several groups with certain features, which are described next.

- **Texture:** The features of damage include:
  - *Raveling:* Loss of aggregate (used in road construction) due to (i) cohesive failure of the bituminous mortar or (ii) adhesive failure in the adhesive zone.
  - *Skidding resistance:* Characterization of the cumulative effects of snow, ice, water, loose material, and the road surface on the traction produced by the wheels of a vehicle.
- **Evenness:** The features of damage include:
  - *Transverse and longitudinal evenness:* Measurement of longitudinal (transverse) profiles for determination of rutting. A rut is a sunken track or a groove made by the passage of vehicles within pavement layers that accumulates over time.
  - *Irregularities:* Something irregular, such as a bump in a smooth surface.

---

[13] This section is based on material from Worm and van Harten (1996).

- o *Roughness:* Deviations of the surface from the true planar surface, with characteristic dimensions that affect vehicle dynamics, ride quality, dynamic loads, and drainage. This is defined using the International Roughness Index (IRI) – deviations (in meters) per kilometer. Roughness is a function of age, strength, traffic loading, potholes, cracking, raveling, rutting, environment, and so on.
- **Soundness:** The features of damage include:
  - o *Longitudinal (transverse) cracking:* Cracks that run along (perpendicular to) the road.
  - o *Craze:* A fine crack in a surface of the road.
  - o *Pothole:* An open cavity in the road surface with at least 150 mm diameter and at least 25 mm depth.
- **Marginal strip:** The features of damage include:
  - o *Edge damage:* Loss of bituminous surface material (and possibly base materials) from the edge of the pavement, expressed in square meters per km.
  - o *Kerb damage.*
- **Miscellaneous:** The features of damage include:
  - o *Water run-off.*
  - o *Damage to the verge:* A strip of grass or other vegetation beside a road.

The quality of a road is often described by a function involving one or more of these features. Figure 3.4 illustrates quality deterioration over time (as defined by some feature such as raveling) with no maintenance actions.

The *quality standards* (also referred to as *norms*) for a road are derived from the lowest acceptable values for these features. They can be (i) local – for segments (for example, 100 m in length) of a road (or lane) or (ii) global – the whole length of the road. In Figure 3.4 at time t (since the construction of the road), the quality is Q_t and Y denotes the time when the quality reaches the minimum acceptable level, Q_{min}, at which instant corrective maintenance (CM) action is needed. The interval $(Y - t)$ provides a window over which preventive maintenance (PM) action can be initiated to avoid the need for CM action.

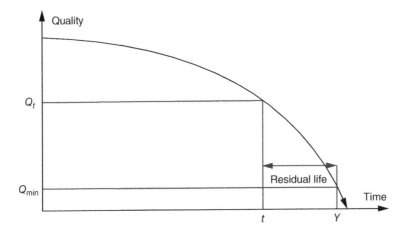

Figure 3.4   Deterioration of road quality over time.

3.6.3 Concrete Infrastructures

Dams, multi-story buildings, bridges, and so on, are massive concrete structures. The degradation of concrete structures over time can be defined, in a broad sense, as the loss of certain properties due to the result of chemical, physical, or mechanical processes, or combinations thereof. Chlorides are present in the concrete and result in chlorine corrosion of rebars (bars used in reinforcing the concrete), leading to spalling (spalls are flakes of a material that are broken off a larger solid body and are a result of water entering and forcing the surface to peel, pop out, or flake off) and delamination. The latter occurs in reinforced concrete structures due to reinforcement corrosion, which oxidizes the metal of the reinforcement. This increases the volume of the metal and causes a wedge-like stress on the concrete. This force eventually overcomes the relatively weak tensile strength of concrete, resulting in a separation or delamination of the concrete above and below the reinforcing bars.

The condition of concrete bridges is based on a rating involving 10 levels, with the level being a function of the degree of spall, delamination, electrical potential, and chloride content, as shown in Table 3.4.

Example 3.10 shows failure mechanisms for bridge collapses.

3.6.4 Pipeline Infrastructures

From a technical point of view, pipelines are structures that include straight pipes, nozzles, pipe-bends, dissimilar welded joints, and so on. Their operating conditions can be quite severe due to internal pressure and cyclic loading (vibration), and to the influence of internal and external environments (such as temperature, corrosion, etc.).

A major cause of pipeline failure is corrosion. External corrosion and cracking are the major threats and the principal mechanisms of buried pipeline deterioration. Internal corrosion is normally caused by chemical reaction between the pipeline material and a fluid or gas, such as carbon dioxide (CO_2), hydrogen sulfide (H_2S), or oxygen (O_2). Corrosion can take the form of general or localized metal loss from a pipe, and it may also give rise to cracking in the

Table 3.4 Concrete bridge condition ratings (FHWA, 1979).

Rating	Condition indicator (% deck area)			
	Spall	Delamination	Electrical potentials	Chloride content (lb/yd^3)
9	None	None	0	0
8	None	None	None > 0.35	None > 1.0
7	None	<2%	45% < 0.35	None > 2.0
6	<2% spall or sum of all deteriorated or contaminated deck concrete <20%			
5	<5% spall or sum of all deteriorated or contaminated deck concrete <20–40%			
4	>5% spall or sum of all deteriorated or contaminated deck concrete <40–60%			
3	>5% spall or sum of all deteriorated or contaminated deck concrete >60%			
2	Deck structural capacity grossly inadequate			
1	Deck repairable by replacement only			
0	Holes in deck – dangers of other sections of deck failing			

Example 3.10   Bridges

Bridges collapse for different reasons and their consequences can be severe, as indicated in Table 3.5.

Table 3.5   Reasons for bridge collapse.

Bridge	Year	Reason	Consequences
Mianus River Bridge, Greenwich, Connecticut, USA	1983	Metal corrosion and fatigue deferred maintenance	Four killed, five injured
Sgt Aubery Cosens VC Memorial Bridge, Latchford, Ontario, Canada	2003	Fatigue fracture of three steel hanger rods	Partial failure of bridge deck
Highway 19 overpass at Laval, Quebec, Canada	2006	Shear failure due to incorrectly placed rebar, low quality concrete	Four killed, six injured, bridge demolished and rebuilt in 2007
Myllisilta Bridge Turku, Finland	2010	Structural failure of both piers	Bridge bent 143 cm, demolished

pipeline material. The rate of corrosion depends on the pipe materials, the type of products being transported through the pipelines, and the corrosion inhibitor.

3.7   Failure Mechanisms and Maintenance

In the life cycle of a product, several failure mechanisms may be activated by different environmental and operational conditions (characterized by parameters and variables) that can change over time. An important example of interactions between degradation processes is corrosive wear, which is mechanical wear that is accelerated by chemical damage to the worn material. However, only a few operational and environmental parameters and failure mechanisms are, in general, responsible for the majority of failures. High-priority mechanisms are those select failure mechanisms that may cause the product to fail earlier than the product's intended life duration. These mechanisms occur during the normal operational and environmental conditions of the product's application. Identifying high-priority failure mechanisms provides effective utilization of resources. Failure Modes, Mechanisms, and Effects Analysis (FMMEA) is a method similar to Failure Modes and Effects Analysis (FMEA) that focuses on failure mechanisms rather than failures and can be used for the purpose of prioritizing failure mechanisms.

Effective decision making in maintenance requires an understanding of the mechanisms leading to failures. This understanding enables proper modeling in order to predict the remaining life and plan maintenance actions. Understanding of failure mechanisms is also crucial for designing high-quality products through proper selection of materials and manufacturing processes.

3.8   Summary

The performance of an engineered object degrades due to the degradation of the materials and components of the object. Understanding the ways in which degradation leads to failure requires understanding the underlying failure mechanisms. This is essential for developing reliable objects at the design stage and for predicting failure for effective maintenance decision making during operation.

Failure mechanisms can be divided into two broad categories: (i) overstress mechanisms and (ii) wear-out mechanisms. In the former case, an item fails only if the stress to which the item is subjected exceeds its strength. If the stress does not exceed the strength, the stress has no permanent effect on the item. In the latter case, however, the stress causes damage that usually accumulates irreversibly. The accumulated damage does not disappear when the stress is removed, although sometimes annealing is possible. The cumulative damage does not cause any performance degradation as long as it is below the endurance limit. Once this limit is reached, the item fails. The effects of stresses are influenced by several factors – geometry of the part, constitutive and damage properties of the materials, manufacturing, and the operational environment.

The stresses that trigger the failure mechanism can be mechanical, electrical, thermal, radiation, and/or chemical. Several examples have been provided in the chapter to illustrate failure mechanisms affecting products, plants, and infrastructures.

Review Questions

3.1   Define failure mechanisms.

3.2   What are the factors that affect a failure mechanism?

3.3   Explain the effect of material on failure mechanisms.

3.4   Explain the effect of manufacturing on failure mechanisms.

3.5   Define stress and strength.

3.6   Explain failure in terms of stress and strength.

3.7   What are the different types of stresses that may be acting on components? Give an example of each.

3.8   Provide at least two classifications of failure mechanisms.

3.9   Give an example of an overstress failure mechanism and describe it briefly.

3.10 Give an example of a wear-out failure mechanism and describe it briefly.

3.11 Explain the importance of understanding failure mechanisms for sound maintenance practice.

Exercises

3.1   Following on from Exercise 2.1, describe the different failure mechanisms for the following two components of a bicycle:
      (a) Tires.
      (b) Wheel bearings.

3.2   Following on from Exercise 2.2, describe the different failure mechanisms for the following components of a diesel engine:
      (a) Bearings of the crank.
      (b) Fuel injector.

3.3 Following on from Exercise 2.3, describe the different failure mechanisms for the following components:
   (a) Valves that control the flow of fluids in a refinery.
   (b) Heating elements to heat the fluid.

3.4 Following on from Exercise 2.4, describe the failure mechanisms of the following components of a wind turbine:
   (a) Bearings.
   (b) Blades.
   (c) Gearbox.

3.5 Following on from Exercise 2.5, describe the different failure mechanisms for the following components of a rail infrastructure:
   (a) Rail track.
   (b) Ballast.
   (c) Concrete bridges.

3.6 Following on from Exercise 2.6, describe the different failure mechanisms for the following components of a road infrastructure:
   (a) Traffic lights at an intersection.
   (b) The road surface on a highway.

3.7 A seat belt in a car is a safety device consisting of a strap that clamps to protect passengers and the driver. The clamping mechanism is housed in a plastic casing. Discuss the degradation of the plastic and how it may lead to the failure of a seat belt and the possible consequences.
   Hint: Failure may be (i) unable to clamp and (ii) unable to unclamp.

3.8 List the different kinds of material used in a passenger vehicle. Describe the mechanisms that lead to degradation of the following components of the vehicle:
   (a) Chassis.
   (b) Paint work.
   (c) Windshield wiper.

3.9 List the different components of a gas-heated boiler used for supplying hot water to a building. Describe the degradation mechanisms of the different components.

3.10 Carry out a literature review of the failures of steel bridges and summarize the different failure causes and mechanisms leading to bridge failure.

3.11 Repeat Exercise 3.10 for concrete dams.

3.12 List the chemical properties of crude oil and how they affect the degradation of pipes in a petroleum-refining plant.

3.13 The failure of composite materials is, unlike metals, a complex, multi-stage process. The failure of a composite sample may be triggered in a certain "mode," but its propagation and final failure modes may be significantly different. Given the widespread use of composites in many applications, describe the common failure mechanisms of these materials.

References

Blischke, W.R. and Murthy, D.N.P (2000) *Reliability: Modelling, Prediction, and Optimization*, John Wiley & Sons, Inc.: New York.

Carter, T.J. (2005) Common failures in gas turbine blades, *Engineering Failure Analysis*, **12**(2): 237–247.

Dasgupta, A. and Pecht, M. (1991) Material failure mechanisms and damage models, *IEEE Transactions on Reliability*, **40**: 531–536.

EN (European Standard) 13306 (2010) *Maintenance – Maintenance Terminology*, European Committee for Standardization, Brussels.

FHWA (1979) Federal High Way Administration, USA.

Koppenhoefer, K.C., Crompton, J.S., and Dydo, J.R. (2006) Prediction of Failure in Existing Heart Valve Designs, in *Fatigue and Fracture of Medical Metallic Materials and Devices* (eds M.R. Mitchell and K.L. Jerina), ASTM, West Conshohocken, PA, pp. 77–86.

Kordonsky, K.B. and Gertsbakh, I.B. (1995) System state monitoring and lifetime scales, *Reliability Engineering and System Safety*, **47**: 1–14.

Richards, F. (2013) Failure analysis of a natural gas pipeline rupture, *Journal of Failure Analysis and Prevention,* **13**(6): 653–657.

Taylor, C.M. (1998) Automobile engine tribology – design considerations for efficiency and durability, *Wear* **221**: 1–8.

Tinga, T. (2012) *Mechanism Based Failure Analysis*, Nederlandse Defensie Academie.

Worm, J.M. and van Harten, A. (1996) Model based decision support for planning of road maintenance, *Reliability Engineering and System Safety*, **51**: 305–316.

Yang, G., Yoon, K.B., and Moon, Y.C. (2013) Stress corrosion cracking of stainless steel pipes for Methyl-Methacrylate process plants, *Engineering Failure Analysis*, **29**: 45–55.

4

Maintenance – Basic Concepts

Learning Outcomes

After reading this chapter, you should be able to:

- Describe different types of maintenance actions and their classification;
- Explain the difference between preventive and corrective maintenance;
- Describe the characteristics of predetermined preventive maintenance;
- Describe the characteristics of condition-based preventive maintenance;
- Differentiate design out maintenance from design for maintenance;
- Classify maintenance actions according to their effect on equipment reliability and failure behavior;
- Identify the maintenance requirement of various engineered objects;
- Describe common maintenance policies;
- Discuss warranty issues related to products;
- Explain the importance of a comprehensive framework for effective maintenance that includes engineering technology and management.

Introduction to Maintenance Engineering: Modeling, Optimization, and Management, First Edition.
Mohammed Ben-Daya, Uday Kumar, and D.N. Prabhakar Murthy.
© 2016 John Wiley & Sons, Ltd. Published 2016 by John Wiley & Sons, Ltd.

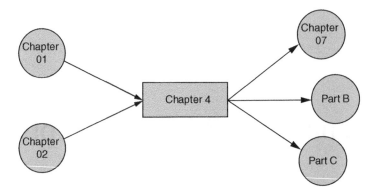

4.1 Introduction

The performance of an engineered object depends not only on its design and operating environment, usage intensity, and so on, but also on the maintenance carried out, since proper functioning over an extended time period requires effective maintenance. The degradation and failure mechanisms vary from object to object and so do the maintenance requirements to control the degradation and/or restore the object to the desired performance should a failure occur. This poses challenging problems for people involved with the maintenance of objects.

This chapter deals with the maintenance of different types of engineered objects. Our focus is on the different types of maintenance actions – the underlying basic concepts and their effect on the object being maintained. We discuss and illustrate the complexities and the unique issues in the maintenance of the different types of objects. In the process we present a bigger picture (and a wider perspective) of maintenance that highlights the relevance of the topics discussed in earlier chapters as well as those that are discussed in later chapters of the book.

The chapter is organized as follows. Section 4.2 presents the different types of maintenance actions. Section 4.3 focuses on preventive maintenance (PM) actions, Section 4.4 on corrective maintenance (CM) actions, and Section 4.5 on design out maintenance (DOM). Section 4.6 focuses on uptime and downtime of engineered objects and their components. Warranty and maintenance and related issues are discussed in Section 4.7. Sections 4.8–4.10 deal with maintenance of products, plants, and infrastructures, respectively. Section 4.11 lists important elements of effective maintenance and provides links to other chapters of this book where these elements are discussed in detail. Finally, Section 4.12 provides a summary of the chapter.

4.2 Types of Maintenance Actions

We use the following terminology proposed by Pintelon and Parodi-Herz (2008):

Definition 4.1   Maintenance Action

Basic maintenance intervention, elementary task carried out by a technician (What to do.)

Definition 4.2 Maintenance Policy

Rule or set of rules describing the triggering mechanism for the different maintenance actions (How is it triggered?)

Definition 4.3 Maintenance Concept

Set of maintenance policies and actions of various types and the general decision structure in which these are planned and supported. (The logic and maintenance solution used.)

In Chapter 1, we discussed maintenance concepts. In this section we discuss maintenance actions, and maintenance policies are discussed in a later section.

Maintenance actions for an item[1] include the following:[2]

- *Inspection:* Check for conformity by measuring, observing, testing, or gauging the relevant characteristics of an item. Generally, inspection can be carried out before, during, or after other maintenance activities.
- *Compliance test:* Test used to show whether or not a characteristic or a property of the item complies with the stated specification.
- *Monitoring:* Activity performed either manually or automatically which is intended to observe the actual state of the item. Monitoring is distinguished from inspection in that it is used to evaluate any changes in some parameters of the equipment with time. Monitoring may be continuous, over an interval of time, or after a given number of operations. Monitoring is usually carried out in the operating state.
- *Routine maintenance:* Regular or repeated elementary maintenance activities which usually do not require special qualification, authorization(s), or tools. Routine maintenance may include, for example, cleaning, tightening of connections, checking liquid level, lubrication, and so on.
- *Overhaul:* A comprehensive set of examinations and actions carried out in order to maintain the required level of availability and safety of the equipment. An overhaul may be performed at prescribed intervals of time or after a number of operations, and may require a partial or complete dismantling of the item.
- *Rebuilding:* Action following the dismantling of the item and the repair or replacement of those components that are approaching the end of their useful lives and/or should be replaced regularly. The objective of rebuilding is normally to provide the item with a useful life that may be greater than the lifespan of the original equipment. Rebuilding differs from overhaul in that the actions may include improvements and/or modifications, understood as follows:
  - *Improvement:* Combination of all technical, administrative, and managerial actions intended to ameliorate the dependability of the item without changing its required function.

---

[1] In the remainder of the chapter, we will often use the term "item" to denote an engineered object, component, or something in between.

[2] Our discussion is based on definitions from the European Standard EN 13306.

- *Modification:* Combination of all technical, administrative, and managerial actions intended to change the required function of the item. Modification, in fact, is not a maintenance action but concerns changing the original function of the item to a new required function. The changes may have an influence on the dependability or on the performance of the item, or both.
- *Repair:* Physical action taken to restore the required function of faulty equipment. Within a repair we can normally find the following actions:
  - *Fault diagnosis:* Actions taken for fault recognition, fault localization, and fault isolation at the appropriate indenture level and cause identification.
  - *Fault correction:* Actions taken after fault diagnosis to put the item into a state in which it can perform a required function.
  - *Function check-out:* Action taken after maintenance actions to verify that the item is able to perform the required function.
- *Turnaround maintenance:* Planned, periodic shutdown (total or partial) of a plant (refinery, chemical plant, power plant, etc.) to perform overhaul and repair activities and to inspect, test, and replace process materials (catalyst in a chemical plant). Turnarounds are expensive, both in terms of lost production (duration of several weeks) and direct costs for the labor, tools, heavy equipment, and materials used to execute the project. Management of turnaround maintenance will be discussed in Chapter 19.

Maintenance actions can be carried out at any level of the decomposition of an engineered object:

- At the plant level (for example, periodic plant shutdown for major maintenance activities);
- At the intermediate level (for example, module, subassembly, etc.) by standard replacement (electronic card, electric motor, etc.);
- At the component level (PM actions to replace filters, lamps, belts, etc., based on age and/or usage).

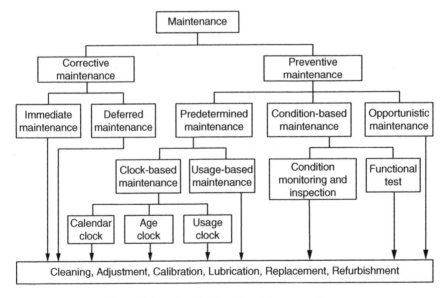

Figure 4.1   Classification of maintenance actions.

4.2.1   Classification of Maintenance Actions

Figure 4.1 shows an effective way of grouping the various maintenance actions (and policies) into a multi-level hierarchy. At the first level we have (i) preventive and (ii) corrective maintenance. Each of these can be subdivided, as indicated in the figure, and we discuss them further in the next three sections.

4.3   Preventive Maintenance Actions

Definition 4.4

Preventive maintenance (PM) actions are carried out according to prescribed criteria of time, usage, or condition and are intended to reduce the probability of failure or the functional degradation of an item.

PM actions usually require taking an operational item out of service and are intended to increase the span of its lifetime and/or its reliability. PM actions are generally carried out at discrete time instants. Actions can range from relatively minor servicing requiring a short downtime, such as visual inspection, lubrication, testing, planned replacement of parts or components, and so forth, to major overhauls requiring a significant amount of downtime and proper planning and adequate resources.

The aims of PM actions are to:

- Prevent failure.
- Detect the onset of failure: Whilst we may not be able to prevent a failure, frequently we do know how to detect the onset of failure. Our knowledge of how to do this is increasing every day, through condition monitoring technology.
- Find hidden failures: Check to see if a failure has occurred before equipment is called into service.

PM actions can be grouped into predetermined, condition-based maintenance, and opportunistic maintenance, as shown in Figure 4.1. They involve some form of intrusion into the object.

4.3.1   Predetermined Maintenance Actions

Definition 4.5

A *predetermined PM action* has the following characteristics:

- The action is carried out at discrete time instants determined by some predetermined rule.
- The action timing involves either a clock or a measure of the usage (for example, output).

Predetermined PM actions can be further divided into two subcategories: (i) clock-based and (ii) usage-based, as indicated in Figure 4.1.

4.3.1.1 Clock-Based Maintenance Actions

One can define three types of clocks and, as a result, we have three further subcategories of predetermined maintenance actions, as shown in Figure 4.1.

- *Calendar clock:* This is the familiar clock with a fixed starting time – start of a new year. As such, maintenance actions are carried out at predetermined time instants based on this clock. One such policy is the block replacement used for a fleet of identical items (such as bulbs in a city suburb) where all of them are replaced periodically.
- *Age clock:* Here, one uses the familiar clock which is set to zero when an item is put to use. PM actions are carried out at predetermined time instants using this clock. In other words, the PM actions are based on an item reaching some age.

The above two clocks run continuously, and they are appropriate for items in continuous operation over time. The next clock is appropriate for items used intermittently.

- *Usage clock:* Here, one uses the age clock which stops when the item is not in use (either due to idling, or undergoing PM or CM action). An example of this is the usage clock which clocks the number of hours flown for an aircraft engine.

4.3.1.2 Usage-Based Maintenance Actions

PM actions are based on usage of the item. Usage can be measured in different ways such as output (number of copies made by a photocopier, number of take-offs and landings for landing gear, number of tons produced, etc.). This is appropriate for items from various sectors (such as manufacturing, transport, etc.) that are used either continuously or intermittently.

Predetermined PM actions based on a calendar clock are easier to administer than those using age- or usage-based clocks. However, the latter reflect the degradation in a more sensible manner and hence are preferred in many cases.

Example 4.1   Aircraft Landing Gear

Landing gear is made up of several components and they need to be maintained and inspected on a regular basis. Listed below are some examples of key inspection areas along with the typical clock used. Note that sometimes a calendar clock and an age clock are combined and action is initiated based on which reaches its allotted time first.

1. After 300 hours or after one year in-service inspection:
   - Shock absorber nitrogen pressure check.

2. After 600 hours inspection:
   • Landing gear hinge points visual inspections;
   • Leak inspection (oil, hydraulic fluid, etc.);
   • Inspection of torque link play.
3. After seven years or 5000 cycles:
   • Landing gear overhaul.   ∎

Definition 4.6

A *condition-based maintenance (CBM) action* is a PM action that has the following characteristics:

• There is a measurable parameter that correlates with the degradation over time and the onset of failure.
• Changes in the measurable parameter are obtained from data collected using appropriate condition-monitoring techniques.
• The data collection usually does not require intrusion[3] into the object.

4.3.2 Condition-Based Maintenance Actions

From this definition, notice that the most important factor is the ability to identify a measurable parameter[4] that provides a correlation between the measurement and the level of degradation. A combination of the understanding of failure mechanisms (Chapter 3), availability of the appropriate measurement technology (Chapter 6), and analysis tools is vital for CBM implementation.[5]

The selected parameter can provide either direct or indirect measurement of degradation. An example of indirect measurement is the monitoring of bearing degradation based on analysis of particle debris in the lubricating oil, and this is discussed further in Chapter 6.

The parameter may be monitored continuously or at discrete points in time, online or offline, and in the case of monitoring at discrete points in time, the monitoring frequency needs to be determined.

Figure 4.2 illustrates the concept of CBM. The periodic monitoring of degradation occurs at time instants $t_1, t_2, \ldots$. These are PM actions to assess the state. The inspection at time t_a (the first time after the alarm threshold is crossed) is the starting point of a CBM window. Note that t_p represents a possible time for CBM action and t_f is a failure time if CBM action has not been taken by that time. Beyond t_f a CM is required which is usually much more expensive than the preventive action to improve the state through some appropriate action.

As shown in Figure 4.1, CBM actions can be further divided into the following two categories:

• Functional tests;
• Condition monitoring and inspection.

---

[3] Action to measure might not require intrusion, but carrying out the PM action subsequent to measurement requires intrusion.
[4] Some authors use the term "variable" instead of parameter. We will use the term parameter in the rest of this book.
[5] The tools for data collection, data analysis, modeling, and so on, are discussed in Part B of the book.

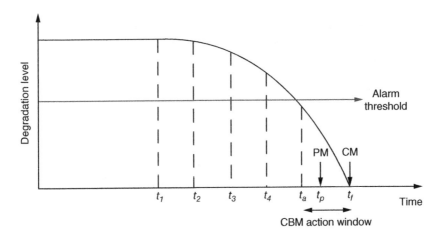

Figure 4.2 Concept of condition-based maintenance.

In both cases the data are analyzed to assess the condition or state of the item. In the former case, the focus is on deciding whether to take immediate action (such as to repair or not) while the latter case involves extrapolating the degradation into the future (through trend analysis) and predicting when action might be needed. This allows some time for proper planning of maintenance activities and the necessary resources needed. A proper understanding of the failure mechanisms (discussed in Chapter 3) is critical for CBM, and proper modeling is needed to predict the residual life to plan effective maintenance strategies.

Predetermined PM actions are unable to avoid many item failures because they do not take into account the condition of the item. The only way of reducing the probability of failure is through early replacements. This results in the discarding of useful remaining life. With CBM the discarding of useful life is reduced and hence CBM is more desirable than predetermined PM. However, this is achieved at the additional expense of technologies (sensors for collecting data, data analysis, model building, etc.). Therefore, for expensive objects, CBM is the preferred option whereas for cheap objects, predetermined PM is more appropriate.

4.3.3   Opportunistic Maintenance Actions

Definition 4.7

Opportunistic maintenance actions are carried out at convenient moments which are unpredictable and they can be categorized broadly into the following two groups:

- *Internal to the object:* Failure of a component providing an opportunity to carry out PM actions on some of the non-failed components.
- *External to the object:* The object being in the idle state due to external factors (such as production ceasing due to running out of inputs or excess inventory resulting in a temporary shutdown of the plant) provides an opportunity to carry out PM actions.

As a result, opportunistic maintenance actions are used extensively in mining, manufacturing, and processing industries.

Example 4.2   Wind Farm

Nowadays, one of the important sources of clean and renewable energy is wind energy. To control operation and maintenance costs, effective maintenance strategies need to be adopted in this context. Opportunistic maintenance is one of four maintenance strategies mentioned in a European wind energy report from 2001 for European offshore wind farms. When an opportunistic maintenance strategy is used, the maintenance team sent on location to perform CM will take this opportunity simultaneously to perform PM on other wind machines and their components which show relatively high risks of failure according to some prescribed criteria. Some of these farms are offshore or in remote locations; therefore, opportunistic maintenance may be more cost-effective as it takes advantage of already allocated resources and time.   ∎

4.3.4   Failure and Fault Detection

In some situations, one is not able to detect the onset of the failure of an item that is used intermittently. However, one can check to see if a failure has occurred before the item is called into service. This is the case, for example, for standby and special purpose items such as protective devices. Unfortunately, these devices are the last line of protection when things go wrong and are often poorly maintained or receive no maintenance at all.

The objective of failure finding (FF) tasks is to ensure that a protective device or a back-up system will provide the required protection when it is called into service. The FF task can be characterized as follows:

Definition 4.8

A *failure-finding task* is a maintenance action carried out at regular intervals that involves checking for a hidden failure.

Many different types of technologies are used for detecting faults and these are discussed in Chapter 6.

4.3.4.1   No Fault Found (NFF)

Definition 4.9

No fault found (NFF) is a term used in the field of diagnostics to describe a situation where an originally reported mode of failure cannot be duplicated by the maintenance engineer/ technician and therefore the defect cannot be detected and fixed.

The NFF phenomenon is a major problem when dealing with complex technical systems. NFF can be attributed to intermittent faults (resulting from defective connections in an electrical circuit, temporary shorts, or opens in a circuit, software bugs, temporary environmental factors, operator error, etc.).

4.3.5   Effect of PM Actions on Item Reliability

Since PM can be viewed as actions to compensate for the unreliability of an item, maintenance actions have a significant effect on reliability and future failures. Based on this effect, PM actions can be classified into the following two categories: (i) perfect PM and (ii) imperfect PM.

4.3.5.1   Perfect Preventive Maintenance

With perfect PM an item is restored to an "as good as new" condition. The replacement of an item with a new one is an example of perfect PM. As such, the system failure rate (defined in Chapter 2) at time t is given by $h_1(t) = h_0(t - \tau)$ for $t > \tau$ where τ denotes the time instant where PM is carried out and $h_0(.)$ is the failure rate of a new system, as indicated in Figure 4.3.

4.3.5.2   Imperfect Preventive Maintenance

Imperfect PM results in the item reliability improving so that it is between an "as bad as old" and an "as good as new" condition. An engine tune-up is an example of imperfect PM, as this action may not make an engine as good as new, but its performance is somewhat improved. Here, the failure rate after PM is lower than it was just before failure, so the item failure rate at time t, $t > \tau$, is given by $h_2(t)$ with $h_0(t) > h_2(t) > h_1(t)$, as shown in Figure 4.3.

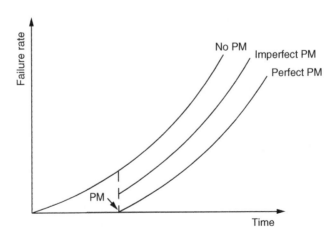

Figure 4.3   Impact of PM action on system failure rate.

4.3.6   Overhaul (Major Shutdown)

An overhaul involves a complete dismantling of the system and replacing all the components that have deteriorated significantly. In general, an overhaul improves the reliability but there is progressive deterioration in the sense that the hazard function after the first overhaul is greater than that for a new system and the hazard function after the jth overhaul is higher than that after the $(j-1)$th overhaul for $j \geq 1$.

In some cases an overhaul can be perfect, so that the system is as good as new after this type of action.

4.4   Corrective Maintenance Actions

Definition 4.10

Corrective maintenance actions are actions carried out after fault recognition and are intended to put a failed item into a working state to perform its normal function.

CM actions are actions to restore a failed item to an operational state through appropriate rectification (repair or replacement) of the failed components. When a repairable item fails, there is an option either to repair or replace it by a new or refurbished item. The optimal decision is usually based on cost considerations and the impact of the actions on future failures of the item involved. These issues will be discussed later in this book.

CM actions can range from relatively minor repairs or replacements requiring a short downtime, such as replacing a lamp, adjusting a machine, and so forth, to major repairs requiring a significant amount of downtime and resources. Although CM is unplanned and happens without notice, the maintenance organization must be prepared to deal with such events to minimize their negative effects and consequences.

Based on the timing of CM actions, they may be classified into two categories: (i) immediate (emergency) CM and (ii) deferred CM, as shown in Figure 4.1.

4.4.1   Immediate CM

Definition 4.11

Immediate CM is corrective maintenance which needs to be carried out immediately after fault detection.

Immediate CM is carried out to rectify failures of critical items or failures having environmental or safety consequences. Maintenance needs to be initiated immediately to restore the failed item to its functional state in order to prevent further damage, avoid injuries or death, prevent environmental damage, or avoid the high cost of lost production.

4.4.2 Deferred CM

If the failure is not critical and does not need immediate action or can be delayed (car windshield wiper failure can be delayed if is not the rainy season), the maintenance action can be deferred to a more convenient time.

Definition 4.12

Deferred CM is corrective maintenance which is not immediately carried out after fault detection but is delayed in accordance with given maintenance rules (EN 13306, 2001).

In many cases, a deliberate decision is made to allow an item to operate until it fails. This practice is known as run to failure (RTF). Such a decision can be taken occasionally for one of the following reasons:

- There is no PM task that will have an effect regardless of how much one is willing to spend.
- The PM cost is more than the RTF cost and there is no safety consequence.
- The failure is too low on the priority list and cannot be addressed within the allocated budget.

However, if, after failure, the item is still needed, then it needs to be restored to an operable status in a timely manner.

4.4.3 Effect of CM Actions on Item Reliability

As with PM actions, CM actions (in the form of repair and/or replacement of failed components) have a significant effect on the reliability and subsequent failures of an item. Based on this effect, repair actions can be grouped into three categories – perfect, minimal, and imperfect – as indicated below.

4.4.3.1 Perfect Repair

Similar to perfect PM, perfect repair means a maintenance action that restores the failed item to an "as good as new" condition. A complete overhaul of an engine with a broken connecting rod is an example of perfect repair. As such, the item failure rate at time t is given by $h_1(t) = h_0(t - \tau)$ for $t > \tau$, where τ denotes the time of first failure and $h_0(.)$ is the failure rate of a new item, as shown in Figure 4.4.

4.4.3.2 Minimal Repair

As the name suggests, minimal repair implies bringing the failed item to an operational state whereby the condition (or reliability) of the item is not affected, so that it is the same as that just before the failure. Changing a flat tire on a car and changing a broken fan belt on an engine are examples of minimal CM. Here, the failure rate of the item is unaffected by the maintenance action. As such, the reliability after repair is the same as that just before failure.

This is an appropriate characterization at the system level if the failure is due to one or few components and either repairing or replacing them has very little impact on the overall item reliability and failure rate. In this case, the item failure rate at time t is given by $h_2(t) = h_0(t)$ for $t > \tau$, as shown in Figure 4.4.

4.4.3.3   Imperfect Repair

There are two types of imperfect repair: Type (i) and Type (ii). In Type (i) imperfect repair the condition of a repaired item is between "as bad as old" and "as good as new." As a result, the failure rate after repair is lower than it was just before failure. This characterizes the situation where the failure is a major failure requiring the replacement of several components by new ones so that the overall failure rate improves. In this case, the item failure rate at time t, $t > \tau$, is given by $h_3(t)$ with $h_2(t) > h_3(t) > h_1(t)$. In Type (ii) imperfect repair the failure rate of a repaired item is higher than that just before failure. This usually is the effect of a poor quality repair that reduces the reliability so that the item failure rate at time t, $t > \tau$, is given by $h_4(t)$ with $h_4(t) > h_2(t)$ Figure 4.4 shows these two cases.

4.5   Design Out Maintenance

Definition 4.13
Design out maintenance (DOM) deals with redesigning a part of an object.

The reason may be to improve the reliability and/or to reduce the likelihood of the failure of the part. DOM deals with strategies and techniques that reduce or eliminate the possibility of failure rather than dealing with its consequences. In fact, failure is viewed in a positive sense, as it presents an opportunity to improve equipment reliability and advance technology.

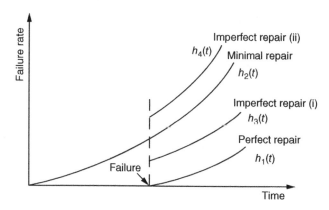

Figure 4.4   Impact of different types of maintenance action on system failure rate.

While the aim of PM is to minimize the number of failures, DOM aims to reduce or eliminate the need for maintenance for a particular failure mode. It is more suitable for items with high maintenance costs so that the redesign effort can be justified. The choice to be made is between the cost of redesign and the cost of recurring maintenance. Notice that DOM targets improving reliability that leads to a reduction in the number of failures. On the other hand, design for maintainability (DFM – discussed further in Chapter 7) focuses on improving maintainability that leads to a reduction in maintenance and repair times.

Example 4.3[6]   Helicopter

The UH-60A Black Hawk utility helicopter requires only 2.5–3 man-hours of maintenance per flight hour at the unit level, as opposed to 12–20 hours for the family of helicopters it replaced. This increase in maintainability is attributed largely to simplification of the main rotor head design. Unlike previous designs which had upper and lower plates tied together with vertical and horizontal hinges, the Black Hawk main rotor head is forged from a single piece of titanium. An elastomeric bearing, which is completely dry and requires no lubrication and has an average life of 3000 hours, eliminates the need for hinges. In the earlier designs, the hydraulic reservoir seal failures led to loss of lubrication oil and faster wear of the main rotor head, which necessitated overhaul. Typically, the mean time between removals (MTBRs) for older designs of such components was less than 500 hours. Simplification has resulted in dynamic component MTBR requirements of at least 1500 hours.   ■

4.6   Uptime and Downtime

Uptime and downtime are concepts that play an important role in effective maintenance.

- *Uptime:* The interval between two downtimes during which the object is available for operation. It consists of two components: (i) operating time and (ii) idle time. If the object is used in continuous mode, then idle time is zero.
- *Downtime:* The time interval where the object (product or plant) is not operational and is undergoing either PM or CM. It consists of three main components: (i) administrative time, (ii) logistic time, and (iii) active (or actual) repair time. Administrative time is the time spent in organizing maintenance activities. Logistic time is the portion of downtime during which the repair activity is waiting for a repair crew, repair facilities, tools, spare parts, replacement instructions, and so on. The actual repair time is also called active repair time. It is the time during which the repair crew is working on the object to carry out the necessary repair activities. The details of these different times are illustrated in Figure 4.5.

---

[6] DOD-HDBK-791.

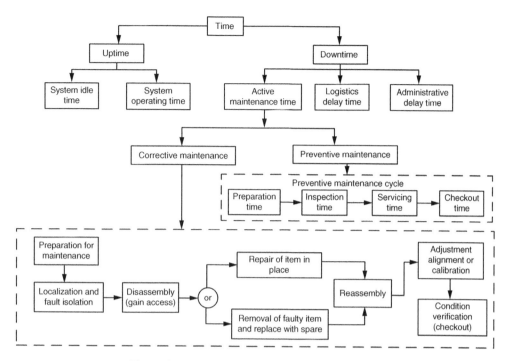

Figure 4.5 Decomposition of uptime and downtime.

4.6.1   Repair Time

Time to repair (D) is a random variable and is characterized by a repair distribution function $G(t)$ with

$$G(t) = P(D \leq t) \tag{4.1}$$

If the function $G(t)$ is differentiable, the repair density function is defined by:

$$g(t) = \frac{dG(t)}{dt} \tag{4.2}$$

Similar to the failure rate, one can define the repair rate as follows:

$$\rho(t) = \frac{g(t)}{1 - G(t)} \tag{4.3}$$

Note that the repair rate is usually a decreasing function of repair time.

The mean time to repair, *MTTR* is the expected value of the random variable D and is given by:

$$MTTR = E[D] = \int_0^\infty t g(t)\, dt. \tag{4.4}$$

Example 4.4   General

Let the repair time distribution for a failed item be given by the two-parameter Weibull distribution with shape parameter $\beta = 0.7$ and scale parameter $\alpha = 2.5$ hours.

(a) Determine the probability of repairing the failed item within three hours.
(b) Determine the repair rate function.
(c) Find the mean time to repair.

Solution:

(a) The probability of restoring the failed item to an operational state within three hours is given by:

$$P(T \leq 3) = G(3) = 1 - e^{-(3/2.5)^{0.7}} = 0.68$$

(b) The repair rate function is given by:

$$\rho(t) = \frac{g(t)}{1 - G(t)} = \left(\frac{t}{2.5}\right)^{0.7}$$

(c) The mean time to repair is given by:

$$MTTR = E(T) = 2.5\Gamma\left(1 + \frac{1}{0.7}\right) = 3.16 \text{ hours.} \qquad \blacksquare$$

4.7   Warranty and Maintenance

4.7.1   Warranty: Concept and Role

A *normal* (or *base*) *warranty* is a written and/or oral manufacturer's (builder's) assurance to customers that the engineered object purchased (built) shall perform satisfactorily. It is a contractual agreement between customer and manufacturer (builder) that is entered into upon sale (delivery) of the object. The contract specifies the performance of the object, the owner's responsibilities, and what the warrantor (generally the manufacturer) will do if the object fails to meet the stated performance.[7]

The manufacturer (builder) carries out remedial actions (at no cost to the customer) to rectify any failures over a specified period subsequent to the purchase (delivery) of the object. This period is called the *warranty period.* Many different types of warranties are offered, depending on the engineered object, the manufacturer (builder), and the buyer. As a result, warranties play an increasingly important role in most consumer and commercial transactions.

4.7.2   Types of Warranties

The type of warranty depends on whether the engineered object is a product or plant, and in the case of a product, whether it is standard or custom-built.[8]

---

[7] A base warranty is integral to the sale and its cost is factored into the sale price. Once the warranty expires, the customer often has the option to buy an extended warranty by paying an additional amount. This is discussed further in Chapter 18.
[8] Warranties for infrastructures (such as roads, pavements, etc.) involve more complex characterizations of performance and when these are not met, the builder has to carry out the required remedial actions.

4.7.2.1 Standard Products

In the case of standard products, because of the wide variety of products, company objectives, and market conditions, a large number of different warranties is offered.[9] The most common type is the non-renewing Free-Replacement Warranty (FRW) – which is available in one- and two-dimensional forms.

- *One-dimensional non-renewing FRW policy:* The manufacturer agrees to repair or provide replacements for failed items free of charge up to a time W from the time of the initial purchase. The warranty expires at time W after purchase.
- *Two-dimensional non-renewing FRW policy:* The manufacturer agrees to repair or provide a replacement for failed items free of charge up to a time W or up to a usage U, whichever occurs first, from the time of the initial purchase. W is called the warranty period and U the usage limit.

If the usage is heavy, the warranty can expire well before W, and if the usage is very light, then the warranty can expire well before the limit U is reached. Should a failure occur at age T with usage X, it is covered by warranty only if T is less than W and X is less than U. If the failed item is replaced by a new item, the replacement item is warranted for a time period $W - T$ and for usage $U - X$. Nearly all auto manufacturers offer this type of policy, with usage corresponding to distance driven.

4.7.2.2 Custom-Built Products and Plants

Warranties for custom-built products or plants are typically part of a service maintenance contract and are used principally in commercial applications and government acquisitions of large, complex items – for example, aircraft or military equipment. Nearly all such warranties involve time and/or some function of time as well as a number of characteristics that may not involve time, for example, fuel efficiency. One such is the *reliability improvement warranty* (RIW).

The basic idea of an RIW is to extend the notion of a basic consumer warranty (usually the FRW) to include guarantees on the reliability of the object and not just on its immediate or short-term performance. This is particularly appropriate in the purchase of complex, repairable equipment that is intended for relatively long use. The aim of an RIW is to negotiate warranty terms that will motivate a manufacturer to continue improvements in reliability after the object is delivered. Under an RIW, the contractor's fee is based on his/her ability to meet the warranty reliability requirements. These often include a guaranteed MTBF (mean time between failures) as a part of the warranty contract.

4.7.3 Maintenance under Warranty

Over the warranty period (when the engineered object is covered under warranty) the cost of CM actions is borne by the manufacturer. In the case of standard products (such as automobiles), the costs of PM actions over the warranty period are paid by the owner. In the case of custom-built products and plants, some or all of the PM costs are paid by the manufacturer (builder).

---

[9] A taxonomy to classify the different types of warranties can be found in Blischke and Murthy (1994).

4.8   Maintenance of Products

Products can be divided into three categories:

- *Type A:* Standard consumer products;
- *Type B:* Industrial and commercial products;
- *Type C:* Defense products.

All products come with some form of warranty. CM actions under warranty are done by the manufacturers/contractors or their agents. PM actions under warranty are carried out by the manufacturer at a cost to the customer unless a contract requires the manufacturer to bear the cost. Table 4.1 highlights the key issues in the maintenance of the three types of products.

Once the base (or extended) warranty expires, the owner is responsible for the maintenance of the product. There are many different policies that have been proposed for maintenance and we now discuss a few of these.

4.8.1   Maintenance Policies

The type of maintenance policy (Definition 4.2) that is appropriate depends on the item. The maintenance policies at the product level are, in general, different from those at the component level. As a result, there is a plethora of maintenance policies that are best characterized by a multi-dimensional characterization.

A maintenance policy is characterized by one or more parameters that can be viewed as decision variables to be selected optimally to optimize some objective function involving the performance of the item over a specified time horizon. The optimal choice depends on several factors such as monitoring, inspection, usage, and so on. Table 4.2 summarizes the different dimensions and their characteristics and this provides the basis for the various maintenance policies defined in this section.

4.8.2   Classification

One needs a two-level approach to classify the different maintenance policies. At the first level, we can categorize them into three groups based on whether the decision variable is: (i) only with PM actions, (ii) only with CM actions, or (iii) with both PM and CM actions. The second level depends on the clock used – calendar, age, or usage, as discussed earlier.

Table 4.1   Key issues in the maintenance of products.

Issue	Product type		
	Type A	Type B	Type C
Type of warranty	FRW	FRW, RIW	RIW
Warranty duration	Short	Short to medium	Medium to long
Maintenance actions	Age/usage	Age/usage/condition	Age/usage/condition
PM actions	Inspect, replace	Monitor, replace	Monitor, replace
CM actions	Repair, replace	Repair, DOM	Repair, DOM
Skill level needed	Low	Low to medium	Medium to high

We define several maintenance policies. Each policy is characterized by one or more parameters (denoted by $\wp$) which are the decision variables to be selected optimally.[10]

Comment: We use the terms parameters and variables interchangeably in the context of these policies.

4.8.3 Maintenance Policies I – Component Level

4.8.3.1 Policies with PM Decision Variables

Here, the PM actions are characterized by parameters to be selected optimally to optimize some objective function. The CM actions are pre-specified.

Based on an age (usage) clock

Policy 4.1   Age Policy

Under this policy, a unit is replaced (PM action) when it reaches age T (a constant) or on failure (CM action) should it occur earlier. The decision variable is the parameter $\wp \equiv \{T\}$.

Table 4.2   Maintenance policies – dimensions and characterization.

Dimensions	Variables	Comments
Item	Single component	—
	Multiple components	Opportunistic maintenance possible
State of item	Two-level	Up or down
	More than two and finite	Different levels of degradation
	Infinite	Continuous degradation (Chapter 3)
Monitoring of state	Continuous	See Chapter 6
	Discrete	
Inspection interval	Constant	—
	Variable	—
Parameters	Known	
	Uncertain	Bayesian approach
Time horizon	Finite	—
	Infinite	—
Item usage	Continuous	Calendar or age clock
	Intermittent	Usage clock (number of times used, usage time)
Repair	Minimal	Always repair
	Imperfect	Different levels of repair
	Perfect	Choice between repair and replace
Repair time	Taken into account	—
	Negligible	Very small compared to time between actions

---

[10] The analysis of these policies and the optimal selection of the decision parameters are discussed in Part C of the book.

This policy can be used with a usage clock with PM replacement occurring when the total usage reaches T.

Based on a calendar clock

Policy 4.2   Block Policy

Under this policy, a unit is replaced (PM action) at fixed time intervals kT, $(k = 1,2,\ldots)$ with T (a constant), and also at intervening failures (CM action).

Policy 4.3   Periodic Policy

Under this policy, a unit is replaced (PM action) at fixed time intervals kT $(k = 1,2,\ldots)$, and all intervening failures are minimally repaired.

In both policies, the decision variable is the parameter $\wp \equiv \{T\}$.

Comment: Policy 4.2 is appropriate for non-repairable components and is often used for repairable components if the cost of repair is high relative to the purchase price of the component. Policy 4.3 is applicable only for repairable components and is appropriate if the cost of repair is low relative to the purchase price.

Based on failure rate

Policy 4.4   Failure Limit Policy

Under this policy, a unit is replaced (PM action) when its failure rate reaches some specified limit ψ. All failures between PM actions are repaired (CM action) through minimal or imperfect repair. The decision variable is the parameter $\wp \equiv \{\psi\}$.

4.8.3.2   Policies with CM Decision Variables

The following two policies are based solely on CM and involve no PM. The CM actions are characterized by one or more parameters that need to be selected optimally. These are appropriate for components which have a constant failure rate. The CM decision (repair or replace) in the first policy is based on the cost to repair and in the second policy on the time to repair.

Policy 4.5   Repair Cost Limit Policy

When an item fails, the repair cost is estimated and repair is undertaken if the estimated cost is less than a predetermined limit v; otherwise, the item is replaced. The decision variable is the parameter $\wp \equiv \{v\}$.

Policy 4.6   Repair Time Limit Policy

An item is repaired at failure: if the repair is not completed within a specified time limit T (called the *repair time limit*), it is replaced by a new one; otherwise, the repaired unit is put into operation again. The decision variable is the parameter $\wp \equiv \{T\}$.

For components with increasing failure rates, the following policy based on the number of times a failed item is repaired is more appropriate.

Policy 4.7   Repair Count Policy

The first $(K - 1)$ failures of an item are minimally repaired and the item is replaced at the Kth failure. The decision variable is the parameter $\wp \equiv \{K\}$.

4.8.3.3   Policies with PM and CM Decision Variables

One can define several maintenance policies with PM and CM decision variables. The one given below combines features of Policies 4.1 and 4.7.

Policy 4.8   Age and Repair Count Policy

An item is replaced at age T (PM action) or at the Kth failure (CM action), whichever occurs first. The first $(K - 1)$ failures are minimally repaired (CM action). The decision variables are the parameters $\wp \equiv \{T,K\}$.

4.8.4   Maintenance Policies II – Product Level

A product can be viewed as a system comprised of several components. When no strong dependence exists between the different components, the component maintenance policies can be independently applied to each component. However, in most cases, interactions exist between the various components of the product. In particular, economic dependence means that savings in costs or downtime can be achieved when several components are jointly maintained. When the system is down (due to failure or external factors), there is an opportunity to carry out PM actions on non-failed components or perform deferred CM.

A large number of policies has been proposed and studied in the literature. The following is an illustrative sample. The components are, in general, not identical. Let n denote the number of components in the product.

Policy 4.9

CM action is performed whenever a component fails. The failed component is replaced by a new one (CM action) and the opportunity is used to replace all non-failed components (PM action) if their age exceeds some specified threshold levels. The decision variables are the parameters $\wp \equiv \{$threshold levels for the different components of the product$\}$.

Policy 4.10

This policy depends on two control limits T_1 and T $(T_1 < T)$: During $(0, T_1]$, individual CM is carried out as soon as a component fails. When the first failure occurs in the interval $(T_1, T]$, the corresponding component is replaced, jointly with all other (non-failed) components. If no failure occurs in $(T_1, T]$, the whole system is preventively replaced at time T. The decision variables are the parameters $\wp \equiv \{T_1, T\}$.

Policy 4.11

This policy depends on two failure rate control limits L and U. A unit is replaced (active replacement) when the failure rate reaches L or at failure with the failure rate in interval $(L - U, L]$. When a unit is replaced due to the failure rate reaching L, all of the operating units with their failure rate falling in $(L - U, L]$ are replaced (passive replacement) at that time. A unit is repaired (minimal repair) at failure with the failure rate in $(0, L - U]$. The decision variables are the parameters $\wp \equiv \{L, U\}$.

4.8.5   Examples

Example 4.5   Aircraft

Aircraft maintenance checks are periodic inspections that have to be done on all commercial/civil aircraft after a certain amount of time or usage. Table 4.3 summarizes the four commonly known types of checks and their frequency.

Table 4.3   Aircraft maintenance checks.

Check type	Frequency	Maintenance man hours[a]	Comments
A	500–800 FH[b]	20–50	Light check performed overnight at an airport gate or hangar
B	4–6 mo	150	Light check performed overnight at a hangar
C	20–24 mo	6000	Heavy check performed overnight at a hangar
D	5 yr	50 000	Major overhaul performed at a maintenance base

[a] Several specialized technicians carry out different tasks concurrently.
[b] FH: flight hours.

Each check type usually incorporates each of the lower check types to take advantage of the benefits of grouping maintenance tasks.   ■

Example 4.6 Automobile

A typical maintenance schedule showing the different maintenance actions and their timing is shown in Table 4.4.

Table 4.4 Car maintenance schedule.

Tasks	Number of miles (thousands) or months																			
	3	6	9	12	15	18	21	24	27	30	33	36	39	42	45	48	51	54	57	60
Lube oil filter	●	●	●	●	●	●	●	●	●	●	●	●	●	●	●	●	●	●	●	●
Engine flush service								●							●					
Rotate tires		●		●		●		●		●		●		●		●				●
Wheel alignment					●					●					●					●
Tune-up										●										●
Fuel system cleaning								●								●				
Power steering flush								●								●				
Transmission flush					●					●										●
Cooling system										●										●
Inspect brakes		●		●		●		●		●		●		●		●		●		●
Brake system flush								●								●				
Battery					●					●						●				

Note that the PM schedule uses different levels of maintenance at periodic intervals of mile usage or months, whichever comes first. ∎

4.9 Maintenance of Plants and Facilities[11]

Plants and facilities can be viewed as systems comprised of several products. As such, the maintenance policies (at product level) described in the previous section are also applicable to plants. In particular, plants can take advantage of opportunistic maintenance policies to lower the total maintenance cost. Additional issues that arise in the case of plants and facilities include dealing with fleets (or groups) of products where each unit operates independently. This offers the scope for additional maintenance policies. The effect of maintenance actions on production and operations implies that maintenance policies need to take this into account. These two issues are discussed in the next two sections.

4.9.1 Maintenance of a Fleet or Group of Products

A fleet is defined as a group of identical or similar items (such as ships, vehicles, aircraft, etc.) that are under the control of one entity (a municipality in the case of buses or an airline company in the case of aircraft, etc.).

---

[11] Plants produce goods whereas facilities produce services.

There are several issues that need to be taken into account in the context of fleet maintenance.[12] Some of these are as follows:

- The age and the condition of units in a fleet can vary significantly so that the units are not statistically similar. The main reasons for this include: (a) the units are purchased at different time points; (b) the usage of each unit can be quite different and hence the degradation levels of units with the same age can be quite different; and (c) the ages of constituent components of a unit can be quite different due to the maintenance history. This raises an issue – how to control the "health level" of the fleet by appropriate maintenance and replacement decisions.
- Fleet maintenance needs to coordinate with production (or service) requirements and needs to take into account resource constraints.
- A failure consequence of a unit strongly depends on the configuration of a fleet and the functional requirements assigned to units within the fleet. This implies that the fleet maintenance needs to consider the priority of each unit and devise appropriate maintenance policies.
- The technological evolution of the unit makes maintenance options multi-dimensional – repair or replacement; if replacement, whether a particular unit should be replaced by a unit with the same technology or one with more advanced technology. This implies that one needs to take into account technological evolution in the decision-making process when retiring (or replacing) old or degraded units.
- Because of the multi-unit nature, group and opportunistic maintenance are appropriate for fleet maintenance. Many different policies have been proposed and we will discuss a few of them.

Below we describe some group maintenance policies for a fleet, and we first look at a simple case where the fleet is monitored continuously.

Policy 4.12

The fleet is monitored continuously and maintenance actions are initiated only when the number of failed units reaches m. At this instant, all failed units are replaced with new ones (CM action) and all functioning units are serviced (PM action) so that they are restored to as-good-as-new condition. The decision variable is the parameter $\wp \equiv \{m\}$.

The following two policies involve the fleet being inspected at discrete time instants.

Policy 4.13

The fleet is inspected at discrete time instants. At an inspection, the failed units are repaired if the number of failed units is greater than or equal to m; otherwise, they are left idle (failed state). The time to the next inspection is decided based on the number of failed units. The decision variables are the parameter $\wp \equiv \{m\}$ and the state-dependent inspection time instants.

---

[12] For more information, see Cassidy *et al.* (1998).

In the following policy, the trigger is both age and the number of failed units and the fleet is monitored continuously.

Policy 4.14

The maintenance actions are initiated when the fleet reaches an age T or at the time instant when the number of failed units reaches m, whichever comes first. All failed units are replaced with new ones (CM action) and all functioning units are serviced (PM action) so that they become as good as new. The decision variables are the parameters $\wp \equiv \{m,T\}$.

In the following policy, the fleet consists of a group of identical units and the trigger is based on both a time limit and the number of failures.

Policy 4.15

Each unit is replaced on failure during the interval $(0, \tau]$. Beyond this interval, failed units are left idle until the number of failed units reaches a specified number m, when a block replacement is performed. The decision variables are the parameters $\wp \equiv \{\tau,m\}$.

In the case of complex products, one often uses maintenance policies involving overhaul (major shutdown maintenance). We list two such policies.

Policy 4.16

The system is subjected to K (≥ 2) overhauls. The first overhaul is done after the system has been in use for a period T_1 and the kth overhaul is done after the system has been in use for a period T_k subsequent to the $(k-1)$th overhaul for $k \geq 2$. All failures in between overhauls are minimally repaired. The system is replaced by a new one after a period T_K subsequent to the last overhaul. The decision parameters to be selected optimally are given by the set $\wp \equiv \{K;T_k, k=1,\ldots,K\}$.

Policy 4.17

The system is subjected to K (≥ 2) overhauls. The first overhaul is done after the system has been in use for a period T_1 or at failure, whichever occurs first, and the kth overhaul is done after the system has been in use for a period T_k since the $(k-1)$th overhaul for $k \geq 2$ or at failure, whichever occurs first. The system is replaced by a new one after a period T_K subsequent to the last overhaul. The decision parameters to be selected optimally are given by the set $\wp \equiv \{K;T_k, k=1,\ldots,K\}$.

4.9.2 Maintenance and Production

Production here is viewed in a general sense to include production of goods as well as services. There is a strong interaction between maintenance and production. The failure of a component can have an effect on the entire plant or facility. Failures of such systems can result in high production losses and societal costs (due to negative effects on the environment and public safety).

Effective maintenance needs to take into account the links between maintenance and production, quality, and sustainability. Plant maintenance, especially in the process industries such as petrochemical plants and refineries, is characterized by the following:

- A well-structured PM program developed through sound methodology such as reliability-centered maintenance (RCM) (discussed in Chapter 17) and based on a good understanding of the functions of the key subsystems of the plant. Such a program is usually a combination of various time-based, condition-based, and opportunistic maintenance policies.
- Periodic shutdown of the plant (every one to five years) to carry out major maintenance work called turnaround maintenance (discussed in Chapter 20). This is a major undertaking that requires months of planning and can take several weeks to carry out. Maintenance work scope includes PM tasks, deferred CM tasks, and opportunistic maintenance.

4.9.3   Examples

We indicate the salient features of the policies used in the maintenance of two machines used in the mining industry: (i) a dragline and (ii) an excavator.

Example 4.7   Dragline

The dragline and its operation were discussed in Example 1.6. The dragline operates continuously. Each day of operation is divided into three 8-hour shifts, giving a total of 21 shifts per week. The different kinds of PM actions are listed below.

- *Minor PM:* One shift out of every 63 shifts (i.e., three weeks) is dedicated to minor PM. The minor PM actions are divided into three groups: Mechanical (M), Electrical (E), and Ropes, Bucket, and Rigging (B). At each minor PM action, the maintenance jobs to be undertaken depend on the service sheets for the day. The mechanical jobs consist of 12 service sheets, which are rotated over a 36-week period. The electrical jobs consist of 24 service sheets, which are rotated over a 78-week period. The ropes, bucket, and rigging are all checked at each minor PM action.
- *Inspection:* There are two types of inspections undertaken – scheduled and unscheduled. Scheduled inspections include inspection of ropes, bucket, and rigging, which are undertaken once a week. Operators also do a quick cursory inspection of the dragline including the bucket (including rigging and dump rope), ropes, machinery, and motors at the start of their shift.

Unscheduled inspections occur when an operator detects a problem which has not caused the machine to fail (for example, motors sparking). The inspection can result in one of the following four actions being taken: (i) no problem was detected and no work has been done; (ii) no work is to be done at the moment but an eye is to be kept on the item; (iii) a problem was detected and work was done immediately (CM action); or (iv) a problem was detected and work has been scheduled for the next minor PM. Table 4.5 shows a list of some of the different maintenance activities that are undertaken.

Table 4.5   Dragline maintenance actions.

Component	Maintenance activity/purpose	Frequency
Bucket, rigging, drag, hoist, and dump ropes	Inspection	7 d
Boom protection system	Inspection	3 mo
Air compressors	Inspection	3 mo
Gear cases/bearings	Vibration analysis	4 mo
Winch rope box	Crack detection	6 mo
Machinery, boom, boom suspension ropes	Crack detection	12 mo
Gear cases	Oil particle analysis	12 mo
Motors	Greasing	40 wk

Shutdown maintenance is carried out for change-outs of major components, modifications, repairs to cracks (for example, in boom), and so on.   ∎

Example 4.8   Excavator

An excavator is a machine used in mining for loading coal or ore onto trucks. Table 4.6 presents a summary of the different maintenance actions carried out on an excavator along with some additional information relating to the duration.

The scheduled oil sampling (SOS) is performed in conjunction with servicing at 500-hour intervals. During servicing, a sample of oil is sent away for analysis as a guide to forewarn of possible failure. The results of the oil (hydraulic fluid) sampling are returned sometime after the scheduled service is completed. Consequently, if the oil sample contains a high content of a particular metal or mineral, indicative of probable oncoming failure, it is not possible to change the pump during that service. Instead, if failure is indeed imminent, as indicated by the SOS, the pump must last until the next scheduled service interval or component failure, which adds to the downtime of the machine.   ∎

Table 4.6   Excavator maintenance actions.

Maintenance type	Policy	Planning	Duration (h)	Adds to downtime?
Preventive	500 h service	Scheduled	18	No
	1000 h service	Scheduled	24	No
	2000 h service	Scheduled	24	No
Corrective/ breakdown	Repairs	Unscheduled	Varies	Yes
Condition-based (predictive)	SOS	Scheduled	Negligible	No
	Visual checks	Scheduled	Negligible	No

N.B. Servicing follows pattern: 500, 1000, 1500, 2000 (i.e., serviced every 500 hours).

4.10   Maintenance of Infrastructures

Infrastructures have both discrete and distributed components such as road and rail networks, electricity networks, and gas, oil, water, and sewage networks. The characterization of the state is more complex than that of products and plants for the following reasons:

- Each type of infrastructure is different;
- The degradation of the state of the infrastructure is influenced by several factors such as weather, state of the system, usage intensity, output of (or throughput through) the system;
- Safety plays an important role, as an asset in poor condition can have dramatic consequences.

The main purpose of maintenance actions (PM and CM) is to control infrastructure degradation due to age, usage, load carried, other environmental factors, and so on, and restore it to a normal operating condition in the case of failure or other faults. PM and CM actions are infrastructure-specific and we discuss these for a few different infrastructures in the next few sections. Maintenance of infrastructures include service/operations that are different from PM and CM actions relating to infrastructure *per se*. They include services such as clearing (snow, sand, and any object hindering the operation in the case of road and rail tracks), cleaning (routine cleaning of buildings, rolling stock, vegetation growth on the sides of roads and rail tracks, etc.), and fixing (damaged road signs) for safe operations.

PM actions include inspections to monitor and assess the condition of the infrastructure. Based on the inspection results and considering, for example the severity of the risk to traffic, the availability of resources, and other business risks, a decision is made as to whether to rectify the fault immediately or plan it for a later stage. Also, since failure is not so well-defined, there is a blurred line between PM and CM actions. In general, PM actions are those tasks that can be carried out in a short time period without too much interruption to the normal operation of the infrastructure. In contrast, CM actions take a longer time to complete (possibly running into months), affect normal operations in a significant manner, and are costly.

In the next two sections, we discuss maintenance requirements of rail and pipeline infrastructures as examples.

4.10.1   Rail Infrastructures

In terms of rail track maintenance, we focus on the maintenance of ballast and rails (see Figure 3.3). Ballast is crushed or moved by the weight of trains passing over it. If unattended, the track may become uneven, leading to swaying, rough riding, and possibly derailments.

4.10.1.1   Maintenance of Ballast

This involves one or more of the following activities:

- *Tamping:* This is conducted to correct the longitudinal profile, cross level, and alignment of track. A number of sleepers at a time are lifted to the correct level with vibrating tamping tines inserted into the ballast.

- *Track stabilization:* Track stabilizers vibrate the track in the lateral direction with a vertical load to give controlled settlement. Tamping and compacting ballast underneath sleepers reduces the lateral resistance of the track. Track stabilization can restore the lateral resistance to the original level.
- *Ballast injection (stone-blowing):* Ballast injection, or stone-blowing, is conducted to correct the longitudinal profile. The process introduces additional stones to the surface of the existing ballast bed whilst leaving the stable compact ballast bed undisturbed.
- *Sleeper replacement:* In almost all types of sleeper defects, remedial action is not possible and the sleeper requires replacement. Defective sleepers can result in the rail losing the correct gauge, which can cause rolling stock derailments.

4.10.1.2 Maintenance of Rails

The activities involved in rail maintenance can be classified broadly into the following areas:

- Rail inspection to assess the state (condition) of the rails. Some inspections are done using an instrumented bogey and others are done manually.
- PM actions to control the rail degradation rate due to the interaction of rail and wheel. This, in turn, involves activities for:
  - Rail wear reduction;
  - Reduction in rolling contact fatigue (RCF) initiated defects through periodic grinding of rails.
- Rail rectification and replacement. Rail rectification consists of "minimal repair" actions (such as small rail section replacement, rail welding, tamping, adjustment of rail sections, etc.). Replacing a substantial part of the rail is known as "rail replacement," and major overhauling of rails is known as "re-railing."

4.10.2 Pipeline Infrastructures

Pipeline infrastructures have been employed as one of the most practical and low-cost methods for large-scale oil and gas transport for decades. Pipeline networks for water distribution and sewage systems are everywhere and they pose critical management and maintenance problems. Aging infrastructure and replacement costs are major challenges for municipal water utilities. With populations increasing and available freshwater resources decreasing, water and wastewater distribution pipelines need to be maintained to prevent water loss and damage to the surrounding environment. Pipeline replacements are bound to get costlier over time, so the burdens, if shifted to the future, are bound to get heavier.

In this section, we focus on pipeline systems in the oil and gas industries to illustrate their maintenance requirements. It is important to ensure the safe utilization of such infrastructure in order to prevent economic, social, and ecological losses. Pipeline integrity monitoring is one of the most important maintenance activities. This includes a variety of measures taken to monitor the condition of the pipeline in order to anticipate any damage to the pipe and its associated equipment, maximize the efficiency and safety of the pipeline, minimize potential accidents and service interruptions due to pipeline neglect, and safeguard company and public interests.

A partial list of measures that should be involved in pipeline integrity monitoring includes leak detection, inspection pigs,[13] visual and underwater inspection of the pipe exterior, and checking of pressure regulators and pressure-relief valves.

Inspection pigs, in particular, are widespread devices employed for maintaining pipeline integrity. Whether a pipeline is onshore or offshore, the only way a complete inspection can be carried out is from inside the pipeline using "pigs." Regular operational pigging serves the following purposes:

- Prevention of scale build-up;
- Cleaning of the pipe wall;
- Removal of internal debris;
- Removal of liquids (condensation and water);
- Enhancement of the performance of corrosion inhibitors.

4.11   Effective Maintenance

Effective maintenance decisions need to be made in a framework that takes into account technical, commercial, social, and managerial issues from an overall business perspective. From a knowledge point of view, a multi-disciplinary approach is required, as the know-how required spans areas across the various fields of knowledge including science, engineering, technology, and management. Some of the key issues in each are shown in Figure 4.6.

Science plays a key role in understanding and modeling failure mechanisms, as discussed in Chapter 3. Understanding failure mechanisms (physics of failure) is the key to sound decision making, as it aids in predicting failures.

CBM relies on several technologies for the condition monitoring of systems, with the aim of confirming health or scheduling maintenance action prior to failure. Also, many technological advances are changing the way engineered objects are maintained. For example, using smart sensors, wireless communication technology, and web technology, experts at service centers can conduct diagnostics and analysis of remote equipment. Therefore, maintenance technology plays a crucial role and is a key enabler of effective maintenance practice. Maintenance-technology-related issues are discussed in Chapters 6, 16, and 23.

The link between science and technology is engineering, as discussed in Chapter 6. Engineering deals with design, manufacture, operation, and maintenance. A key to controlling maintenance costs is to design systems that are reliable and easy to maintain, so the link between product design and maintenance is very strong. It requires a close collaboration between designer and operators and maintainers of systems to improve their maintainability. These efforts may lead sometimes to maintenance-free components. These issues are discussed in detail in Chapter 7.

An effective maintenance system provides supporting decision-making techniques, models, and methodologies, and enables maintenance personnel to apply them in order to set the global production costs at a minimum and to ensure high levels of customer service. Therefore, effective maintenance management needs to be based on quantitative business models that integrate maintenance with other decisions, such as those involving production, and so on. Modeling and maintenance models are presented in Parts B and C.

---

[13] Pig: any device made to pass through a pipeline driven by the pipeline fluid.

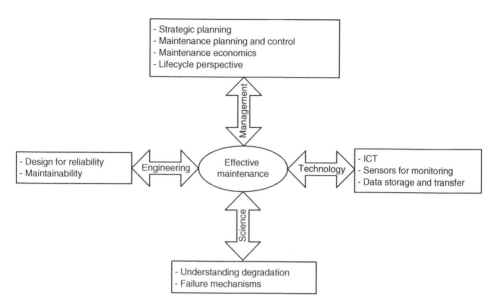

Figure 4.6   Elements of effective maintenance.

Engineering and technology are not enough to deal with the complexities of the maintenance of various engineering objects without effective maintenance management. Various maintenance management issues are discussed in Part D including strategic planning, maintenance control, maintenance planning, and scheduling, maintenance logistics, maintenance outsourcing, maintenance economics, and maintenance performance management.

4.12   Summary

The performance of an engineered object depends not only on its design and operation, but also on its maintenance during its operational lifetime. Thus, proper functioning over an extended time period requires proper maintenance activities.

Maintenance actions can range from visual inspection of an object to the overhaul of very complex and large systems such as turbines, chemical reactors, or the turnaround maintenance of a complete plant. Typical maintenance activities include inspection, compliance testing, monitoring, routine maintenance, overhaul, rebuilding, repair, and turnaround maintenance.

Maintenance actions can be divided into two broad categories: (i) preventive maintenance (PM) and (ii) corrective maintenance (CM). PM can be divided into predetermined PM tasks based on a clock or usage and CBM (based on equipment condition). CM actions are maintenance activities that are carried out after a failure has occurred. CM must be initiated immediately to restore critical systems to their functional state or can be deferred to a more convenient timing if the failure is not critical and does not need immediate action. Maintenance improvement is possible by using maintenance data and feedback information to design out maintenance (removing the need for maintenance) or design for maintenance (ease of maintenance).

Since maintenance can be viewed as actions to compensate for the unreliability of an engineered object, maintenance actions do have an effect on the reliability and future failures of such objects and they can be classified into the following three categories: (i) perfect maintenance (item as good as new after maintenance), (ii) minimal repair (item restored without change to failure rate), and (iii) imperfect maintenance (item state between as good as new and as bad as old after maintenance).

Engineered objects have different maintenance requirements, different levels of complexities, and make use of different maintenance policies. Understanding these differences shows clearly that maintenance is multifaceted and there are several aspects involved including technical, commercial, social, and management viewpoints. This implies that effective maintenance decisions need to be made in a framework that takes into account these issues from an overall business perspective. In particular, a comprehensive maintenance system is needed, combining science, engineering, technology, and management.

Review Questions

4.1   What are the characteristics of condition-based maintenance actions?

4.2   What is the difference between design out and design for maintenance?

4.3   What are the main issues in the maintenance of consumer products?

4.4   What are the main issues in the maintenance of plants?

4.5   Explain the complexity of infrastructure maintenance issues.

4.6   What are the main elements of an effective maintenance framework? Can effective maintenance be achieved without one of these elements?

4.7   Justify the need for a multi-disciplinary approach to maintenance.

Exercises

4.1   Following on from Exercise 2.1 for a bicycle:
  (a) Describe the kinds of PM actions to ensure that the bicycle is safe to ride and that the ride is satisfactory.
  (b) Make a table to indicate the frequency of the various PM actions.
  (c) List the PM actions for the tires.
  (d) Which of these may be carried out by a lay person and which require a skilled mechanic?

4.2   Following on from Exercise 2.2 for a truck diesel engine:
  (a) Describe the PM requirements for the engine.
  (b) The truck is involved in a head-on collision and the engine is damaged. What kind of damage can be fixed through CM?
  (c) What damage will require the engine to be scrapped?

4.3   Following on from Exercise 2.3 for a refinery plant:
  (a) List the components which are critical and require continuous inspection.
  (b) Many valves may remain in a failed state and this may lead to serious consequences. What kind of testing will allow the detection of a failed valve?

4.4   Following on from Exercise 2.4 for a wind turbine:
    (a) Identify the unique maintenance issues, especially for those turbines installed off-shore.
    (b) Explain why opportunistic maintenance may be appropriate for this type of engineered object.

4.5   Following on from Exercise 2.5:
    (a) List the different types of PM actions for the maintenance of rail track.
    (b) List the different types of CM actions to restore a degraded track.

4.6   Following on from Exercise 2.6:
    (a) List the different types of PM actions for the maintenance of the road surface.
    (b) List the different types of CM actions to restore a degraded road surface.

4.7   Discuss the reasons why CM is more expensive than PM, with reference to resource utilization, equipment preservation, and environmental impact, amongst other dimensions. Explain how the understanding of failure mechanisms helps in deciding the type of appropriate maintenance actions for a particular engineered object.

4.8   The number of grid-connected solar photovoltaic (PV) systems is expected to increase dramatically over the coming decades. This increase in the number of PV units leads to an increased focus by utility companies and other solar-generating firms on achieving the highest level of performance and reliability from the solar asset. What maintenance policies do you think will be appropriate for solar panels?

4.9   List two engineered objects where opportunistic maintenance is appropriate. For each object explain the advantages of such a maintenance policy.

4.10   Identify two engineered objects where group maintenance policies are appropriate. For each object explain the advantages of such a maintenance policy.

4.11   The maintenance of many public infrastructures (bridges, sewage and water networks) is the responsibility of government organizations and is often subcontracted to a third party. Discuss the economic and social implications of delaying or having poor maintenance of such infrastructures.

4.12   Explain how the maintenance of products, plants, and infrastructures differs in terms of requirements and types of maintenance actions.

References

Blischke, W.R. and Murthy, D.N.P. (1994) *Warranty Cost Analysis*, Marcel Dekker: New York.
Cassidy, C.R., Murdock, W.P, Nachlas, J.A., and Pohl, E.A. (1998) Comprehensive fleet maintenance management. *Proceedings of the IEEE International Conference on Systems, Man, and Cybernetics*, **5**: 4665–4669.
European Standard EN 13306 (2001) *Maintenance Terminology*, European Committee for Standardization.
Pintelon, L. and Parodi-Herz, A. (2008) Maintenance: An evolutionary perspective. In *Complex System Maintenance Handbook*, Kobbacy, K.A.H. and Murthy, D.N.P. (eds), Springer-Verlag, London, pp. 21–48.

5

Life Cycle of Engineered Objects

Learning Outcomes

After reading this chapter, you should be able to:

- Explain the concept of the life cycle of engineered objects;
- Differentiate between standard and custom-built objects from the maintenance and reliability points of view;
- Describe the standard engineered object life cycle from a manufacturer's/builder's perspective;
- Describe the standard engineered object life cycle from a customer's/owner's perspective;
- Describe the custom-built engineered object life cycle from a manufacturer's/builder's perspective;
- Describe the custom-built engineered object life cycle from a customer's/owner's perspective;
- Explain the difference between design reliability, inherent reliability, reliability at sale, and field reliability in the context of products;
- Explain the difference between design reliability, inherent reliability, reliability at the start of operation, and field reliability in the context of plants;
- Explain the importance of considering life cycle cost in making investment decisions in engineered objects.

Introduction to Maintenance Engineering: Modeling, Optimization, and Management, First Edition.
Mohammed Ben-Daya, Uday Kumar, and D.N. Prabhakar Murthy.
© 2016 John Wiley & Sons, Ltd. Published 2016 by John Wiley & Sons, Ltd.

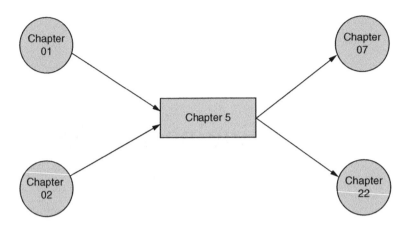

5.1   Introduction

The life cycle concept plays a very important role in the decision-making processes of both customers (owners) and manufacturers (builders) of engineered objects (products, plants, or infrastructures). There are several different notions of life cycles and a proper understanding of these is critical in the context of product reliability and maintainability.[1] In this chapter we focus on reliability from the life cycle perspective.[2]

The outline of the chapter is as follows. In Section 5.2 we start with a general discussion of life cycles. The life cycle from the owner's (customer's) perspective is different from that of the manufacturer (builder). Section 5.3 deals with the life cycle of standard objects from the manufacturer's and customer's perspectives, and Section 5.4 looks at the life cycle of custom-built objects from the manufacturer's and customer's/owner's perspectives. Reliability of engineered objects from the new product development (NPD) life cycle perspective is the focus of Section 5.5, where different notions of reliability are introduced. The concept of Life Cycle Cost (LCC) is presented in Section 5.6. We conclude with a summary of the chapter in Section 5.7.

5.2   Life Cycle Concept and Classification

5.2.1   Life Cycle Concept

> **Definition 5.1**
>
> The life cycle refers to the entire spectrum of activity for a given engineered object, commencing with the identification of a consumer need and extending through design and development, production and/or construction, operational use, sustaining maintenance and support, and finally the retirement and phase-out (Blanchard and Fabrycky, 1997).

---

[1] Maintainability deals with maintenance issues during the design of the engineered object, as opposed to maintenance, which covers actions carried out after the object is put into operation. Chapter 7 deals with maintainability.
[2] The study of maintenance from a life cycle perspective is discussed in Chapter 7.

Table 5.1 Life cycle duration.

Engineered object		Perspective	
		Manufacturer (builder) (yr)	Customer (owner) (yr)
Product	Computer	3–5	1–3
	Automobile	10–15	5–7
	Cell phone	2–7	2–3
Plant	Mining	3–7	20–30
	Oil refinery	3–5	20–40
Infrastructure	Road	3–5	>30
	Rail	5–7	>40
	Dam	5–7	>100

Thus, the life cycle of an engineered object is the duration (or time period) that is relevant and appropriate for making decisions with regard to the object.[3] The duration depends on the engineered object and the perspective (customer or manufacturer). Table 5.1 shows typical durations for a small sample of engineered objects.

The duration is comprised of several phases (or stages) and there is no universal agreement on the number of phases that is most appropriate. Decisions need to be made continuously in each phase – whether to continue with the activities of the phase or move to the subsequent phase or terminate the process completely.

5.2.2 Classification of Life Cycles

One can classify the life cycles of engineered objects based on (i) whether they are standard or custom-built and (ii) the perspective – customer/owner or manufacturer/builder. We will discuss these in the next two sections.

5.3 Standard Objects

Standard objects are mainly consumer, commercial, and industrial products.

5.3.1 Manufacturer/Builder Perspective

The life cycle for standard objects can be classified into two types: (i) the marketing perspective and (ii) the NPD perspective.

5.3.1.1 Marketing Perspective

From a marketing perspective, the product life cycle is the sales rate or volume (characterized either as a number or as a fraction of the total market volume or share) over time. In the product life cycle, the marketing perspective plays an important role, in the sense that it helps in integrating

---

[3] A similar concept exists for technologies. The technology life cycle is discussed in Chapter 6.

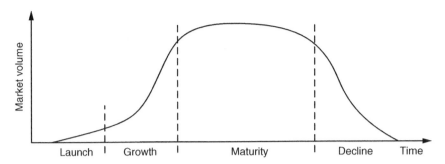

Figure 5.1 Product life cycle – marketing perspective.

a customer's requirements and also helps in assessing the volume of products or after-sales services (such as spare parts, etc.) and their management over the product life cycle. The life cycle has four distinct phases, as indicated in Figure 5.1 and a brief description of these follows.

Phase 1 (Launch): Customers for new products can be grouped into innovators – the customers who are looking for new technologies and products and are ready to try them (around 5–15%) – and imitators – the consumer group that follows innovators. During the launch phase, the innovators are the customers and their evaluation and assessment of the product (in terms of the different characteristics, with product reliability being one of the most dominant) play a significant role in attracting the imitators to buy the product. The growth rate is slow. The innovators are attracted mainly by the novelty whilst other factors, such as cost, play a less important role in the purchase decisions.

Phase 2 (Growth): This phase marks the imitators buying the product, leading to a sharp increase in the growth rate. The strategy used is to offer better warranty terms in the case of standard products to attract more customers.

Phase 3 (Maturity): During this phase, competitors (producing products which act as substitutes to the new product) appear on the market, leading to vigorous competition. The sale price is an important factor in ensuring market share. Sales are mostly repeat purchases by customers who bought the product in Phase 2.

Phase 4 (Decline): This phase is characterized by declining sales or market share for the product. It is usually due to the appearance of a superior product on the market. As a result, this phase overlaps with Phase 1 of the life cycle for the newer product.

Example 5.1   Boeing 707

The Boeing 707 was a mid-sized, long-range, narrow-bodied, four-engine jet airliner that was built by Boeing Commercial Airplanes from 1958 to 1979. Developed as Boeing's first jet airliner, the 707 dominated passenger air transport in the 1960s and remained common through the 1970s; the 707 is generally credited with ushering in the Jet Age. Boeing produced and delivered more than 1000 aircraft all over the world. Figure 5.2 illustrates the life cycle of the Boeing 707 from the marketing perspective. The early years, from 1956 to 1960, correspond to the growth phase; the period from 1960 to 1970 is the maturity phase (this phase was marked by decline due to conflicts and global recession), and the period from 1970 onwards (after the launch of the newer 7 × 7 series – especially the Boeing 767) corresponds to the decline phase.   ∎

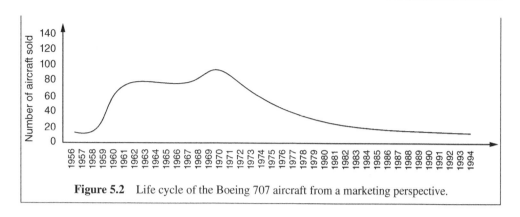

Figure 5.2   Life cycle of the Boeing 707 aircraft from a marketing perspective.

5.3.1.2   New Product Development (NPD) Perspective

The life cycle from the NPD perspective is important in the context of deciding the reliability characteristics of a new product based on experiences from an existing product and/or the use of new technologies to improve product reliability. It has six phases, as indicated in Figure 5.3. We discuss each of these phases briefly.

Phase 1 (Front-end): In this phase, an evaluation of a new product idea from a business perspective is made to decide whether the idea is feasible or not. This is done in terms of identifying the desired characteristics of the product from the customer perspective, the implications for development and manufacturing (technical aspects such as technology acquisition, research, and development effort needed, etc.), and for marketing (commercial aspects such as pricing, demand, etc.). Such evaluations result in go or no-go decisions.

Phase 2 (Design): This phase involves two subphases: (i) conceptual design and (ii) detailed design. The first subphase looks at alternative design concepts for the new product. Once the choice of the appropriate concept is finalized, one proceeds to the next subphase, which deals with detailed drawings and specifications such as components, materials, and so on.

Phase 3 (Development): This phase deals with converting the detailed design into a physical product – a prototype of the new product. The performance is assessed based on testing. If the performance does not meet the required performance, additional development efforts will be needed to close the gap, with the possibility of going back to Phase 2 to make changes to the design.

Phase 4 (Production): This phase deals with the actual manufacturing of the product. Raw materials, components from external vendors, human resources, machines, and other resources needed for the production of the new product are sorted out and organized. Procedures and standards to ensure quality of conformance are set up so that items produced conform to specifications.

Phase 5 (Marketing): Marketing is the process of communicating the value of a product or service to customers. It involves making decisions regarding distribution channels, promotion media, pricing, and so on to bring the product to potential customers.

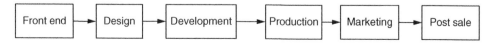

Figure 5.3   Product life cycle – NPD perspective.

Phase 6 (Post-sale servicing): Customers buying a product need assurance that the product will perform satisfactorily over its useful life. Manufacturers provide this assurance through warranties which require the manufacturer to rectify any failures occurring within the warranty period, as discussed in Chapter 4.

Beyond the warranty period, customers need adequate post-sale support for maintenance. These include spare parts if maintenance is done in-house and spare parts as well as trained personnel to carry out the maintenance actions if maintenance is outsourced. Extended warranties and maintenance service contracts are two options available to customers when maintenance is outsourced, and these are discussed further in Chapter 18.

Example 5.2   Dreamliner 787

The Dreamliner 787 is the direct descendant of the 707. The experiences from earlier models of Boeing passenger and cargo aircraft were used to develop the new aircraft. The Dreamliner designers had the goal of developing an aircraft which was faster, better, and cheaper to run. This involved extensive use of composite materials and state-of-the-art electrical systems to achieve a high level of performance and an efficiency of 70% in fuel consumption. Apart from improved performance, the aircraft requires larger time intervals between maintenance, leading to higher availability.[4]   ∎

5.3.2   Customer/Owner Perspective

The life cycle from the customer perspective involves three phases, as shown in Figure 5.4, and we describe each of these briefly.

Phase 1 (Purchase): This involves choosing the best object from a set of similar objects on the market. Often, a multi-dimensional criterion is used for evaluation that will maximize the return on investment and minimize the total LCC. Due consideration is given to factors such as cost, manufacturer's reputation, object features, and post-sale warranty and service.

Phase 2 (Operation and Maintenance): Operation and maintenance of standard objects rely on well-defined, tested, and verified procedures that are made available by the manufacturer. However, owners have the choice to either develop an in-house maintenance strategy

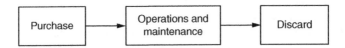

Figure 5.4   Life cycle for standard objects – customer perspective.

---

[4] Availability is discussed in Chapter 7.

or outsource the maintenance service to the OEM (original equipment manufacturer) or to some external (third party) qualified maintenance contractor. One advantage of standard objects is that spare parts are normally readily available. This is not the case for custom-built objects, where spare parts are engineered to order and require long lead times.

Phase 3 (Discard): Engineered objects are usually discarded either because of a reduction in their efficiency or because a new engineered object with better features, often with lower operational costs, enters the market. Owners must have a plan for discarding old and obsolete items. Options include scrapping the old items, trading them in for new items, or selling them on the second-hand market.

5.4 Custom-Built Objects

5.4.1 Manufacturer/Builder Perspective

Custom-built objects include plants and infrastructures as well as some specialized products. These are produced in small volumes – a single unit (for example, a bridge) – to tens (for example, a fleet of ships for the navy) or hundreds (commercial aircraft), as opposed to standard products which are produced in large volumes (thousands to millions). Custom-built objects often use the latest and untried technologies, so there is a degree of risk to both parties in terms of the final outcome.[5] The life cycle for custom-built objects differs slightly from that of standard products, but it also involves six phases, as shown in Figure 5.5.

Phase 1 (Contract): Here, the customer's requirements regarding the object and the ability of the builder (also referred to as the *contractor*) to deliver the object (within the time and cost limits) are the main issues that are jointly decided leading to a legal contract (covering all aspects – technical, financial, risk, penalties, etc.) between the two parties. This can be a very involved process for complex infrastructures.

Phase 2 (Design): This phase is similar to that for standard products discussed earlier. The important difference is that the customer is involved in the process.

Phase 3 (Development): This phase is similar to that for standard products and it is critical if the design involves using new technologies. Extensive testing is needed to ensure that the final object meets the desired requirements.

Phase 4 (Fabrication/Construction): In the case of complex plants and infrastructures, the fabrication/construction needs to be done on site. This involves efficient project management – proper coordination of resources (material, human, equipment, etc.) – so that the schedules are met. If not, the builder/manufacturer can incur heavy penalties.

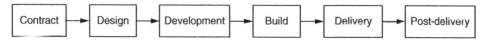

Figure 5.5 Life cycle for custom-built objects – builder perspective.

---

[5] Chapter 6 deals with technology, and this includes materials, processes, techniques, and so on.

Phase 5 (Delivery): Once the object is built, it is rigorously tested to ensure that it meets the performance requirements stated in the contract. If not, the contractor needs to make the appropriate changes to the design and/or fabrication to achieve the agreed performance requirements. Once these are resolved, the object is formally handed over to the customer. The parties may resort to some settlement that might involve legal systems if the requirements cannot be met.

Phase 6 (Post-sale servicing): This phase is similar to that for standard products.

Example 5.3   The Bothnia Line – Railway Infrastructure

Railway infrastructure is built as per customer (government authority – ministry of railways, the private sector such as a mining company transporting ore or steel, or the public sector) requirement. The process involves (i) invitation of tenders, (ii) selection of the best offer and negotiation of terms and conditions, (iii) contract signing, (iv) detailed design and development, (v) construction, and (vi) handing over of the railway infrastructure.

The Bothnia Line (Swedish: Botniabanan) is a high-speed railway line in northern Sweden built as per Swedish Transport Administration (formerly Banverket – now Trafikverket) specification by the company Botniabanan AB, a company owned 91% by the Swedish state and 9% by the regional governments.

The construction budget was SEK 15 billion (approximately $2.2 billion). The work began on the line in 1999 and was completed in 2010. After completion, the line was leased to the Swedish Transport Administration – Trafikverket. The 190-km long Botniabanan, from Kramfors Airport to Umeå, was opened in 2010 and was designed for travel speeds up to 250 km/h. When Botniabanan AB recovers its investment (estimated to be around the year 2050), ownership of the line will pass to Trafikverket.   ∎

5.4.2   Customer/Owner Perspective

The life cycle from the customer perspective consists of the five stages shown in Figure 5.6.

Phase 1 (Contract): In custom-built objects, the first step is to have a formal contract between the customer and manufacturer (builder). The customer invites tenders for the engineered object from the local, national, or international market. This depends on the type of object needed and the level of expertise required. It is always advisable to make the contract as detailed as possible to include specifications, quality control checks, schedules, unforeseen circumstances, compensations, maintenance, and so on, so that ambiguities and sources of conflict are reduced.

Phase 2 (Fabrication): Once the contract is signed and payment details sorted out, the builder starts the fabrication process. The object could be built at the manufacturer's site

Figure 5.6   Life cycle for custom-built objects – customer perspective.

and transported to the customer's premises, or it could be built on site as a stand-alone system, as in the case of bridges, tunnels, railway lines, space rockets, and so on.

Phase 3 (Delivery/Commissioning): When the custom-built object is completely ready, it is tested rigorously for all its intended functions. Any defects or deviations from the planned design are fixed. When the object is completely ready, it is delivered to the customer.

Phase 4 (Operation and Maintenance): The operation and maintenance of custom-built objects are complex tasks that are equally as important as their design and manufacturing. The builder (contractor) trains the customer's team regarding operation and maintenance. Spare parts are provided in case of emergency. Depending on the type of engineered object, the builder may be offered a maintenance contract to provide the expertise needed keep the object within acceptable reliability limits.

Phase 5 (Discard): As discussed previously, once the system completes its intended life cycle, several options are available. It may be overhauled completely to extend its useful life if it is economical to do so, sold to a third party, or traded in as part of a new purchase. However, often such objects are just discarded or scrapped (for example, an oil platform). Furthermore, discarding a custom-built system is a sensitive issue in today's world that is concerned with environmental regulations, pollution control measures, and hazardous waste materials.

5.5   Reliability: Product Life Cycle Perspective

The reliability of products and plants from a life cycle perspective is of interest from both manufacturers' and customers' perspectives and this leads to different notions of reliability.

5.5.1   NPD Perspective

Figure 5.7 describes a typical scenario of the reliability of a new product over the different phases of its life cycle.

During the feasibility phase, a study is carried out using various target values for product reliability, such as customer satisfaction, warranty costs, profits, market share, and so on.

The design stage is carried out in two phases. In the conceptual design phase, reliability targets for the product are defined in order to ensure that reliability-related targets such as customer satisfaction (measured by no failures over the warranty period) and warranty costs (specified and defined in the feasibility phase) are met.

During the detailed design phase, the architecture of the product is established through a sequential process proceeding down to component level. The product reliability is determined from component reliabilities using the structure function (discussed in Chapter 2). Deciding on the desired component reliabilities to achieve the reliability target at the system (object) level is called *reliability allocation*. If the actual reliability of components does not meet the desired reliability, then one needs to make design changes (such as using redundancy – if appropriate) or use a development process to understand the cause and mechanisms of failure (discussed in Chapter 3) and then redesign the component, perhaps using a different material. This is the reliability improvement process (which is costly and time-consuming) and it stops once the final reliability of the prototype matches the target reliability.

In the pre-production phase, the reliability of items produced is usually below the target reliability for a variety of reasons which are process and/or supplier related. This is mainly

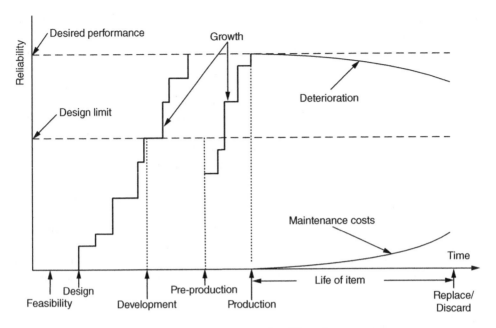

Figure 5.7   Reliability from a product life cycle perspective.

caused by variations resulting from the manufacturing process. These are sorted out (through proper quality control of the process and input materials and components) so that the reliability of items produced meets the target reliability. Once this is achieved, full-scale production starts and manufactured products are rechecked (by random sampling), approved, and released for sale.

During the operation phase, the reliability of a product decreases with age and/or usage, and maintenance (discussed in Chapter 4) is used to ensure that it functions properly. The frequency of failures increases with age (for most of the mechanical components) and this results in maintenance costs going up and availability going down.

When it is no longer economical to continue using the item, it either needs to be discarded or replaced by a new item. The reliability characteristics of a product over its life cycle have implications for both the manufacturer and customers.

5.5.2   Different Reliability Notions (Product)

When viewed from the product life perspective, one can define four different notions of product reliability. Figure 5.8 shows these notions, the relationships between them, and the main factors that affect the various reliabilities. We discuss these briefly:

* *Design reliability* is based on component reliability and is decided at the end of the design/ development phase, as discussed earlier.
* *Inherent reliability* is based on the reliability of the product at the end of the manufacturing and assembly process. In spite of proper quality control, errors may be made in manufacturing

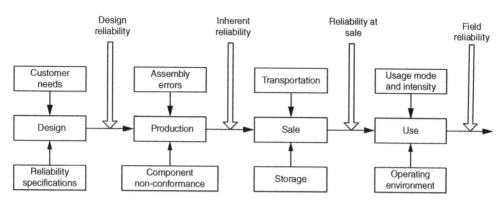

Figure 5.8 Different notions of product reliability.

or assembly, or some of the components may be non-conforming. As a result, a fraction of the items produced may not meet the reliability target. The production output is a mixture of conforming and non-conforming items and the inherent reliability is a composite of the reliabilities of these two types of item.

- *Reliability at sale* represents the reliability attributes that a customer gets. Items need to be transported to markets and are often stored for some period before they are sold. Items can experience mechanical shocks if transported without proper packaging. The storage conditions can also have an impact on the items (for example, trucks left in open lots are exposed to the harsh influence of weather). As a result, the reliability at sale is different and lower than the inherent reliability.
- *Field reliability* is a notion of product reliability that a customer achieves in day-to-day operations. Errors in commissioning of engineered objects and different operating or usage environments (good roads as opposed to poor roads in the case of automobiles) and usage intensity (the number of times a washing machine is used per week or the amount washed per week) create loads (mechanical, thermal, electric, etc.) that differ from the loads under nominal usage. When the loads exceed the nominal values, degradation occurs at a faster rate (discussed in Chapter 3) and as a result, the field reliability is the composite of how these factors affect the reliability of the product. In practice, field reliability is often lower than the reliability at sale, mainly due to poor maintenance, a non-favorable operating environment, or intensity of usage.

5.5.3 Different Reliability Notions (Plant)

The reliability notion for a plant differs from that of a product in the sense that plant capacity and performance are discussed at equipment or infrastructure level and reliability attributes are jointly decided by the manufacturer/supplier and the customer. When viewed from the plant life cycle perspective, one can define four different notions of plant reliability: (i) design reliability; (ii) inherent reliability; (iii) reliability at the start of operation; and (iv) field reliability. Figure 5.9 shows these notions, the relationships between them, and the main factors that affect the various reliabilities.

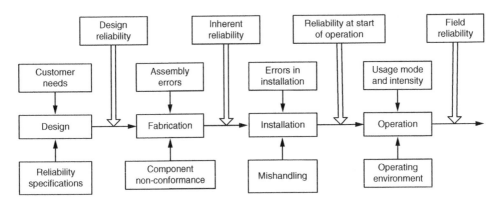

Figure 5.9 Different notions of plant reliability.

As can be seen, this is very similar to the different notions of product reliability discussed earlier. The main differences between the two are as follows:

- In the case of a plant, customers and the builder (contractor) jointly decide on design reliability (whereas in the case of products, the decision is made by the manufacturer based on customer needs).
- Poor field reliability of a plant requires the contractor to make design changes to improve reliability.
- Poor field reliability of a plant necessitates increased maintenance and service preparedness.
- Poor reliability also necessitates effective and efficient spare parts and inventory management.
- Often, poor reliability leads to non-conformance of product delivery schedules and an increased focus on quality checks.
- Poor field reliability has major implications for plant owners as it increases maintenance costs and lowers availability, as discussed in Chapter 4, and it also results in accidents and injuries.

5.6 Life Cycle Cost

LCC refers to the total costs associated with an object (product, plant, or infrastructure) over its useful life. It is comprised of several cost elements and these are discussed in more detail in Chapter 21.

The LCC from the manufacturer's perspective is different from that of the customer. They differ in terms of cost elements included in the LCC, as will be indicated later in the section.

Life cycle costing refers to the evaluation of alternative products, system design configurations, operational, or maintenance solutions. The objective is to assist in deciding on the best way to use scarce resources (monetary and human) to achieve the desired business goals. Life cycle costing will be discussed in more detail in Chapter 21.

5.6.1   Manufacturer's Perspective

LCC from the manufacturer's perspective (LCC_M) includes all the costs associated with all the items produced over the product life cycle. The period of LCC starts with the feasibility phase of NPD (see Figure 5.3) and ends when the post-sale service (as part of a sale) for the last item sold expires. It can be expressed as:

$$LCC_M = \text{design and development cost} + \text{manufacturing cost} + \text{marketing}$$
$$+ \text{post-sale service support costs}$$

If N units are produced, then the LCC per unit is given by LCC_M/N.

5.6.2   Customer's Perspective

LCC from a customer's perspective (LCC_C) for an item purchased can be expressed as:

$$LCC_C = \text{acquisition cost} + \text{operating costs} + \text{disposal cost}$$

The operating costs include the maintenance cost (either done in-house or outsourced) and the acquisition cost is the LCC cost per unit + the profit margin for the manufacturer.

5.6.3   The Importance of LCC

The bulk of the LCC (up to 80%) is the result of decisions made during the design, development, and manufacturing phases. As such, proper decision making is needed to ensure the lowest possible LCC. Figure 5.10 shows the role and importance of LCC in the context of the continuous evolution of products. LCC-related data from earlier generations are used in the design of the next generation of products.

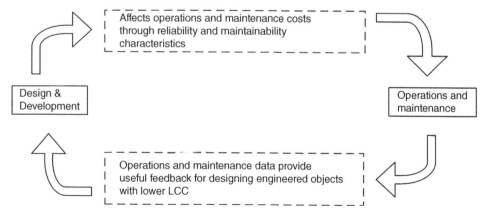

Figure 5.10   LCC as part of a new product development process.

LCC analysis may be applied for (IEC60300-3-3, 2004):

- The evaluation and comparison of alternative designs;
- The assessment of economic viability of projects/products;
- The identification of cost drivers and cost-effective improvements;
- The evaluation and comparison of alternative strategies for product use, operation, test, inspection, maintenance, and so on;
- The evaluation and comparison of different approaches for replacement, rehabilitation/life extension, or disposal of aging facilities;
- The optimal allocation of available funds to activities in a process for product development/ improvement;
- Long-term financial planning.

5.7 Summary

The life cycle refers to the entire spectrum of activity for a given engineered object, commencing with the identification of a consumer need and extending through system design and development, production and/or construction, operational use, sustaining maintenance and support, and system retirement and phase-out. This chapter looked at the product life cycle from reliability and maintenance points of view.

The life cycle duration is comprised of several phases (or stages) and there is no universal agreement on the number of phases that is most appropriate. The number and nature of these phases depends on (i) whether the engineered object is standard or custom-built and (ii) the perspective – the customer/owner or the manufacturer/builder.

From a manufacturer's perspective, the life cycle of standard objects can be viewed from a marketing perspective (launch, growth, maturity, and decline), an NPD perspective (front end, design, development, production, marketing, and post-sale), or a functional product perspective. The customer's perspective has three phases (purchase, operations and maintenance, and discard).

From a manufacturer's perspective, the life cycle of custom-built objects has six phases (contract, design, development, fabrication, delivery, and post-sale). The customer's perspective has five phases (contract, fabrication, delivery, operations and maintenance, and discard).

For the various stages of a product's life cycle, different reliability notions have been identified and linked, and the factors that affect them have been discussed. For products, design reliability, inherent reliability, reliability at sale, and field reliability follow the design, production, sale, and use phases of the life cycle, respectively. For plants, design reliability, inherent reliability, reliability at the start of operations, and field reliability follow the design, production, installation, and operations phases of the life cycle, respectively.

Review Questions

5.1 What is the difference between standard and custom-built objects?

5.2 What is the definition of a life cycle?

5.3   Identify the phases of the life cycle of a standard object from a manufacturer's point of view using a marketing perspective.

5.4   Identify the phases of the life cycle of a standard object from a manufacturer's point of view using a new product development perspective.

5.5   Identify the phases of the life cycle of a standard object from a customer's perspective.

5.6   Identify the phases of the life cycle of a custom-built object from a manufacturer's perspective.

5.7   Identify the phases of the life cycle of a custom-built object from a customer's perspective.

5.8   What is the difference between design reliability, inherent reliability, reliability at sale, and field reliability in the context of products?

5.9   For the notions of reliability mentioned in Question 5.8, identify the factors affecting each one of them.

5.10   What is the difference between design reliability, inherent reliability, reliability at the start of operations, and field reliability in the context of plants?

5.11   For the notions of reliability mentioned in Question 5.10, identify the factors affecting each one of them.

5.12   Define life cycle cost.

5.13   Explain the statement "LCC is an effective engineering approach for a new product development."

5.14   Explain "LCC facilitates selection of cost-effective design alternatives."

Exercises

5.1   Discuss why LCC is not important for most consumer products and purchase price is an important variable in the decision-making process.

5.2   For some consumer products (such as a car), the operating cost and the resale value are important. What factors determine the salvage value?

5.3   For expensive industrial/commercial products, LCC is very important. Consider the case where an airline operator is planning on buying a new fleet of aircraft. List the variables that need to be taken into account to determine the LCC.

5.4   Redo Exercise 5.3 for a dragline that a mining company is planning to buy for its open cut mine.

5.5   Redo Exercise 5.3 for an offshore wind farm (a collection of wind turbines to generate electricity from wind).

5.6   Redo Exercise 5.3 for a new rail infrastructure system.

5.7   Redo Exercise 5.3 for a new road infrastructure system.

5.8 With good quality control procedures, the inherent reliability of an item is very close to the design reliability. What control methods are needed to ensure the following?
(a) The quality of components supplied by a vendor conforms to the required specifications.
(b) The quality of assembly is high.

5.9 Field reliability depends on several factors, as indicated in Figure 5.7. Characterize these in the context of a vehicle.

5.10 Characterize the factors that affect the field reliability for a rail infrastructure.

References

Blanchard, B.S. and Fabrycky, W.J. (1997) *Systems Engineering and Analysis*, 3rd edition, Upper Saddle River, NJ: Prentice Hall Engineering.
IEC IEC60300-3-3 (2004) *Life Cycle Costing*, International Electrotechnical Commission, Geneva.

6

Technologies for Maintenance

Learning Outcomes

After reading this chapter, you should be able to:

- Explain the link between science, engineering, and technology;
- Describe the elements of the maintenance decision-making process;
- Explain the role of technology in the maintenance decision-making process;
- Define sensors and explain their role in assessing the state of engineered objects;
- Explain the scientific basis of commonly used sensors;
- Describe the technologies used for condition-monitoring methods;
- Describe the application of condition-monitoring technologies, for example of products, plants, and infrastructures.

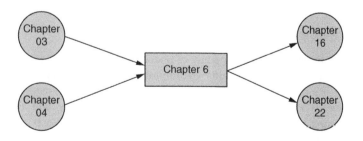

Introduction to Maintenance Engineering: Modeling, Optimization, and Management, First Edition.
Mohammed Ben-Daya, Uday Kumar, and D.N. Prabhakar Murthy.
© 2016 John Wiley & Sons, Ltd. Published 2016 by John Wiley & Sons, Ltd.

6.1 Introduction

Technologies (such as food, health, communication, transport, materials, manufacturing, etc.) have played a dominant role in the evolution of mankind. Two important features of technologies are: (i) they have evolved from *handed-down craft skills* to those that are more *science-based* and (ii) they have been changing over time at an ever-increasing pace. Technologies for maintenance have evolved in a similar manner and have impacted significantly on how engineered objects are maintained. One such example is the increasing use of *Condition-Based Maintenance* (CBM) policies (based on the condition or state of the object) as opposed to age- or usage-based policies. CBM uses sensor- and data-related technologies to assess the physical condition (or state) of an item (usually at the component level of an object but also at the system or some intermediate level, as discussed in Chapter 2).[1] The dynamic characterization of item state (discussed in Chapter 2) and the underlying degradation mechanisms (discussed in Chapter 3) play an important role in the CBM approach.[2] In this chapter our focus is on sensor- and data-related technologies for assessing item state.

The outline of the chapter is as follows. We start with a brief introduction to technology in Section 6.2, where we discuss the links between science, engineering, and technology. Section 6.3 looks at the issues involved in assessing the state of an item and the role of technology in the process. Section 6.4 deals with sensors and looks at the various issues and aspects. Section 6.5 looks at condition inspection and monitoring methods. These provide the data to assess the state of the item. Section 6.6 looks at data collection, storage, and analysis, and the technologies that have evolved. Sections 6.7–6.9 deal with technologies for maintenance of products, plants, and infrastructures, respectively. We conclude with a summary in Section 6.10.

6.2 Technology – An Overview

Technology is practical knowledge comprising goods, tools, processes, methods, techniques, procedures, and services that are invented and put into some practical use. As such, it is embedded in every engineered object (product, plant, or infrastructure) and includes both hardware and software.

6.2.1 Science, Engineering, and Technology Link

Science deals with the investigation or study of phenomena, and is aimed at discovering the principles of the underlying phenomena by employing a scientific approach or method. In other words, it is a method of discovering reliable knowledge about nature. For example, reliability science is concerned with the properties of materials and the causes for deterioration, leading to the degradation and failure of engineered objects. Chapter 3 dealt with the various failure mechanisms, and an understanding of these phenomena is critical when making CBM decisions.

---

[1] In the remainder of the chapter we use the terms *condition* and *state* interchangeably.
[2] CBM policies are more effective than age-based or usage-based policies and maintenance decisions based on CBM are discussed in Chapter 16.

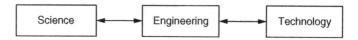

Figure 6.1   Science, engineering, and technology link.

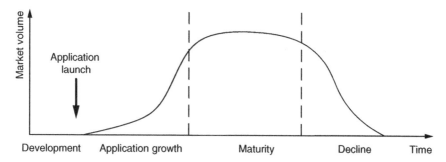

Figure 6.2   Technology life cycle.

Engineering is the process by which the scientific understanding of nature is used to produce engineered objects. This involves various activities such as design, manufacture, operation, and so on.

The link between science and technology is engineering, as shown in Figure 6.1. Engineering uses scientific principles and prevalent technology to build, operate, and maintain engineered objects. In general, it is difficult to say whether science precedes technology or vice versa.

An important feature of technology is its evolutionary nature – current technologies become obsolete with the appearance of new technologies. As such, each technology has a life cycle – a concept similar to the product life cycle (PLC) discussed in Chapter 5. The technology life cycle (TLC) has five stages:

- *Stage 1:* Development;
- *Stage 2:* Application launch;
- *Stage 3:* Application growth (through adoption and diffusion);
- *Stage 4:* Maturity;
- *Stage 5:* Decline (due to obsolescence and substitution).

Figure 6.2 shows the typical shape of the TLC. The duration of the Development stage is getting longer and the length of the remaining stages (from Launch to Decline) is getting smaller for all technologies.

6.3   Assessing the State (Condition) of an Item

The state (condition) of an item characterizes the effect of degradation with age and/or usage. In Chapter 2 we characterized the degradation by a variable $X(t)$ and this defines the state of the item at age t (or time t based on an age clock). It can be characterized as a discrete variable (with a finite set of values) or a continuous variable (with an infinite set of values), as discussed in that chapter. We assume $X(t)$ to be a non-decreasing function of t with higher values indicating a more degraded state. In general, $X(t)$ evolves over time in an uncertain manner and is influenced by operating environment, usage intensity, and maintenance history over the interval $[0,t)$.

6.3.1 Methods for Assessing Item State

The appropriate method with which to assess the state of an item depends on several factors, such as the material of the item, the mechanism causing the degradation and its effect, and so on. There are many different methods and they can be grouped broadly into two categories: (i) direct methods and (ii) indirect methods.

6.3.1.1 Direct Methods

Direct methods assess the item state by direct measurement of the level of degradation. For example, degradation of a bearing can be measured directly in terms of wear in the bearing by directly using a device such as a measurement scale or laser scanner.

In many cases, direct methods require removing the item from service and dismantling it (for example, inspection of bearing wear in a turbine requires the disassembly of the turbine) and in a few cases the testing results in the destruction of the item (for example, cutting a pipe in a boiler to examine internal cracks), so that it cannot be put into service again. In the past, these direct methods were performed manually by trained specialists. There is a growing trend towards the use of technologies and automated processes to replace such specialists.

Technology has played a very important role in devising testing methods which are not destructive, and these are discussed in more detail in a later section.

6.3.1.2 Indirect Methods

The indirect methods involve a measurable variable, $Y(t)$, that is related to $X(t)$ and whose values can be obtained using a measurement device. This produces an output, $Z(t)$, which provides the data with which to assess the state of the item. This involves proper data collection and data analysis to yield an estimate of $X(t)$, and the overall process is shown in Figure 6.3. We will discuss briefly some of the elements of the process.

Item State
The variable characterizing the item state can be viewed as an indicator of some form of energy – mechanical, electrical, thermal, electromagnetic (including light), chemical, or acoustic (for example, temperature is an indicator of how hot an item is).

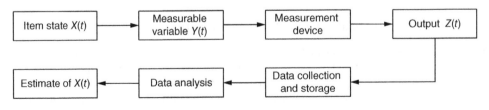

Figure 6.3   Indirect methods for assessment of an item's state.

Measurable Variable

The relationship between $X(t)$ and $Y(t)$ is given by an algebraic equation of the form:

$$Y(t) = \Phi(X(t)) \tag{6.1}$$

Very rarely can the expression $\Phi(\cdot)$ be obtained from some scientific theory. Often it is obtained empirically by carrying out controlled experiments in a laboratory and the relation is nonlinear.

The measured variable $Y(t)$ can also be viewed as an indicator of some form of energy, which can be the same as or different from that for $X(t)$.

Measurement Device

The words *sensor* and *transducer* are both widely used in the description of measurement devices.[3] We discuss these two forms and another called an *actuator* as these are all important in the context of maintenance.

Definition 6.1

A transducer is a device that receives energy from one object and transmits it, in the same or different form, to another object.

Devices which perform an "input" function are commonly called *sensors* and devices which perform an "output" function are called *actuators*. This is best illustrated by a smoke detector. It consists of a sensor (to sense smoke – the input energy is chemical and the output energy is electrical) and an actuator (a loudspeaker that produces a high-pitched sound – the input energy is electrical and the output energy is acoustic) to warn people about a fire.

A more precise definition of a sensor, appropriate in the context of maintenance, is the following.

Definition 6.2

A sensor is a device that consists of a sensing element (to sense some form of input energy) and some other element to produce a signal.

In most cases, the signal is an electrical signal and the other element is the circuitry to convert the output of the sensing element to an electrical signal.

Example 6.1   Thermocouple

A thermocouple is used for measuring the temperature of a liquid. The sensing element consists of two metallic strips (of different material) joined together in a "V" shape. When the tip is immersed in the fluid, it generates a voltage across the non-connected ends and this is the output of the sensing element. The voltage is then digitized and displayed on a screen. ∎

---

[3] The former is popular in the USA whereas the latter has been used in Europe for many years. The word "sensor" is derived from the meaning "to perceive" and "transducer" from the meaning "to lead across".

A more precise definition of an actuator is the following.

Definition 6.3

An actuator is a device that accepts energy (usually a signal) and produces some physical action for motion or control of an object.

The signal can be from a sensor, human, or other source. The actuator needs an external energy source to carry out the action. This can be hydraulic, pneumatic, electrical, or mechanical. The following are commonly used actuators:

- A *hydraulic actuator* consists of a cylinder or fluid motor that uses hydraulic power to facilitate mechanical operation. The mechanical motion gives an output in terms of linear, rotary, or oscillatory motion.
- A *pneumatic actuator* converts energy formed by a vacuum or compressed air at high pressure into either linear or rotary motion.
- An *electric actuator* is powered by a motor that converts electrical energy to mechanical torque. The electrical energy is used to actuate equipment such as multi-turn valves.
- A *mechanical actuator* functions by converting rotary motion into linear motion to execute movement. It requires gears, rails, pulleys, chains, and other devices to operate.

In the context of maintenance, actuators play an important role. They are used to initiate actions to prevent further damage when a fault is detected. An example is the shutting off of a valve if a leak is detected in a pipe. The input signal is from a sensor that detected the leak and the external energy is electrical (for example, an electric motor or solenoid).

As can be seen from the above discussion, sensors and actuators are transducers, but the reverse is not the case. A *combination transducer* is a device which has both a sensor and an actuator. The smoke detector discussed earlier would be a combination transducer in our terminology.

Example 6.2   Turbine Bearings

The item is the bearing of a turbine and its state $X(t)$ is the wear of the bearing. As the bearing wears, the vibration of the turbine, the temperature of the bearing, and the noise generated all increase. In addition, the composition of the wear debris in the lubricant oil changes. As such, one or more of these can be used as the measurable variable, $Y(t)$, to assess the wear. They all require different types of sensors. The outputs of the sensors are electrical signals (electrical energy) whereas the inputs (quantity sensed) and the form of energy are all different, as indicated in Table 6.1.   ∎

Table 6.1   Measurement variables and forms of energy for turbine bearings.

Measurable variable	Energy
Vibration	Mechanical
Noise	Acoustic
Temperature	Thermal
Wear debris	Chemical

Output

The output of most sensors is usually an electrical signal. When the output signal of a sensor is very small, the unwanted signals or voltages (from the circuitry and referred to as *noise*) can have an impact – the observed output signal $Z(t)$ differs significantly from the true signal $Y(t)$. The two signals are related as follows:

$$Z(t) = Y(t) + V(t) \tag{6.2}$$

Technology has played a critical role in how data are collected and stored, and this is discussed further in a later section.

6.3.2 Evolution of Technologies for Assessing Item State

Technology plays an important role in the different elements of the process shown in Figure 6.3. These technologies have been evolving over time, as shown in Figure 6.4 and new technologies will continue to evolve in the future at a faster rate.

6.4 Sensors

There are many types of sensors and many aspects to a sensor – sensor technologies, classification, selection, and so on. We discuss these briefly in this section.

Sensors have two basic functions in the maintenance context:

- *Measure condition:* In this case, the output from the measurement device is used to determine the rate of degradation and to assess the remaining useful life of an item.
- *Protection:* Here, a sensor is used to trigger an alarm when the variable characterizing the state of an item crosses some pre-specified threshold, so that appropriate action can be initiated to avoid a catastrophic failure. An example of this is the pressure building up in a boiler – unless action is taken, the boiler may explode.

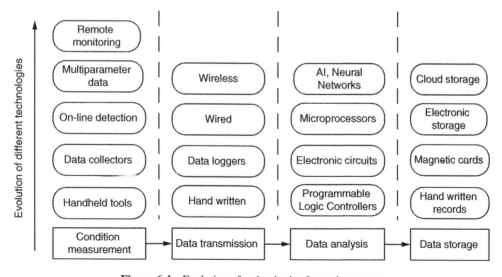

Figure 6.4 Evolution of technologies for maintenance.

Sensors constitute the basis not only for assessing item state but also for fault diagnosis. Some sensors measure quantities that are directly related to fault modes identified as candidates for diagnosis. Among them are strain gauges, ultrasonic sensors, proximity devices, acoustic emission sensors, electrochemical fatigue sensors, and so on. Other sensors are of the multipurpose kind, such as temperature, speed, flow rate, and so on, and are designed to monitor process variables for control and/or performance assessment in addition to diagnosis. Some sensors are built into the engineered object. This allows for online monitoring of the object.[4] In other cases, they are external devices (usually handheld) used by maintenance technicians and operators to record the signals.

6.4.1   Sensor Technologies

Figure 6.5 shows the role of science and engineering in the evolution of sensor technologies. Scientific knowledge (relating to materials and their behavior) has been increasing at an exponential rate. This, combined with the rapid advances in computer technologies and nanoscale manufacturing, has led to dramatic changes in sensor technology. New sensors are appearing at a faster rate to replace current ones.

6.4.2   Classification of Sensors

There are many different types of sensors and there are many different methods of classification. Sensor classification schemes range from very simple to complex.

6.4.2.1   Simple Classification Schemes

- *Classification 1: Active versus passive sensors.* An active sensor requires an external energy source to operate and produce the output signal. An example of this is a strain gauge used for measuring strain or force. It is basically a pressure-sensitive resistive bridge network.

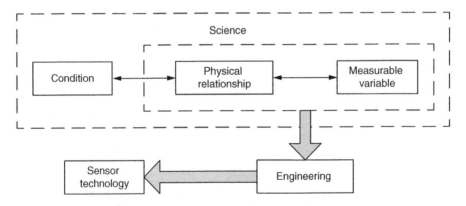

Figure 6.5   Sensors: science, engineering, and technology link.

---

[4] Monitoring is discussed in the next section.

It does not generate an electrical signal itself, but by passing a current through it (from an external source such as a battery), its electrical resistance can be measured by detecting variations in the current and/or voltage across it. In contrast, a passive sensor does not need any external energy source and directly generates an electrical signal in response to an external stimulus. An example of this is a thermocouple. Passive sensors are direct sensors which change their physical properties, such as resistance, capacitance or inductance, and so on.

- *Classification 2: Analog versus digital sensors.* The classification is based on the output of the sensor. An analog sensor produces an output signal which is a continuous real variable and is proportional to the quantity being measured. Physical quantities, such as temperature, speed, pressure, displacement, and so on, are all continuous in nature. In contrast, a digital sensor produces a discrete digital output signal that is a digital representation of the quantity being measured using an analog-to-digital converter.

6.4.2.2 Complex Classification Schemes

A proper way to look at a sensor is to consider all of its properties, such as the stimulus (input variable that is being measured), the underlying physical principles (phenomena) and mechanisms involved, the materials used, and the fields of application.

6.4.2.3 Classification Based on Underlying Physical Principles

Here, the classification of sensors is based on the underlying physical principles (phenomena). Table 6.2 lists some of the many underlying principles.

Other underlying physical principles of sensors include photoelectric, thermo-magnetic, thermo-elastic, magneto-electric, and thermo-optic effects. The following example illustrates the underlying physical principles of common household products.

Table 6.2 Underlying physical principles of sensors.

Piezoresistive effect	Converts an applied strain to a change in resistance that can be sensed using electronic circuits
Piezoelectric effect	Converts an applied stress (force) to a charge separation or potential difference
Magnetoresistive effect	Based on the fact that the conductivity varies as the square of the applied flux density
Thermoresistive effect	Based on the fact that the resistance changes with temperature
Thermocouples	Based on the fact that if a circuit consists of two different materials joined together at each end, with one junction hotter than the other, a current flows in the circuit
Photoconductive sensors	Photons generate carriers that lower the resistance of the material
Chemiresistors	Have two electrodes coated with specialized chemical coatings that change their resistance when exposed to certain chemical challenge agents
Electrochemical transducers	Rely on currents induced by oxidation or reduction of a chemical species at an electrode surface

---

Example 6.3   Smoke Detector

A smoke detector is a combination transducer device that senses smoke (typically an indicator of fire). Smoke detectors used in households (also referred to as smoke alarms) are typically housed in a disk-shaped plastic enclosure about 150 mm in diameter and 25 mm thick. Most smoke detectors work either by optical detection (photoelectric) or by a physical process (ionization), whilst others use both detection methods to increase sensitivity to smoke. They produce a local audible or visual alarm using an actuator which is powered by a single disposable battery.   ■

Table 6.3 provides a classification of sensors based on the underlying scientific principle and the various faults they detect for different materials.

6.4.3   Sensor Selection

Factors important in choosing a sensor for a particular application are the following:

- *Ruggedness:* The ability to withstand overloads beyond the limit with safety stops for overload protection;
- *Linearity:* For the input–output characteristics of the sensor;
- *Repeatability:* The ability to reproduce the output signal exactly when the same input is applied under the same environmental conditions;
- *Convenient instrumentation:* A sufficiently high analog output signal with a high signal to noise ratio;
- *Robustness:* Minimal error in measurement, unaffected by temperature, vibrations, and environmental variations;
- *Reliability:* No degradation in performance over time;
- *Resolution:* The smallest change that can be detected in the quantity it is measuring.

6.4.4   Sensors Used in Maintenance

A brief description of some of the commonly used sensors for condition-monitoring purposes, their underlying principles, and their application areas is given below.

6.4.4.1   Accelerometers

A piezoelectric accelerometer is the sensor of choice for vibration-monitoring applications. These types of sensors exhibit many desirable properties including: vibration measurement in a wide range of environmental conditions, usability over a wide range of frequencies, self-generating power supply, excellent linearity over a wide dynamic range, and they are extremely compact. Piezoelectric accelerometers are used to measure all types of vibrations as long as the accelerometer has the correct frequency and dynamic ranges.

Piezoelectric accelerometers rely on the piezoelectric effect of quartz or ceramic crystals to generate an electrical output that is proportional to applied acceleration. Using Newton's law of motion that force is equal to mass times acceleration, it is possible to determine acceleration by measuring the force and dividing it by the mass. Figure 6.6 is an illustration of the working of one type of accelerometer that uses a shear force.

Table 6.3 Classification based on the underlying scientific principles.

Material	Degradation mechanism	Mechanical				Electrical				Optical		Thermal	Chemical	
		Force	Displacement	Ultrasonic testing	Energy/duration/counts/maximum amplitude	Heat/resistance/temperature	Strain/resistance/temperature	Electrical resistance	Thickness/eddy currents	X-ray	Laser	Flux/magnetic field	Number of particles/shape size	Sulfide/cyanides
		Strain gauges	Accelerometer											
Metal	Wear	Strain gauges	Accelerometer	—	—	—	—	—	—	—	—	Infrared imaging technology	SOAP	—
	Crack	—	—	✓	—	—	—	—	Eddy current	✓	—	Magnetic particle testing	—	—
	Erosion	—	—	✓	Acoustic emission	—	—	—	Eddy current	✓	—	—	—	—
	Corrosion	—	—	—	—	Electro magnetic	—	Galvanic probe	Eddy current	✓	—	—	—	Hydrogen penetration probe
	Fatigue	—	—	✓	—	—	—	—	Eddy current	—	Laser thermal vibration	Electro magnetic aging	—	—
Concrete	Crack	—	—	✓	—	—	Piezoelectrical	—	—	✓	—	—	—	—
Wood	Crack	—	—	✓	—	—	—	—	—	—	—	—	—	—
Composites	Crack	—	—	✓	—	—	—	—	—	—	—	Thermography	—	—
	Fatigue	—	—	✓	—	—	—	—	—	✓	—	—	—	—
Ceramic	Erosion	—	—	—	—	—	—	—	—	—	Laser scanning	Thermography	—	—

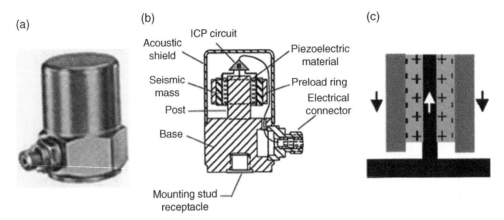

Figure 6.6   Shear-mode accelerometer. (a) Sensor, (b) sensor components, and (c) piezoelectric effect.

6.4.4.2   Strain Gauges

A strain-gauge sensor is based on the principle that the resistance of a conductor is directly proportional to its length and resistivity and inversely proportional to its cross-sectional area. When a stress or strain is applied to the metal transduction element, its length and cross-sectional area vary, causing a change in resistance that can be measured as an electrical signal.

6.4.4.3   Ultrasonic Sensors

Ultrasonic sensors are used for monitoring the state of critical structures such as aircraft and bridges. This device has a piezoelectric element that produces an ultrasonic pulse when it is excited by an extremely short electrical discharge. When the pulse is reflected by a surface or flaw in the scanned material, the same element generates an electrical signal.

6.4.4.4   Temperature Sensors

Temperature sensors such as resistance temperature detectors (RTDs), thermistors, and thermocouples are used widely to monitor temperature levels and variations.

Thermistors
A *thermistor* is a thermally sensitive resistor in which the electrical resistance varies directly with temperature. Thermistors are ceramic semiconductors exhibiting a negative temperature coefficient (NTC)[5] or positive temperature coefficient (PTC), and typically they are used within the −200 to 1000° C temperature range for NTCs and within the range 60–180° C for PTCs.

Thermocouples
Thermocouples are the most widely used sensors in industry. They are very rugged and can be used from sub-zero temperatures to temperatures over 2000° C. The underlying principle is based on the Seebeck, or thermoelectric, effect in which if two dissimilar materials are joined, and if the two junctions are placed at different temperatures, a current flows around the circuit.

---

[5] An NTC occurs when a physical property (such as thermal conductivity or electrical resistivity) of a material lowers with increasing temperature.

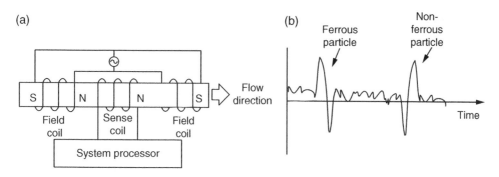

Figure 6.7 Oil debris sensor. (a) Sensor components and (b) output signal.

6.4.4.5 Oil Debris Sensors

An oil debris sensor is a device installed in-line with the oil system and it operates as follows (see Figure 6.7). As a particle passes through the magnetic coil assembly in the device, it causes a disturbance to the alternating magnetic field, resulting in a characteristic output signature. The amplitude of the signal produced by a passing particle is proportional to the mass of the particle for ferromagnetic materials and to the surface area of the particle for conductive non-ferromagnetic materials. A distinction between ferromagnetic and non-ferromagnetic materials needs to be made, since the phases of their signals are opposite to each other. Signal conditioning using a threshold algorithm is used to detect the particles that pass through the sensor. Each time a particle passes through the sensor, an electrical pulse is generated which can be used to count particles.

6.4.4.6 Eddy Current Proximity Sensors

Eddy current sensors are based on the physics of electromagnetic induction. When an alternating current flows in a coil in close proximity to a conducting surface, the magnetic field of the coil will induce circulating (eddy) currents in that surface. These currents set up a magnetic field that tends to oppose the original magnetic field. The impedance of the coil in the probe is affected by the presence of the induced eddy currents in the object being tested. The interaction of the magnetic fields is dependent on the distance between the probe and the target. As the distance changes, the electronics sense the change in the field interaction and produce a voltage output which is proportional to the change in distance between the probe and the target. Figure 6.8 illustrates the eddy current sensor principle.

6.4.4.7 Micro-Electromechanical System (MEMS) Sensors

MEMS sensors are microstructures produced using MEMS technology.[6] Micro-machined MEMS sensors in silicon or other materials are fabricated in a batch process with the potential for integration with electronics, thus facilitating on-board signal processing and other smart

---

[6] MEMS technology is a process technology used to create tiny integrated devices or systems that combine mechanical and electrical components. Whilst the device electronics are fabricated using integrated circuit (IC) technology, the micromechanical components are fabricated using micromachining of silicon and other substrates. These tiny devices can range in size from a few micrometers to millimeters and have the ability to sense, control, and actuate on the micro scale, and generate effects on the macro scale.

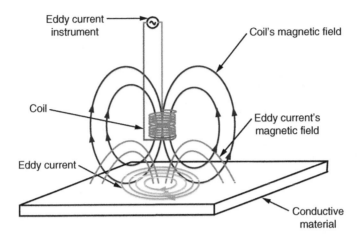

Figure 6.8 Eddy current sensor.

functions. A number of MEMS sensors are available commercially, and these are capable of monitoring such critical variables as temperature, pressure, acceleration, and so on (see Venktasubramaniam, 2003).

6.4.5 Smart Sensors/Transducers

Definition 6.4

A smart transducer is a transducer that provides functions beyond those necessary for generating a correct representation of a sensed or controlled quantity. This functionality typically simplifies the integration of the transducer into applications in a networked environment (IEEE 1451).

Smart sensors are basic sensing elements with embedded intelligence. The fundamental idea of a smart sensor is that the integration of silicon microprocessors with sensor technology, described earlier, can significantly improve sensor system performance and capabilities. In addition, many sensors may be included in a single smart sensor.

A smart sensor should have:

- Integrated intelligence close to the point of measurement;
- Basic computation capability;
- The ability to communicate data and information in a standardized digital format.

The output of a sensor is conditioned and scaled, then converted to a digital format through an analog-to-digital converter. The digitized sensor signal can then be easily processed by a microprocessor. Any of the measured or calculated parameters can be passed on to any device or host in a network by means of the network communication protocol.

Smart sensors represent a new generation of devices with high sensing capability and self-awareness, which will be essential components of future intelligent systems. Bringing intelligence down to the component level through the design of smart sensor systems will automate data collection and analysis and have a profound impact on effective condition monitoring of various engineered objects.

6.4.6 Wireless Sensor Networks

Definition 6.5

A wireless sensor network (WSN) consists of a number of smart sensors spread across a geographical area. Each sensor has wireless communication capability and some level of intelligence for signal processing and networking of the data.

In a WSN for condition monitoring, smart sensors are moved to sensor nodes. Their architecture provides the basic hardware and software that enable wireless modules to carry out the following:

1. *Intelligent sensing:* Decide when to send information relating to an object's condition. This is done by local evaluation of the significance and confidence in reaching a first-level decision.
2. *Network self-organization:* Given the large number of nodes and their potential placement in hostile locations, it is essential that the network is able to self-organize, since manual configuration is not feasible. Since nodes may fail (either from a lack of energy or for other reasons) and new nodes may join the network, the network must be able to reconfigure itself periodically so that it can continue to fulfill its function. Also, a high degree of connectivity must be maintained even if individual nodes become disconnected.

6.5 Testing Technologies

6.5.1 Testing

Definition 6.6

Testing is performed either to identify or diagnose a problem with an engineered object or to confirm that repair measures taken to rectify the problem have been effective.[7]

Since an engineered object can be decomposed into many levels, testing can be performed at any level – system level, component level, or something in between.Testing uses system

---

[7] Testing in the context of maintenance is often referred to as *maintenance testing*.

performance requirements as the basis for identifying the appropriate components for further inspection or repair. As part of in-service inspection, an engineered object is subjected to a test or series of tests performed on a frequency established by the manufacturer.[8] We will illustrate using a few examples.

Example 6.4   Components of a Nuclear Power Plant

Periodic inspections of nuclear power plant components are important for the safe operation of the facility. Such inspections include a range of tests such as:

- Leak-rate tests for containment structures (the reactor boundary, the containment vessel, etc.);
- Calibration tests to ensure shutdown systems will activate in over-power or over-temperature conditions.   ∎

Example 6.5   Fire Safety Systems in a Multi-Story Building

In most countries there are stringent laws relating to the maintenance and testing schedule for fire safety systems. These involve the following:

- Testing and tagging of electrical equipment;
- Visual inspections of all electrical equipment;
- Testing RCDs (residual current devices) in the switchboard;[9]
- Testing emergency and evacuation lighting;
- Testing the fire alarm and evacuation system.   ∎

Example 6.6   Lead–Acid Batteries

IEEE recommended practice for maintenance, testing, and replacement of vented lead–acid batteries for stationary applications involves the following tests:

- *Acceptance test:* A constant-current or constant-power capacity test made on a new battery to determine if it meets specifications or manufacturer's ratings;
- *Capacity test:* A controlled constant-current or constant-power discharge of a battery to a specified terminal voltage;
- *Performance test:* A constant-current or constant-power capacity test made on a battery after it has been in service to detect any change in the capacity;
- *Service test:* A test of a battery's capability during service to meet the battery duty cycle.   ∎

---

[8] The frequency and type of tests is based on prior experience with the item (system, component, or something in between) and upon engineering analysis of the probable failure rate for the equipment.

[9] An RCD is a safety device that monitors electrical current flowing within a circuit from the switchboard. It works on the principle that the electricity current flowing in must be equal to the current flowing out of the circuit. If the RCD detects an imbalance in the electrical current, indicating a leakage to earth, for example, current flows through someone's body to earth, the RCD immediately cuts the electricity supply to prevent electrocution. RCDs are extremely sensitive, disconnecting within 10–50 ms of detecting 30 mA or more of current leakage.

A good testing program maintains a record of test results and maintenance actions taken. These data are used to extract information to see trends and make decisions on appropriate testing frequency, replace or upgrade, performance improvement opportunities, and so on.

Finally, testing can be destructive or non-destructive.

6.5.2 Destructive Testing

In a destructive test, the item is damaged during testing so that it is not serviceable after the test. These tests are used to assess the state when there are a large number of items (for example, pipes in a boiler or cables in a mine shaft for hauling ore) by testing only a few of them. The form of the test depends on the item and the two commonly used destructive tests are the following.

6.5.2.1 Stress Tests

This type is used to assess the strength of a component which decreases with age and/or usage. The test involves increasing the load until the item fails and provides information on the residual strength.

6.5.2.2 Metallography Tests

The item is cut to reveal its cross-section and this allows one to observe internal cracks not visible from the outside.

6.5.3 Non-Destructive Testing (NDT)

NDT (non-destructive testing) is a wide group of analysis techniques used in science and industry to evaluate the properties of a material, component, or system without causing damage. It can be divided into several *methods* based on the underlying scientific principle and can be further subdivided into various *techniques*. The various methods and techniques, due to their particular natures, may lend themselves especially well to specific applications.

Test method names often refer to the type of penetrating medium or the equipment used to perform that test. Current NDT methods include: Acoustic Emission Testing (AE), Electromagnetic Testing (ET), Laser Testing Methods (LMs), Leak Testing (LT), Magnetic Flux Leakage (MFL), Liquid Penetrant Testing (PT), Magnetic Particle Testing (MPT), Neutron Radiographic Testing (NR), Radiographic Testing (RT), Thermal/Infrared Testing (IR), Ultrasonic Testing (UT), Vibration Analysis (VA), and Visual Testing (VT). The six most frequently used test methods are MPT, PT, RT, UT, ET, and VT. We discuss briefly the principles of some of the commonly used methods.

6.5.3.1 Radiographic Testing

RT is an NDT method of inspecting materials for hidden flaws by using the ability of short-wavelength electromagnetic radiation (X-rays or gamma rays) to penetrate various materials.

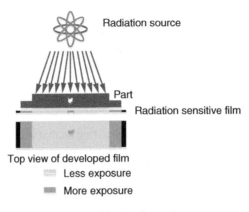

Top view of developed film
Less exposure
More exposure

Figure 6.9   Radiography.

The three basic elements of RT are: (i) a radiation source, (ii) the test piece being evaluated, and (iii) a recording medium. These elements are combined to produce a radiograph, as depicted in Figure 6.9. The intensity of the radiation that penetrates and passes through the material is either captured by a radiation-sensitive film (Film Radiography) or by a planar array of radiation-sensitive sensors (Real-time Radiography).

The underlying principle can be described as follows. The radiation source emits energy that travels in straight lines and penetrates the tested object, producing an image on the recording medium opposite to the X-ray source. This image is used to evaluate the condition of the object being examined. The image is produced on film or electronically for real-time systems.

RT offers a number of advantages over other NDT methods, including the ability to detect surface and internal discontinuities, the production of a permanent test record, and good portability, especially for gamma-ray sources. However, one of its major disadvantages is the health risk associated with the radiation.

Radiographic testing can be used for detecting internal defects in thick and complex shapes in metallic and non-metallic materials, structures, and assemblies. Applications include detection of surface and subsurface features in welded pipes, corrosion mapping, detection of blockages inside sealed equipment, and so on.

6.5.3.2   Ultrasonic Testing

The ultrasonic principle is based on the fact that solid materials are good conductors of sound waves. Sound waves are not only reflected at the interfaces but also by internal flaws such as cracks. Because the interaction effect of sound waves with the material is stronger for high frequencies, ultrasound wave frequencies used are in the range 0.5–25 MHz.

Ultrasonic sensors are discussed in Section 6.4. The essential element is the probe. When the pulse produced by the sensor is reflected by a surface or flaw in the scanned material, the same element generates an electrical signal that can be registered on a display screen (see Figure 6.10). The probe is coupled to the surface of the scanned object with a liquid or coupling paste, so that the sound waves from the probe are able to be transmitted into the

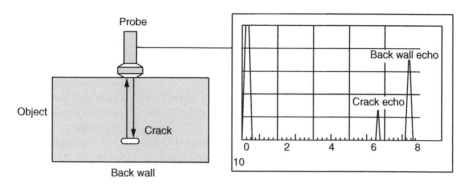

Figure 6.10 Ultrasonic principle.

object. The object is scanned by moving the probe evenly to and fro across the surface. Any signals caused by reflections from internal discontinuities can be observed on an instrument display. The three signals shown in the display in Figure 6.10 correspond to the first surface of the object, the crack, and the back surface or wall, respectively.

Radiography and ultrasonic testing are probably the most frequently used methods of testing different objects for internal flaws. NDT ultrasonic testing is used widely to test welded gas and oil pipelines, railway track, and other shaped engineered object pieces.

6.5.3.3   Liquid Color Penetrant Testing

This is mainly a surface crack detection technique. It may be applied to all non-porous materials, but only cracks open to the surface can be detected. The surface to be inspected is coated with a film of a special liquid or penetrant. This is drawn into any surface-breaking cracks by capillary action. Then the surface is cleaned with a suitable remover. After this, a layer of developer (chalk powder or similar absorbent material) suspended in a suitable solvent is applied. As this layer dries, it draws the liquid out of the crack and spreads it over a larger area of the surface, making it more visible. The liquid used as the penetrant is frequently colored to contrast with the chalk developer. It is also possible to use a fluorescent penetrant that can be inspected under ultra-violet light. Liquid penetrant inspection is an extension of visual inspection and is used for detecting surface-breaking flaws, such as cracks, laps, and folds on any non-absorbent material's surface.

6.5.3.4   Eddy Current Testing

Eddy current testing uses sensors based on the physics of electromagnetic induction discussed in Section 6.4. When the eddy currents in the tested object are distorted by the presence of flaws or material variations such as cracks, the impedance in the coil is changed. This change is measured and displayed in a manner that indicates the type of flaw or material condition. This method is used to detect small surface flaws in an item and for measuring the thickness of a non-conductive coating of an item.

Table 6.4 NDT methods used for detection of defects and condition monitoring.

NDT methods	Degradation	Limitations	Remarks
Acoustic emissions	To identify propagating surface or subsurface crack	Not suitable for stable cracks	Long-term data collection for decision making
Eddy currents	Voids in thin layers, head checks	Maximum depth 2 mm, local resolution 2 mm	Commonly used in railways for head check detections
Ultrasonics	Internal voids or measuring the thickness of surface coatings	Defects below 4–5 mm test surface are difficult to detect	Useful for rapid and automated inspection. Used for railway track inspections
Radiography	Internal voids	Hazardous in use. Maximum investigated plate thickness 70 mm	Can be used for most materials
Color penetration test	Surface cracks	Only surface-breaking defects can be detected	Useful for detecting small surface discontinuities

6.5.3.5 Acoustic Emission Testing

AE refers to the generation of transient elastic waves during the rapid release of energy from localized sources within a material. The dislocation movement accompanying plastic deformation of material and the initiation and extension of cracks under stress are examples of sources of AE.

AE piezoelectric sensors are used for monitoring defects in real time while the phenomenon is taking place. Anomalies that can be detected include corrosion, fatigue cracks, fiber breakage, and cracking in concrete or reinforced concrete structures.

Table 6.4 summarizes the salient features of the NDT methods discussed above along with their limitations.

6.6 Data-Related Technologies

Modern-day maintenance technology requires addressing the following issues relating to maintenance data: acquisition, transmission, storage, and analysis for proper and effective maintenance of an engineered object. The amount of maintenance data is increasing day by day, posing new challenges for storage and analysis. We focus on maintenance-related data.

6.6.1 Data Collection

Data collection may be either manual or automatic and the collection is either performed continuously or at discrete time instants (also referred to as *sampled data* – data sampled over time). Different technologies are used for data collection and they can be classified broadly into the categories indicated below:

- Technologies for manual data collection using portable devices.
- Technologies for automatic data collection: For critical machines, when the failure development process is such that it demands daily or even hourly monitoring, implementation for online automatic or continuous monitoring devices is preferred. Depending on the criticality of the machines, a continuous data-collection system may be implemented.

6.6.2 Data Transmission

In the past, data were collected manually using handheld devices by technicians who had to physically walk from machine to machine. Currently, most critical objects are monitored continuously and data can be transferred using wired physical infrastructure or a wireless network. Data transfer using traditional cabling is very expensive and in some cases impractical (remote bridge). In the past, data had to be downloaded onto a USB stick/memory card to be transferred to an office computer for analysis if cable connections were not possible. In a modern condition-monitoring system, data are often transferred by wireless network, or the data may be stored in Internet clouds to be used on demand from anywhere. A customer can access the data using his/her smartphone or any Internet-enabled device with Wi-Fi modules. Slow transfer and crowded wireless net security are critical issues in wireless data transfer in the maintenance context, where data collection is too large and decisions need to be made in real time due to safety-related functions (fault identification and protection). Security is another issue that is of major concern for many companies operating in a competitive environment, where information regarding the physical status of machines is critical.

6.6.3 Data Storage

In many applications, the volume of both structured and unstructured data is so large that it's difficult to use local computers for storage, and the data are being collected for different purposes. Cloud storage seems to be the solution for managing such huge amounts of data. Cloud storage is simply a term that refers to online space that can be used to store data.

6.6.3.1 Cloud Storage

Cloud storage is a form of service model in which data are maintained, managed, and backed up remotely and made available to users over a network or Internet. In cloud storage, data are stored in virtualized pools of storage which are generally hosted by third parties. Companies hosting storage operate large data centers, and customers who require their data to be hosted buy or lease storage capacity from these companies. Data storage in cloud computing can save companies the capital expenditure required for investment in storage infrastructure, as they can buy storage space from vendors or hosting companies. However, there are data-security-related risks that need to be addressed to ensure the integrity of the data stored in the cloud.

Example 6.7   Cloud Storage

SKF launched its cloud solution in November 2012. It connects all SKF Remote Diagnostic Centers allowing SKF to monitor and diagnose millions of machines all over the world. Recently, the global manager of SKF Asset Diagnostic Services made the following statement about the SKF cloud system: "The new cloud-based system lets us monitor and diagnose millions of machines. We already have half a million machines in the cloud. This shared knowledge helps us better serve customers. In addition, cloud-based software services give customers 24/7 access to an easy-to-understand diagnostics dashboard and regular reporting. And direct access is available to a knowledge network to collaborate with SKF machine-health experts around the globe, day or night. In this way we help customers increase machine availability whilst decreasing maintenance costs." ∎

6.7 Technologies for Maintenance of Products

6.7.1 Automobile: On-Board Diagnostics for Automotive Applications

Automotive engine faults can lead to increased emissions and fuel consumption or even damage to the engine. These negative effects can be prevented or reduced if these faults can be detected and isolated. On-board diagnostic (OBD) systems consist of computers and sensors built into the vehicle to monitor emission systems on a continuous basis. They were developed in the 1980s in the US to help technicians diagnose and service the computerized engine systems of modern vehicles. A new generation of these systems (OBD II) is found in 1996 models and newer vehicles.

OBD II systems monitor virtually every component that can affect the emission performance of the vehicle. If a problem is detected, the OBD II system illuminates a warning light on the vehicle instrument panel to alert the driver. This warning light typically contains the phrase "Check Engine" or "Service Engine Soon." The system will also store important information about the detected malfunction so that a repair technician can accurately find and fix the problem.

6.7.1.1 Diagnostic Technologies Used

Various sensors are used to collect data that are fed to the computer, which then diagnoses the faults using appropriate diagnostic models and algorithms (see Chapter 16) and produces a code for each identified malfunction. The Society of Automotive Engineers (SAE) established the Diagnostic Trouble Codes (DTCs) used for OBD II systems. The DTCs are designed to be identified by their alphanumeric structure. DTCs are stored whenever the "Check Engine" light is illuminated. A standardized tester (scan tool) can be used to access the stored DTCs.

6.7.1.2 Future Developments: OBD III

With current OBD systems, vehicle owners can delay having emission problems fixed. OBD III, however, would force owners to have the faults repaired. An on-board transmitter may send out information about a vehicle's emissions system. These data will be collected by roadside readers, a local station network, or satellites. The vehicle's owner will be notified of a problem by mail and will be given time to have it fixed. The concept involves the use of wireless techniques to query the OBD computer installed in a vehicle.

6.8 Technologies for Maintenance of Plants

6.8.1 Wind Turbines

Wind turbines (WTs) are usually located in remote locations and sometimes offshore. Access to these machines may be difficult and requires mobilization of many resources (ships for offshore WTs, cranes, etc.). Maintenance technologies play a crucial role in managing these machines. In this section we describe key technologies and the role of remote condition monitoring in the maintenance of WTs.

6.8.1.1   Condition-Monitoring Techniques and Sensors

Turbine components and their main failure modes are shown in Figure 6.11.

Vibration monitoring is the most popular technology employed in wind turbines. Position sensors are used for low frequencies, velocity sensors for medium-range frequencies, and accelerometers for high frequencies. Table 6.5 summarizes the main CM (corrective

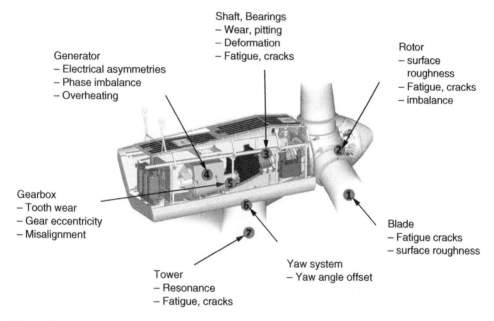

Figure 6.11   Wind turbine hierarchical decomposition and failure modes. Source: de Novaes, Alencar, and Kraj (2012).

Table 6.5   Condition-monitoring methods used for various WT components.

CM method	Wind turbine component					
	Blades	Rotor	Gearbox	Generator	Bearing	Tower
Vibration	•	•	•		•	•
Acoustic emissions	•		•		•	
Ultrasonics	•					
Radiography	•					
Thermography	•	•		•		

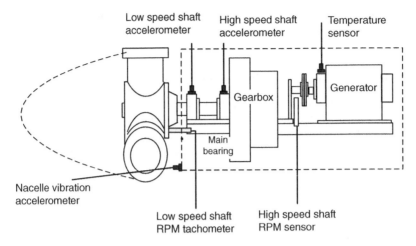

Figure 6.12   Wind turbine sensor configuration.

maintenance) techniques used and the WT components on which they are used. Examples of sensors used in WTs and their configuration are shown in Figure 6.12.

6.8.1.2   Remote Condition Monitoring

Since WTs are located in remote locations and dispersed over wide geographical areas, regular inspection visits to collect condition-monitoring data are almost impossible. Modern technologies, through remote condition monitoring, offer an effective solution, as illustrated in Example 6.8.

Example 6.8   WT Gearbox

With no condition monitoring implemented, gearbox damage occurs without warning, running to approximately $130 000 for replacement costs, and four to eight weeks of downtime. With condition monitoring in place, gearbox damage is detected early, maintenance can be scheduled around the wind, and it will cost approximately $70 000 for repair/ overhaul and one or two weeks of downtime.   ∎

Remote condition monitoring of WTs allows an operator to monitor a single turbine or an entire wind farm from a remote location. The system consists of a number of sensors connected to an intelligent monitoring unit (IMU) located in each wind turbine. The IMU can be configured remotely using a graphical interface and the signals and measurements can also be analyzed and trended from a remote location. Measurements can be taken from the main bearing, the gearbox, the generator, the rotor and blades, and the tower. Condition-monitoring methods used include vibration, strain, oil, and temperature.

Many original equipment manufacturers (OEMs) are offering required remote condition-monitoring equipment and service to their customers. Typically, data are collected on a local server at each wind farm and transmitted through wireless technologies to the OEM server in a Remote Condition Monitoring Center. Customers can access the data through Web services and receive assistance from the OEM in case of problems. Since the OEM is typically managing hundreds of machines, this enables increased productivity and the necessary feedback to improve technology.

6.8.2 Bearings

6.8.2.1 Vibration Monitoring

Vibration monitoring is the most widely employed technique for condition monitoring of bearings. Bearings vibrate even if they are geometrically perfect and have no defects. The number of rolling elements and their position in the load zone change with bearing rotation. This leads to vibrations commonly known as *varying compliance vibrations*. The presence of defects, such as spalling[10] of the races or the rolling elements, causes a significant increase in the level of vibration. Sensors, mainly accelerometers (see Section 6.4), are used to record these vibrations, as explained in Section 6.4.

6.8.2.2 Ultrasonics

Bearing vibration produces a broad range of sounds. The high-frequency ultrasonic components of these sounds are extremely shortwave in nature. Because a shortwave signal tends to be fairly directional, it is relatively easy to detect its exact location by separating these signals from background plant and operating equipment noises. In addition, as changes begin to occur in mechanical equipment, the subtle, directional nature of ultrasound allows these potential warning signals to be detected early, before actual failure, often before they are detected by vibration or infrared. Ultrasonic detectors are capable of accurately interpreting the sounds created by under-lubrication, over-lubrication, and early signs of wear.

6.8.2.3 Acoustic Emissions (AE)

Sources of AE in rotating machinery include impacting, cyclic fatigue, friction, material loss, and so on. For instance, the interaction of surface asperities and impingement of the bearing rollers over a defect on an outer race will result in the generation of acoustic emissions. These emissions propagate on the surface of the material as Rayleigh waves and the displacement of these waves is measured with an AE sensor, as explained in Section 6.4. AE signals can identify faults before they appear in the vibration acceleration range. Signals detected in the AE frequency range usually represent bearing defects rather than other defects such as imbalance, misalignment, looseness, and shaft bending.

6.8.2.4 Oil Debris Analysis

As described in Section 6.4, online sensors allow one to monitor debris count without taking oil samples. The following sensor technology allows for inspection and analysis as well as not

---

[10] Fatigue cracks begin below the metal and move to the surface until a piece of metal breaks away to leave a small pit or spall.

interrupting the machine. The device has two chambers, each with its own debris sensor. Only one chamber is used to monitor oil debris. If the oil sensor captures bearing debris, a remote signal alerts the operator to switch the flow to the alternative chamber. The first chamber, which has been isolated from the oil, may then be depressurized and opened to retrieve the debris particles for inspection and analysis. Monitoring the shape, count, size, and nature of wear particles can point to the source of the fault and assist in taking action in time before failure occurs.

6.8.2.5 Smart Bearings

Smart bearings are bearings equipped with smart sensors and have the following characteristics:

- *Miniaturization:* MEMS sensor technology enables measurement of several critical condition parameters such as temperature, velocity, vibration, load, rpm, and so on.
- *Self-powered:* They can generate their own power using the application environment.
- *Wireless communication:* They make use of intelligent wireless communication technology packaged inside the bearing to communicate in environments where Wi-Fi cannot operate.
- *Smart networks:* They communicate through each other and via a wireless gateway to form a network that can send information about their condition for analysis to remote locations.

6.9   Technologies for Maintenance of Infrastructures

6.9.1   Pipelines

Although pipelines have failure rates that are much lower than other modes of transportation, failures, when they occur, can sometimes have catastrophic consequences in terms of loss of life and damage to the environment.

NDT techniques are used widely for detecting faults in pipelines and very often smart pigs are used for this purpose. Pigs are autonomous devices sent into a pipeline for a certain length of time to collect condition-monitoring data. They are usually propelled by the pipeline product flow. Smart pigs are equipped with many sensors of different varieties to carry out their condition-monitoring task. A typical smart pig anatomy is shown in Figure 6.13.

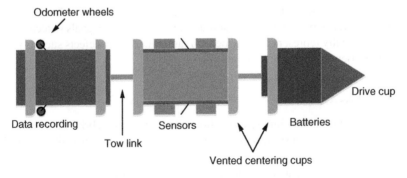

Figure 6.13   Smart pig components.

The main internal inspection technologies used for many types of pipes are MFL, ultrasonic technology, and geometry tools. The following is a brief description of these tools and the types of fault they can detect.

MFL devices identify and measure metal loss by inducing a magnetic flux along the axial length of the pipe wall. Near a pipe wall, most magnetic flux is carried by the steel. Flux leaks when metal loss is present. The strength and shape of leaking flux helps determine anomaly dimensions. Sensors mounted on the tool detect and measure the amount of flux leakage, which can be attributed to anomalies such as corrosion, cracks, or other forms of metal loss. High-resolution MFL in-line inspection pigs have increased sensor density and sampling frequency and may be equipped with tri-axial sensors to detect MFL in the axial, radial, and circumferential directions.

Different types of ultrasonic tools with sensors that emit ultrasonic signals through a liquid-coupling medium such as oil or water are used to measure pipe wall thickness, metal loss, cracks, and weld defects. Pipeline cleanliness is important for the effective use of ultrasonic tools.

Geometry tools use mechanical arms to measure the pipeline bore. Geometry tools are used to identify dents, deformations, and other changes in the pipe circumference. These tools can be used to identify the orientation, location, and depth measurement of each dent or deformation.

Another important issue in pipeline condition monitoring is leak detection. For gas pipelines, leaks are a major concern due to their catastrophic consequences in terms of human injuries and environmental impact. The main causes of gas pipeline accidents are external interference, corrosion, construction defects, material failure, and ground movement.

A wide variety of leak-detecting techniques is available for gas pipelines and can be classified broadly into three categories:

- Non-technical leak-detection methods are those that do not make use of any device and rely only on the natural senses (i.e., hearing, smelling, and seeing) of humans and/or animals.
- Hardware-based methods rely mainly on the usage of special sensing devices in the detection of gas leaks, including acoustic sensors, optical sensors, cable sensors, soil monitoring, ultrasonic flow meters, and vapor sampling. We briefly discuss acoustic and optical methods.
  - *Acoustic methods:* Escaping gas, as it flows out of a pipe, generates an acoustic wave. Acoustic sensors, integrated in handheld detection devices employed by personnel patrolling the pipeline or in smart pigs, can be used to detect a leak signal. Continuous monitoring is sometimes done by installing acoustic sensors outside the pipeline at certain distances from each other.
  - *Optical methods:* A common approach is to survey the natural gas pipeline networks using aircraft-mounted optical devices for leak detection. The absorption or scattering of the emitted radiation caused by natural gas molecules is monitored. Significant absorption or scattering detected above a pipeline is an indication of a leak. Thermal imaging detects a leak by utilizing the differences in temperature between the leaked gas and the surrounding environment.
- Software-based methods monitor pipeline parameters such as the state of pressure, temperature, flow rate, and so on continuously. The software methods use different approaches to detect leaks: mass/volume balance, real-time transient modeling, acoustic/negative pressure waves, pressure point analysis, statistics, or digital signal processing.

6.9.2 Railway Track

Rail track components were introduced in Chapter 3 and are shown in Figure 3.3. In this section, we focus on the condition-monitoring methods and technologies for rail track and ballast.

Rail tracks are inspected at regular intervals for internal and surface defects, as well as rail profile irregularities and wear, missing fastenings, failed sleepers, and abnormal variations in rail gauge (distance between rails). Rail tracks are inspected either visually by appropriately trained personnel walking along the tracks or by using a number of common rail-inspection techniques, including ultrasonics, magnetic induction (or MFL), eddy current sensing, and automated visual inspection. Condition-monitoring instrumentation is usually mounted on various types of inspection vehicle.

6.9.2.1 Assessing Rail Condition

Rails are systematically inspected for (i) internal and surface defects using various NDT techniques and (ii) geometric irregularities by testing several geometric parameters of the track.

Condition monitoring of rails makes use of several techniques; the commonly used ones are ultrasonics, MFL, and eddy current testing. Ultrasonic inspection is carried out by a variety of different instruments ranging from handheld devices, through dual-purpose road/track vehicles to test fixtures that are towed or carried by dedicated rail cars. Eddy current testing is applicable for the inspection of the surface and near-surface areas of the rail head. The application of MFL sensors is mainly focused on the detection of near-surface or surface-breaking transverse defects, such as rolling contact fatigue (RCF) cracking. Automated vision systems can operate at very high speeds and are typically used to measure the rail profile and percentage of wear of the rail head, rail gauge, corrugation, and missing bolts. The machine vision system generates three-dimensional (3-D) images and detects surface defects in the rails. The cameras can capture images at high speeds.

Track geometry involves the measurement of the relativity of the two running rails. This information is critical in order to maintain a safe interface between wheels and rail. Track geometry deterioration is an indication of an underlying problem with some aspect of the track. For example, a wide gauge could be due to rail wear, poor fastening restraint, or poor sleeper condition, whilst a large twist error could be due to formation subsidence, poor ballast condition, a dipped weld, or a broken joint. Some of the parameters generally measured include position, curvature, alignment of the track, smoothness, and the cross level of the two rails. Inspection vehicles use a variety of sensors, measuring and data management systems to create a profile of the track being inspected. Because track geometry cars are full-sized rail cars, they provide a better picture of the geometry of the track under loading compared to manual methods.

Ballast Condition Monitoring

Ballast refers to the upper layer of the track substructure upon which the rail and sleepers are placed (see Figure 3.3). It is made of a permeable granular material placed under and around sleepers to maintain the stability and geometry of the rail. An inadequate ballast section leads to track geometry degradation that may result in catastrophic types of failure.

Non-destructive inspection technologies such as the laser imaging detection and ranging (LIDAR) inspection system for profile measurement and ground penetrating radar (GPR) for ballast depth measurement provide good information about the ballast profile and depth and can be used to determine the amount of missing ballast as well as the condition of the ballast itself. These two technologies are described briefly below.

LIDAR is an inspection technology used to measure and map the ballast profile. LIDAR uses optical remote-sensing technology to produce a three-dimensional track profile surface. A LIDAR unit is a combined laser source and detector that rotates continuously to determine the distance to the nearest solid object by detecting its laser light reflection. As the measurement device travels along the track, the collection and storage of data is continuous, and so aggregating a series of consecutive cross-sections produces a three-dimensional track profile surface. This profile is used to identify ballast anomalies.

GPR is used to measure the depth of the ballast layer and facilitates the mapping of the ballast layer. GPR is an NDT method that uses electromagnetic radiation in the microwave band (UHF (ultra high frequency)/VHF (very high frequency)) of the radio spectrum, and detects the reflected signals from subsurface structures. GPR devices transmit radio into the ground and when the wave hits a boundary with different dielectric constants, the receiving antenna records variations in the reflected return signal. The change in material from ballast to soil represents a detectable boundary layer, as do locations where moisture is present. Clay-laden soils and soils with high electrical conductivity are also readily detectable.

6.10 Summary

This chapter dealt with technologies for maintenance. Science, engineering, and technology are linked. In fact, engineering provides the link between the other two, since it uses scientific principles and prevalent technology to build, operate, and maintain engineered objects.

In the past, maintenance actions were either performed after a clear indication of impending failure (PM (preventive maintenance) action) based on manual inspection or after failure (CM action). Advances in science and engineering have led to a rapid evolution in technologies for maintenance over time. In most industrial contexts, the purpose and role of maintenance technologies are to identify the state of the item so that maintenance actions can be planned properly.

Sensors constitute the basis for fault diagnosis and prognosis. They are the primary source of data collected for assessing an item's state. The chapter provided a definition of sensors, the science underlying them, and the types of sensors with applications in the area of condition monitoring to assist in the effective maintenance of various engineered objects.

There are many classifications of sensors. Some sensors measure quantities that are directly related to fault modes identified as candidates for diagnosis. Among these are strain gauges, ultrasonic sensors, proximity devices, acoustic emission sensors, electrochemical fatigue sensors, and so on. Other sensors are of the multi-purpose kind, such as temperature, speed, flow rate, and so on, and are designed to monitor process variables for control and/or performance assessment in addition to diagnosis.

Smart sensors are basic sensing elements with embedded intelligence. They allow remote condition monitoring of engineered objects through WSNs.

Condition-monitoring methods can be direct or indirect. Indirect methods are dependent on establishing a cause–effect relationship between the condition of the item and some variable such as temperature, force/pressure, position, velocity, acceleration, chemical composition, and so on, that forms the basis of the data used for assessing an item's state via the different types of sensors.

There are many condition-monitoring methods and the most commonly used ones were discussed in this chapter, including vibration, thermography, oil analysis, and several NDT techniques such as ultrasonics, eddy current testing, radiographic testing, and acoustic emissions.

Technologies for data collection, transmission, and storage were discussed, including wireless communication and cloud systems.

Applications of maintenance technologies for products, plants, and infrastructure were presented to highlight the role of technology in effective maintenance decision making.

Review Questions

6.1   What is the link between science, engineering, and technology?

6.2   Verbalize Figure 6.1 highlighting the role of technology in effective maintenance decision making.

6.3   Define a sensor and provide some examples.

6.4   Describe three physical principles used to make sensors.

6.5   Explain the physical basis of an accelerometer and identify areas of application.

6.6   Explain the physical basis of a strain gauge and identify areas of application.

6.7   Explain the physical basis of an ultrasonic sensor and identify areas of application.

6.8   What is an acoustic emission?

6.9   Describe two thermographic sensors.

6.10 Explain the physical basis of an eddy current sensor and identify areas of application.

6.11 Explain the physical basis of an online oil debris sensor and identify areas of application.

6.12 What is a smart sensor and what are its elements?

6.13 What is a wireless sensor network and what are the advantages of using one?

6.14 Contrast direct and indirect condition-monitoring methods.

6.15 What are the main requirements for indirect condition-monitoring methods?

6.16 What are the commonly measured vibration signals? Provide the type of sensor used for each.

6.17 Describe the measurements collected from oil debris testing. What is the advantage of online oil debris monitoring?

6.18 Explain how the following NDT methods are used for condition monitoring and provide an application for each:
      (a) Ultrasonics.
      (b) Acoustic emissions.

(c) Eddy current testing.

(d) Liquid color penetrant testing.

(e) Radiography.

Exercises

6.1 Following on from Exercise 3.1, suggest condition-monitoring techniques for the failure mechanisms identified.

6.2 Following on from Exercise 3.2, suggest condition-monitoring techniques for the failure mechanisms identified for the diesel engine components discussed there.

6.3 There is a growing trend in most modern cities to have separate bicycle paths and an infrastructure that allows people to hire bicycles for moving within the city. There are special centers where cyclists can collect and leave the bicycles. This, combined with technological advances in the bicycle itself, has resulted in an ever-increasing number of cyclists.
   (a) List the technological advances in the bicycle since it was first built.
   (b) List the different technologies involved in creating the cycle infrastructure.[11]

6.4 A turbine in a thermal power plant is a rotating object that converts energy from steam into mechanical rotational energy.
   (a) Describe two condition-monitoring techniques that are appropriate for monitoring the wear in the bearings of the turbine.
   (b) Describe sensors that may be used to assess the state.

6.5 The maintenance of turbines in an offshore wind farm can be costly, as it involves specialist equipment such as support vessels and cranes and the maintenance may have to be carried out during unfavorable weather and wave conditions. Explain the benefits of condition monitoring and, in particular, monitoring via remote access using state-of-the-art communications media. Describe the necessary technologies needed.

6.6 Following on from Exercise 3.4, suggest appropriate sensors and condition-monitoring techniques for the failure mechanisms identified for the wind turbine components discussed there.

6.7 Following on from Exercise 3.5, suggest condition-monitoring techniques for the failure mechanisms identified for the components of rail infrastructure discussed there.

6.8 Suggest condition-monitoring techniques for rail wagon wheels and rail track and explore the possibility of online monitoring of rail and wheel condition. Describe the variables that need to be monitored for this purpose and the technologies needed.

6.9 In a railway system, a significant number of all rail accidents are due to track degradation. Describe a condition-monitoring system that may be used to detect defects in track geometry and cracks resulting from RCF (rolling contact fatigue).

---

[11] See Taylor and Fairfield (2010).

References

de Novaes, G., Alencar, E., and Kraj, A. (2012) Remote condition monitoring system for a hybrid wind diesel system – application at Fernando de Naronha Island, http://www.ontario-sea.org (accessed 5 August 2015).

IEEE 1451. Standard for a Smart Transducer Interface for Sensors and Actuators. Institute of Electrical and Electronics Engineers, Inc., Piscataway, New Jersey.

Taylor, M. and Fairfield, C.A. (2010) IntelliBike: Condition monitoring of our cycling infrastructure. Bicycle and Motorcycle Dynamics 2010, Symposium on the Dynamics and Control of Single Track Vehicles, Delft, The Netherlands, October 20–22, 2010.

Venktasubramaniam, V. (2003) A review of the process fault detection and diagnosis: part I – quantitative model-based methods, *Computers and Chemical Engineering*, **27**: 293–311.

7

Maintainability and Availability

Learning Outcomes

After reading this chapter, you should be able to:

- Define maintainability;
- Differentiate between maintenance and maintainability;
- Explain the effect of maintainability on life cycle cost;
- Describe the benefits of maintainability;
- Identify various maintainability attributes and explain how they contribute to the ease and economy of maintenance;
- Define and use quantitative measures of maintainability;
- Define availability and differentiate between point availability and interval availability;
- Explain how reliability and maintainability affect availability;
- Define and use different types of availability measures;
- Explain why a sound maintainability process must be an integral part of the object's overall system design approach;
- Describe the maintainability process in the context of the system design development cycle.

Introduction to Maintenance Engineering: Modeling, Optimization, and Management, First Edition.
Mohammed Ben-Daya, Uday Kumar, and D.N. Prabhakar Murthy.
© 2016 John Wiley & Sons, Ltd. Published 2016 by John Wiley & Sons, Ltd.

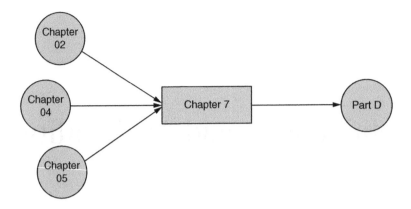

7.1 Introduction

Some objects need to last for a specified time with no maintenance (for example, satellites, offshore items on the seabed, etc.). For these items, mission reliability is important. High reliability is achieved by greater expenditure on research and development (R&D) and/or higher production costs. For other items this is either technically not feasible or uneconomical. As the reliability decreases, greater maintenance (preventive maintenance – PM – and corrective maintenance – CM) effort is needed and this, in turn, increases the downtime and results in higher maintenance costs. As a result, to achieve a suitable tradeoff, maintenance issues need to be addressed at the design stage. Maintainability deals with this topic.

This chapter deals with maintainability and availability, and is organized as follows. The definition of maintainability, the difference between maintenance and maintainability, and the relation with reliability amongst other basic concepts are discussed in Section 7.2. In Section 7.3, important maintainability attributes are presented. Availability definition measures and how availability relates to maintainability and reliability are discussed in Section 7.4. Section 7.5 is devoted to maintainability planning and its requirements. Well known maintainability standards are listed in Section 7.6. Section 7.7 describes the relationship of maintainability to other disciplines, while Section 7.8 provides a summary of the chapter.

7.2 Maintainability – An Overview

Maintainability is an inherent characteristic of the design of an engineered object. It relates to the ease, economy, and safety of maintenance. Most engineered objects require maintenance throughout their life cycle. Therefore, they should be designed so that their maintenance can be carried out safely, without large investment in time, resources (labor, materials, tools, facilities, etc.), and with minimal impact on the environment. Maintainability can be defined as follows:

Definition 7.1

Maintainability is defined as the relative ease and economy of time and resources with which an item can be retained in, or restored to, a specified condition when maintenance is performed by personnel having specified skill levels, using prescribed procedures and resources, at each prescribed level of maintenance and repair and in this context, it is a function of design. (Department of Defense, 1997)

A more quantitative definition of maintainability is as follows:

Definition 7.2

Maintainability is the probability that an item will be retained in, or restored to, a specified condition within a given time period if prescribed procedures and resources are followed.

It is important to note that maintenance and maintainability are different. Maintainability is a design characteristic and is a prime responsibility of the maintainability engineer working in cooperation with the designer, whilst maintenance consists of the actions necessary to retain an engineered object in, or restore it to, a serviceable condition. Maintainability deals with maintenance issues at the design stage, whilst the maintenance required is a result of the design decision.

Achieving the desired performance for an engineered object requires treating reliability and maintenance issues together during the design phase of the life cycle discussed in Chapter 5.

7.2.1 Historical Perspective

The historical background of the concept of Reliability, Availability, and Maintainability (RAM) is very similar to that of reliability-centered maintenance (RCM) (see Chapter 17). Its origin was in the 1970s, and it was developed to ensure the availability of military systems. Whilst RCM originated in the aerospace industry, RAM was developed by the Department of Defense (DoD) for its weapon systems. The main difference between RAM and RCM is that the former concentrates more on the design and construction phases than RCM. Consequently, RAM tends to be more of a life cycle concept.

Reliability and maintainability are often considered complementary disciplines, as both contribute to system availability. Due attention to reliability at the design stage results in a lower frequency of failures, and due attention to maintainability helps to restore the system quickly if it fails. This idea is illustrated in Figure 7.1.

Reliability and maintainability impact on availability, and this relationship will be explored in more detail later in this chapter.

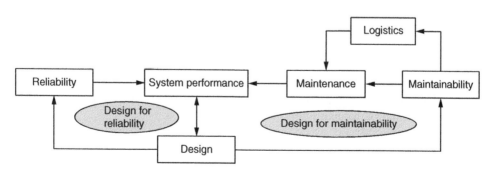

Figure 7.1 Complementary roles of reliability and maintainability on system performance.

7.2.2   Maintainability: A Life Cycle Perspective

Maintainability is very important to operations or mission accomplishment, as it directly affects system availability. Products that never fail and are always available are rare. Most engineered objects do fail and must be restored quickly to the functioning state to avoid negative effects on operations or mission accomplishment.

Maintainability affects life cycle costs.[1] The cost at the R&D stage is relatively high, especially for complex systems. Depending on system complexity, the maintainability engineer may need to implement design approaches that could account for 10% of the development cost, especially if extensive built-in tests and diagnostics are involved to meet maintainability requirements.[2] However, it is important to note that maintainability investments made in this phase can significantly reduce operation and maintenance costs later on, and thus minimize the total life cycle costs.

Maintainability costs incurred in the production and construction phase are attributed to initial tests and evaluation and demonstration testing. Again, tests at this stage can be viewed as an investment to ensure the required system availability and a reasonable total cost of ownership.

Maintainability influences operation and maintenance costs because it directly affects the ease and economy of maintenance. Ease and economy affect the number and qualifications of personnel needed to support the system, the time required to perform maintenance actions, the number and types of support equipment and tools needed, and the level of safety for both product and personnel with which maintenance jobs can be carried out. This discussion highlights the importance of logistics support and the fact that it is affected by maintainability. In particular, even if the inherent maintainability exceeds design requirements, the observed operational maintainability will only be achieved if the required logistical support is provided. Logistics support issues are discussed in Chapter 20.

In many cases, the maintainability engineer should also be concerned with the costs associated with the disposal of a system's components and materials in the retirement phase. Issues that need to be anticipated include material durability, regulations, adequate disposal of hazardous and radioactive materials, environmental concerns, and so on.

In addition to impacting life cycle costs, as discussed above, maintenance and support issues are also vital for enhancing customer satisfaction throughout the life cycle of the engineered object. A key element in achieving the desired level of satisfaction is to include maintenance and support personnel on the integrated product team(s) early in the design phase. The expertise of maintenance professionals can provide very useful feedback that enhances maintainability. Also, companies with an effective maintainability program can design maintainability into their products and use this attribute to distinguish themselves from the competition, making their product more attractive to customers.

7.3   Elements of Maintainability

There are several elements that characterize maintainability and which are used in the design stage to enhance the ease of maintenance of an engineered object. In this section we describe some of the well-known maintainability elements.

---

[1] Life cycle costs are discussed in Chapter 21.
[2] Department of Defense (1997).

7.3.1   Accessibility

PM and CM tasks can be performed safely and easily if the system has features that ensure adequate access. "Access" means enough space for the component, tools, hand, arm, and possibly head or head and body of the maintenance engineer. Examples of such features include, but are not limited to, the following:

* Adjustment should not require the removal of components to access the adjustment point, the exception being where an entire module is easily removed for adjustment on the workbench;
* Where a tool is required to remove a component, there must be access for the tool and the engineer's hand, in normal grasp;
* Access holes and spaces should be designed for the full range of human body shapes and limb sizes, not just their averages.

7.3.2   Simplification

System simplicity relates to the number of subsystems, the number of parts in a system, and the types of parts – whether they are standard or special purpose. Simplification of engineered objects enhances the ease and effectiveness of maintenance diagnostics, reduces the number of spare parts, and leads to reduced training costs. A complex design solution is often easier than a simple solution – until one has to maintain it. Designers should go for simpler designs whenever feasible.

7.3.3   Standardization and Interchangeability

Interchangeability refers to a component's ability to be replaced with a similar component without a requirement for recalibration. This flexibility in design reduces the number of maintenance procedures and consequently reduces maintenance costs.

7.3.4   Modularization

Modularization is the division of a system into physically and functionally distinct modules (units) to allow easy removal and replacement. The degree of modularization is dictated by cost practicality and functionality. Modular designs should be used whenever practical as they have many benefits, including the simplification of product design, shorter design time, easy removal and replacement of modules, independent inspection of modules, lower skill levels and fewer tools required in the field when it comes to replacement, and so on.

7.3.5   Identification and Labeling

It is important to identify critical components whose failure can have a serious impact on a system's performance. These components should be labeled properly so that they can be identified quickly. In addition, components with low life expectancies should be easily accessible and components requiring frequent routine maintenance should be at the outer edge of the engineered object in a position suitable for convenient access. This is especially important, for example, for components requiring routine lubrication or visual inspection.

7.3.6 Testability and Diagnostic Techniques

Diagnosability is a measure of how easily a particular problem (such as the location of papers jamming a photocopier) can be identified. Diagnosing a problem is often a time-consuming and difficult job. In order to maintain a high level of system readiness, designers should explicitly consider ways to speed up problem identification and diagnosis. This can be done by including functional testing capabilities or built-in test facilities in key components or system modules.

Testability is a measure of the ability to detect system faults and to isolate them at the lowest replaceable component(s). The speed at which faults are detected and isolated can greatly influence downtime and maintenance costs.

As technology advances continue to increase the capability and complexity of systems, the use of automatic diagnostics as a means of fault detection, isolation, and recovery is substantially reducing the need for highly trained maintenance personnel and can decrease maintenance costs by reducing the erroneous replacement of non-faulty equipment.

7.3.7 Human Factors

Human factor design requirements should also be applied to ensure proper design consideration. The human factors discipline identifies structure and equipment features that impede task performance by inhibiting or prohibiting maintainer body movement, and it also identifies requirements necessary to provide an efficient workspace for maintainers. Normally, the system design must be well specified and represented in drawings or sketches before detailed anthropometric evaluation can be effective. However, early evaluation during concept development can ensure early application of anthropometric considerations. Use of these evaluation results leads to improved designs largely in the areas of system provisions for equipment access, arrangement, assembly, storage, and maintenance task procedures. The benefits of the evaluation include less time to effect repairs, lower maintenance costs, improved supportability systems, and improved safety.

7.3.8 Environmental Factors

The environment in which maintenance activities are carried out is very important and should be controlled to facilitate maintenance actions. Whilst simplicity, accessibility, and other attributes described earlier refer to the actual design of components, environmental factors refer to the condition under which maintenance is carried out. The best design possible will be difficult to maintain under harsh environmental conditions such as excess heat or cold, poor lighting, excess humidity, dust, and so on.

7.3.9 Maintainability Index

Maintainability evaluation of engineered objects is important for several reasons. It can be used as part of the design process for improvement. It can also be used to select among competing alternatives. In an existing plant, maintainability evaluation can help determine if it would be cost effective to introduce modifications that can improve maintainability characteristics.

The available maintainability indices tend to be product- or plant-specific. Several indices have been developed, for example, by the Department of Defense (1966) and J817a by the Society of Automotive Engineers (SAE). The Bretby Maintainability Index (Rushworth and Mason, 1992) was developed for the mining industry based on the SAE index.

7.4 Availability

Uptimes and downtimes are two important concepts discussed in Chapter 4 and which play a role in defining availability – an important system effectiveness measure for engineered objects. The uptimes and downtimes (see Figure 4.5) are random variables and are character-ized by failure time and repair time distributions. Availability depends on the times to failure and the times to repair. This is best illustrated graphically by the functioning profile of a repairable object used in a continuous mode. The state of the object (denoted by $X(t)$) alter-nates between two states: (i) functioning (up) and (ii) failed (down), as shown in Figure 7.2.

Let

$$X(t) = \begin{cases} 1 & \text{if the system is functioning,} \\ 0 & \text{otherwise.} \end{cases} \tag{7.1}$$

Let (T_i, D_i), $i = 1, 2, \ldots$, be the successive uptimes and downtimes, as indicated in Figure 7.2. There are several different notions of availability, as discussed next.

7.4.1 Point Availability

Definition 7.3

Point availability, $A(t)$, is the probability that an item is in an up state at some future time t.

Point availability is a measure of the item being in the operational state at some future time instant. This measure is important in the context of safety or protective devices that can fail but the failure is not detected immediately (due to the item not being monitored continuously). On detection, the failed item is restored to the operational state.

Note that:

$$E\left[X(t)\right] = 1 \times P\{X(t) = 1\} + 0 \times \Pr\{X(t) = 0\} = P\{X(t) = 1\} \tag{7.2}$$

and, as a result, we have $A(t) = P\{X(t) = 1\} = E[X(t)]$.

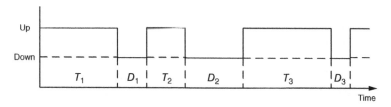

Figure 7.2 Functioning profile of a repairable system.

7.4.2   Interval Availability

Define $U(t)$ as follows:

$$U(t) = \int_0^t X(t)\,dt \tag{7.3}$$

Then, the availability over an interval $[t_1, t_2)$ with $t_2 > t_1$ is a random variable given by:

$$\tilde{A}(t_1, t_2) = \frac{U(t_2) - U(t_1)}{t_2 - t_1} \tag{7.4}$$

Definition 7.4

Average interval availability is the expected value of interval availability over the period $[t_1, t_2)$.

As a result, the average interval availability is given by:

$$A(t_1, t_2) = E\left[\tilde{A}(t_1, t_2)\right] = \frac{E\left[U(t_2) - U(t_1)\right]}{t_2 - t_1} \tag{7.5}$$

Definition 7.5

Asymptotic average interval availability is the expected value of availability over the period $[0, \infty)$.

The asymptotic availability (obtained as the limit with $t_1 = 0$ and $t_2 = \infty$) is given by:

$$A_\infty = A(0, \infty). \tag{7.6}$$

7.4.3   Special Case

The uptimes $(T_i, i = 1, 2, \ldots)$ are independent and identically distributed with distribution function $F(t)$ and the average uptime is given by the mean time to failure (MTTF). The downtimes $(D_i, i = 1, 2, \ldots)$ are also independent and identically distributed with distribution function $G(t)$ and the average downtime is given by the mean time to repair (MTTR). In this case, $X(t)$ is an alternating renewal process (see Appendix B.6.3) and each uptime (T_i) + downtime (D_i), $i = 1, 2, \ldots$, defines a cycle for the alternating renewal process. Then, from the Renewal Reward Theorem (Appendix B.6), we have:

$$A_\infty = \frac{MTTF}{MTTF + MTTR} \tag{7.7}$$

Equation (7.7) provides an interesting relationship between availability, reliability (measured by MTTF), and maintainability (measured by MTTR). One of the insights that we can deduce is that the availability of a system can be improved by either increasing the MTTF (improving its reliability) or decreasing the MTTR (improving its maintainability).

The above insight clearly highlights the effect of maintainability on the availability of an engineered object. This is a clear motivation for giving due attention to maintainability at the design stage.

Example 7.1

A product has an exponential time to failure distribution with probability density function given by:

$$f(t) = \frac{1}{50} e^{-(t/50)}.$$

The failed item is repaired to as good as new, and the repair time distribution is lognormal with probability density function:

$$g(t) = \frac{1}{t\sqrt{2\pi}} e^{-\frac{1}{2}(\ln t - 2)^2}.$$

In order to compute the asymptotic availability, we need the MTTF and the MTTR. Since the failure distribution is exponential, we have $MTTF = 50$ (see Equation (A.21)). The repair time follows a lognormal distribution with mean 2 and standard deviation 1. Its mean is given by $MTTR = e^{2+1/2} = 12.18$ (see Equation (A.28)). Hence, the asymptotic availability is given by:

$$A_\infty = \frac{MTTF}{MTTF + MTTR} = \frac{50}{50 + 12.18} = 0.80 \qquad \blacksquare$$

7.5 Maintainability Process

7.5.1 Process Steps

From the new product development (NPD) life cycle point of view (see Chapter 5), the decisions made at the design and development stage have implications on the operation and maintenance of engineered objects.

The objective of an effective maintainability program is to design and produce/build an engineered object that is easily and economically maintained. As maintainability is a design characteristic, it has to be considered early on in the design stage. The more it is delayed, the more expensive and inefficient it becomes to attempt to improve the

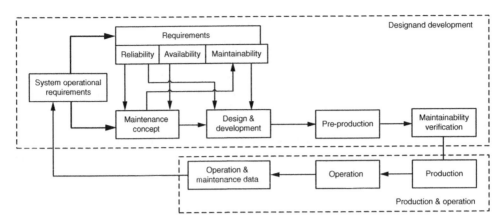

Figure 7.3   Maintainability process as part of the NPD process.

object's inherent maintainability. The maintainability process, as part of overall system development, is depicted in Figure 7.3.

A sound maintenance process requires the following six steps,[3] which also shed more light on the processes shown in Figure 7.3.

1. *Translate customers' needs into maintainability requirements:* Determine the required level of maintainability, as measured by the user during actual use of the product. This is derived from system operational requirements. Data and experience from operation and maintenance on prior or similar systems play an important part in understanding customer needs. By knowing past problems, the maintainability engineer can suggest approaches that design out these problems. As shown in Figure 7.3, maintainability requirements are based on operational requirements and feedback from operations and maintenance.
2. *Integrate maintainability with the systems engineering process:* It is important to make the maintainability activities conducted during design and manufacturing an integral part of the product and process design effort. This systems approach is essential, as maintainability is related to other system characteristics such as reliability and availability requirements. System operational requirements help in determining reliability and availability requirements, and together form the basis for the object's maintenance concept. The maintenance concept covers all aspects of the object's maintenance program throughout its life cycle and is an important input in maintainability implementation in the design. These relationships are depicted in the left side of the upper box of Figure 7.3.
3. *Thoroughly understand the design:* Thoroughly understanding the design is essential to making the final product maintainable. Accessibility, diagnostic capability, and repair times must be known with as much certainty as knowledge allows.
4. *Design for desired level of maintainability:* A conscious and dedicated effort is required to incorporate design features that make preventive and corrective maintenance easy, safe, and economical in terms of time and resources.

---

[3] Department of Defense (1997)

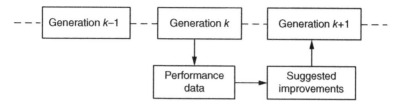

Figure 7.4 Data for design and maintainability improvement.

5. *Validate the maintainability through analysis and development testing:* As shown in Figure 7.3, after pre-production, maintainability verification is carried out through analyses, simulation, and testing to uncover maintainability problems, revise the design, and validate the effectiveness of the redesign. As such, the process is iterative in nature until an acceptable design solution meeting the targeted maintainability requirements is achieved.

6. *Monitor and analyze operational performance:* One must assess the operational maintainability of the product in actual use to uncover problems, identify necessary improvements, and provide "lessons learned" for incorporation in handbooks and for refining modeling and analysis methods. This is the closed loop shown in Figure 7.3, which is needed for several reasons:
   - To identify possible improvements in policy, procedures, or design to address performance problems;
   - Despite all the efforts to design properly and to validate the design through development testing, products are seldom perfect. Some problems may be overlooked and may only become evident when the object is put in the field.

The use of performance data derived from operation and maintenance experience is shown in Figure 7.4.

7.6 Maintainability Standards

Due to the importance of maintainability and its effect on the life cycle cost of engineered objects, many military and civilian organizations have developed maintainability standards to provide guidelines for various maintainability issues, these include:

- US Department of Defense (DoD);
- International Electrotechnical Commission (IEC);
- Society of Automotive Engineers (SAE);
- National Aeronautics and Space Administration (NASA).

The IEC standards, in particular, are gaining wide acceptance and are briefly presented in this section. The purpose of these and other standards is to make recommendations for the standardization of maintainability practices.

Table 7.1 Maintainability standards (IEC-706).

Part	Section	Title
Part 1 (IEC 706–1, 1982)	Section 1	Introduction to Maintainability
	Section 2	Maintainability Requirements in Specifications and Contract
	Section 3	Maintainability Program
Part 2 (IEC 706–2, 1992)	Section 5	Maintainability Studies During the Design Phase
Part 3 (IEC 706–3, 1987)	Section 6	Maintainability Verification
	Section 7	Collection, Analysis and Presentation of Data Related to Maintainability
Part 4 (IEC 706–4, 1992)	Section 8	Maintenance and Maintenance Support Planning
Part 5 (IEC 706–5, 1982)	Section 4	Diagnostic Testing
Part 6 (IEC 706–6, 1982)	Section 9	Statistical Methods in Maintenance Evaluation

The IEC 706 maintainability standard consists of nine sections divided into six parts (six separate publications), as indicated in Table 7.1. A brief summary of the various sections of the standards is given below.

- *Sections 1 and 2:* These provide an introduction to maintainability and the elements important to defining maintainability specifications and requirements.
- *Section 3:* This provides an outline and description of the elements of an overall maintainability program plan. It covers all phases of program development, and outlines information to be provided by the contractor and by the customer when there is a contractual relationship.
- *Section 4:* This provides guidance for the early consideration of testability aspects in design and development, and determining effective test procedures as an integral part of operation and maintenance.
- *Section 5:* This outlines the maintainability studies that need to be carried out in the preliminary and detailed design phases and their relationship to other maintenance and maintainability support tasks.
- *Section 6:* This describes the aspects of verification needed to ensure that the specified maintainability requirements have been met.
- *Section 7:* This provides an overview of issues related to data collection, analysis, and presentation of maintainability-related data.
- *Section 8:* This describes the tasks required for planning maintenance and maintenance support. The interfaces between reliability, maintainability, and the maintenance support planning program and their tasks are also presented.
- *Section 9:* This covers some quantitative aspects of maintainability engineering in various phases of the system life cycle. It is applicable to the tasks of maintainability allocation, maintainability demonstration, and maintainability data evaluation, as described in Sections 5–7 of the guide.

7.7 Relationship with Other Disciplines

The relationship between maintainability, reliability, and life cycle costs has been discussed earlier in this chapter. However, maintainability is also related to other design and support disciplines, as depicted in Figure 7.5.

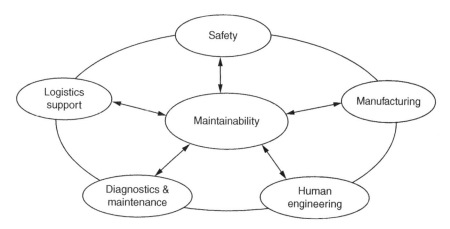

Figure 7.5 Relation of maintainability to other disciplines.

- *The manufacturing/construction processes* used to transform the design into an engineered object can impact the inherent maintainability achieved. Maintainability design features should not make the product difficult or too expensive to build.
- *Human engineering* (HE) is the discipline that deals with the safe and effective integration of people in the use, operation, and maintenance of an engineered object. Its relation to maintainability was discussed in Section 7.3.7.
- *Safety* dictates designing systems and maintenance procedures that minimize harm to maintenance personnel and avoid damage to system components. Warning labels, precaution information, and procedures for disposing of hazardous materials are all examples of safety considerations that can be brought to the table.
- *Diagnostic tools* (software, hardware, procedures, etc.) help determine the status of an item and detect and isolate faults quickly and efficiently. A highly maintainable design will make use of these tools efficiently to reduce downtime.
- *Logistics support* requirements are effected by maintainability design decisions. This includes manpower, support and test equipment, facilities, training, spares and materials, and technical manuals. In turn, the logistics support provided will determine the degree to which inherent maintainability is realized in the field. Maintenance logistics will be discussed in detail in Part D.

7.8   Summary

Maintainability is an inherent characteristic of the design of an engineered object. It relates to the ease, economy, and safety of maintenance. Maintainability deals with maintenance issues at the design stage, whilst maintenance is a result of the design decision. Maintainability investments made in the early design phase can significantly reduce operation and maintenance costs and later help to minimize total life cycle costs.

Several features can have a positive impact on the maintainability characteristics of a system. These include:

- Design attributes such as simplicity, accessibility, standardization and interchangeability, diagnosability, identification and labeling, and so on.

- Human engineering attributes such as weight, size, shape, and position of equipment from an ergonomic point of view, and insulation, protection against leaks, and warning systems from a safety point of view.
- Logistics support considerations such as standardization of components and tools, clear instructions and procedures, and so on.

High availability of many engineered objects is a very important objective. It is affected by both the reliability and maintainability of the system. Reliability impacts on the frequency of maintenance actions, as a more reliable object fails less often. The ease of maintenance or maintainability characteristics of the object and the skill of maintenance personnel and other support resources govern the duration of the maintenance actions. The optimal levels of reliability and maintainability need to be determined based on overall business objectives and life cycle cost.

A sound maintainability process must be an integral part of the object and design process effort. This system approach is essential, as maintainability is related to other system characteristics such as reliability, diagnostics and the maintenance concept, human engineering, logistics, safety, and manufacturing/construction.

Review Questions

7.1   Define maintainability.

7.2   What is the difference between maintenance and maintainability?

7.3   Explain the effect of maintainability on operations.

7.4   Explain how maintainability affects life cycle cost.

7.5   Describe three design maintainability attributes and explain how they enhance maintainability.

7.6   Define two quantitative measures of maintainability.

7.7   Give two benefits of a maintainability index.

7.8   Define availability.

7.9   Explain how availability can be improved through reliability and maintainability.

7.10 Explain the difference between inherent availability, achieved availability, and operational availability.

7.11 Describe the relationship of maintainability to each of the following disciplines:
      (a) Human engineering.
      (b) Safety.
      (c) Logistics.
      (d) Manufacturing.

Exercises

7.1   It is often said that simplicity, reliability, and maintainability are the core virtues of a good engineered product. Using the list of attributes discussed in this chapter, identify those maintainability attributes that apply to bicycles and discuss their benefits.

7.2   For some engineered objects, the simple notion of availability is not appropriate to the item produced due to external factors. Wind turbines are a good example, as there is no output when there is no wind. In this case, the steady-state availability cannot be defined in the manner suggested in the chapter, as the unit may be either in the operational state or the failed state when there is no wind. Propose an alternative definition for asymptotic availability in this case.

7.3   List some engineered objects for which point availability is appropriate.

7.4   Several objects used on a farm (such as harvesters, crushers, etc.) are needed only for a short period each year. The reliability of such objects is one measure of significance as well as the interval availability. How do these change from year to year as the object ages? Indicate what maintenance actions will ensure the desired reliability and availability.

7.5   One way of increasing the asymptotic availability of an engineered object is to have more built-in sensors to detect failure and to indicate the actions needed to rectify the failure. This is true if the detectors are very reliable. On the other hand, if they are not very reliable, the availability goes down (since a specialist mechanic is needed to replace a failed detector) and the maintenance cost goes up. How would one account for this as part of maintainability?

7.6   When standardization is carried out in system design, certain major advantages are gained by the support activities required for the system. Can you list some of these advantages with justification?

7.7   In the context of an aircraft, discuss the different notions of reliability and availability and their significance.

7.8   Consider the maintainability attributes discussed in Section 7.3. Consider an engineered object with which you are familiar and identify how some of these attributes have been implemented to simplify maintenance activities.

7.9   Consider the maintainability attributes discussed in Section 7.3 and identify those attributes that enhance the different stages of a corrective maintenance activity cycle.

7.10   Compare inherent availability, achieved availability, and operational availability.

References

Department of Defense MIL-HDBK-472 (1966) Maintainability Prediction, Department of Defense, Washington, DC.

Department of Defense MIL-HDBK-470A (1997) Designing and Developing Maintainable Products and Systems, Department of Defense, Washington, DC.

Rushworth, A. and Mason, S. (1992) The Bretby maintainability index – A method of systematically applying ergonomic principles to reduce costs and accidents in maintenance operations, *Maintenance* 7(2): 7–14.

Part B

Reliability and Maintenance Modeling

Chapter 8: Models and the Modeling Process
Chapter 9: Collection and Analysis of Maintenance Data
Chapter 10: Modeling First Failure
Chapter 11: Modeling CM and PM Actions
Chapter 12: Modeling Subsequent Failures

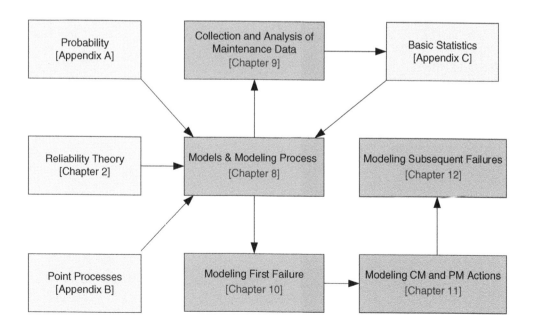

8

Models and the Modeling Process

Learning Outcomes

After reading this chapter, you should be able to:

- List and describe the different types of models;
- Show a proper understanding of descriptive and mathematical models;
- Define mathematical formulation and explain its role in mathematical modeling;
- List and describe the different types of mathematical formulations;
- Describe different mathematical modeling approaches in the context of maintenance;
- Describe the steps of the mathematical modeling process;
- Explain the role of data in model building.

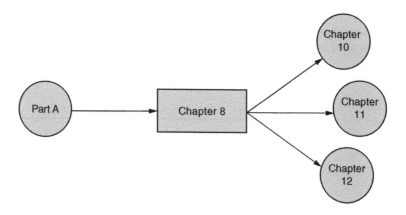

Introduction to Maintenance Engineering: Modeling, Optimization, and Management, First Edition.
Mohammed Ben-Daya, Uday Kumar, and D.N. Prabhakar Murthy.
© 2016 John Wiley & Sons, Ltd. Published 2016 by John Wiley & Sons, Ltd.

8.1 Introduction

Models play an important role in understanding and solving real-world problems. In the context of decision problems they are used to (i) carry out analysis such as the effect of changes to decision variables on objectives of interest (for example, the effect of different preventive maintenance (PM) actions on system failures) and (ii) decide on the optimal decision variables to achieve some specified objectives (for example, optimal PM to minimize total maintenance costs). There are several different types of models (physical, mathematical, simulation, etc.). Many of these are used in solving maintenance-related problems. Mathematical modeling is the process of building a mathematical model that is adequate for solving a given problem. Modeling is both an art and a science – the science aspect deals with methodology and techniques for modeling and the art aspect involves making decisions based on intuitive judgments. In this chapter we look at models and the modeling process. This will be used in later chapters to build a variety of models to assist in maintenance decision making.

The outline of the chapter is as follows. We start with a general discussion on models in Section 8.2 and briefly discuss the different types of models. Two types of models – descriptive and mathematical models – are of particular importance in solving maintenance decision problems. Mathematical modeling involves linking a descriptive model to an abstract mathematical formulation. This requires a good understanding of the different types of mathematical formulations. Section 8.3 looks at the different mathematical formulations for modeling and this forms the basis for the classification of mathematical models. Section 8.4 deals with the different approaches to modeling and Section 8.5 looks at the modeling process. Data play a critical role in modeling and are the focus of Section 8.6. Section 8.7 provides a brief discussion on the modeling of maintenance-related decision problems, and we conclude with a summary of the chapter in Section 8.8.

8.2 Models

A model is a simplified representation of the real world that is adequate for solving the problem under consideration. There are several different types of models. They can be grouped broadly into two categories with several subcategories, as shown in Table 8.1.

Table 8.1 Types of models.

Categories	Subcategories	Examples
Physical	Scaled models	Scaled wing of an airplane for testing in a wind tunnel
	Analog models	Electrical network representation of a water pipe network
Abstract	Descriptive models	Description of the world using some natural language or special symbols and icons
	Mathematical models	Model involving a mathematical formulation
	Simulation models	Simulating the real world relevant to the problem on a computer

8.2.1   Physical Models

Physical models are physical entities and can be categorized into two types: (i) scaled models and (ii) analog models. The former involve using a scaled version to study some aspect whereas the latter use the notion of analogy between the model and the real world, as illustrated in Table 8.1. Physical models are used to train pilots (to fly aircraft), operators of complex plants (such as processing plants), and so on. They are also used by engineers in design and analysis to study the effect of operating and environmental conditions on the performance of a new object.

8.2.2   Abstract Models

Abstract models, in contrast, are non-physical entities and mental constructs. The three types of abstract models are as indicated in Table 8.1.

8.2.2.1   Descriptive Models

A descriptive model is a representation which describes the variables and parameters of significance and the interactions between them, either by using some natural language or a diagrammatic representation. Characterization of a descriptive model uses terms and concepts from systems theory, and these are discussed briefly below.

System
The real world relevant for solving a problem can be viewed as a system consisting of several interconnected elements. The system boundary separates the elements from the outside world and the system interacts with the outside world through input and output variables.

Parameters and Variables
Parameters are attributes intrinsic to an element. Variables are attributes needed to describe the interaction between elements. Variables and parameters are terms from some theory or theories, if such exist, and are used to explain the state and behavior of the system. When no such theory exists, they are words from a natural language with the usual meaning.

A parameter or variable is said to be time invariant (*static*) if it does not change with time. If not, it is time varying (*dynamic*). In a *continuous time* description, the dynamic variables are characterized as continuous functions of time (i.e., for all points along the time axis). In contrast, in a *discrete time* description, the dynamic variables are characterized by changes at discrete time instants (they may be either equally or unequally spaced points along the time axis).

Relationships
The interactions between elements of a system are described by relationships. Of particular importance are *causal* (also referred to as *cause–effect*) relationships. Causal relationships are often displayed using either *graph-theoretic* (*network*) or *matrix tabular* displays.[1]

---

[1] The graph-theoretic representation uses nodes to represent variables and directed arcs to indicate the cause–effect relationship. The reliability block diagram (discussed in Chapter 2) is a graph-theoretic representation of a system from a reliability perspective.

Another type of relationship is the *correlation* between two or more variables. Two variables X and Y are said to be correlated if there is a third variable Z which is the cause and this is related to both X and Y in a cause–effect relationship. As a result, changes in X and Y occur due to changes in Z.

Example 8.1 Automobile Suspension Assembly

The suspension assembly of an automobile is there to ensure satisfactory ride characteristics over bumpy roads. The performance of the suspension assembly over time is important from a reliability and maintenance perspective. As the automobile hits a pothole (or hump), the tires (and wheels) follow the profile of the road and this results in a vertical motion, and possibly a rotational motion along the direction of travel, leading to passenger discomfort. The suspension assembly acts as a spring to absorb the impact and to damp out the energy, and in the process reduce the vertical and rotational motions. A poorly designed assembly will lead to more frequent maintenance actions. The suspension assembly can be characterized in terms of two parameters: spring stiffness (k) and the damping coefficient (q).

If we focus only on the vertical motion, then the factors that are important are as follows:

- Mass of the automobile (m);
- Gravitational effect (g);
- The profile of the road along the direction of travel (as it determines the vertical displacement of the wheel);
- The velocity of travel.

Let $X(t)$ denote the vertical displacement of the car chassis. If the car is traveling on a smooth road, then $X(t) = 0$. The car starts deviating once it hits a bump (or pothole). Some of the key parameters and variables are shown diagrammatically in Figure 8.1a.[2]

The vertical displacement of the wheel (denoted by a circle in Figure 8.1a) depends on the road profile and the velocity of travel, $V(t)$, of the automobile (not shown in the figure). This acts as the input to the system (the automobile is viewed as a system and the road is viewed as external to the system), and the interaction between the wheel and the automobile is defined through the road profile.

If the roll motion is significant, then an additional parameter is the moment of inertia (for roll motion along the direction of travel) and an additional variable $\varphi(t)$ is needed to characterize the angle of rotation from a vertical line passing through the center of gravity of the automobile. In this case the descriptive model is given by Figure 8.1b, where we have assumed that the spring characteristics for the front and rear suspension are identical. If not, the spring stiffness and damping coefficient would be different, resulting in extra parameters.

---

[2] The symbols (or icons) used are from the theory of dynamics – automobile mass is represented as a box, spring characteristics as a coil, and damping characteristics as a dashpot.

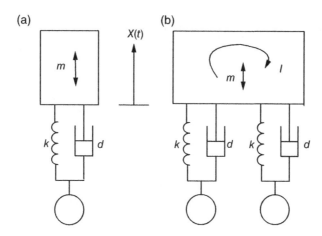

Figure 8.1 Descriptive model of an automobile suspension assembly. (a) vertical motion; (b) vertical motion and rotation.

Note that the spring is in compression when the automobile hits a hump and in tension when it hits a pothole. As a result, the suspension assembly is subjected to a cyclical loading, leading to fatigue failure of the spring.[3] If one is looking at a short time interval (order of months), the spring stiffness is time invariant. On the other hand, over a long time interval (order of years), the spring stiffness and damping characteristics will change.   ∎

8.2.2.2 Mathematical Models

A mathematical model involves the linking of a descriptive model with a mathematical formulation, as indicated in Figure 8.2. There are many different types of formulations and these are discussed in the next section. The linking involves a one-to-one correspondence between variables, parameters, relationships, and so on, of the descriptive model and of the formulation. We will discuss this further later in the chapter.

8.2.2.3 Simulation Models

A simulation model is a computer representation of a system that mimics the time history of changes taking place in the system. In the context of maintenance it involves simulating a time history of events such as failures, maintenance (CM and PM) actions, and evaluating their impact on the system performance (costs, availability, number of spares needed, etc.).

Simulation to generate a time history involves the following steps:

Step 1: Generating the next event. This is done using random number generators when events occur in an uncertain manner.

---

[3] Fatigue failure is discussed in Section 3.3.2. Suspension assemblies are designed such that 90% of the units survive some specified number of such cycles (usually several million).

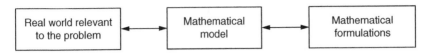

Figure 8.2 Mathematical model.

Step 2: Updating the clock and other counts (such as the number of PM and CM actions, etc.). This generates a time history of events.

When uncertainty (discussed in the next section) is a significant factor, each simulation run provides a sample outcome of a stochastic process.[4] The outcome changes when the simulation is repeated. In this case, one needs to carry out the simulation several times to obtain proper estimates and confidence limits (for the variables, performance measures, and so on, of interest).

8.3   Mathematical Modeling

Mathematical modeling is the building of mathematical models to solve real-world problems. As mentioned earlier, it involves linking a descriptive model to an abstract mathematical formulation. For a given problem there can be several descriptive models (due to the pluralistic view of the world), and similarly there are many types of mathematical formulation that can be used. As a result, there can be many different mathematical models for a given problem, depending on the model builder. However, not all models will be appropriate and this leads to the notion of an *adequate mathematical model.* The selection of an appropriate model is referred to as *model validation* and is discussed in the next section.

8.3.1   *Mathematical Formulations and Classification*

A mathematical formulation is an abstract structure and mental construct. It involves variables, parameters, relationships, and operations (such as integration, differentiation, and so on) using symbols with precise mathematical meaning. The manipulation of symbols is dictated by the rules of logic and mathematics. A formulation makes no sense outside of mathematics. However, formulations play a very important role in building mathematical models (discussed further in later chapters of the book).

The variables of a mathematical formulation can be grouped broadly into the following two categories:

- *Independent variables:* Usually representing time and/or spatial coordinates when the formulation is used in modeling.
- *Dependent variables:* These are functions of the independent variables.

Each of these can be further divided into two subgroups:

- *Discrete variables:* Assume a discrete set of values – finite or infinite.
- *Continuous variables:* Assume any value over an interval – finite or infinite.

---

[4] Stochastic processes are discussed in Appendix B.

Table 8.2   Classification of mathematical formulations.

		Dependent variable	
		Discrete	Continuous
Independent variable	Discrete	A	B
	Continuous	C	D

Based on the variables of the mathematical formulation, the formulations can be grouped broadly into four categories, as indicated in Table 8.2. Many other subcategories can be defined based on the nature of the formulation (deterministic or not) and the independent variables (static or dynamic).

8.3.1.1   Deterministic Formulations

Static Formulations
Here, the variables and parameters do not change with time, so that the formulations are algebraic equations of the form $y = g(x, \theta)$ where x is the independent variable, y is the dependent variable, and θ is the parameter of the formulation. There are several subcases based on the form of the function $g(\cdot, \cdot)$ – for example, linear, polynomial, exponential, and so on.

Dynamic Formulations
Here, one of the independent variables represents time (denoted by t), so that in the case of one independent variable, the dependent variables are functions of the form $g(t)$. If there are two or more independent variables, one represents time and the others (such as x, y, z) represent spatial coordinates, so that the dependent variables are functions of the form $f(t, x)$ or $g(t, x, y)$, and so on.
Based on Table 8.2, dynamic formulations can be grouped into four main types with several subtypes. For example, the two subtypes for type D are the following:

- *Ordinary differential equations:* The formulation has one independent variable (t) and derivatives (first or higher order) of the dependent variables ($X(t)$, $Y(t)$, and so on). A first-order differential equation with one dependent variable is $dX(t) / dt = g(X(t))$ with the initial condition given by $X(t_0) = X_0$.
- *Partial differential equations:* The formulation has two or more independent variables (t, x, and so on) and various partial derivatives (first or higher order) of the dependent variables ($Q(t, x)$, and so on).

Differential equations can be subdivided into linear versus nonlinear; time-varying versus time-invariant, and so on.

8.3.1.2   Uncertainty

An event is said to be uncertain if its outcome cannot be predicted with certainty before the event occurs. However, once the event has occurred, the uncertainty disappears and the outcome is known exactly. Uncertainty basically means non-repeatability – in other words, not always getting the same outcome. That leads to the concept of variability. A few illustrative examples are given below.

Example 8.2   NPD Process

The new product development (NPD) process (discussed in Chapter 5) involves research and development (R&D) to achieve some reliability goal. The outcome is uncertain in the sense that it can lead to success (goal achieved) or failure (goal not achieved). A formulation to model this would involve a binary variable X that can take a value 1 (indicating success) or 0 (indicating failure) and its values are unknown before the event occurs.   ∎

Example 8.3   Item Life

The life of a new item is uncertain before it is put into use. Once it is put into operation and operated till failure, the uncertainty disappears. The time until failure of the item can be characterized by a continuous-valued variable that can take any positive value in the interval zero to infinity. (The latter is to avoid nominating an upper limit and facing the embarrassing situation where the item survives beyond its upper limit.)   ∎

Example 8.4   Tire Wear

When two physical objects rub against each other, a wear mechanism is in operation. The wear on a component (for example, an automobile tire) is a function of usage, which, in turn, can be a function of age. As a result, the wear on a new tire as a function of the age, $X(t)$, changes in an uncertain manner, so that it is not possible to predict its exact value at any future time instant.   ∎

Note that in Example 8.2, the variable X is discrete and is not a function of time (as the time needed for development is ignored and the focus is only on the outcome), whereas the variable $X(t)$ in Example 8.4 is a continuous function of time. The modeling of the former involves probabilistic formulations and the latter stochastic formulations.

8.3.1.3   Probabilistic Formulations

Probabilistic formulations involve random variables (see Appendix A) whose outcomes are uncertain before the event occurs.[5] Once the event occurs, the uncertainty disappears. In a sense, these are *static formulations* with uncertainty. In the case of a scalar variable, the values that the variable can assume are characterized by a probability distribution function $F(x)$ defined over an interval, with the end points representing the minimal and maximal values that the variable can assume.

---

[5] The concepts also apply for the case where the variable is a vector.

- *Discrete probability distribution:* The probability distribution is discrete if the variable assumes values from a discrete set. In this case, the probability distribution is a staircase function. The distribution is non-decreasing, starting with a value zero (implying that the variable cannot assume a value lower than the minimum) and reaching a value 1 (implying that the variable must always assume a value less than the maximum). The jumps indicate the probability of the variable assuming the different values in the interval.
- *Continuous probability distribution:* The probability distribution is continuous if the variable can assume any value over an interval, and in this case the distribution is a smooth function (non-decreasing with value zero at one end and one at the other end).

Appendix A describes a few discrete and continuous distribution functions that will be used in modeling failure times, failure costs, and so on, in later chapters.

8.3.1.4 Stochastic Formulations

Stochastic formulations involve stochastic processes (see Appendix B). A stochastic process, $X(t)$, can be viewed as an extension of a random variable X in the following sense – t represents time (an independent variable). As a result, a stochastic process can be viewed as a dependent variable whose value (at any time instant) is uncertain, so that it is like a random variable before the event and is a deterministic function after the event has occurred.

There are several types of stochastic processes and they can be classified broadly into four categories depending on whether t is discrete or continuous and whether the values assumed by $X(t)$ are discrete or continuous. Table 8.3 indicates the four categories with illustrative applications of relevance in the context of maintenance. Each of these can be subdivided, and a brief introduction to some of the formulations is given in Appendix B.[6]

8.3.2   Classification of Mathematical Models

There are several different ways of classifying models based on:

1. Discipline (physics, biology, engineering, economics, reliability, maintenance, etc.);
2. Problem (component failure, maintenance cost, reliability growth, etc.);
3. Formulation used (as discussed in the previous section);
4. Purpose and/or techniques used (optimization, analytical, computational, etc.).

Table 8.3   Classification of stochastic processes.

Time (t)	Values of $X(t)$	Type	Examples
Discrete	Discrete	A	Count of the number of machine breakdowns each week
Discrete	Continuous	B	Wear of item measured on a weekly basis
Continuous	Discrete	C	Time to failure for a component in continuous use
Continuous	Continuous	D	Wear of item monitored continuously

---

[6] Some of the formulations will be used in later chapters for modeling various maintenance decision problems.

In later chapters of the book we will be discussing a variety of models (mostly based on probabilistic and simple stochastic formulations) in the context of maintenance.

8.4   Approaches to Modeling

The real world relevant to a problem is a complex system. Each of its elements can be decomposed into several levels, with the degree of detail increasing with the number of levels. In Chapter 2 we discussed how a product can be decomposed into several different levels. The modeling can be done at any level – from the lowest to the highest and anywhere in between. Two approaches are used in the selection of the mathematical formulation to model a system: (i) empirical and (ii) theoretical (physics-based).

8.4.1   Empirical and Physics-Based Approaches

8.4.1.1   Empirical Approach

In the empirical approach, the selection of the model formulation is based solely on the data available. This approach is also referred to as *data-based* or *black-box* modeling and is used when the knowledge and understanding of the underlying processes is not sufficient to provide a foundation for derivation of a theoretical model. The parameters of the formulation do not necessarily lend themselves to explicit interpretation in the context of the original problem. Only rarely would one be able to relate the estimated parameter values obtained to underlying failure causes, environmental conditions, and so on in the context of modeling for maintenance.

8.4.1.2   Physics-Based Approach

In the theoretical approach, the selection of the model formulation is based on some relevant theory (for example, theories of corrosion, fatigue, etc.). This approach is also referred to as the *theoretical* or *white-box* approach. One of the key consequences of the theoretical under-pinning is that parameters of the formulation relate to important parameters of the descriptive model. This provides extra information in finding the solution to the problem and the implementation of the solution. As an example, one can predict how the material properties of a pipe and the chemical composition of the fluid being pumped affect failure due to corrosion, so that better decisions can be made in the building of pipe networks.

Often, some elements of a system are modeled using the empirical approach and others using the theoretical approach.

So far the discussion on modeling has been general, whereas the next few subsections deal with maintenance modeling at the component, system, and business levels.

8.4.2   Component-Level Modeling

Consider a component in which failure is due to a crack exceeding some specified limit. Let $X(t)$ denote the state (crack length) at time t, with $t = 0$ corresponding to the time at which the component is put into operation. Here, one can use four different mathematical formulations to model the crack growth and failure.

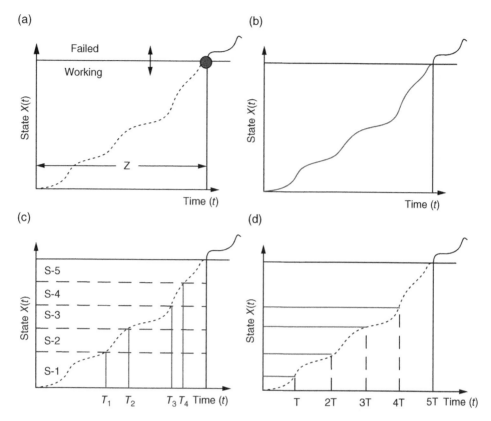

Figure 8.3 Alternative formulations for modeling component failure for four different models, shown in parts (a) to (d).

Model 1: The characterization (see Figure 8.3a) is done through the time to failure (Z) without explicitly characterizing the evolution of $X(t)$. Since time to failure is uncertain, Z is a random variable that is modeled using a *probability distribution function.*

Model 2: The characterization (see Figure 8.3b) involves treating both t and $X(t)$ as continuous variables, so that the modeling involves a Type D (see Table 8.2) formulation.

Model 3: The state is characterized as a discrete variable (see Figure 8.3c where five levels are used) and the transition times (T_i, $i = 1,2,...$) are random variables. The modeling involves a Type C (see Table 8.2) formulation.

Model 4: The state is characterized as a continuous variable for discrete values of the time variable (see Figure 8.3d). In this case, the model involves a Type B (see Table 8.2) formulation.

The formulation to use in modeling depends on the problem context. Consider a non-repairable component that is maintained using an age-based policy (discussed in Chapter 4). To determine the optimal replacement age and for optimal spare part management, one would use Model 1. For condition-based maintenance, one would use Model 2 (with continuous time monitoring) or Model 4 (with discrete time monitoring).

8.4.3   System-Level Modeling

System-level modeling is often used for plants and infrastructures comprised of several elements, and it involves the modeling of each of the elements. One can use different types of formulations to model each element, as indicated in the previous subsection.

Figure 8.4 is a very simple descriptive model (graph-theoretic representation) of maintenance costs and their impact on the profits of a manufacturing plant. As mentioned in Chapter 1, there are two types of maintenance actions – PM and CM. The production rate depends on market demand and impacts on system state (which characterizes the degradation of the plant). Note that the production rate, in turn, depends on the system state.

In the case of infrastructure, the modeling of distributed elements requires two-dimensional (2-D) formulations (one representing time and the other a spatial coordinate). By discretizing the spatial coordinate, one has several sections and each is treated as a separate lumped element, so that only 1-D formulations are needed.

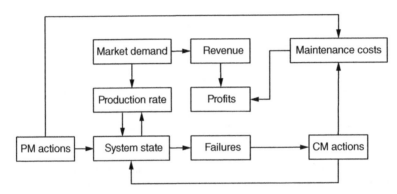

Figure 8.4   Simple descriptive model for modeling maintenance costs.

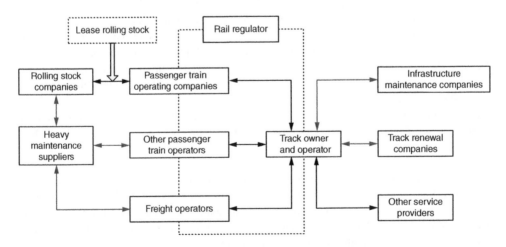

Figure 8.5   Different parties in the British Rail system.

8.4.4   *Business-Level Modeling*

At the business level, an engineered object (product, plant, or infrastructure) is just one element of the business, which also has many other elements such as commercial (economy, marketing, etc.), technical (new acquisitions, new technologies, etc.), legal (contracts, regulatory requirements, etc.), and other issues. Maintenance decisions at the business level (such as in-house versus outsourcing) need to take into account these elements and the interests of the different parties involved. Figure 8.5 shows the various parties involved in the British Rail system after the privatization.

The track owner and operator is responsible for the maintenance of the track. The track condition is affected by the condition of the rolling stock and vice versa. The rolling stock is owned by different independent businesses (running passenger or freight services) and they, in turn, might outsource some or all of the maintenance. There are contracts between different parties to ensure the smooth operation of the rail system. Game-theoretic formulations are appropriate for building models to assist in the decision making for maintenance outsourcing contracts, and this approach is discussed further in Chapter 18.

8.5   Mathematical Modeling Process

The use of models to solve real-world problems involves several interconnected steps, and these are indicated in Figure 8.6. Below we discuss briefly the eight important steps.

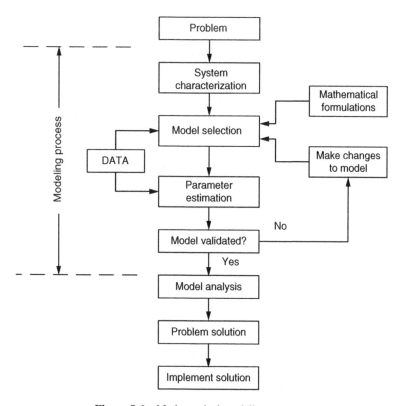

Figure 8.6   Mathematical modeling process.

Step 0: Problem definition

A good understanding of the problem is essential for its proper formulation and modeling and useful interpretation of results at the end.

Step 1: System characterization

Characterization of a system details the salient features of the system that are relevant to the problem under consideration. This generally involves a process of simplification. The variables used in the system characterization and the relationships between them are problem dependent. This corresponds to the descriptive model discussed earlier in Section 8.3.

Step 2: Model selection

The kind of mathematical formulation to be used depends on the system characterization and the approach used. In the theoretical approach, the theory being used provides the basis for model selection.

Example 8.5   Automobile Suspension Assembly

The model formulation needed for modeling the spring characteristics of the suspension assembly depends on the type of spring. For hard and soft springs, the relationship between force and displacement (given by the well-known Hooke's Law) is as shown in Figure 8.7. Note that in both cases, the relationship is nonlinear and the formulation is nonlinear and static. However, if the forces are small (implying the humps are not too high and the potholes not too deep), then one can view the spring as a linear spring characterized by a linear relationship.

The theory of Newtonian mechanisms is appropriate for modeling the vertical and rotational motions. If one confines oneself only to the vertical motion, then the system characterization (descriptive model) is given by Figure 8.1a. In this case, the model formulation is a second-order ordinary differential equation. In other words, the vertical motion can be modeled by the following differential equation:

$$m\frac{d^2X(t)}{dt^2} + k\frac{dX(t)}{dt} + qX(t) = \psi\left(V(t), Z(t), g\right)$$

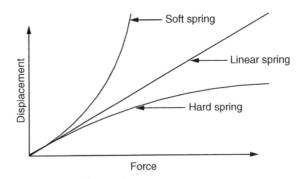

Figure 8.7   Mathematical modeling of a spring.

where $V(t)$ represents the velocity of the car in the direction of travel, $Z(t)$ the vertical displacement of the wheel (depends on the road profile and $V(t)$), and g the gravitational constant.

If one is interested in both the vertical motion and the rotational motion, the system characterization is given in Figure 8.1b. In this case, we need two sets of ordinary differential equation formulations to model the motion of the automobile.   ∎

In the empirical approach, analysis of the data is the starting point for model formulation. This is often referred to as *descriptive statistics* (see Appendix C). Here, one tries to extract information from the data to assist in the selection of the appropriate formulation. Various kinds of techniques and plots have been devised for this purpose. These are discussed in Chapter 9 and are used in the context of modeling first failure (Chapter 10) and subsequent failures (Chapter 12).

Step 3: Parameter estimation
The parameters of the model are the parameters of the mathematical formulations used in the modeling. In the theoretical approach, there is often a one-to-one correspondence between the parameters of the mathematical formulation and the parameters of the descriptive model. In Example 8.5, the parameter m corresponds to the mass of the automobile and can be estimated by weighing the automobile. In the empirical approach, the parameters need to be estimated from the data available. For models involving probabilistic and stochastic formulations there are a variety of methods for estimating the model parameters and these can be grouped into (i) non-statistical and (ii) statistical. These are discussed in Chapter 10.

Non-statistical methods: Here, one uses plots (based on data) to estimate the parameters. This involves plotting the data and selecting a formulation (for example, a distribution function) that fits the data; estimates are obtained from the fitted plot.
Statistical methods: Here, one uses techniques from inferential statistics (see Appendix C). These provide a rigorous basis for evaluating the estimates and providing confidence limits (discussed in Appendix C).

Both of these methods are used in estimating model parameters in the context of modeling first failure (Chapter 10) and subsequent failures (Chapter 12).

Step 4: Model validation
Validation involves testing whether or not the model selected (along with the assigned parameter values) mimics system behavior reasonably well to yield meaningful solutions to the problem of interest. Validation requires data that are different from the data used for parameter estimation in Step 3. When the data set is large, one can divide the data into two parts – one for estimation and the other for validation. With small data sets this is not possible. One needs to look at the fit between data and model and then decide whether the model is valid or not.

In the case of probabilistic and stochastic models there are a variety of methods for model validation and they can be grouped into (i) non-statistical and (ii) statistical.

Non-statistical methods: These involve the following two steps:

Step 1: Comparing graphical plots obtained from the data and the model selected to assess visually if the fit between the two is reasonable or not.

Step 2: Use of quantitative measures such as the sum of error squares. Any model with a measure value below some specified limit is accepted as being adequate and the one that yields the lowest value is viewed as the best fit.

Statistical methods: Here, one uses statistical tests to judge the adequacy of the fit. This provides a rigorous framework for the analysis and comparing of fits. It involves (i) hypothesis testing and (ii) goodness-of-fit tests. These are discussed further in Chapter 10.

In general, obtaining an adequate model requires an iterative approach, where changes to the system characterization and/or the mathematical formulation are made in a systematic manner until an adequate model is obtained.

Step 5: Model analysis
Once an adequate model has been developed, a variety of techniques may be used to carry out the analysis and obtain solutions. These techniques can be grouped broadly into three categories, as indicated below.

Analytical: Analytical solutions are mathematical expressions involving the variables and parameters of the model formulation. They allow one to study the effect of model parameters in an explicit manner. Often, this is only possible for a few special formulations – linear ordinary differential equations, failures based on an exponential failure distribution, and so on.

Computational: When analytical solution is not possible, computer analysis based either on numerical methods or Monte Carlo simulation needs to be used to generate approximate solutions. In the case of ordinary differential equations, numerical methods involve approximating the derivatives by finite differences and the solution is obtained for discrete values of the independent variable. A study of the effect of model parameters requires re-solving of the equation using the new parameter values.

Monte Carlo simulation is similar to numerical methods for model analysis. The process is simulated several times to generate different time histories and solutions are obtained by a statistical analysis of the simulated results. The number of times the simulation is replicated determines the confidence limits for the solution.

Step 6: Problem solution
Problem solution often requires mathematical techniques from optimization theory. Here again, a variety of techniques has been developed and these are discussed briefly in Appendix D.

In Steps 5 and 6, the analysis and optimization are purely mathematical exercises and the link to the descriptive model is of no significance. It is this that makes the use of models so effective. Once the solution is obtained, the link to the descriptive model is re-established so that it can be viewed in terms of the physical variables of the real world.

Step 7: Implement solution
The implementation of the solution requires careful evaluation of the solution. This is important, as the decision-maker needs to keep in mind that a model is a simplified representation of the

real world and must be viewed as a tool to assist in decision making. As such, intuitive judgment combined with the results from model analysis forms the basis for implementation.

8.5.1   Model Complexity and Selection

The complexity of the model depends, to some extent, on the mathematical formulation used in the modeling – for example, two-dimensional distribution functions are more complex than one-dimensional ones, a non-Markov process formulation is more complex than a Markov process formulation, and so on.[7]

In modeling one takes a pluralistic view, as one can have several descriptive models and often one could choose more than one formulation. The final choice depends on the data available (if one is using the empirical approach) or on the appropriate theory (if one is using the theoretical approach). The final choice also involves a trade-off between model complexity and tractability for model analysis. The most important thing to remember is best summarized by the following quotes from two well-known statisticians:

> All models are wrong, but some are useful
> G.E.P. Box

> No model is correct. But some are useful.
> G. Kempthrone

8.5.2   Pitfalls in Modeling

A pitfall can be defined as a conceptual error into which, because of its specious plausibility, one can easily fall. There are several pitfalls that a model builder should be aware of, and we list a few of them below.[8]

- Failure to understand the problem properly. Models based on poor understanding have limited or no value.
- Forgetting that the modeling (Steps 1–4) is not an end in itself but is only a means toward solving real-world problems (Steps 1–7).
- Forgetting that the model is only a simplification of the real world associated with the problem under consideration.
- Failure to check the validity of assumptions and/or the incorrect use of mathematical techniques.
- Failure to interpret properly the results of the mathematical analysis in the physical terms of the problem.

8.6   Statistics versus Probability Perspectives

Data play a critical role in the modeling process. In the empirical approach, the execution of Steps 2–4 (of the modeling process) requires data, whereas in the theoretical approach, it is needed only for Steps 3 and 4. Appropriate and adequate data and other information must be

---

[7] See Appendix B for Markov and non-Markov formulations.
[8] These are from Chapter 13 of Murthy, Page, and Rodin (1990), where a more complete list and additional references can be found.

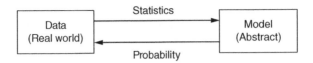

Figure 8.8   Probability and statistics.

available so that various candidate models can be built, evaluated, and compared. Additional data (i.e., data not used in the model building) are needed for validation.

When uncertainty is a significant feature of the real world, one needs concepts and formulations from probability theory (and the theory of stochastic processes) as well as data from the real world in the modeling process. Another discipline of relevance is the theory of statistics. We have tried to highlight their roles in the context of modeling. Probability and statistics are linked, and the relationships between them are as shown in Figure 8.8.

Attempting to understand and deal rationally with uncertainty or randomness is the basic underlying goal of both probability and statistics. These two related fields arose, in part, out of two quite different contexts – the analysis of games of chance and the analysis of experimental data – and the two look at randomness from essentially opposite perspectives.

In probability, the objective, simply stated, is to construct mathematical models of randomness. Given the model, one can then make statements about the nature of the data that may result in realizations of the random phenomenon. In statistics, the objective is to use the observed data to make meaningful statements about the nature of the probability model. Thus, probability and statistics are, in a very real sense, inverses of one another.

Chapter 9 deals with descriptive statistics where we do the preliminary analysis of data to decide on the mathematical formulation that is appropriate for modeling based on data from the real world.

8.7   Modeling of Maintenance Decision Problems

Decision making in the context of maintenance of a system (product, plant, or infrastructure) involves the modeling of the different elements of the system. The modeling of system degradation and failure, in terms of the degradation and failure of its various elements, is a challenging task. Models based on the black-box approach are much simpler than those based on the white-box approach.

For products and plants, modeling based on the black-box approach uses probability distribution functions to model first failure and stochastic point processes to model subsequent failures. Chapter 10 deals with the modeling of first failure. The subsequent failures depend on maintenance actions (see Chapter 4) and the modeling of these actions is discussed in Chapter 11. The results of Chapter 11 are used in Chapter 12 to model subsequent failures. Models for condition-based maintenance (CBM) use a variety of stochastic process formulations to model degradation, and a few of these are discussed in Chapter 16.

For infrastructures, the distributed elements are discretized into segments and the state of each segment is modeled by a discrete-state, discrete-time stochastic process formulation.

Modeling to decide on optimal PM actions is the focus of Part C, and here models from Chapter 12 are used in conjunction with optimization techniques (Appendix D) to derive optimal solutions.

In Part D, one needs more complex models for solving problems relating to spare part management, maintenance planning and scheduling, capital replacement, maintenance outsourcing, and so on.

8.8   Summary

Models play an important role in understanding and solving real-world problems. There are several different types of models (physical, mathematical, simulation, etc.). Many of these are used in solving maintenance-related problems. In this chapter the focus has been on mathematical modeling, which is the process of building a mathematical model that is adequate for solving a given problem. It involves linking a descriptive model to an abstract mathematical formulation.

A mathematical formulation is an abstract structure and mental construct. It involves variables, parameters, relationships, and operations (such as integration, differentiation, and so on) using symbols with precise mathematical meaning. The manipulation of symbols is dictated by the rules of logic and mathematics. The variables can be dependent or independent and they can be discrete or continuous. Mathematical formulations can be deterministic, dynamic, probabilistic, or stochastic.

There are several approaches for modeling maintenance problems. Two approaches are used in the selection of the mathematical formulation to model a system: (i) empirical and (ii) theoretical. In Chapter 2 we discussed how a product can be decomposed into several different levels. The modeling can be done at any level – from the lowest to the highest and anywhere in between. This includes component-level, system-level, and business-level modeling.

The mathematical modeling process involves several interconnected steps that include: problem definition, system characterization, model selection, parameter estimation, model validation, model analysis, problem solution, and solution implementation.

The complexity of the model depends, to some extent, on the mathematical formulation used in the modeling. One can choose more than one formulation. The final choice depends on data availability and involves a trade-off between model complexity and tractability for model analysis.

Data play a critical role in the modeling process. Appropriate and adequate data and other information must be available so that various candidate models can be built, evaluated, and compared. Concepts from probability and statistics are needed to understand and deal rationally with uncertainty.

Review Questions

8.1   What are the different types of models?

8.2   What is a descriptive model?

8.3   What is a mathematical model? How does it differ from a descriptive model?

8.4   Provide a definition for a mathematical formulation and a classification for it.

8.5   What is involved in probabilistic formulations?

8.6   What is involved in stochastic formulations?

8.7   Describe briefly the steps of the mathematical modeling process.

8.8   Give three pitfalls of modeling.

8.9   Explain the role of data in modeling and identify the disciplines needed for dealing with data.

8.10   Describe the complementary roles of probability and statistics in dealing with data.

Exercises

8.1   Carry out a characterization of the activities from the time a failed item arrives at a workshop to the time it leaves after being repaired.

8.2   XYZ is a maintenance service provider which maintains several elevators in a city. Carry out a characterization of the activities from the time the provider receives a call for service (to fix a failure of an elevator) to the time the service is completed.

8.3   A city council operates a fleet of buses to provide public transport for the people living in the city. Carry out a system characterization of the daily operation of the fleet.

8.4   The city council in Exercise 8.3 has a workshop to maintain the fleet. Buses arriving can be classified into the following three groups:
       (a) Regular PM maintenance.
       (b) Minor CM.
       (c) Major CM.
       Carry out a system characterization that is appropriate for planning the maintenance activities on a daily basis.

8.5   The cost of the fuel needed to operate the buses is a significant portion of the annual operating costs. Fuel efficiency (defined as km/l of fuel) is a variable of importance to the city council. This efficiency degrades with age and usage and is affected by several factors. Make a list of the different factors as part of building a descriptive model for the degradation of engine fuel efficiency. Discuss the different formulations that may be used to model the degradation in fuel efficiency.

8.6   The workshop in Exercise 8.4 needs spare parts for use in PM and CM. Carry out a system characterization needed to manage the inventory process.

8.7   The kind of model to describe changes in inventory level depends on whether the component is fast or slow moving. What kind of formulation is appropriate to model each of these cases?

8.8   The profits of the service provider XYZ of Exercise 8.2 have been decreasing for various reasons – for example, inflation (leading to higher wages, fuel costs, labor costs, etc.). The manager needs to decide on the pricing policy for new service contracts. Estimating the cost of servicing a call requires a characterization of the various costs involved. List these costs. There is variability in the costs. How would you characterize this?

8.9   Following on from Exercise 8.8, the manager needs to decide on the pricing of service contracts. Build a descriptive model for this task.

8.10   Hospitals use many different types of equipment (such as X-ray machines and MRI scanners to name a couple) for diagnostic purposes. The demand (defined in terms of number of times used on a daily basis) is uncertain. What type of formulation is appropriate for modeling this?

8.11   When a complex piece of diagnostic equipment used in a hospital fails, it requires an external service agent (certified by the OEM – original equipment manufacturer) to fix the problem. Build a descriptive model of this process.

Reference

Murthy, D.N.P., Page, N.W., and Rodin, I. (1990) *Mathematical Modelling*, Pergamon Press, Oxford.

9

Collection and Analysis of Maintenance Data

Learning Outcomes

After reading this chapter, you should be able to:

- Distinguish between data, information, and knowledge;
- Describe the different types of data;
- Explain the difference between structured and unstructured data;
- List the sources, types, and classification of maintenance data;
- Differentiate between failure and censored data;
- Discuss data-related problems;
- Describe the different technologies used for data collection, transmission, and processing;
- Conduct data analysis and provide numerical and graphical summaries of maintenance data using statistical and graphical methods;
- Explain the role of data in effective decision making and continuous improvement.

Introduction to Maintenance Engineering: Modeling, Optimization, and Management, First Edition.
Mohammed Ben-Daya, Uday Kumar, and D.N. Prabhakar Murthy.
© 2016 John Wiley & Sons, Ltd. Published 2016 by John Wiley & Sons, Ltd.

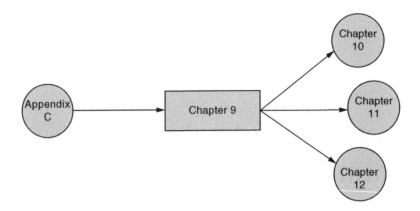

9.1   Introduction

Data are the source for generating the information and knowledge needed to make proper operational and strategic maintenance management decisions. Maintenance data comprise (i) data that are collected during the execution of preventive maintenance (PM) and corrective maintenance (CM) actions for an engineered object and (ii) various kinds of supplementary data (for example, the age of machines, suppliers, the make of spare parts, waiting time, travel time for repair and service crews, etc.). Data analysis is the process of extracting information from the data collected for this purpose and it can be either qualitative or quantitative. In this chapter, we deal with the collection and analysis of maintenance data.

The outline of the chapter is as follows. In Section 9.2 we discuss data, information, and knowledge, the types of data and some other aspects of data. Section 9.3 looks at maintenance data and some related topics as well as data sets that will be used as illustrative examples in this and later chapters. In Section 9.4 we look at data analysis. This involves two topics – descriptive and inferential statistics – and these are discussed in Sections 9.5 and 9.6, respectively. Sections 9.7–9.9 deal with maintenance data collection in the context of products, plants, and infrastructures, respectively. Finally, we conclude with a summary of the chapter in Section 9.10.

9.2   Data, Information, and Knowledge

Data represent a collection of realizations of a measurable quantity such as component failure time, component material properties, load on the component, and so on. *Information* is extracted from data through analysis to understand possible relationships (such as cause and effect) between pieces of data. Often, data and information are used either interchangeably as synonyms or with only slight differences. Data are raw facts that have not been organized or cannot possibly be interpreted. Information is data that are understood. *Knowledge* is the ability of humans to understand the information and how it can be used in a specific context (such as prediction). It includes theories, models, tools and techniques, technical standards, and so forth.

The link between data, information, and knowledge can be characterized through the DIKW (Data, Information, Knowledge, and Wisdom) hierarchy, a term attributed to (Ackoff, 1989)

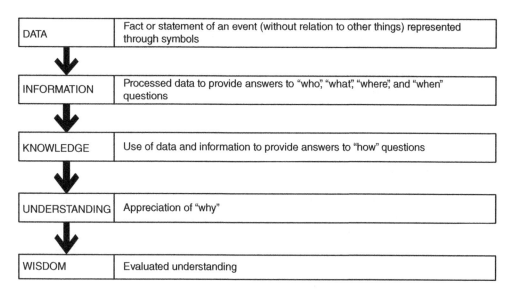

Figure 9.1 DIKW. [Adapted from Ackoff (1989)].

and shown in Figure 9.1. It indicates the role of information and knowledge in providing answers to different types of questions. The two other terms *understanding* and *wisdom* involve evaluation and a value-based framework.

9.2.1 Data Types

Data can be one of four different types and the categorization is based on *type* or *level of scales* used in the measurement and recording of data. The four scales are: nominal, ordinal, interval, and ratio. Nominal and ordinal data are inherently qualitative in nature, representing item attributes; interval and ratio data are inherently quantitative, representing amounts rather than types. Their definitions and allowable calculations on each scale are as follows:

1. Nominal scale
   * Data are categorical.
   * *Examples:* The failure modes of an engineered object, variables (such as male or female gender) in a study regarding customer satisfaction with an automobile service center, the brand name of the automobile serviced.
   * *Allowable operations:* Counts only; no ranking or numerical operations.
2. Ordinal scale
   * Data are categorical with a rank-order relationship.
   * *Examples:* Rating scales (severity of damage on a scale of 1–4; quality of maintenance service on a scale of 1–7).
   * *Allowable operations:* Counts and ranking; no numerical operations.

3. Interval scale
   - Data are numerical values on an equal-interval scale. (*Note*: There is no true zero.)
   - *Example:* Temperature.
   - *Allowable operations:* Ranking; addition and subtraction (and therefore averaging); multiplication and division are not meaningful.
4. Ratio scale
   - Data are numerical values on an equal-interval scale with a uniquely defined zero.
   - *Examples:* Time to failure of an item, cost of repair, number of replacements on a yearly basis.
   - *Allowable operations:* All ordinary numerical and mathematical operations.

The level of scale is separate from the distinction between discrete and continuous scales. Except for the nominal scale, which is always discrete, data on the remaining scales may be either discrete or continuous. For example, time to failure for an item is continuous data based on a ratio scale, whilst the number of failures in a sample of size n is discrete data based on a ratio scale.

9.2.2   Structured versus Unstructured Data

Structured data have a well-defined format that requires closed-ended answers – a choice from a finite set of choices. In contrast, unstructured data are usually in the form of a text with no specified set of choices. Two common sources of unstructured data in the context of maintenance of engineered objects are customer descriptions of problems and technicians' comments, as illustrated by the examples given below.

1. *Customer statements describing problems:*
   - The brakes are sluggish.
   - The air conditioner is not cooling adequately.
   - The pump is making some noise.
2. *Technicians' comments on fault identification:*
   - Brake pedal stiff.
   - Compressor in the air conditioner not functioning properly.
   - Bolt not tightened sufficiently.

For analysis (either qualitative or quantitative), it is necessary to convert unstructured data into structured data. This involves a natural language-processing technique called *named-entity extraction* (also referred to as *text tagging and annotation*).

9.2.3   Data Collection

Data collection is any process of preparing and collecting data and it is expensive. Each data item collected requires time, effort, and money to collect, store, retrieve, and use. The purpose of data collection is to obtain information to keep on record, to make decisions about important issues such as, for example, evaluating different maintenance strategies or trying to understand the degradation process.

9.3 Maintenance Data

9.3.1 Classification of Data

Maintenance data for an engineered object (product, plant, or infrastructure) can be grouped broadly into two categories:

1. Data collected during the servicing of (PM and CM) maintenance actions
   These data can be divided into two subcategories:
   - Data relating to the object (for example, condition at inspection);
   - Data relating to maintenance actions (material, labor, costs, etc.).
2. Supplementary data
   These are other data that are relevant for proper maintenance decision making and they can be grouped into several subcategories:
   - *Object-related data:* Detailed drawings, decomposition, failure, and censored data (discussed later in the chapter), and so on.
   - *Logistics-related data:* Spare parts, repair personnel and facilities, component suppliers, and so on.
   - *Production-related data:* Production is used in a general sense and refers to both products and services. Some examples: tonnage moved in the case of a rail network, output in the case of a production or processing plant, and so on.
   - *Business-related data:* Service contracts, maintenance costs, and so on.

For the proper collection of maintenance data, one needs a good understanding of the maintenance process. The maintenance process depends on whether the maintenance is done in-house or outsourced. There are several different scenarios depending on the engineered object (product, plant, or infrastructure) and the maintenance (done on site or at a service center; in-house or outsourced).

9.3.2 Data Sources

The data sources depend on the engineered object (product, plant, or infrastructure), as indicated below:

- *Products:* These can be consumer, industrial, commercial, or defense.
  - Original equipment manufacturer (OEM);
  - Customers (usage, operating environment, etc.);
  - Service centers (owned by the OEM, retailers, or independent companies);
  - Component suppliers.

- *Plants:* A plant consists of a collection of several elements and the data sources include:
  - The OEM;
  - The production department;
  - The maintenance department (for in-house maintenance);
  - External maintenance service agents (for outsourced maintenance);
  - Customers;
  - Component suppliers.

- *Infrastructures:* Infrastructures are more complex and involve many parties. As such, the data sources include:
  - ○ The owner;
  - ○ The builder;
  - ○ Users (for infrastructures like road, rail, etc.);
  - ○ The maintenance department (for in-house maintenance);
  - ○ External maintenance service agents (for outsourced maintenance);
  - ○ Regulatory authorities (directives and laws relating to safety, etc.).

9.3.3   Failure and Censored Data

Failure data are the ages (and possibly usage at failure) of the items that have failed over the data collection period.

Censored data are the data for items that were put into operation but did not fail over the data collection period. Censored data arise due to (i) items being replaced before failure (for example, due to PM actions), in which case the data refer to the age at replacement and (ii) items still being in operation at the end of the data collection period, in which case the data refer to the ages of the working items.[1] Figure 9.2 shows a time history of replacements under an age-based policy (defined in Chapter 4), showing the failure and censored data over a data collection period $[0, L)$.

Two other terms that will be used often in later chapters are the following:

- *Complete data:* A data set is said to be complete if it contains no censored data.
- *Incomplete data:* A data set is said to be incomplete if it contains both failure and censored data.

9.3.4   Ordered Data

Here and in the ensuing chapters, we assume that we have a sample of size n and denote the generic sample values by $Y_1, Y_2,\ldots, Y_n$ if the values are considered to be random variables, and by $y_1, y_2,\ldots,y_n$ if the values are observed values of the random variables. For certain types of

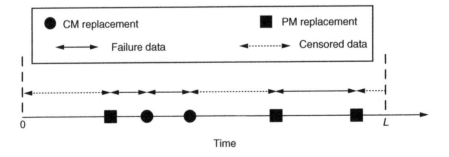

Figure 9.2   Time history of failure and censored data.

---

[1] Censored data of this type are also referred to as *service times.*

data analysis it is more convenient to order the observations from smallest to largest. The ordered set of observations will be denoted $y_{(1)}, y_{(2)}, \ldots, y_{(n)}$ with $y_{(1)} \leq y_{(2)} \leq \ldots \leq y_{(n)}$. Ordered data play an important role in the selection of distribution functions to model first failure and this is discussed further in the next chapter.

9.3.5 Data-Related Problems

A number of problems may be encountered in the collection of data. In the case of product maintenance, the most important of these are the following:

- Delays in reporting;
- Unreported data;
- Data reported incorrectly (resulting from fraud, accidental error, etc.);
- Incomplete data;
- Aggregated data (for example, the total number of failures over an interval instead of the failure times for each item);
- Pooled data (pooling of data for more than one item).

9.3.5.1 Loss of Information

Ideally, the data that should be collected in connection with each CM action are the following:

- Age of item at failure;
- Usage of item at failure (if relevant);
- Reason for failure – failure mode (component causing the failure, assembly errors, etc.);
- Usage mode, intensity, operating environment (unstructured data from customers);
- Symptoms prior to failure (unstructured data from customers);
- Actions taken to rectify the failure (replacement or repair of failed components, condition of other components, etc.);
- Repair data – part used, time to repair, and so on.

There is a loss of information when some of the data and information on the above list are not collected. There is also a loss of information with aggregation and pooling of data.

9.3.6 Illustrative Data Sets

Appendix E contains several data sets (mainly relating to failure and censored data associated with products) and these will be used later in this chapter and subsequent chapters for illustrative purposes.

9.4 Data Analysis

Consider a set of data where either time plays no role (for example, the cost of repair) or one is dealing with a single event (for example, the time to first failure). The analysis involves first deciding whether there is a significant variability in the data or not. One way of deciding this is to estimate the sample mean and sample standard deviation (these are discussed in the next section). If the sample standard deviation is very small compared to the sample mean, one can ignore the variability and the data can be viewed as deterministic. If not, then one needs to use a probabilistic formulation to model the data.

If the data have a time dimension (for example, monthly sales of service contracts or annual maintenance costs in a plant) then a plot of the data versus time can reveal certain patterns such as trend and seasonality. These can be removed from the data, leaving one with the random element of the data. If the sample mean of the data is very close to zero and the standard deviation is very small relative to the smallest value in the data set, then one can ignore the variability and model the data by a deterministic dynamic formulation (such as a difference or differential equation). If not, one needs to use stochastic formulations.

If the data relate to events over time (such as failures over time for one or several items), one plots an event plot (discussed in Section 9.5.2) to see if the variability is significant or not. One can also plot the time between failures and see if the data exhibit variability that is significant or not. If it is significant, one needs point process formulations to model the data.

As discussed in Chapter 2, failures occur in an uncertain manner. As such, failure data exhibit significant variability and we focus on the analysis of such data. The following two topics from the theory of statistics deal with such analysis:

- Descriptive statistics;
- Inferential statistics.

We discuss these topics briefly in the next two sections.

9.4.1   Statistical Packages for Data Analysis

One can either do the statistical analysis by hand (a rather tedious chore) or use computer packages. There are many statistical software packages that can be used for carrying out data analysis. They produce estimates of different summary statistics and plots for descriptive statistics and present the results for inferential statistics.

In this chapter and in the other chapters in Part B we use Minitab and present the computer outputs for several illustrative examples.

9.5   Descriptive Statistics

The first step in getting an insight into the data is to use a graphical presentation of the data and carry out some analysis to obtain summary statistics. These are referred to as *preliminary analysis of data* or *descriptive statistics*. They provide a foundation for building models for failure and maintenance and are discussed further in the next three chapters.

The techniques used in descriptive statistics can be grouped into the following two categories:

- Numerical (summary or sample) statistics;
- Graphical plots.

9.5.1 Numerical Statistics

One begins with a sample, that is, a set of observations (measurements, responses, etc.), and various calculations and operations are performed in order to focus and understand the information content of the sample data. The word "statistic" is used to refer to any quantity calculated from the data – average, range, percentile, and so forth. This section looks at a number of statistics that are intended to describe the sample and summarize the sample information.

The procedures in this section are appropriate for complete data. The data set consists of observed values $y = \{y_1, y_2, \ldots, y_n\}$ of the random variables $Y = \{Y_1, Y_2, \ldots, Y_n\}$ and we discuss very briefly the various types of statistics and expressions used to derive the sample statistics (more details are given in Appendix C).

9.5.1.1 Fractiles

The p-fractile of a sample is defined as that value y_p such that at least a proportion p of the sample lies at or below y_p and at least a proportion $1 - p$ lies at or above y_p. The expression for computing this statistic is given by Equation (C.1). The value may also not be unique and there are several alternative definitions that may be used. Of particular interest are the 0.25-, 0.50-, and 0.75-fractiles, called the *quartiles*, and denoted Q_1, Q_2, and Q_3.

9.5.1.2 Measures of Central Tendency

The most common measures of the center of a sample (also called measures of location, or simply averages) are the sample mean and median. The sample median is the 0.50-fractile or Q_2 and is a natural measure of the center, since at least one-half of the data lie at or above it and at least one-half lie at or below Q_2. The sample mean of Y, denoted $\overline{y}$, is the simple arithmetic average (given by Equation (C.2)).

An approach for dealing with the distortion caused by outliers (discussed later in the section) is to calculate a trimmed mean – remove a fixed proportion of both the smallest and largest values from the data and calculate the average of the remaining values.[2]

9.5.1.3 Measures of Dispersion

A second descriptive measure commonly used in statistical analysis is a measure of dispersion (or spread) of the data. These measures reflect the variability in the data and are important in understanding the data and in properly interpreting many statistical results. The most important measures of dispersion for most purposes are the sample variance s^2 (given by Equation (C.3)) and the sample standard deviation (s), which is the square root of the sample variance.

A few other measures are sometimes used. These include the mode, which is not of much use in statistical inference, and various measures that can be defined as functions of fractiles, for example $(Q_3 - Q_1)/2$, $(y_{0.90} - y_{0.10})/2$, and so forth. In analyzing failure and other data, we will use the mean and median as measures of center, and occasionally look at the trimmed mean[3] as well.

---

[2] Minitab removes the smallest and largest 5% (using the nearest integer to $0.05n$). This usually removes the values causing the distortion and provides a more meaningful measure.

[3] The trimmed mean is calculated by deleting a few of the very small and/or the very large values in the data set.

Another measure of variability sometimes used is the *interquartile range (IQR)*, *I*, given by $I = Q_3 - Q_1$. An advantage of the IQR is that it is not affected by extreme values. A disadvantage is that it is not readily interpretable, unlike the standard deviation.

Finally, a useful measure of dispersion in certain applications is the *coefficient of variation*, given by c.v. $= s / \bar{y}$.

9.5.1.4 Measures of Skewness

Although both are measures of center $\bar{y}$ and Q_2 measure this differently; a comparison of the two provides additional information about the sample (and, by inference, about the population from which it was drawn). If the sample is perfectly symmetrical about its center, the mean and median are identical. If the two differ, this is an indication of skewness. If $Q_2 < \bar{y}$, the data are skewed to the right; if $Q_2 > \bar{y}$, the data are skewed to the left. Failure data and the distributions used to model them are often skewed to the right, which results from (usually) small numbers of exceptionally long-lived items.

Example 9.1 Automotive Engine Repair Costs

Table E.2 gives kilometers driven and repair costs for $n = 32$ automotive engine failures. The ordered values for repair cost, say *y*, are given in Table 9.1. Thus, $y_{(1)} = 7.75$, and so forth.

We calculate the quartiles and $y_{0.05}$. For the first quartile, we have $k = [0.25(33)] = 8$, and $d = 0.25$, so from Equation (C.1) $Q_1 = \$34.68 + 0.25(42.71 - 34.68) = \$36.69$. Similarly, $Q_2 = \$88.66$, and $Q_3 = \$788.30$.

The sample mean (obtained using Equation (C.2)) is $\bar{y} = \$536.09$. Note the very large difference between the mean and the median, indicating skewness to the right. In fact, this is apparent in the data – there are several large values and one very large value. The trimmed mean, eliminating the largest and smallest observations (i.e., trimming about 5%), is \$371. This is still considerably larger than the median, indicating real skewness, beyond the influence of a few unusually large observations.

The variance (using Equation (C.3)) is 941 463; the standard deviation is the square root of this value, or \$970.30. These large values reflect the significant amount of variability in the data. Both are influenced by the large outliers and the interpretation given above is not valid in this case because of these and the overall skewness of the data. In fact, 28 observations (86%) lie within one standard deviation of the mean and 31 (97%) lie within two and three standard deviations. For these data, the IQR is $I = \$788.30 - 34.68 = \$753.62$.

The results obtained from Minitab (rounded to the nearest dollar) are given in Table 9.2. ∎

Table 9.1 Ordered repair costs for Data Set E.2.

Repair costs (ordered)							
7.75	11.70	24.60	26.35	27.58	27.78	29.91	34.68
42.71	42.96	48.05	60.35	77.22	77.24	77.57	78.42
98.90	127.20	149.36	253.50	388.30	432.89	556.93	658.36
831.61	918.53	1007.27	1101.90	1546.75	1638.73	1712.47	5037.28

Table 9.2 Minitab results rounded to the nearest dollar.

Variable	N	Mean	Median	TrMean	StDev
Repair cost	32	536	89	371	970
Variable	Minimum		Maximum	Q1	Q3
Repair cost	8		5037	37	788

9.5.1.5 Measures of Relationship

When the data include two or more variables, measures regarding the relationship between the variables are of interest. A measure of the strength of the relationship for two variables, the sample correlation coefficient r is introduced in Appendix C. r lies in the interval $[-1,1]$, with the values -1 and $+1$ indicating that the variables are co-linear, with lines sloping downward and upward, respectively. The general interpretation is that values close to either extreme indicate a strong relationship and values close to zero indicate very little relationship between the variables.

Example 9.2 Automotive Engine Repair Costs

For the Data Set E.2 the correlation between the repair cost (y) and kilometers driven (x) until the first engine failure (using Equation (C.4)) is found to be $r = 0.254$ and this is relatively small, indicating little relationship between the two variables. This is also seen from Figure 9.3, which is plot of (x_i, y_i), $i = 1,2,\ldots,32$. Note that the apparent outlier is quite evident in the plot. The weak relationship between kilometers driven and cost as measured by the correlation coefficient is reflected in the flat pattern of the data in the figure. ∎

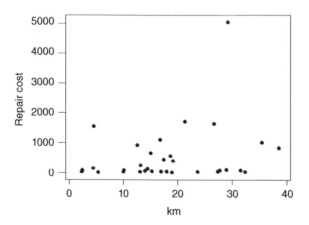

Figure 9.3 Plot of repair cost versus kilometers driven.

9.5.2   Graphical Plots

9.5.2.1   Pareto Charts

A Pareto chart is a graphical representation of qualitative or categorical data. The chart is formed simply by ranking categories by frequency of occurrence, from highest to lowest, and then forming a histogram-type graph, with frequencies as heights and category names as identifiers.

The usefulness of a Pareto chart is that it gives an immediate identification of the most important categories. In typical reliability and quality applications, the charts are used to determine defects, failure modes, and so forth, that are most in need of attention; that is, those whose rectification (for example, through engineering or production changes) would have the greatest impact on increasing overall reliability and quality and hence reducing warranty costs.

Example 9.3   Photocopier

The data given in Table E.3 represent failures of a photocopier due to the failure of 17 different components. A list of the components and their frequency of failure is given in Table 9.3.

Figure 9.4 is a Pareto chart of the failure modes of the copier as identified by the failed component. The chart includes a curve indicating the cumulative failure proportion. Note that there is not a predominant failure mode or even a few that account for a majority of the failures. The first five account for a total of only 49% of the failures. The conclusion is that significant improvements to many items are needed to reduce service costs.   ∎

Table 9.3   Frequency distribution of copier component failures.

Failed component	Frequency
Cleaning web	15
Toner filter	6
Feed rollers	11
Drum blade	2
Toner guide	7
Cleaning blade	7
Dust filter	6
Drum claws	5
Crum	6
Ozone filter	8
Upper fuser roller	5
Upper roller claws	5
TS block front	2
Charging wire	6
Lower roller	2
Optics PS felt	3
Drive gear D	2

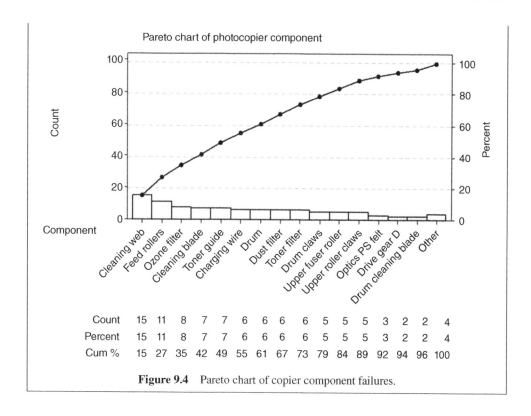

Count	15	11	8	7	7	6	6	6	6	5	5	5	3	2	2	4
Percent	15	11	8	7	7	6	6	6	6	5	5	5	3	2	2	4
Cum %	15	27	35	42	49	55	61	67	73	79	84	89	92	94	96	100

Figure 9.4   Pareto chart of copier component failures.

9.5.2.2   Histograms

The most commonly used graphical presentation of quantitative data is the *histogram*. To form a histogram, the observations are grouped into intervals (usually contiguous intervals of equal length) and counts are made of the number of observations falling into each interval. These counts are called *frequencies*, and the set of intervals and associated counts is a *frequency distribution*. A *histogram* is simply a plot of the frequency distribution. A second type of frequency distribution is a table of counts of occurrence of events, for example, failures of components or modes of failure of a system.

Example 9.4   Automotive Engine Repair Costs

A histogram of the repair costs of Data Set E.2 is given in Figure 9.5. Note that an extreme outlier is apparent in the histogram. Even without the outlier, the distribution of repair cost is skewed, as noted previously. The histogram also indicates that the majority of cases are clustered at the lower end of the scale, with most being below $500. There is, however, a case in the class centered on $5000. This high value for only a single case has a significant effect on the mean but little effect on the median, making the median a better indicator of central tendency for the repair cost data.   ■

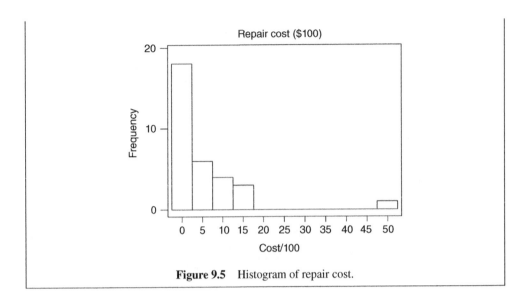

Figure 9.5   Histogram of repair cost.

Example 9.5   Photocopier

The data given in Table E.3 represent the failure data of a photocopier; time between failures can be measured by the number of copies made (column 1 of Table E.3 and referred to as *count*) as well as the days of service (column 2 of Table E.3 and referred to as *days*). We look first at the 14 times between failures of the cleaning web. The data are given in Table 9.4.

Table 9.4   Time in days and number of copies run between failures (copier cleaning web).

Days	Copies
99	71 927
269	232 996
166	61 981
159	74 494
194	96 189
100	78 102
95	40 795
245	183 726
56	33 423
36	4 315
66	56 497
69	51 296
26	22 231
31	9 413

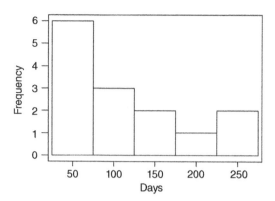

Figure 9.6   Histogram of days to failure of copier cleaning web.

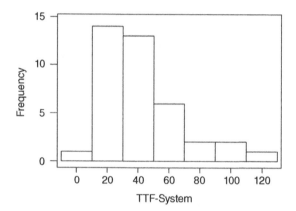

Figure 9.7   Histogram of days to failure of copier.

Figure 9.6 is a histogram of the data on days to web failure. Note that for such a relatively small sample, the histogram is not very informative. In fact, small samples in general provide limited information (which may be adequate for some purposes).

As a further illustration, a histogram of all 39 times to failure (service calls) of the copier system is given in Figure 9.7. Here, we see a bit more of a pattern. There are some early failures, a high point at about 20 days, nearly as high a frequency at about 40 days, and then progressively fewer instances of longer times to failure. This pattern of a skewed-to-the-right distribution is typical of the failure patterns of many items.

For the 14 times between failure of the copier cleaning web, the sample mean and median of times to web failure are 115.1 and 97.0 days, respectively. Note that the mean is well in excess of the median, suggesting the right-skewness shown in Figure 9.6.

For days to failure of the cleaner web, the range is 269 − 26 = 243. The sample standard deviation, obtained from Minitab, is 79.2. For the system, these values are 112 and 27.1, respectively. The system mean is 41.5, so we would expect that approximately 95% of the failures would occur between 0 and 41.5 + 2(27.1) = 95.7. In fact, 36 out of 39 or 92.3% of the inter-failure times were in this range. (We note that these were the first, third, and fourth observations in the sample, suggesting that perhaps the failure rate is increasing.) This is reasonably close to the expected 95%, given the relatively small sample size and the skewness of the frequency distribution.   ■

9.5.2.3   Detecting Outliers

Extreme observations, relatively large or relatively small values that appear to be far removed from the bulk of the data, are occasionally encountered in data analysis. In truly messy data, not uncommon in dealing with warranty claims, unusual observations may occur with disconcerting frequency. The extreme observations, called *outliers*, may occur for a number of reasons, including at least the following:

- Errors in measurement;
- Errors in recording, transcription, and so on;
- Valid measurements of items from a distribution having very long tails;
- Valid measurements of items from a different population.

Example 9.6   Automotive Engine Repair Costs

Figure 9.8 shows boxplots of the usage and repair costs for the data of Table E.2. For kilometers driven, the boxplot looks quite reasonable, although the median is below the center of the box. The boxplot for repair costs, however, shows a very skewed distribution, with the median near the lower edge of the box, the lower whisker so small that it does not appear in the plot, and the extreme outlier indicated by a "*". (Note: Minitab does not distinguish between mild and extreme outliers in its notation.)

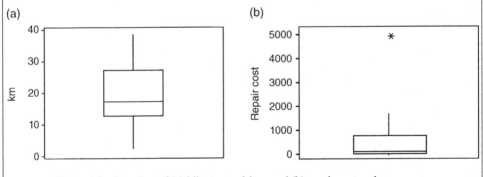

Figure 9.8   Boxplots of (a) kilometers driven and (b) repair cost under warranty.

For a better picture of the data, the extreme observation may be removed. A boxplot of the remaining data is given in Figure 9.9. Here, the skewness of the distribution remains obvious and two additional possible outliers are identified. ■

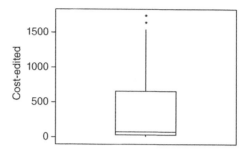

Figure 9.9 Boxplot of repair cost after removal of the maximal value.

Whatever the cause, outliers can significantly affect the outcome of tests and other statistical procedures.

A frequently used graphical device for identifying outlying observations is the boxplot (originally called a *box-and-whisker plot*). This is a plot in which the middle 50% of the data, that is, all observations between the quartiles Q_1 and Q_3, is represented by a rectangle (the "box") and the remaining data are indicated by lines outside the box (the "whiskers") or by points beyond the lines (the outliers). The length of the lines is calculated as 1.5 times the IQR, where IQR = $Q_3 - Q_1$. Observations between 1.5(IQR) and 3(IQR) are designated "mild" outliers; those beyond 3(IQR) are "extreme" outliers. The box also shows a horizontal line for the median, Q_2.

9.5.2.4 Event Plots

An event plot is a plot of events over time for an item. As such, it is a record of the time history for an item, as illustrated by the following example.

Example 9.7 Valve Seat Replacement Data for Diesel Engines

We consider the data set given in Table E.5. The data include 46 instances of valve replacement for several items. Figure 9.10 is an event plot of the data that was created by using Splida[4] (Meeker and Escobar, 1998). The event plot may also be created by Minitab. The plot shows when the failures occurred for each system. Each line extends to the last day of observation. Note that service times range from just under 400 days (System 409) to nearly 700 days (System 251). ■

---

[4] Splida (S-Plus Life Data Analysis), see www.public.iastate.edu/~splida.

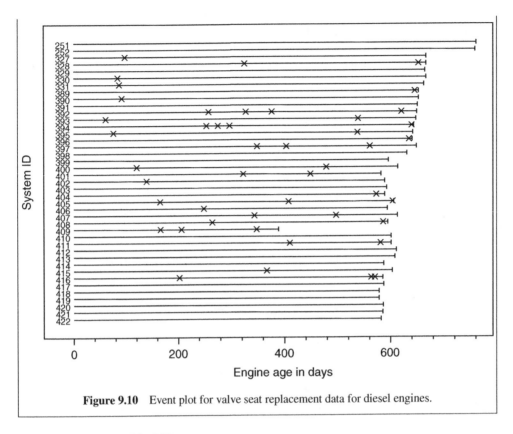

Figure 9.10   Event plot for valve seat replacement data for diesel engines.

9.5.2.5   Other Graphical Plots

Many other techniques are used to display both quantitative and qualitative data. These include the Q-Q plot, scatter plot, stemplot, control chart, and so on.[5] In the context of first failure, there are many plots to assist in model building – some non-parametric (for example, the empirical distribution function (EDF) plot) and others parametric (the Weibull probability plot (WPP)).[6]

9.6   Inferential Statistics

In statistics, statistical inference is the process of drawing conclusions from data subject to random variation. Statistical inference is generally distinguished from descriptive statistics. In simple terms, descriptive statistics can be thought of as being just a straightforward presentation of facts.

Any statistical inference requires some assumptions. A statistical model is a set of assumptions concerning the generation of the observed data and similar data.[7] Descriptions of statistical models usually emphasize the role of population quantities of interest, about which

[5] For more details, see Blischke and Murthy (2000).

[6] EDF and WPP are discussed in Chapter 10.

[7] Section 8.6 discussed such data and models.

one wishes to draw inference. Descriptive statistics are typically used as a preliminary step before more formal inferences are drawn.

The three levels of modeling assumptions are:

- *Fully parametric:* The process generating the data is assumed to be fully described by a family of formulations (probability distributions, stochastic formulations, etc.) involving only a finite number of unknown parameters.
- *Non-parametric:* The assumptions made about the process generating the data are far fewer than in parametric statistics and may be minimal.
- *Semi-parametric:* This term typically implies assumptions "between" the fully parametric and the non-parametric approaches. For example, one may assume that a population distribution has a finite mean and also depends in a linear manner on some covariate (a parametric assumption) but not make any parametric assumption describing the variance around that mean.

Whatever level of assumption is made, correctly calibrated inference, in general, requires these assumptions to be correct; in other words, that the data-generating process has been specified correctly.

In Chapters 10 and 12 we will look at the parametric approach to modeling first and subsequent failures, which involves the following three topics:

- Parameter estimation;
- Hypothesis testing;
- Goodness-of-fit tests.

A brief discussion of the basic concepts and expressions is given in Appendix C.

9.7   Collection of Maintenance Data for Products

There are two different scenarios:

Scenario 1

Here, the maintenance is outsourced and needs to be done on site by an external service agent. This requires the owner (customer) to report a failure (to the maintenance service provider) and the maintenance technician to travel to the site to carry out the CM. The object can be:

- A consumer product (a washing machine, dishwasher, refrigerator, etc.);
- A commercial product (an elevator in a building);
- Components of a plant (such as pumps, boilers, turbines).

Table 9.5 lists the sequence of events, activities, and types of maintenance data that can be collected.

Table 9.5 Maintenance service process for maintenance on site.

Events	Activities	Typical maintenance data
Customer reports failure	Logging the request for service	Unstructured description of symptoms
Dispatch of service technician	Manuals, spares, tools to be taken	Time to travel to site
Start of diagnosis	Testing	No fault detected
	Trouble shooting	Fault detected
		Cause of failure
		Age (usage) of failed component
		Action initiated
Ordering spares (if not in store)	Order spares	Cost of spares ordered
		Time to delivery
Start of repair	Repair	Time to travel to site
		Components replaced
		Time for repair
		Further action regarding disposal of failed components
Invoicing	Check if the object is covered by service contract or not	Costing of parts used
	Amount to be billed to customer (depending on the service contract)	Costing of labor
		Other overheads
Follow-up action with customer	—	Customer satisfaction
		Need for further action

Comments:

- Some of the events and actions might not be relevant in some cases.
- The last column includes some supplementary data.
- More detailed data would include the technical details of parts ordered, suppliers of parts, and so on.

Scenario 2

Here, the owner (customer) needs to bring the failed object to a service center to get it fixed. This is appropriate for consumer products where the owner does not have the expertise to fix the failure and the failed item is brought to a service center. The maintenance process is very similar to the earlier case, with some slight differences, and it is shown in Table 9.6.

Table 9.6 Maintenance service process for maintenance at a service center.

Events	Activities	Typical maintenance data
Customer brings failed item	Logging the request for service	Unstructured description of symptoms
Start of diagnosis	Testing	No fault detected
	Trouble shooting	Fault detected
		Cause of failure
		Age (usage) of failed component
		Action initiated
Ordering spares (if not in store)	Order spares	Cost of spares ordered
		Time to delivery
Start of repair	Repair	Components replaced
		Time for repair
		Further action regarding disposal of failed components
Invoicing	Amount to be billed to customer	Costing of parts used
		Costing of labor
		Other overheads
Follow-up action with customer	—	Customer satisfaction
		Need for further action

9.8   Collection of Maintenance Data for Plants

Various types of data are needed for effective maintenance of plants. The data can be grouped broadly into the following categories:

- *Technical data:* These relate to the plant, its decomposition down to component level, failure modes, and so on.
- *Operational data:* These relate to the process data such as the production rate (which depends on demand), efficiency, resource (such as water, electricity) consumption, and so on.
- *Maintenance servicing data:* These relate to inspection, PM, and CM actions carried out. When some or all of the maintenance actions are carried out by external agents, these data are collected by the service agents.
- *Cost data:* These relate to materials, labor, administration, and so on.
- *Other data:* These relate to regulatory requirements in terms of safety, protection of the environment, and so on.

9.8.1   *Dragline*[8]

The PM actions are discussed in Example 4.7. Here, we focus on the data collection and analysis at a mine site.

---

[8] This subsection is based largely on material from Townson (2002).

9.8.1.1   Data Collection

Each maintenance job is recorded on a database and has the following information:

- A job number;
- A job title;
- The date the job was raised;
- The start time for the job;
- The end time for the job;
- The elapsed time for the job;
- The usage hours of the machine;
- The component group assigned by the mine based on its dragline characterization chart;
- The action carried out (for example, inspection, repair, replacement, or modification);
- Job instructions;
- Job comments.

9.8.1.2   Recording of Maintenance Work

Maintenance work is carried out at the component level, but the level at which the maintenance action is recorded is usually higher. Different subsystems are recorded at different levels. Component/system codes for the maintenance system do not go down to the level at which work is undertaken. The comments made by maintenance personnel sometimes contain necessary information. Also, at the lower levels, no distinction is made between different components such as:

- Brushes for motors and generators;
- Seals on gear cases;
- Bolts on gear cases or on the swing rack;
- The left or right propel-lube components.

Data are only recorded in the database for PM maintenance when maintenance work is carried out.

Various job codes are used for recording. A sample of job code descriptions for "Physical Cause" is given below:

BB: Bent/Broken/Damage/Crack	NF: No Fault
CL: Calibration	NP: No Physical Cause (Preventive)
CN: Contamination	OC: Over Current
CR: Corrosion/Cavitation	OR: Open/Short Circuit
FB: Fusion/Burnt/Overheated	OV: Over/Under Voltage
FO: Flash Over	RP: Rupture
LS: Loose/Slipping/Tracking	SZ: Seized/Jammed/Blocked
ML: Misalignment/Vibration/Noise	WN: Worn Out/Worn

A sample of code descriptions for "Job Results" is given below:

AD: Adjusted/Calibrated/Charged	RE: Relocated
AR: Add/Remove cable	RO: Rotation
CA: Canceled	RP: Repaired
FM: Fabricated/Machined	RT: Reset
IN: Inspected/Tested	SB: Serviced – Maintenance and Lubrication
IS: Isolated/Disconnect/Reconnect	SL: Serviced – Lubrication
MD: Modified/New Installation	SM: Serviced – Maintenance
RC: Replaced/PEX	

9.8.1.3   Operational (Process) Data

These are stored in another database and include the following:

- Cycle time;
- Swing angle;
- Net and gross bucket load;
- Dig mode;
- Event code;
- Reasons for downtime periods;
- Beginning and end times of a cycle;
- Beginning and end times of operational modes.

9.8.1.4   Data Analysis

The analysis of data depends on the type of management decision problem. At the business level, the analysis would involve computing availability on a periodic basis (daily, weekly, monthly, and yearly), production volumes, and so on. Also, total costs would be a variable of interest to the top management. At lower levels, the analysis would be used to evaluate the performance of the main- tenance plan, of the external service agents involved, and so on. Also, the data may be used for design modification (design out maintenance) and improving the maintenance operations.

9.8.2   Ore-Dressing Plant

Ore, mineral, or coal taken from a mine needs to be processed. For example, in a typical iron ore mine, the ore, after crushing, is transported to an ore-dressing plant consisting of screens, secondary crushers, tertiary crushers, classifiers, cyclones, a filter, a thickener, and so on. The ore is processed to separate it from waste and sort it into the required sizes and quality to meet cus- tomers' requirements on its way to becoming a different material via proper concentration of the ore. The maintenance data collection practices in a typical iron-ore-dressing plant are as follows:

- *Process data:* In general, ore-dressing plants are identical to process plants. Any change in the physical state of the item will affect the process parameters, such as flow rate, energy consumption, water consumption, production rate, and so on. All of these parameters are monitored continuously.

- *Failure/stoppage reporting:* Whenever there is a plant stoppage due to failure, it is reported at the system level (crusher), the subsystem level (crusher mantle), or the component level (gearboxes, etc.).
- *Potential failure reporting:* Often, the plant operators-cum-mechanics, maintenance or safety inspectors report such failures. These engineers meet at the end of a shift and record their findings regarding the state of an item in the equipment logbook, to be followed by engineers on the next shift. Such reports are detailed in terms of the state of the item. If possible, an estimated degradation rate or even an estimated number of hours of remaining useful life is reported by the operators.
- *Condition monitoring data:* Modern ore-dressing plants are fitted with many sensors and measurement devices to monitor the performance and assess the health of components continuously.
- *Periodic inspection reports by maintenance engineers:* Periodic inspection reports are the results of visual inspection and/or handheld instruments and are done on a periodic basis (weekly or bi-weekly depending on the criticality of items based on past experience or the recommendation of suppliers). Such data are compared with condition-monitoring data, especially when safety-related issues are involved, and instructions are issued to the maintenance planning department for scheduling of maintenance actions/tasks as a priority on such equipment.
- *Periodic safety inspection reports by safety engineers:* Inspection reports concerning the maintenance of safety-instrumented systems or safety-critical equipment also form important feedback to maintenance planners.

Apart from condition-monitoring data stored in a computerized maintenance management system (CMMS), the maintenance department continuously monitors the consumption of spare parts and consumables. The pattern of consumption triggers alarms and indicates a need for additional focus on items and consumables consumed in high numbers or quantities.

- *Repair/service history reports:* Such reports give details of repairs performed, resources used versus planned, the true state of the item recommended for replacement, or additional work/service/repair performed due to the detection of additional faults.

The age of equipment, hours run, maintenance history, and contractors used also form part of the maintenance data. The failure data collected for individual items often provide the basis for monitoring of the reliability and availability of items.

9.9   Collection of Maintenance Data for Infrastructures

9.9.1   Road Infrastructure

A road infrastructure is a network of pavements (distributed elements) with several other lumped (discrete) elements such as bridges, traffic signals, signposts, and so on. One can define three levels of data, as indicated below:

- *Network-level data:* These are needed for general planning, programming, and policy decisions. The data relate to growth of volume in traffic and this, in turn, depends on growth of the population, urban development, transport of goods, and so on.

- *Project-level data:* These assist in making maintenance decisions for a selected section of road. As the data are collected, they can be stored to create a more complete database over time. However, a method must be established to keep the data current.
- *Research-level data:* These relate to specific attributes of an element, for example, degradation of a concrete bridge, the effect of de-icing on the road characteristics, and so on.

A report by the World Bank (Bennet *et al.*, 2007) proposed five levels of road management with different information quality levels (IQLs), as indicated below:

- *IQL-1* represents fundamental, research, laboratory, theoretical, or electronic data types, where numerous attributes may be measured or identified.
- *IQL-2* represents a level of detail typical of many engineering analyses for a project-level decision.
- *IQL-3* is a simpler level of detail, typically two or three attributes, which might be used for a large production like a network-level survey or where simpler data-collection methods are appropriate.
- *IQL-4* is a summary or a key attribute which has use in planning, senior management reports, or in low-effort data collection.
- *IQL-5* represents top-level data such as key performance indicators (discussed in Chapter 17), which typically might combine key attributes from several pieces of information. Still higher levels can be defined as necessary.

At IQL-1, pavement conditions are described by twenty or more attributes. At IQL-2, these would be reduced to six to ten attributes, one or two for each mode of distress. At IQL-3, the number of attributes is reduced to two to three, namely roughness, surface distress, and texture or skid resistance.[9] At IQL-4, all of the lower-level attributes may be condensed into one attribute, "Pavement Condition" (or "state" or "quality"), which may be measured by class values (good, fair, poor) or by an index (for example, 0–10). An IQL-5 indicator would combine pavement quality with other measures such as structural adequacy, safety aspects, and traffic congestion – representing higher-order information, such as "road condition." ■

9.9.2 Rail Infrastructure

Railway track is a complex system consisting of components, such as rails, sleepers, ballast, sub-grade, fasteners, turnouts, and so on. Often, maintenance data for railway track comprise measurement data about the track geometry, traffic type, traffic density, axle load, the age of rails, defect history, rail material, curvature, yearly and total accumulated million gross tons (MGT), and so on. Data collected may be at the component level (rails, sleepers), the sub-system level (turnouts), or the system level (track), depending on the functionality and criti-

---

[9] Roughness: Deviations of a surface from the true planar surface with characteristic dimensions that affect vehicle dynamics, ride quality, dynamic loads, and drainage, expressed by the International Roughness Index, IRI (m/km).
Texture depth: The average depth of the surface of a road expressed as the quotient of a given volume of standardized material (sand) and the area of that material spread in a circular patch on the surface being tested.
Skid resistance: Resistance to skidding expressed by the sideways force coefficient (SDF) at 50 km/h measured using the Sideways Force Coefficient Routine Investigation Machine (SCRIM).

cality of the item. In Sweden, the collected data and information are stored in the centralized databases, BIS, BESSY, and Ofelia.

- *BIS:* This is Trafikverket's infrastructure register (computerized database) for the railway system, containing information about infrastructure and facilities, and arranged geographically in accordance with Trafikverket's facility structure. In BIS, for example, information is collected prior to work on train timetables and before work in connection with inspections. Apart from this, information about agreements, accident reports, the history of tamping and grinding, and curve information can also be obtained.
- *BESSY:* This is an inspection system in which comments are registered for each facility on completion of an inspection. Data are also registered directly during the course of an inspection with the aid of a palm computer.
- *Ofelia:* This is a database containing information on all the faults in the infrastructure that have been registered for a particular component. The faults are sorted on the basis of the structure used in BIS. Here, faults are classified into three types:
  - o *Type A:* This type of fault needs to be corrected immediately and the action taken is noted in the BESSY system. These types of faults are safety related.
  - o *Type B:* This type of fault should be rectified and recorded in the BESSY system within a period of two weeks.
  - o *Type C:* This class of fault should be rectified in the near future at the convenience of infrastructure managers. These are non-safety- and non-functionality-related failures, often influencing the lifespan of the track.

Apart from the data collection and its verification within the above-mentioned databases, information may also be collected through discussions and consultations with outside experts.

All data recorded from the different systems are further analyzed by an expert. Historical data and information stored in a centralized database are also used to correlate failure patterns. Finally, a decision is made to prioritize these defects. Priority of defects is based on several factors but is mainly governed by functionality, performance, and safety issues. The consequential costs and risks associated with a particular defect are also taken into consideration, bearing in mind derailment scenarios due to such defects.

9.10 Summary

Data are the source for generating the information and knowledge needed to make proper operational and strategic maintenance management decisions. Maintenance data comprise: (i) data that are collected during the execution of PM and CM actions for an engineered object and (ii) various kinds of supplementary data. Data analysis is the process of extracting information from the data collected for this purpose and can be either qualitative or quantitative.

Data are raw facts that have not been organized or cannot possibly be interpreted. Information is data that are understood. *Knowledge* is the ability of humans to understand the information and how it can be used in a specific context.

Data can be one of four different types and the categorization is based on the *type* or *level of scales* used in the measurement and recording of data. The four scales are: nominal, ordinal, interval, and ratio. Data can be structured or unstructured. Structured data have a well-defined format that requires closed-ended answers whereas unstructured data are usually in the form of a text with no specified set of choices.

Failure data exhibit significant variability and we have focused on the analysis of such data. Data analysis provides a foundation for building models for failure and maintenance. Descriptive and inferential statistical methods can be used for this purpose.

Descriptive statistics provide numerical and graphical summaries of data. Numerical summaries include fractiles, measures of central tendency, measures of dispersion, measures of skewness, and measures of relationship. Graphical summaries include Pareto charts, histograms, and event plots.

Inferential statistics is the process of drawing conclusions from data subject to random variation and requires some assumptions. There are three levels of modeling assumptions:

- *Fully parametric:* The process generating the data is assumed to be fully described by a family of formulations involving only a finite number of unknown parameters.
- *Non-parametric:* The assumptions made about the process generating the data are far fewer than in parametric statistics and may be minimal.
- *Semi-parametric:* This term typically implies assumptions "between" the fully parametric and non-parametric approaches.

The collection of maintenance data for various engineered objects was discussed in this chapter.

Review Questions

9.1 What is the difference between data, information, and knowledge?

9.2 What are the two main categories of maintenance data?

9.3 What are the different types of data based on type or level of scales used? Give an example of each type of data.

9.4 What is the difference between structured and unstructured data? Provide examples.

9.5 What are the main sources of maintenance data?

9.6 What is the difference between censored and uncensored failure data?

9.7 Describe some problems encountered in maintenance data collection.

9.8 What are the main statistical methods used in maintenance data analysis?

9.9 List some numerical methods for summarizing data.

9.10 List some graphical methods that can be used to summarize data.

9.11 Explain the importance of data for effective decision making for different engineered objects.

Exercises

9.1 Following up on Exercise 2.2, what type of data needs to be collected about the diesel engine for effective maintenance?

9.2 Following up on Exercise 2.3, what type of data needs to be collected about the pump motor unit for effective maintenance?

9.3 Following up on Exercise 2.4, what type of data needs to be collected about the wind turbine for effective maintenance?

9.4 Following up on Exercise 2.5, what type of data needs to be collected about the rail track for effective maintenance?

9.5 The bus workshop in Exercise 8.4 has been buying batteries from two different battery manufacturers (A and B). The life times (in months) for 25 batteries bought from manufacturer A are given below:

2.06	1.04	4.33	6.45	2.28
26.16	1.02	9.29	1.21	13.27
1.37	7.12	6.47	1.26	4.94
7.34	5.78	3.17	3.62	15.23
8.39	2.42	16.71	2.09	5.08

Carry out an analysis of the data using Minitab to obtain the summary statistics (descriptive statistics). Interpret the results of the analysis.

The life times (in months) for 25 batteries bought from manufacturer B are given below:

1.33	4.09	2.22	13.61	0.06
7.83	6.04	6.34	7.24	8.54
0.16	1.47	5.82	6.85	3.47
3.41	2.99	11.62	5.34	0.99
0.51	12.01	14.93	11.94	0.33

Carry out an analysis of the data using Minitab to obtain the summary statistics (descriptive statistics). Compare the results of the analysis for the two sets of batteries. What inferences can you draw – for example, are the batteries from the two manufacturers similar (in a statistical sense)?

9.6 The data given below are the failure times (in days) of 50 (non-repairable) components subjected to accelerated testing until all the items failed.

0.73	1.44	3.39	5.50	3.26	4.03	8.08	5.91	6.12	1.73
3.18	1.96	5.16	2.85	6.96	2.65	7.06	3.15	8.25	7.02
5.88	1.68	4.14	7.87	3.29	4.34	1.09	3.53	2.96	8.09
5.20	3.17	2.37	2.12	5.74	2.67	4.17	3.79	4.34	7.31
5.33	5.87	1.55	3.06	3.71	7.68	6.84	1.72	8.34	2.29

Carry out an analysis of the data using Minitab to obtain the summary statistics (descriptive statistics). Interpret the results of the analysis.

9.7   You are working for a small consulting firm that specializes in helping small mainte-
      nance service businesses improve their operations. You have been asked by the owner of
      a small computer repair facility to help to improve the operations. What data should the
      owner collect relating to customers, processes, and so on?

9.8   Carry out a preliminary analysis of the data set in Table E.4. Interpret the results of the
      analysis.

9.9   Carry out a preliminary analysis of the data set in Table E.6. Interpret the results of the
      analysis.

References

Ackoff, R.L. (1989) From data to wisdom, Journal of Applied System Analysis, **16**: 3–9.

Bennet, C.R., Chamorro, A., Chen, C., de Solminihac, H., and Flintsch, G.W. (2007) *Data Collection Technologies for Road Management, Version 2.0*, East Asia Pacific Transport Unit, The World Bank, Washington, DC.

Blischke, W.R. and Murthy, D.N.P. (2000) *Reliability*, John Wiley & Sons, Inc., New York.

Meeker, W.Q. and Escobar, L.A. (1998) *Statistical Methods for Reliability Data*, John Wiley & Sons, Inc., New York.

Townson, P.G.A. (2002) *Load-maintenance interaction: Modelling and optimization*. Unpublished doctoral thesis. The University of Queensland, Brisbane, Australia.

10

Modeling First Failure

Learning Outcomes

After reading this chapter, you should be able to:

- Distinguish one-dimensional modeling of first failure from two-dimensional modeling;
- Provide a classification of the types of distribution functions that may be used as mathematical formulations to model the time to first failure;
- Identify properties of distribution functions that are relevant in the context of modeling;
- Explain the role of transformed plots in the modeling of first failure;
- Determine expressions of the transformed plots of: (i) exponential; (ii) Weibull; and (iii) lognormal distributions;
- Explain the role of the empirical distribution function (EDF) in the modeling of first failure;
- Describe methods for calculating the EDF for complete and incomplete data;
- Describe methods for the selection of an appropriate mathematical formulation (distribution function);
- Use graphical and statistical methods for parameter estimation;
- Describe various validation methods and explain the importance of model validation.

Introduction to Maintenance Engineering: Modeling, Optimization, and Management, First Edition.
Mohammed Ben-Daya, Uday Kumar, and D.N. Prabhakar Murthy.
© 2016 John Wiley & Sons, Ltd. Published 2016 by John Wiley & Sons, Ltd.

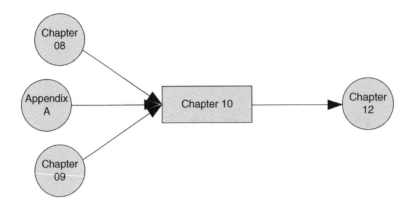

10.1   Introduction

As discussed in Chapter 8, modeling involves the linking of the descriptive model (of the real world relevant to the problem under consideration) with an appropriate mathematical formulation. The most important issue in modeling is that the model formulation must mimic the real world. In this chapter we focus on modeling the time to first failure of an item using the empirical (or data-based) approach. We consider the following two cases:

- *Case (i):* The data consist of the failure/service times of several items.
- *Case (ii):* The data consist of the failure/service times and usage levels.

Since variability is a significant factor in first failure, the empirical approach to modeling requires concepts and techniques from probability theory and the theory of statistics.[1] In Case (i) the mathematical formulations appropriate for modeling are one-dimensional distribution functions, and in Case (ii) they can be either one- or two-dimensional distribution functions.

The outline of the chapter is as follows. Section 10.2 deals with one-dimensional formulations and the focus is on one-dimensional distribution functions appropriate for modeling the age at first failure. Section 10.3 deals with two-dimensional formulations and presents three different approaches to modeling age and usage at first failure. Section 10.4 looks at some properties of distribution functions – in particular the various plots and their characterization. Section 10.5 deals with the data for modeling, and the focus is on data analysis and various plots. We give the formulae for doing the calculations and illustrate by hand calculations for some of the data. This is followed by a plot for the complete data using the Minitab software package. Section 10.6 deals with model selection – deciding whether a distribution function is suitable or not for modeling the given data set. This involves comparing the theoretical plots of Section 10.4 with the empirical plots of Section 10.5. Section 10.7 deals with the estimation of model parameters and Section 10.8 looks at model validation. In Section 10.9 we illustrate

---

[1] A brief introduction to these topics is given in Appendices A and C, respectively.

the complete process (model selection, parameter estimation, and validation) by looking at three examples involving data from the real world. We conclude with a summary of the chapter in Section 10.10.

10.2   One-Dimensional Formulations

The time to first failure is uncertain and, as a result, one-dimensional distribution functions with random variable T have been used as mathematical formulations for modeling. When the formulation is viewed as a model, T denotes the time to first failure.

10.2.1   Basic Concepts

The failure distribution function $F(t; \theta)$ characterizes the probability that the time to failure T (a random variable) is less than or equal to t and is given by Equation (2.1). The dependence on the set of model parameters is being suppressed temporarily in what follows. The *failure density* function, denoted $f(t)$ is given by Equation (2.2). The *survivor* (*reliability*) function $\bar{F}(t) = 1 - F(t)$ (sometimes denoted by $R(t)$),[2] is given by Equation (2.3). The *hazard function* (also often referred to as the *failure rate*) $h(t)$ is given by Equation (2.7).

The *cumulative hazard function, $H(t)$*, is defined as:

$$H(t) = \int_0^t h(\tau)d\tau \tag{10.1}$$

$H(t)$ is also referred to as the *cumulative failure rate function*.
It is easily shown that:

$$F(t) = 1 - e^{-H(t)}, f(t) = h(t)e^{-H(t)}, \quad \text{and} \quad H(t) = -\log(1 - F(t)). \tag{10.2}$$

The last expression follows from integrating Equation (2.7), the first expression is obtained from this by exponentiation (and rearrangement) and finally, the middle expression follows from differentiating the first expression.

10.2.2   Classification of Distribution Functions

There are many different distribution functions that may be used as mathematical formulations to model the time to first failure. They can be categorized broadly into the following three groups:

- Standard distributions;
- Derived distributions;
- Complex distributions.

---

[2] We will use both notations throughout the book.

The first two are appropriate when the data available for modeling are failure data and censored data and nothing else. The third is suitable when there is additional information such as quality variations, more than one failure mode, and so on.

10.2.2.1 Standard Distributions

Some of the standard distribution functions that have been used in modeling failure times are the following:[3]

- *Exponential distribution:* The distribution function is given by:

$$F(t;\theta) = 1 - e^{-\lambda t}, \quad 0 \le t < \infty. \tag{10.3}$$

The parameter set is $\theta = \{\lambda\}$ with $\lambda > 0$.

- *Two-parameter Weibull distribution:* The distribution function is given by:

$$F(t;\theta) = 1 - e^{-(t/\alpha)^{\beta}}, \quad 0 \le t < \infty. \tag{10.4}$$

The parameter set is $\theta = \{\alpha, \beta\}$ with $\alpha > 0$ and $\beta > 0$. α is called the scale parameter and β the shape parameter.

- *Normal distribution:* The density function is given by:

$$f(t;\theta) = \frac{e^{-(t-\mu)^2 / 2\sigma^2}}{\sigma\sqrt{2\pi}}, \quad -\infty < t < \infty. \tag{10.5}$$

The parameter set is $\theta = \{\mu, \sigma\}$ with $\sigma > 0$ and $-\infty < \mu < \infty$.

Comment: Note that here, $F(t) > 0$ for $t < 0$. If $\sigma \ll \mu$ then $F(0) \approx 0$ and the normal distribution function can be used to model failure time.

10.2.2.2 Derived Distributions

Let $G(t;\theta)$ denote the derived distribution. T is a random variable from a standard distribution $F(t;\theta)$ and Z is a random variable from $G(t;\theta)$. We have two cases:

- **Case (1):** Z and T are related by a transformation. The transformation may be either (a) linear or (b) nonlinear. Two examples of this are as follows:

  Three-parameter Weibull distribution: $Z = T + \gamma$ with $\gamma > 0$ and $F(t;\theta)$ is the two-parameter Weibull given by Equation (10.4). As a result

$$G(t;\theta) = 1 - e^{-\{(t-\gamma)/\alpha\}^{\beta}}, \quad \gamma \le t < \infty. \tag{10.6}$$

---

[3] There are several other distributions that have been used in reliability modeling. Some of these are given in Appendix A. Johnson and Kotz (1970a, b) are good sources from which to obtain more information about these distributions.

The parameter set is $\theta = \{\alpha, \beta, \gamma\}$ with $\alpha > 0$, $\beta > 0$, and $\gamma \geq 0$. γ is called the *location parameter*. Note that when $\gamma = 0$ the distribution reduces to the two-parameter Weibull.

Lognormal distribution: $Z = e^T$ and $F(t; \theta)$ is the normal distribution function with density function given by Equation (10.5). As a result, the density function is given by:

$$g(t;\theta) = \frac{e^{-\left\{(\log(t)-\mu)^2 / 2\sigma^2\right\}}}{\sigma t \sqrt{2\pi}}, \quad 0 \leq t < \infty. \tag{10.7}$$

The parameter set is $\theta = \{\mu, \sigma\}$ with $\sigma > 0$ and $-\infty < \mu < \infty$.

- **Case (2):** Here, $G(t; \theta)$ is related to $F(t; \theta)$ by some relationship. An example of this is the following:

  Exponentiated Weibull distribution: This is given by $G(t) = [F(t)]^\nu$ where $F(t)$ is the two-parameter Weibull distribution (given by Equation (10.4)) and ν (>0) is a parameter. As a result:

$$G(t) = \left\{1 - \exp\left[-(t/\alpha)^\beta\right]\right\}^\nu, \quad 0 \leq t < \infty. \tag{10.8}$$

The parameter set is $\theta = \{\alpha, \beta, \nu\}$ with $\alpha > 0$, $\beta > 0$ and $\nu > 0$. Note that when $\nu = 1$ the distribution reduces to the standard two-parameter Weibull distribution given by Equation (10.4).

10.2.2.3 Complex Distribution Functions

These involve two or more standard and/or derived distributions. Let $F_i(t)$, $1 \leq i \leq k$, $k \geq 2$, denote k standard distributions and $F(t)$ the complex distribution. We confine our discussion to the following three situations.

Competing Risks
The distribution function is given by:

$$F(t) = 1 - \prod_{i=1}^{k} \{1 - F_i(t)\} \tag{10.9}$$

A real-world application of this is the modeling of item failure with k different failure modes. Let T_i (an independent random variable) denote the time to failure if mode i was the only mode of failure with distribution function $F_i(t)$, $1 \leq i \leq k$. When all failure modes are present, the time to failure is given by $T = \min_i \{T_1, T_2, \cdots, T_k\}$. Then the distribution function for T is given by Equation (10.9).

Competing Risk Weibull ($k = 2$) is a special case of Equation (10.9) with $F_i(t)$ being the two-parameter Weibull distribution given by Equation (10.4).

Multiplicative
The distribution function is given by:

$$F(t) = \prod_{i=1}^{k} F_i(t) \qquad (10.10)$$

A real-world application of this is as follows. The item is an assembly of k identical components connected in parallel and the item is working if at least one of the components is working. Let T_i (an independent random variable) denote the failure time of component i with distribution function $F_i(t), 1 \leq i \leq k$. The item failure time is given by $T = \max_i \{T_1, T_2, \ldots, T_k\}$ and the distribution function for T is given by Equation (10.10).

Multiplicative Weibull ($k = 2$) is a special case of Equation (10.10) with $F_i(t)$ being the two-parameter Weibull distribution given by Equation (10.4).

Mixture Formulation

$$F(t) = \sum_{i=1}^{k} p_i F_i(t) \qquad (10.11)$$

where $0 < p_i < 1, 1 \leq i \leq k$, and $\sum_{i=1}^{k} p_i = 1$. The p_i's are called the *mixing parameters*.

A real-world application of this is when items are from a pool of items manufactured by k different component manufacturers. Items from manufacturer i have a failure distribution function $F_i(t), 1 \leq i \leq k$. If items manufactured by different manufacturers cannot be differentiated, then the failure time of an item is given by a mixture of these distributions where p_i is the fraction of items supplied by manufacturer $i, 1 \leq i \leq k$.

A *Weibull Mixture* ($k = 2$) is a special case of Equation (10.11) with $F_i(t)$ being the two-parameter Weibull distribution given by Equation (10.4).

10.3   Two-Dimensional Formulations

For many items the failure is a function of both the age (T) and usage (U) at failure and these are random variables. The notion of usage depends on the item:

- *Automobile:* The distance traveled until the first failure;
- *Photocopier:* The number of copies made until the first failure;
- *Machine tool:* The number of components machined until the first failure.

In this case, one can define the *usage rate* ($Z = U / T$) as the output per unit time until failure. For an automobile, this could represent the distance traveled per week, month, year, and so on.

The time to first failure is a random point in a two-dimensional plane, with age and usage being the two coordinates. The data available for modeling can be either complete or incomplete. In the case of complete data, the age and usage at first failure for all n items are known. In the case of incomplete data, for failed items we have the age and usage and for non-failed items we might or might not know the service time and/or usage.

There are three different approaches to modeling such data:

- *Approach 1:* The time to first failure is modeled by a bivariate distribution function $F(t, u)$. The density and hazard functions associated with this are given by $f(t, u)$ and $h(t, u)$ respectively.[4]
- *Approach 2:* In this approach the two scales, usage u and time t, are combined to define a composite scale v and the time to first failure is modeled by a distribution function $F_v(v)$.
- *Approach 3:* This approach assumes a constant usage rate for an item with the rate varying from item to item. The usage rate is modeled as a random variable Z with distribution function $G(z) = P\{Z \leq z\}$ and density function $g(z)$.[5] The time to first failure, conditional on the usage rate, is given by a conditional failure distribution function $F(t \mid z)$.

We will confine our discussion to Approach 3, as this is the one that has been used extensively in modeling failures involving age and usage. Normally, products are designed for some nominal usage rate, z_0. As the usage rate increases (decreases), the rate of degradation increases (decreases) and this, in turn, accelerates (decelerates) the time to failure. As a result, the reliability decreases (increases) as the usage rate increases (decreases).

Let $F_0(t)$ denote the baseline failure distribution function when the usage rate is the nominal value z_0. Conditional on the usage rate, the time to first failure is modeled by a survivor function:

$$F(t \mid z) = F_0(t \tilde{z}^\gamma) \tag{10.12}$$

where $\tilde{z} = z / z_0$ and $\gamma > 1$.[6] It is easily shown that for $z < z_0$, $F(t \mid z) < F_0(t)$ and for $z > z_0$, $F(t \mid z) > F_0(t)$ for all t.

One can use the available usage rate data to obtain the density function, $g(z)$. An alternative approach is the following. Let u_{max} denote the upper limit for the usage rate. The interval $[0, u_{max})$ is divided into K non-overlapping groups with a usage rate (based on the mean) for each group, as illustrated by the example given below.

Example 10.1[7]   Automobile Component

The data relate to age and usage at first failure of an automobile component for 498 items. For the 498 customers, usage rates $(z_i, 1 \leq i \leq 498)$ were first computed with units of $10\,000$ km per month. The usage rate for one item was excluded from the analysis as it was exceptionally high compared to other items and so was assumed to be an outlier.

---

[4] For more on two-dimensional distributions, see Johnson and Kotz (1972) and Hutchinson and Lai (1990).
[5] For notational ease, we omit the parameters of the distribution functions.
[6] This is also commonly referred to as the *accelerated failure time*(AFT) model in the reliability literature.
[7] This example is based on Example 14.1 from Blischke, Karim, and Murthy (2011).

The usage rates (minimum = 0.012, maximum = 2.10) were divided into three groups as follows:

- *Group 1 (Low usage rate):* $0 \leq z_i < 0.75$;
- *Group 2 (Medium usage rate):* $0.75 \leq z_i < 1.50$;
- *Group 3 (High usage rate):* $1.50 \leq z_i < 2.25$.

The numbers of customers in Groups 1, 2, and 3 were 396, 91, and 10, respectively. The modeling of the data is discussed in Example 10.8. ∎

The modeling involves the following two steps:

- *Step 1:* Deciding on the distribution for the usage rate $G(z)$.
- *Step 2:* Deciding on the distribution function $F_0(t)$.

If the usage rate is divided into several groups then one estimates the failure distribution for each. Then the parameters of the distribution function are obtained by modeling each group separately and then using regression analysis to obtain the parameters of $F_0(t)$ as functions of the usage rate. We illustrate this in Example 10.7. ∎

10.4   Properties of Distribution Functions

There are several properties of distribution functions that are very relevant in the context of modeling. These include the shapes of the density and hazard functions and the transformed plots of distribution functions.

10.4.1   Shapes of Density and Hazard Functions

The shapes of the density and hazard functions depend on the parameter values. Table 10.1 lists the possible shapes for the distributions discussed earlier. These play a role in the selection of a distribution function to model failure times.

Figure 10.1 shows the plots of the density and hazard functions for the Weibull distribution for a range of β values.

10.4.2   Transformed Plots

The relationship between $F(t)$ and t is highly nonlinear. One can obtain a linear relationship by transforming $t \rightarrow x$ and $F(t) \rightarrow y$ so that y is a linear function of x. The transformation depends on the failure distribution function. The plot of y versus x is called a *transformed plot*.[8]

---

[8] In the early 1970s, special papers were developed for plotting data under different transformations. The plotting paper was referred to as exponential plotting paper (EPP) and the plot was called the EPP plot in the case of the exponential distribution; WPP and WPP in the case of the Weibull distribution, and so on. At present, most computer reliability software packages and many statistical program packages contain programs to produce these plots automatically, but the plots continue to be called by their earlier names – EPP, WPP, and so on.

Table 10.1 Shapes of density and hazard functions.

Distribution	Shapes of density function	Shapes of hazard function
Exponential	Decreasing	Constant
Two-parameter Weibull	Decreasing, unimodal	Increasing, decreasing, constant
Normal	—	—
Three-parameter Weibull	Decreasing, unimodal	Increasing, decreasing, constant
Exponentiated Weibull	Decreasing, unimodal	Decreasing, increasing, unimodal, bathtub
Lognormal	Unimodal	Increasing, unimodal
Weibull competing risk $(k = 2)$	Decreasing, unimodal, decreasing followed by unimodal, bimodal	Decreasing, increasing, unimodal followed by increasing, decreasing followed by unimodal, bimodal followed by increasing
Weibull multiplicative $(k = 2)$	Decreasing, bimodal	Decreasing, increasing, decreasing followed by unimodal, bimodal
Weibull mixture $(k = 2)$	Decreasing, unimodal, decreasing followed by modal, bimodal	Decreasing, increasing, bathtub

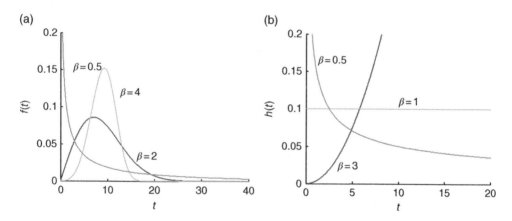

Figure 10.1 (a) Density and (b) hazard function plots for the Weibull distribution.

10.4.2.1 Exponential Probability Plot

Under the transformation

$$y = \ln\left(1 / \left(1 - F(t)\right)\right) \quad \text{and} \quad x = t \tag{10.13}$$

the exponential distribution function (given by Equation (10.3)) is transformed into

$$y = \lambda x \tag{10.14}$$

This is an equation for a straight line with slope λ which passes through the origin. A plot of y versus x is called the *theoretical EPP plot*.

10.4.2.2   Weibull Probability Plot (WPP)

Two-Parameter Weibull Distribution
The two-parameter Weibull distribution function is given by Equation (10.4). Under the transformation

$$y = \ln\left(-\ln\left(1 - F(t)\right)\right) \quad \text{and} \quad x = \ln(t) \tag{10.15}$$

the distribution is transformed into

$$y = \beta\left[x - \ln(\alpha)\right] \tag{10.16}$$

This is an equation for a straight line with slope β (the shape parameter). The intercept with the y axis is $-\beta \ln(\alpha)$ and with the x axis the intercept is $\ln(\alpha)$. A plot of y versus x is called the *theoretical* Weibull probability plot (WPP).

The Weibull transformation transforms the two-parameter Weibull into a linear relationship between y and x, as indicated above. When this transformation is applied to other derived or complex distributions involving the two-parameter Weibull distribution, this results in a non-linear relationship between y and x, as will be discussed later.[9] These plots are called WPPs and play an important role in model selection. We look at such plots for a few distributions.

Exponentiated Weibull
The distribution function is given by Equation (10.8). Under the Weibull transformation (given by Equation (10.15)) the distribution is transformed into:

$$y = w(x) = \ln\left\{-\ln\left[1 - \exp\left(-\left(\exp(x)/\alpha\right)^{\beta}\right)\right]^{\nu}\right\} \tag{10.17}$$

Note that the relationship is not linear. The plot of y versus x is concave for $\nu > 1$ and convex for $\nu < 1$, as indicated in Figure 10.2.

Competing Risk Weibull (k = 2)
The distribution function is given by Equation (A.40). Under the Weibull transformation (given by Equation (10.15)) the distribution is transformed into a nonlinear curve which is convex, as shown in Figure 10.3a.

Multiplicative Weibull (k = 2)
The distribution function is given by Equation (A.45). Under the Weibull transformation (given by Equation (10.15)) the distribution is transformed into a nonlinear curve which is concave, as shown in Figure 10.3b.

---

[9] Further details of the WPPs for these and many other derived and complex distributions involving the two-parameter Weibull distribution can be found in Murthy, Xie, and Jiang (2004).

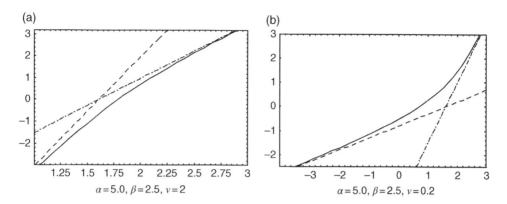

Figure 10.2 WPPs for the exponentiated Weibull distribution (a) $\alpha = 5.0$, $\beta = 2.5$, $v = 2$ and (b) $\alpha = 5.0$, $\beta = 2.5$, $v = 0.2$.

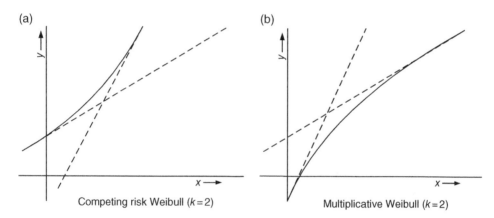

Figure 10.3 WPP for (a) the competing risk Weibull ($k = 2$) and (b) the multiplicative Weibull ($k = 2$) models.

Weibull Mixture (k = 2)

The distribution function is given by Equation (A.48). Under the Weibull transformation (given by Equation (10.15)) the distribution is transformed into a nonlinear curve. The shape of the plot depends on the shape parameters of the two subpopulations $-\beta_1$ and β_2. Figure 10.4a and 10.4b show the shapes for the two cases.

We have only discussed a few distribution functions and the shapes of the plots under the Weibull transformation. Murthy, Xie, and Jiang (2004) give a detailed characterization of the plots shown in Figures 10.2–10.4 as well as the plots for many distributions derived from the two-parameter Weibull. An understanding of shapes plays an important role in modeling, as will be indicated in a later section.

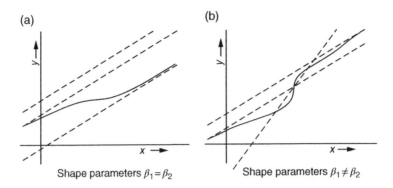

Figure 10.4 WPP for a Weibull mixture model ($k = 2$). (a) Shape parameters $\beta_1 = \beta_2$ and (b) shape parameters $\beta_1 \neq \beta_2$.

10.4.2.3 Lognormal Probability Plot

The lognormal density function $f(t; \theta)$ is given by Equation (10.7). Under the transformation

$$y = \Phi^{-1}(p) \quad \text{and} \quad x = \log(t_p) \tag{10.18}$$

where the function $\Phi^{-1}(\cdot)$ is the inverse of the standard normal distribution function and t_p is the p-fractile (obtained from the relationship $p = F(t_p)$), the distribution gets transformed into:

$$x = \mu + \sigma y \tag{10.19}$$

This is an equation of a straight line with slope $1/\sigma$.

10.5   Preliminary Data Analysis and Plots

The available data consist of failure and service times of n identical items, and times to failure are independent and identically distributed.[10] We have two scenarios: complete data and incomplete data.

- *Complete data:* The failure times (t_i, $i = 1, 2, \ldots, n$) are known for all items. In other words, there are no censored data (service times) and the set has only uncensored data (failure times).
- *Incomplete data:* Here, the failure time for some items is known and for the remaining items the information available is their service times. In this case, the data can be divided into two subsets. Let D denote the set of items which have failed ($T_i = t_i$) and C denote the set of

---

[10] A simple nonparametric test of randomness or independence is the run test. A run is a set of consecutive observations that are all either less than or greater than a specified value. The test statistic is based on the count of the number of runs of like symbols (positive or negative deviations). The run test is available in some software packages including Minitab.

items that have not failed ($T_i > t_i^*$). The data available are the uncensored data (t_i, $i \in$ D) and the censored data (t_i^*, $i \in$ C). As a result, the available data for item i are either t_i (if the item has failed) or t_i^* (if the item has been in operation for a period t_i^* and has not yet failed). Let n_1 [n_2] denote the number of data in set D [C] with $n_1 + n_2 = n$.

Comment: $n_2 = 0$ implies complete data and $n_2 \geq 1$ implies incomplete data.

Given a set of data (t_i, $i = 1, 2, \ldots, n$) we can reorder the set so that the values are in ascending order and this new (ordered) set is denoted by($t_{(i)}$, $i = 1, 2, \ldots, n$). In the case of complete data, one can use descriptive statistics to compute various sample statistics and various plots (for example, a histogram), as discussed in Chapter 8.

In most cases the data are incomplete. In this case, two plots that are very useful in the modeling of first failure are the *empirical distribution function* (EDF) and transformed plots based on the EDF.

10.5.1   Empirical Distribution Function

The EDF, $\hat{F}(t)$, is the sample equivalent of the failure distribution function. The calculation of $\hat{F}(t)$ for complete data is different from that for the incomplete case. Section C.2.6 of Appendix C gives the procedures to compute the EDF. In the case of complete data, it is given by Equation (C.5) and is fairly straightforward. When the data are incomplete, it is a bit more complicated and is given by Equations (C.6) and (C.7). Note that in the latter case, the EDF has jumps only at failure (or uncensored) times.

Most statistical software packages have programs to calculate the EDF. In the case of Minitab, the output includes the details of the computation of the EDF (Kaplan–Meier estimates) as well as a graphical plot of the EDF, and this is illustrated by the following example.

Example 10.2   Battery

Incomplete failure data on a sample of 54 batteries are given in Table E.1. The data include failure times for 39 items that failed and service times (censored data) for 15 items that had not failed (right censored).

The ordered array, including both failures and censored observations (marked with a "*") is as follows: 64, 66, 131*, 162*, 163*, 164, 178, 185, 202*, 232*, 245*, 286*, 299, 302*, 315*, 319, 337*, 383, 385, 405, 482, 492, 506, 548, 589, 599, 619, 631, 639, 645, 656, 681, 722, 727, 728, 761, 765, 788, 801, 845*, 848, 852, 929, 948, 973, 977, 983*, 1084, 1100, 1100, 1259*, 1350, 1384*, 1421*.

The y_i''s are the set of non-marked values in the array (64, 66, 164, 178, etc.). Here, $n = 54$ and $m = 39$. The numbers at risk are $n_1 = 54$, $n_2 = 53$, $n_3 = 49$, $n_4 = 48$, and so on. The d_i are all 1 except for the case of the two values tied at 1100, for which we have $n_{38} = 6$ and $d_{38} = 2$. Thus, $S_1 = 1 - 1/54 = 0.9815$, $S_2 = 0.9815(1 - 1/53) = 0.9630$, $S_3 = 0.9630(1 - 1/49) = 0.9433$, and so on, with the EDF being the complement of these values. The EDF is given by one minus the survivor function.

Figure 10.5 shows the EDF plot using Minitab. The Minitab output is given in Table 10.2. It contains some descriptive statistics such as an estimate of the mean time to failure

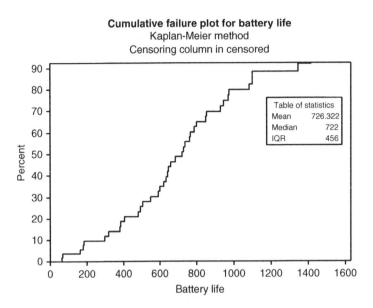

Figure 10.5   EDF of battery life.

(MTTF) along with a confidence interval and some other statistics. These are derived from the EDF. (Since the data are incomplete, the sample mean cannot be obtained from the data directly.) The Kaplan–Meier estimates for the survival probability $(1 - \hat{F}(t))$ for t_i, $i \in \mathrm{D}$ (the uncensored data subset) along with the confidence limits are also given.   ∎

10.5.2   Transformed Plots Based on the EDF

In the previous section we discussed the transformations to produce the theoretical EPP and WPPs. One can use these same transformations to produce empirical EPP and WPPs using the EDF. Since the EDF is a staircase function, the plots are dots and fitting them by a smooth curve yields the empirical plot.

10.5.2.1   Complete Data

EPP plot:
The procedure for plotting the empirical EPP is as follows:

1. Compute $y_i = \ln\left(1 / \left(1 - \hat{F}\left(t_{(i)}\right)\right)\right)$ for $1 \le i \le n$.

2. Compute $x_i = \ln\left(t_{(i)}\right)$ for $1 \le i \le n$.

3. Plot y_i versus x_i for $1 \le i \le n$.

A smooth curve to fit the plotted data yields the *empirical EPP plot.*

Table 10.2 Minitab output for Kaplan–Meier estimation of the EDF.

Nonparametric estimates

Characteristics of variable (mean)

Mean (MTTF)	Standard error	95.0% Normal CI	
		Lower	Upper
726.3222	51.7679	624.8590	827.7855

Characteristics of variable (Fractiles)

Median	IQR	Q1	Q3
722.0000	456.0000	492.0000	948.0000

Kaplan–Meier estimates

Time	Number at risk	Number failed	Survival probability	Standard error	95% Normal CI	
					Lower	Upper
64	54	1	0.9815	0.0183	0.9455	1.0000
66	53	1	0.9630	0.0257	0.9126	1.0000
164	49	1	0.9433	0.0318	0.8810	1.0000
178	48	1	0.9237	0.0367	0.8517	0.9956
185	47	1	0.9040	0.0409	0.8239	0.9841
299	42	1	0.8825	0.0452	0.7939	0.9711
319	39	1	0.8599	0.0494	0.7631	0.9566
383	37	1	0.8366	0.0532	0.7323	0.9410
385	36	1	0.8134	0.0566	0.7024	0.9243
405	35	1	0.7901	0.0596	0.6734	0.9069
482	34	1	0.7669	0.0622	0.6450	0.8888
492	33	1	0.7437	0.0645	0.6172	0.8701
506	32	1	0.7204	0.0665	0.5900	0.8508
548	31	1	0.6972	0.0683	0.5633	0.8311
589	30	1	0.6739	0.0699	0.5370	0.8109
599	29	1	0.6507	0.0712	0.5111	0.7903
619	28	1	0.6275	0.0724	0.4856	0.7693
631	27	1	0.6042	0.0733	0.4605	0.7480
639	26	1	0.5810	0.0741	0.4357	0.7262
645	25	1	0.5577	0.0747	0.4113	0.7042
656	24	1	0.5345	0.0751	0.3873	0.6817
681	23	1	0.5113	0.0754	0.3636	0.6590
722	22	1	0.4880	0.0754	0.3402	0.6359
727	21	1	0.4648	0.0753	0.3171	0.6124
738	20	1	0.4415	0.0751	0.2944	0.5887
761	19	1	0.4183	0.0746	0.2720	0.5646
765	18	1	0.3951	0.0740	0.2500	0.5401
788	17	1	0.3718	0.0732	0.2283	0.5153

(Continued)

Table 10.2 (*Continued*)

Time	Number at risk	Number failed	Survival probability	Standard error	95% Normal CI Lower	Upper
801	16	1	0.3486	0.0722	0.2070	0.4902
848	14	1	0.3237	0.0712	0.1841	0.4633
852	13	1	0.2988	0.0700	0.1616	0.4359
929	12	1	0.2739	0.0684	0.1398	0.4080
948	11	1	0.2490	0.0666	0.1185	0.3795
973	10	1	0.2241	0.0644	0.0978	0.3503
977	9	1	0.1992	0.0619	0.0779	0.3205
1084	7	1	0.1707	0.0592	0.0547	0.2868
1100	6	2	0.1138	0.0514	0.0131	0.2145
1350	3	1	0.0759	0.0462	0.0000	0.1664

WPP plot:

The procedure for plotting the empirical WPP is as follows:

1. Compute $y_i = \ln\left(-\log\left(1 - \widehat{F}\left(t_{(i)}\right)\right)\right)$ for $1 \le i \le n$.

2. Compute $x_i = \ln\left(t_{(i)}\right)$ for $1 \le i \le n$.

3. Plot y_i versus x_i for $1 \le i \le n$.

10.5.2.2 Incomplete Data

The procedure is the same as for the complete data but the plotting is only done for each failure (uncensored data), as the EDF is calculated only for these points.

Example 10.3 Battery

The EDF for the battery lifetime data of Table E.1 is given in Figure 10.5. The transformed plots obtained from Minitab for the exponential and the Weibull transformation (along with two other transformations – lognormal and log-logistic) are shown in Figure 10.6.[11] ■

---

[11] Minitab gives the plots for 11 different distributions. They are two- and three-parameter exponential, two- and three-parameter Weibull, normal, lognormal, three-parameter lognormal, logistic, log-logistic, three-parameter log-logistic and smallest extreme value. Details of some of the distributions can be found in either Section 10.4.1 or Appendix A.

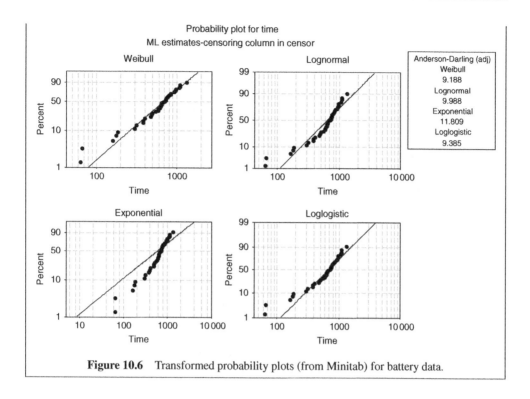

Figure 10.6   Transformed probability plots (from Minitab) for battery data.

10.6   Selection of a Mathematical Formulation

The selection of a mathematical formulation to model a given data set requires knowledge of the theoretical plots of different formulations (requiring concepts and techniques from probability theory) and generating empirical plots (requiring concepts and techniques from the theory of statistics). Figure 10.7 shows the model selection process and how it involves comparing the theoretical and empirical plots. Formulations whose theoretical plots do not match the empirical plots must be rejected as not being suitable for modeling the data set. When a match looks reasonable, the formulation may be accepted (or more correctly, not rejected) for further study to determine whether it yields an adequate model or not.

The most commonly used plots are the transformed plots. They can be done by hand using special plotting paper – a tedious process. Minitab produces the transformed plots for 11 different distributions along with straight-line fits generated using the maximum likelihood estimates (MLEs) of the parameters of the distributions.[12] A visual inspection will reveal if the plotted data are scattered close to the fitted straight line or not. If the fit looks reasonable, then the distribution under consideration can be accepted as a possible formulation to model the data set. If the fit is poor, then one needs to look at more complex distributions.[13]

---

[12] The method of maximum likelihood is discussed in the next section.
[13] This is discussed further in a later section using the WPP and complex distribution functions involving two Weibull distributions.

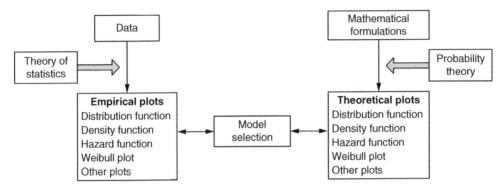

Figure 10.7   Model selection process.

Example 10.4   Battery

This is a continuation of Example 10.3 – the data consist of failure and service times for 54 batteries. The probability plots for the data for four distributions are given in Figure 10.6. The distributions are: exponential, Weibull, lognormal, and log-logistic. The plots were generated by use of the Minitab "Distribution ID" option. The first of the probability plots is the WPP. This plot is nearly linear, except for the first five points, indicating that the Weibull distribution itself may be a reasonable choice for the failure distribution. The first five points, in fact, appear to be out of the pattern of the remaining data in most of the plots, and may either be outliers or observations from a distribution different from that of the remaining data (possibly resulting from a different failure mode).

As noted at the outset, we have not reached a definitive conclusion here with regard to the probability distribution for modeling time to failure. More data are needed (and, if possible, more information, for example on failure modes) in order to make the selection. For purposes of further analysis, one may tentatively select one of the distributions that fits reasonably well, at least visually (for example, the Weibull) and carry out further analysis (discussed in the next two sections) to decide whether to accept or reject the distribution as a model. In the latter case, one needs to look at more complex distributions to obtain one which is adequate.   ■

10.7   Parameter Estimation

The model parameters are the parameters of the distribution (see Section 10.2.1). Many different estimation methods have been developed for estimating model parameters. They can be grouped broadly into two categories:

• Graphical methods;
• Statistical methods.

With graphical methods, the estimates are obtained from plotting the data. The plot depends on the model selected and hence each needs to be treated separately. The main drawback of these methods is that there is no well-developed statistical theory for determining the small sample or asymptotic properties. However, they are useful in providing an initial estimate for statistical methods of estimation.

The statistical methods, in contrast, are more general and applicable to all kinds of models and data types. The asymptotic properties of the estimators are well understood. We first discuss some of the common statistical methods for point estimation and then briefly discuss interval estimation.

10.7.1 Graphical Methods

Here, the estimates are obtained by a straight-line fit to the transformed plots based on the EDF and depend on the distribution involved.

Exponential distribution:

1. Determine the best straight line (passing through the origin) to fit the empirical EPP plot.
2. The slope of this line yields $\hat{\lambda}$.

Weibull distribution:

1. Determine the best straight-line fit using regression or the least squares method to fit the empirical WPP.
2. The slope of this line yields $\hat{\beta}$, the estimate of β.
3. Compute y_0, the y-intercept of the fitted line. $\hat{\alpha}$, the estimate of α, is given by $\hat{\alpha} = \exp\left\{-y_0 / \hat{\beta}\right\}$.

Example 10.5   Shock Absorber

We now look at the shock absorber data given in Table E.10. These are comprised of 38 values with 11 pieces of failure (uncensored) data and 27 pieces of censored data. The EDF is estimated using a procedure similar to Example 10.3 where the distribution function has jumps at times corresponding to uncensored data and is given in Table 10.3. Also shown in the table are the empirical WPP points corresponding to the 11 failure data values.

The WPP is shown in Figure 10.8. Also shown is the straight line fit to the data and, as can be seen, the two-parameter Weibull distribution is an appropriate model formulation to model shock absorber failure.

The parameters can be obtained by fitting a straight line $y = a + bx$ which minimizes the sum of errors squared (in other words, a straight-line regression). The estimates are $\hat{a} = -25.79$ and $\hat{b} = 2.51$. From this we obtain the estimates of the two-parameter Weibull as follows: $\hat{\beta} = \hat{b} = 2.51$ and $\hat{\alpha} = \exp(\hat{a} / \hat{\beta}) = 28996$. ∎

Table 10.3   EDF and WPP for shock absorber data.

	t	F(t)	Ln t	Ln Ln (1/(1 − F(t)))
1	6700	0.0263	8.8099	−3.6249
5	9120	0.0550	9.1182	−2.8723
13	12 200	0.0913	9.4092	−2.3461
15	13 150	0.1292	9.4842	−1.9780
19	14 300	0.1727	9.5680	−1.6629
20	17 520	0.2162	9.7711	−1.4122
27	20 100	0.2816	9.9085	−1.1065
31	20 900	0.3714	9.9475	−0.7673
32	22 700	0.4612	10.0301	−0.4806
34	26 510	0.5689	10.1853	−0.1727
36	27 490	0.7126	10.2216	0.2206

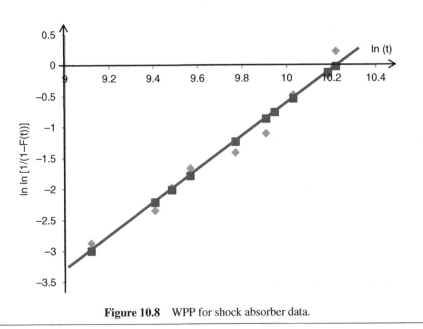

Figure 10.8   WPP for shock absorber data.

10.7.2   Method of Maximum Likelihood

The method of maximum likelihood (ML) involves the *likelihood function,* which is basically the joint distribution of the data, expressed as a function of the parameters of the distribution and the data. The idea is that the likelihood function, in a sense, represents the joint probability of the data and we are choosing parameter values that maximize the probability of obtaining the sample actually observed. (This is often called the *likelihood principle.*) The likelihood function is different for complete and incomplete data.

10.7.2.1   Complete Data

The likelihood function is given by Equation (C.10).

- *Exponential distribution:*
  Here, there is only one parameter ($\theta \equiv \lambda$) and the likelihood function (using Equation (10.3) in Equation (C.10)) is given by:

$$L(\lambda) = \lambda^n \prod_{i=1}^{n} e^{-\lambda t_i} = \lambda^n e^{-\lambda n \bar{t}} \tag{10.20}$$

where $\bar{t} \left(= \sum_{i=1}^{n} t_i \right)$ is the sample mean. The ML estimate, $\hat{\lambda}$, is the value which maximizes this likelihood function and can be obtained from the first order condition (setting the derivative $(dL(\lambda)/d\lambda)$ to zero).

- *Weibull distribution:*
  Here, there are two parameters ($\theta \equiv \alpha, \beta$). The likelihood function (using Equation (10.4) in Equation (C.10)) is given by:

$$L(\alpha, \beta) = \prod_{i=1}^{n} \left(\frac{\beta t_i^{(\beta-1)}}{\alpha^\beta} \right) \exp\left(-\left[\frac{t_i}{\alpha} \right]^\beta \right) \tag{10.21}$$

The ML estimates are obtained by solving the equations resulting from setting the two partial derivatives of $L(\alpha, \beta)$ to zero. As a result, $\hat{\beta}$ is obtained as the solution of:

$$\frac{\sum_{i=1}^{n} \left(t_i^{\hat{\beta}} \log t_i \right)}{\sum_{i=1}^{n} t_i^{\hat{\beta}}} - \frac{1}{\hat{\beta}} - \frac{1}{n} \sum_{i=1}^{n} \log t_i = 0 \tag{10.22}$$

Although an analytical solution of this equation is not available, we can easily solve for $\hat{\beta}$ using a computational approach. Once the shape parameter has been estimated, the scale parameter can then be estimated as follows:

$$\hat{\alpha} = \left(\frac{1}{n} \sum_{i=1}^{n} t_i^{\hat{\beta}} \right)^{1/\hat{\beta}} \tag{10.23}$$

10.7.2.2   Incomplete Data (Right Censoring)

From Equation (C.11) the likelihood function is given by:

$$L(\theta) = \left[\prod_{i \in D} f(t_i; \theta) \right] \left[\prod_{i \in C} \left[1 - F(t_i^*; \theta) \right] \right] \tag{10.24}$$

and the ML estimates are obtained by maximizing this function.

Comment: Minitab (and other packages) has programs to give the ML estimates.

10.8   Model Validation

Validation is the process of determining the degree to which a selected model (model formulation along with the assigned or estimated parameter values) is an accurate representation of the real-world problem of interest. A poor fit of model may occur for two reasons: (i) the model formulation is incorrect or (ii) the model formulation is correct, but the parameter values specified or estimated may differ from the true values by too great an amount.

A basic principle in validating a model is an assessment of the predictive power of the model, as this provides a basis for generalization. Assessment of the predictive power of a model is basically a statistical problem. The idea is to use an estimated model to predict outcomes of other observations, which may be data set aside for this purpose or future observations, and to evaluate the closeness of the predictions to observed values. A key requirement for credibility is that data used for validation are not a part of the data set used in model selection and estimation. When there are several models, one needs a procedure to discriminate between them and choose the final model.

Several approaches may be used for model validation and they can be grouped broadly into three categories: (i) non statistical, (ii) information theoretic, and (iii) statistical.

10.8.1   Non-Statistical

One of the methods is based on comparing the graphical plot of the model predictions and the data set aside for validation. This can be done visually (a very subjective basis) or by gauging the closeness of fit by evaluating the sum of errors squared (a more objective basis). However, models can vary in terms of the number of parameters – a model with more parameters will yield a smaller value for the sum of errors squared.

10.8.2   Information Theoretic

Akaike (1973; 1974) proposed an information theoretic interpretation of the likelihood function and proposed it as a model selection criterion, which is known as the *Akaike information criterion* (AIC). The AIC is given by:

$$\text{AIC} = -2(\text{maximum loglikelihood}) + 2(\text{number of model parameters}) \qquad (10.25)$$

When there are several models that look reasonable (based on non-statistical methods), the one with the minimal AIC is considered to be the best model.

10.8.3   Statistical (Goodness-of-Fit Test)

Here, one uses statistical tests to judge the adequacy of a model (distribution function) to model given data. It starts with a null hypothesis:

$$\text{H}_0 : \text{The given data comes from a distribution function } F(t,\theta)$$

with the form and parameters of the distribution function specified.

The basic idea in testing H_0 is to look at the data set aside for validation and evaluate the likelihood of occurrence of this sample *given that H_0 is true*. If the conclusion is that this is a highly unlikely sample under H_0, then H_0 is rejected. It follows that there are two types of errors that can be made: rejecting H_0 when it is true, and failing to reject it when it is not true. These are called Type I and Type II errors. The level of significance (denoted by α) is the maximal probability of making a Type I error and the power of the test is the probability of not making a Type II error. The test statistic is a summary of the given data and the test is usually based on some such statistic. The test involves comparing the computed values of the test statistic with some critical values (these depend on the level of significance), which are usually available in tabulated form. The null hypothesis is rejected if the test statistic exceeds the critical value.

A test is said to be non-parametric if the distribution of the test statistic does not depend on the distribution function $(F(t))$. Many tests have been developed and Appendix C discusses two of them briefly – the K–S and A–D tests.

In applying goodness-of-fit tests, one encounters two types of problem:

1. A specific distribution is suggested by theoretical considerations, past experience, and so forth, and we wish to determine whether or not a set of data is from this distribution.
2. We have only a vague idea of the correct distribution (for example, that it is skewed to the right) and we want to do some screening of possible candidate distributions.

In the former case, it is usually appropriate to test at a low level of significance, say $\alpha = 0.01$, the idea being that we do not want to reject the theoretical distribution unless there is strong evidence against it. In the second case, it may be more appropriate to test at a much higher level of significance, say $\alpha = 0.1$ or 0.2, in order to narrow the list of candidates somewhat. The choice is, to a great extent, subjective and depends on the particular application.

Most statistical packages give the results for the A–D test through the "*p*-value," which is the probability of obtaining the observed value of the test statistic, given that H_0 is true. If the test of significance gives a "*p*-value" smaller than the significance level α, the null hypothesis is rejected. Such results are informally referred to as being *statistically significant*.

10.9 Examples

In this section we look at the modeling of first failure for a few of the data sets from Appendix E.

Example 10.6   Battery

This is a continuation of Example 10.3. The A–D statistics (obtained as an output from Minitab) and the AIC values (computed using Equation (10.33)) for 11 distributions (listed in footnote 11) are given in Table 10.4. As can be seen, the AIC is smallest for the Weibull distribution, indicating that it models the data better than the remaining 10 distribution functions. All the *p*-values are greater than 0.1, implying that none of the distributions should be rejected.

Table 10.5 gives the ML estimates for the different distributions listed in Table 10.4 obtained from Minitab.   ∎

Table 10.4 A-D and AIC values for battery data.

Distribution	A–D*	AIC
Weibull	9.188	**587.296**
Lognormal	9.988	597.072
Exponential	11.809	606.398
Log-logistic	9.385	591.524
Three-parameter Weibull	9.143	589.180
Three-parameter lognormal	9.090	589.580
Two-parameter exponential	11.300	599.844
Three-parameter log-logistic	9.048	589.378
Smallest extreme value	9.792	601.498
Normal	9.127	590.152
Logistic	9.126	590.200

* The Anderson–Darling test.

Table 10.5 MLEs of the parameters of the different distributions.

Distribution	MLEs of the parameters
Weibull	{Shape = 1.96620, Scale = 836.344}
Lognormal	{Location = 6.45703, Scale = 0.748909}
Exponential	{Mean = 852.949}
Log-logistic	{Location = 6.51589, Scale = 0.382719}
Three-parameter Weibull	{Shape = 2.09463, Scale = 870.239, Threshold = −29.6094}
Three-parameter lognormal	{Location = 7.32693, Scale = 0.248427, Threshold = −822.129}
Two-parameter exponential	{Scale = 764.342, Threshold = 63.9936}
Three-parameter log-logistic	{Location = 7.10774, Scale = 0.179736, Threshold = −524.225}
Smallest extreme value	{Location = 921.296, Scale = 352.415}
Normal	{Mean = 737.358, St Dev = 363.556}
Logistic	{Location = 721.459, Scale = 207.952}

MLE: maximum likelihood estimate.

Example 10.7   Throttle Failures

The data presented in Table E.2 are from Carter (1986) and give the distance traveled (in thousands of kilometers) before failure, or the item being suspended before failure, for a pre-production general-purpose, load-carrying vehicle. Hence, the independent variable is the distance traveled, with 1000 km being the unit of measurement.

The empirical WPP is not linear and hence the two-parameter Weibull distribution is not appropriate to model the data. A smooth fit to the empirical WPP indicates a curve which has a shape similar to the multiplicative Weibull ($k = 2$), as can be seen in Figure 10.9 or the Weibull mixture ($k = 2$), as can be seen in Figure 10.10.

The parameter estimates (obtained by selecting the parameters to minimize the sum of errors squared [$J(\theta)$] between the theoretical and the empirical WPPs) are given in Table 10.6

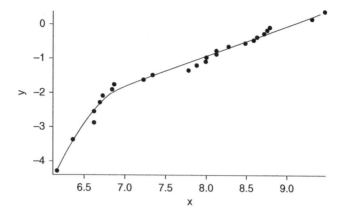

Figure 10.9 WPP for multiplicative Weibull ($k = 2$).

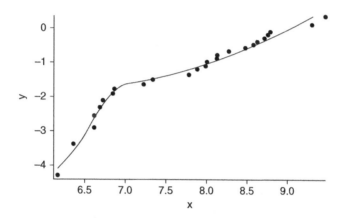

Figure 10.10 WPP for a Weibull mixture ($k = 2$).

Table 10.6 Parameter estimates based on a WPP.

Formulation	$J(\theta)$	Parameter estimates
Two-parameter Weibull ($k = 2$)	2.77	$\hat{\alpha} = 7470, \hat{\beta} = 1.13$
Weibull mixture	0.412	$\hat{\alpha}_1 = 865, \hat{\alpha}_2 = 9160, \hat{\beta}_1 = 6.86, \hat{\beta}_2 = 1.38, \hat{p} = 0.136$
Weibull multiplicative ($k = 2$)	0.37	$\hat{\alpha}_1 = 8700, \hat{\alpha}_2 = 723, \hat{\beta}_1 = 0.87, \hat{\beta}_2 = 3.93$

for the above two distributions as well as for the two-parameter Weibull distribution. As can be seen, $J(\theta)$ for the two-parameter Weibull is large compared to that for the other two distributions and is the smallest for the multiplicative Weibull ($k = 2$).

The ML estimates for the Weibull Multiplicative model are given below:

$$\alpha_1 = 8920, \ \alpha_2 = 779, \ \beta_1 = 0.836, \ \beta_2 = 2.42, \ \ln(\theta) = -246.1$$

Note that they differ slightly from the estimates obtained by the curve-fitting process shown in Table 10.6. ∎

Example 10.8   Automobile Component

This is a continuation of Example 10.1. Figure 10.11 shows probability plots of the usage rates for four distributions (normal, lognormal, exponential, and Weibull). From the plots, one can see that the Weibull distribution appears to provide a very good fit to the data.

The Weibull distribution overview plot shown in Figure 10.12 gives the following ML estimates: shape parameter $\hat{\beta} = 1.695$, scale parameter $\hat{\alpha} = 0.603$, mean mileage accumulation rate = 0.538 (equivalent to 5380 km/month or 179 km/day).

Consider the case where the usage rates are divided into three groups, as indicated in Example 10.1. Figure 10.13 shows four probability plots of age in months (Weibull, lognormal, exponential, and normal) for the three groups of customers. Among these four choices, the Weibull distribution is the best fit for each of the three groups.

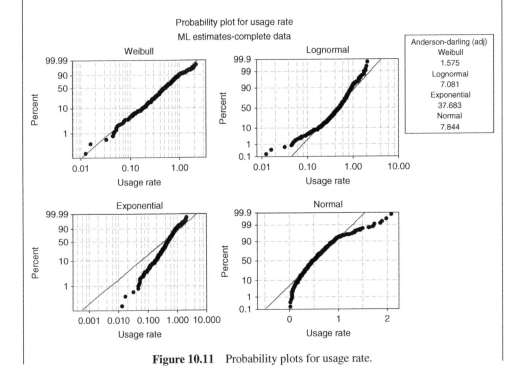

Figure 10.11   Probability plots for usage rate.

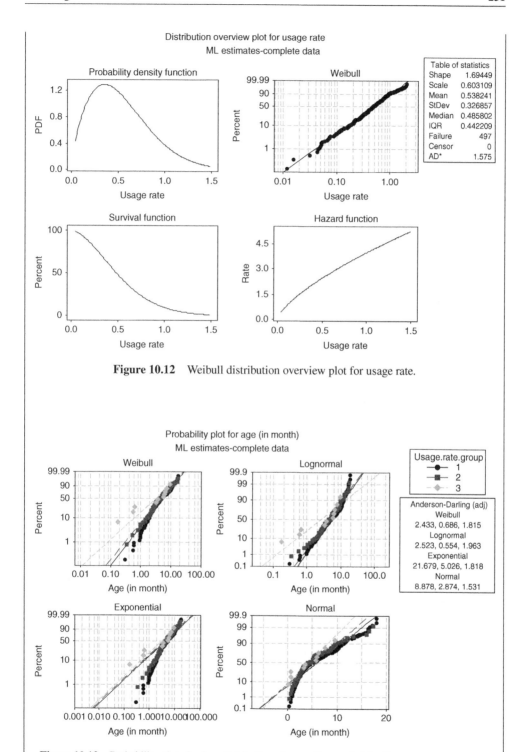

Figure 10.12 Weibull distribution overview plot for usage rate.

Figure 10.13 Probability plots for four distributions for age (in months) for three usage-rate groups.

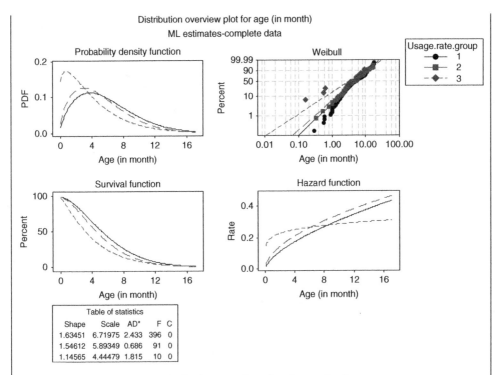

Figure 10.14   Weibull distribution overview plots for age for three usage-rate groups.

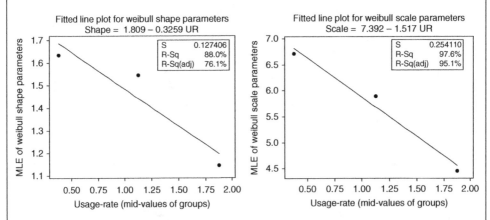

Figure 10.15   Plots of fitted shape and scale parameters of the Weibull distribution versus the usage rate.

The Weibull distribution overview plots for age with the ML estimates of the parameters for each of the three groups are shown in Figure 10.14.

We next examine the relationships between the ML estimates of the parameters and the usage rates. Figure 10.15 shows plots of the ML estimates of the shape and scale parameters of Weibull distributions against the mid-values of each usage rate group. It shows a linear relationship is appropriate. From the figure, the shape and scale parameters of the Weibull distribution can be expressed (using linear regression) as a function of usage-rate, z, and are as follows:

$$\beta(z) = 1.809 - 0.3859z \quad \text{and} \quad \alpha(z) = 7.392 - 1.517z.$$

Comment: In this example, the number of items and the number of groups are small and the results given are preliminary and intended only to demonstrate the techniques. More items with a larger number of groups are necessary for an adequate study of the relationships. ∎

10.10   Summary

This chapter dealt with modeling the time to first failure using the empirical approach considering the one-dimensional case (data consist of the failure/service times of several items) and the two-dimensional case (data consist of the failure/service times and the usage levels). One-dimensional distribution functions are used for the first case and two-dimensional distribution functions are used for the second case.

There are many different distribution functions that can be used as mathematical formulations to model the time to first failure in the one-dimensional case. They can be categorized broadly into the following three groups: (i) standard distributions; (ii) derived distributions; and (iii) complex distributions.

For the two-dimensional case, the time to first failure is a random point in a two-dimensional plane, with age and usage being the two coordinates. There are three different approaches to modeling such data: (i) time to first failure is modeled by a bivariate distribution function $F(t, u)$; (ii) the two scales, usage u and time t, are combined to define a composite scale v and the time to first failure is modeled by a distribution function $F_v(v)$; and (iii) time to first failure, conditional on the usage rate, is given by a conditional failure distribution function $F(t \mid z)$.

There are several properties of distribution functions that are very relevant in the context of modeling. These include the shapes of the density and hazard functions and the transformed plots of the distribution functions. These play a role in the selection of a distribution function to model failure times.

Two plots that are very useful in the modeling of first failure are the EDF (the sample equivalent of the failure distribution function) and the transformed plots based on the EDF.

Many different estimation methods have been developed for estimating model parameters. They can be grouped broadly into two categories: (i) graphical methods and (ii) statistical methods. In the former, the estimates are obtained from plotting the data. Statistical methods, in contrast, are more general and applicable to all kinds of models and data types and use methods such as ML.

An important aspect of modeling is validation, which is the process of determining the degree to which a selected model (model formulation along with the assigned or estimated parameter values) is an accurate representation of the real-world problem of interest. Several approaches may be used for model validation and they can be grouped broadly into three categories: (i) non-statistical, (ii) information theoretic, and (iii) statistical.

Review Questions

10.1   What is the difference between the one-dimensional and the two-dimensional approach to modeling time to first failure?

10.2   What are the different types of distribution functions used as mathematical formulations to model time to first failure?

10.3   What are the different approaches for modeling data in the two-dimensional case?

10.4   Determine the expressions of the transformed plot for the following distributions: (i) exponential; (ii) two-parameter Weibull; and (iii) lognormal.

10.5   Describe how you may estimate the parameter of the exponential distribution using graphical methods.

10.6   Describe how you may estimate the parameters of the Weibull distribution using graphical methods.

10.7   Describe how you may estimate the parameter of the exponential distribution using the maximum likelihood method.

10.8   Describe how you may estimate the parameters of the Weibull distribution using the maximum likelihood method.

10.9   Is there a difference between the expressions for the maximum likelihood function for censored and uncensored data?

10.10   Outline three different approaches for model validation.

Exercises

10.1   Compile the EDF and hazard plots of the two data sets in Exercise 9.5. What can you infer from these plots?

10.2   Assume that the batteries from the two manufacturers in Exercise 9.5 are roughly similar (in a statistical sense). Pool the two data sets to create a new data set with 50 data points. Plot the data on exponential probability paper (EPP) (i.e., plotting the data after the exponential transformation). What can you infer? From the results of Minitab, obtain the ML estimate of the model parameter. Can one accept the model with 95% confidence?

10.3   Following on from Exercise 10.2, plot the data on Weibull probability paper. What can you infer? From the results of Minitab, obtain the ML estimates of the model parameters. Can one accept the model with 95% confidence?

10.4   An object consists of two components with failure distribution $F_i(t), i = 1, 2$. When component 1 [2] fails, it may induce a failure of the other component with probability $p_1 [p_2]$ or not. If there is no induced failure, the reliability of the non-failed component is not affected. In this case, the failed component is minimally repaired with repair time negligible. The object is replaced when both components fail. Let Z denote the time between replacements.
(a) Derive an expression for the distribution function for $\tilde{Z}$.
(b) How do the two probabilities of induced failures affect the failure of the object?

10.5 Following on from Exercise 10.4, discuss the object failure distribution when p_1 and/ or p_2 reach their limits. Describe a real-world situation where the failure may be modeled by this.

10.6 How do the two probabilities of induced failure affect the failure of the object in Exercise 10.5?

10.7 Assume that the data in Exercise 9.6 may be modeled by a Weibull distribution function. Estimate the model parameters using the method of maximum likelihood.

10.8 If the test in Exercise 9.6 was terminated after five days, what data are available for estimating the model parameters? How would one obtain the ML estimates? How do the estimates compare with those in Exercise 10.7?

10.9 If the test in Exercise 9.6 was terminated after the 25th item failure, what data are available for estimating the model parameters? How would one obtain the ML estimates? How do the estimates compare with those in Exercises 10.7 and 10.8?

10.10 Locomotive wheel wear is a linear function of the mileage. A locomotive truck travels at constant speed and a wheel is classified as having failed when the wear exceeds some specified limit. A wheel may also fail due to derailments, which occur randomly. Build a model to compute the expected life of a wheel.

References

Akaike, H. (1973) Information theory and an extension of the maximum likelihood principle. In *Second International Symposium on Information Theory*, B.N. Petrov and F. Csaki (eds), Budapest: Academiai Kiado, pp. 267–281.

Akaike, H. (1974) A new look at the statistical model identification, *IEEE Transactions on Automatic Control*, **AC-19**: 716–723.

Blischke, W.R., Karim, M.R., and Murthy, D.N.P. (2011) *Warranty Data Collection and Analysis*, Springer-Verlag, London.

Carter, A.D.S. (1986) *Mathematical Reliability*, 2nd edition, John Wiley & Sons, Inc., New York.

Hutchinson, T.P. and Lai, C.D. (1990) *Continuous Bivariate Distributions, Emphasizing Applications*, Rumsby Scientific Publishing, Adelaide.

Johnson, N.L. and Kotz, S. (1970a) *Distributions in Statistics: Continuous Univariate Distributions — I*, John Wiley & Sons, Inc., New York.

Johnson, N.L. and Kotz, S. (1970b) *Distributions in Statistics: Continuous Univariate Distributions — II*, John Wiley & Sons, Inc., New York.

Johnson, N.L. and Kotz, S. (1972) *Distributions in Statistics: Continuous Multivariate Distributions*, John Wiley & Sons, Inc., New York.

Murthy, D.N.P., Xie, M., and Jiang, R. (2004) *Weibull Models*, John Wiley & Sons, Inc., New York.

11

Modeling CM and PM Actions

Learning Outcomes

After reading this chapter, you should be able to:

- Describe the different types of repairs;
- Distinguish between the modeling of CM for repairable and non-repairable items;
- Describe the effect of CM and PM actions on failure rate;
- Describe two methods of modeling imperfect repair;
- Describe three kinds of formulations for modeling PM actions;
- Identify the components of downtime;
- Describe the modeling of repair times and downtimes;
- Provide a classification of maintenance costs;
- Explain the modeling issues related to repair–replace decisions;
- Explain the modeling issues related to fleet and infrastructure maintenance.

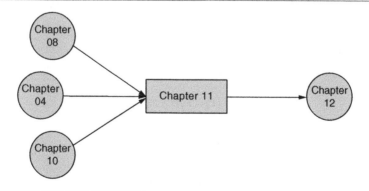

Introduction to Maintenance Engineering: Modeling, Optimization, and Management, First Edition.
Mohammed Ben-Daya, Uday Kumar, and D.N. Prabhakar Murthy.
© 2016 John Wiley & Sons, Ltd. Published 2016 by John Wiley & Sons, Ltd.

11.1   Introduction

As discussed in Chapter 4, maintenance actions can be grouped into (i) preventive maintenance (PM) and (ii) corrective maintenance (CM). There are several types of CM and PM actions. The time to first failure is influenced only by PM actions whereas subsequent failures are impacted by both CM and PM actions. In this chapter we discuss the impact of PM actions on first failure and the effect of CM actions after the first failure. We focus on the characterization of these impacts as well as the cost implications of maintenance actions. The study of subsequent failures is the focus of Chapter 12.

The outline of the chapter is as follows. We start with maintenance for products and plants. We discuss the modeling of CM actions in Section 11.2 and this is followed by the modeling of PM actions in Section 11.3, where we also look at various other issues. Section 11.4 deals with repair and downtime. In Section 11.5 we look at maintenance costs. Section 11.6 looks at the modeling of repair versus replace decisions. We discuss briefly the modeling of fleet and infrastructure maintenance in Section 11.7 and conclude with a summary in Section 11.8.

11.2   Modeling CM Actions

We need to differentiate between repairable and non-repairable items (products, components, or anything in between). The failures may be either *hard* or *intermittent.* In the case of a hard failure, some form of CM action is needed to make the item operational. In the case of an intermittent failure, the item returns to the operational state with no CM action but the underlying cause still persists.

Let $F_0(t)$ denote the failure distribution of a new item and $h_0(t)$ the associated hazard function with $t = 0$ when an item is first put into operation. Let x denote the age at the first failure and $F(t)$ and $h(t)$ the failure distribution function and hazard function after the repair/replacement. We assume that the time to repair or replace is small, so that it can be treated as being zero.[1]

11.2.1   Non-Repairable Items

In this case there are two options: replace a failed item by a new one or by a used item.

11.2.1.1   Replace by a New Item

In this case, the time to the next failure, using a calendar clock,[2] has distribution function:

$$F(t) = F_0(t - x) \tag{11.1}$$

or

$$F(t) = F_0(t) \tag{11.2}$$

using a local clock (which is reset to zero after the replacement).

---

[1] Later on we look at repair times that are not negligible and the modeling of these.
[2] We will be using a calendar clock unless some other clock (such as local, age, etc.) is specifically indicated. With the calendar clock, $t = 0$ corresponds to the state of the clock when the first item is put into operation.

11.2.1.2 Replace by a Used Item

If the failed item is replaced by a used item of age y, then the conditional probability of failing by time $t + y$, given that the item survived up to time y, using the local clock (which is reset to zero after replacement) is given by:

$$\frac{1}{1-F_0(y)} \int_y^{t+y} f_0(\tau) d\tau = \frac{F_0(t+y)-F_0(y)}{1-F_0(y)} \tag{11.3}$$

so that

$$F(t) = \frac{F_0(t+y)-F_0(y)}{1-F_0(y)}, \ t > 0. \tag{11.4}$$

Based on the calendar clock, the distribution function for the time to next failure after replacement is given by:

$$F(t) = \frac{F_0(t)-F_0(y)}{1-F_0(y)}, \ t > y. \tag{11.5}$$

11.2.2 Repairable Items

We assume that the times taken to carry out CM actions are negligible (relative to the times between CM actions), so that they can be ignored.[3] We consider the following three types of repair.

11.2.2.1 Minimal Repair[4]

Under minimal repair, the reliability of the item is unaffected by the repair action. As a result, the hazard function after repair is the same as that just before failure and is given by:

$$h(t) = h_0(t), \ t > x. \tag{11.6}$$

This type of repair is depicted graphically in Figure 11.1.

11.2.2.2 Perfect Repair

Here, the hazard function after repair is the same as that for a new item. In other words, the repair makes the failed item as good as new. In this case, the hazard function after repair is given by:

$$h(t) = h_0(t-x), \ t > x. \tag{11.7}$$

---

[3] One can extend this to the case where the repair time is not negligible. The time to repair is usually uncertain and this is discussed later in the chapter.
[4] The concept of minimal repair was first proposed by Barlow and Hunter (1961).

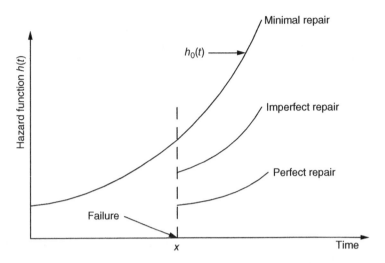

Figure 11.1   Effect of different CM actions on the hazard function.

This type of repair is depicted graphically in Figure 11.1.

Note that usually one replaces the failed item by a new one so that it is a replacement rather than a repair. However, in some rare cases, it is possible to repair a failed item to new, as illustrated by the following example.

Example 11.1   Aircraft Windshield

The windshield on a large aircraft is a complex piece of equipment, comprised basically of several layers of material, including a very strong outer skin with a heated layer just beneath it, all laminated under high temperature and pressure. Failures of this item are not structural failures. Instead, they typically involve damage or delamination of the non-structural outer ply, or failure of the heating system. These failures do not result in damage to the aircraft, but do result in replacement of the windshield. The two common failure modes are: (i) failure of the heating element, and (ii) delaminating of the sheets. In either case, the pilot's visibility is affected. When a failure occurs, the windshield is heated in an oven to melt the glue and it is then re-glued (with the heating element replaced in case it has failed). The repaired windshield is as good as new.   ∎

11.2.2.3   Imperfect Repair

In some situations, the reliability characteristic of a repaired item is better than that under minimal repair but not as good as that for a new item. This type of repair is referred to as *imperfect repair* and the hazard function after repair satisfies the inequality $h_0(t-x) < h(t) < h_0(t)$, $t > x$, as shown in Figure 11.1.

Two ways of modeling imperfect repair are as follows:[5]

---
[5] For more on imperfect repair, see Pham and Wang (1996).

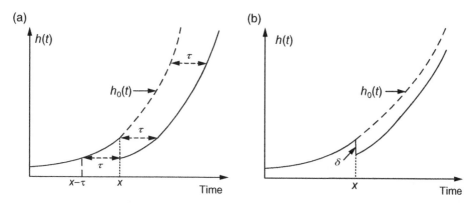

Figure 11.2 Two models for imperfect repair. (a) Reduction in age and (b) reduction in the hazard function.

Reduction in Age (Virtual Age)

The effect of a CM action is modeled by a reduction τ $(0 < \tau < x)$ in the age.[6] The virtual age of the item at time t is given by $(t - \tau)$ for $t > x$. As a result, the hazard function after repair is given by:

$$h(t) = h_0(t - \tau), \ t > x, \tag{11.8}$$

as shown in Figure 11.2a.

Reduction in Hazard Function

Here, a CM action results in a reduction in the hazard function. The effect of CM on the hazard is given by $h(x^+) = h_0(x^-) - \delta$ where δ is the reduction resulting from the CM action at time x. As a result, the hazard function after repair is given by:

$$h(t) = h_0(t) - \delta, t > x, \tag{11.9}$$

as indicated in Figure 11.2b. δ depends on the level of CM effort and needs to be constrained to satisfy the following inequality:

$$0 \le \delta \le h_0(x^-) - h_0(0). \tag{11.10}$$

Note: $\delta = 0$ corresponds to minimal repair and $\delta = h_0(x^-) - h_0(0)$ to perfect repair.

11.2.3 Intermittent Failures

Some items (especially electronic ones) experience intermittent failures. When the item fails, often it becomes operational again by restarting. When the owner takes it to a repair facility (after experiencing the failure one or more times), the maintenance technician might or might not detect a

---

[6] See Kijima (1989) and Doyen and Gaudion (2004) for more details.

fault. In the latter case, the item is returned with a statement *no fault found* (NFF) and in the former case, the fault is rectified. The number of times before a fault is fixed is a random variable, and one way of modeling it is by a random variable with a discrete probability distribution function.

11.2.4 Quality of Repair

The quality of repair is an issue of concern to owners. One way of defining a poor quality of repair is that the repaired item fails again very soon after repair. There are several reasons for poor-quality maintenance – the maintenance technician may lack experience, the repair may have been rushed without exercising due care and diligence, and so on. One way of modeling quality of repair is by a binary random variable which has values 1 (failed item repaired properly) or 0 (failed item not repaired properly). Let p $(0 \leq p \leq 1)$ denote the probability that a failed item is repaired properly. A higher value of p implies a better quality of repair.

11.3   Modeling PM Actions

There are several different ways of modeling PM actions, depending on the kind of formulations used, as indicated in Figure 11.3.

11.3.1   Static Formulations

In the static formulation, PM effort is modeled as a parameter which captures the different actions (or level of PM action) in an aggregated manner. The original equipment manufacturer (OEM) recommended PM level is denoted u_0 and the actual level that the owner decides is u, with $0 \leq u \leq u_0$.[7]

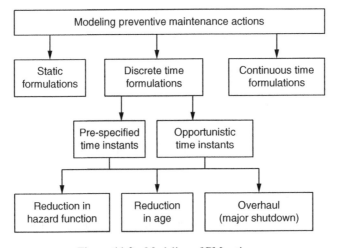

Figure 11.3   Modeling of PM actions.

___
[7] The case where $u > u_0$ implies that the owner is wasting money due to *over maintenance* – carrying out tasks that are not of relevance

11.3.1.1 Impact on Hazard Function

The hazard function is affected by the level of PM, as indicated in Figure 11.4. The hazard function (over the useful life of the item) with the OEM-recommended level of PM is given by $h(t, u_0)$ and with PM level u $(0 < u \leq u_0)$, it is given by $h(t, u)$.

It is important to note the following:

- $\partial h(t,u) / \partial t > 0$, implies an aging effect.
- $\partial h(t,u) / \partial u < 0$, implies a higher hazard function with reduced PM.

11.3.1.2 Impact on Failure Severity

In the simplest characterization, one can classify failures as minor (not too costly to fix) and major (costly to fix).[8] As before, the hazard function with the OEM-recommended level of PM is given by $h(t, u_0)$. With maintenance level u $[0 \leq u \leq u_0]$, the probability that a failure is major [minor] is given by $p(u) [1 - p(u)]$. $p(u)$ decreases as the PM level increases, as indicated in Figure 11.5.

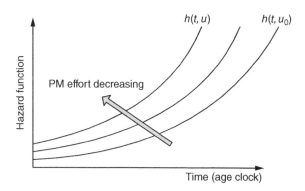

Figure 11.4 Impact of PM on the hazard function.

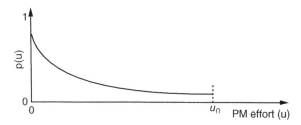

Figure 11.5 Effect of PM on the severity of failure.

---

[8] One can have more than two levels of severity, with different costs for each type.

11.3.1.3   Impact on Salvage Value

The salvage value depends on the age as well as the condition of the item, which, in turn, depends on the level of PM. The salvage value of an item of age t and PM level u can be modeled in many different ways depending on the form of the mathematical formulation. A deterministic model involving a linear ordinary differential equation is as follows:

$$\frac{dS(t)}{dt} = -a(u)S(t), \quad S(0) = S_0 \text{ (purchase price)} \tag{11.11}$$

The rate of decrease in the salvage value as a fraction of the salvage value is given by $[dS(t)/dt]/S(t) = -a(u)$, with $a(u)$ a decreasing function of the PM level, as indicated in Figure 11.6. This model implies that an item maintained with a higher PM level has a higher salvage value over time.

11.3.2   Dynamic Formulations – I

Here, PM actions are carried out at discrete time instants so that they occur at points along the time axis. The time instants can be either deterministic (for example, based on a calendar or age clock) or uncertain (also referred to as opportunistic PM actions).

The effect of a PM action is to improve the reliability of the item after PM. PM actions can be one of the following two types:

- Replace by new;
- Overhaul (major shutdown PM) – the level of reduction is dependent on the components replaced and is modeled by reduction in age or in the hazard function.

The modeling of the effect of PM action on the hazard function is similar to that for the CM case discussed earlier and hence it will not be discussed further.

11.3.2.1   Overhaul (Major Shutdown Maintenance)

As mentioned in Chapter 4, an overhaul rejuvenates a system. The hazard function after the jth ($j \geq 1$) overhaul is given by $h_j(t)$, with $h_0(t)$ being the function for a new system. Typical shapes for these functions are shown in Figure 11.7, where we use a local clock which is reset to zero after each overhaul.

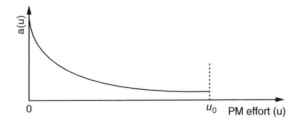

Figure 11.6   Effect of PM level on the salvage value.

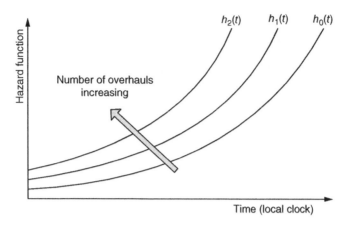

Figure 11.7 Modeling the effect of overhauls (major shutdowns).

Of note are the following points:

1. $h_j(t)$, $j \geq 0$ is an increasing function of t, implying that the system degrades with age.
2. $h_{j+1}(t) > h_j(t)$ for $j \geq 0$, implying that each overhaul improves the intensity function but there is progressive deterioration.

Comment: If the system is restored to as good as new after each overhaul then $h_j(t) = h_0(t)$, $j \geq 1$.

Example 11.2   Dragline

A dragline is subjected to an overhaul after being in operation for a specified number of years. The overhaul involves dismantling, checking (for example, cracks in the components of the boom), replacing all worn-out components and components with cracks, and so on. As a result, the dragline is as good as new after an overhaul. Typically, the life of a dragline is around 20 years and the dragline is subjected to overhaul after being in operation for four to five years. The cost of the overhaul is usually around 30–40% of the price of a new dragline. ■

11.3.2.2   Opportunistic PM

Opportunistic PM deals with carrying out PM actions as a result of opportunities that occur in an uncertain manner. An example of this is a plant with several components. The failure of a component provides an opportunity to carry out PM action on one or more of the non-failed components at the same time. This implies that the time instants of PM for these components are moved forward to exploit opportunities and to reduce the PM cost.

11.3.3   Dynamic Formulations – II

Many complex plants require PM actions to be carried out at fairly short intervals. A discrete time formulation results in a complex model (the curse of dimensionality). In this case, PM actions are better modeled by a continuous function $u(t)$ (representing the PM effort) which changes over time with $0 \leq u(t) \leq u_m$ where u_m is the upper limit. In this case, the impact on salvage value can be modeled by:

$$\frac{dS(t)}{dt} = -aS(t) + b\left(\frac{u(t)}{u_m}\right), \quad S(0) = S_0 \tag{11.12}$$

instead of Equation (11.11). If L is the item's useful life then one needs to constrain b so that even with $u(t) = u_m$ the right-hand side of Equation (11.12) is negative for $0 \leq t \leq L$.

11.4 Repair Times and Downtimes

Downtime is characterized by two events: (i) failure of the item and (ii) the item being put back into operation after repair/replacement. This is usually longer than the actual repair time, which is also characterized by two events: (i) start of the repair and (ii) end of the repair. As shown in Figure 11.8, downtime = repair time + $z_1 + z_2$, where z_1 and z_2 are the times for administrative and logistical activities.

11.4.1 Modeling Repair Times

Repair time is comprised of several time periods – investigation time (time needed to locate the fault), the time needed to carry out the actual repair, and testing time after the repair. It can also include the waiting times that can result from a lack of spares or from other failed items awaiting rectification actions. This time is dependent on the inventory of spares and the staffing of the repair facility.

Some of these times can be predicted precisely whilst others (for example, the time to carry out the actual repair) can be highly variable, depending on the item and the type of failure. The easiest approach is to aggregate all the above-mentioned times into a single repair time X, modeled as a random variable with a distribution function $G_1(x) = P\{X \leq x\}$. We assume that $G_1(x)$ is differentiable and let $g_1(x) = dG_1(x)/dx$ denote the density function and

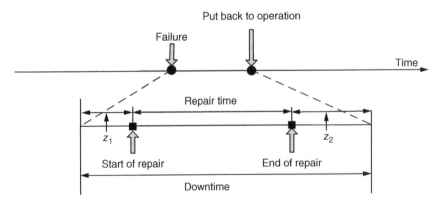

Figure 11.8 Downtime and repair time.

$\bar{G}_1(x) = 1 - G_1(x)$ the probability that the total repair time will exceed x. Analogous to the concept of the failure rate function, one can define a repair rate function $\rho(x)$ given by:

$$\rho(x) = \frac{g_1(x)}{\bar{G}_1(x)} \tag{11.13}$$

$\rho(x)\delta x$ is interpreted as the probability that the repair will be completed in $[x, x + \delta x)$, given that it has not been completed in $[0, x)$. In general, $\rho(x)$ will be a decreasing function of x (see Mahon and Bailey, 1975), indicating that the probability of a repair being completed in a short time increases with the duration that the service has been going on. In other words, $\rho(x)$ has a "decreasing repair rate," a concept analogous to that of a decreasing failure rate.[9]

If the variability in the repair time is small in relation to the mean time for repair, then one can approximate the repair time as being deterministic.

11.4.2 Modeling Downtimes

As indicated in Figure 11.8, the downtime is comprised of three components. If maintenance is to be carried out on site, then z_1 is the travel time and z_2 is negligible. If a failed item has to be transported to a central repair facility, then both of these variables are non-zero. Again, these times can be either predicted precisely or be uncertain. One can aggregate all the three times and model the downtime by a distribution function similar to that for repair time.

Example 11.3   Maintenance of a Heavy Vehicle

Table E.6 contains the repair time and downtime data (in hours) for a heavy vehicle. The descriptive statistics (using Minitab) are given in Table 11.1.

Note that downtime is missing for one failure whilst the downtime for another failure is extremely large (96 hours) so it can be viewed as an outlier. The selection of a distribution to model the data set follows along the lines discussed in Chapter 10, noting that the data are complete data. The probability plots using Minitab indicate that the lognormal distribution is an appropriate distribution to model the repair time data. The various plots (along with the parameter estimates obtained using the method of maximum likelihood) for the lognormal model are given in Figure 11.9. Note that the repair rate is initially increasing and then decreasing.

Table 11.1   Descriptive statistics (using Minitab) for the data in Table E.6.

Variable	N	Mean	St Dev	Q1	Median	Q3
Repair time	69	2.623	2.435	1.0	1.5	3.5
Downtime	68	4.40	11.64	1.0	2.0	4.88

---

[9] Kline (1984) suggests that the lognormal distribution is appropriate for modeling the repair times for many different products.

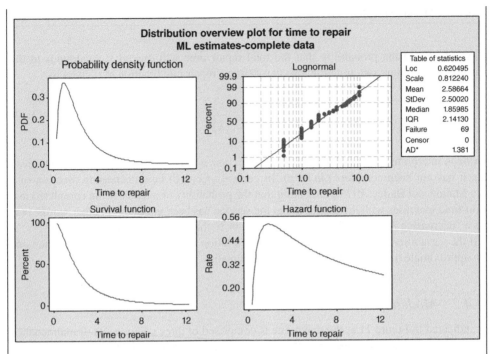

Figure 11.9 Lognormal fit to repair time data.

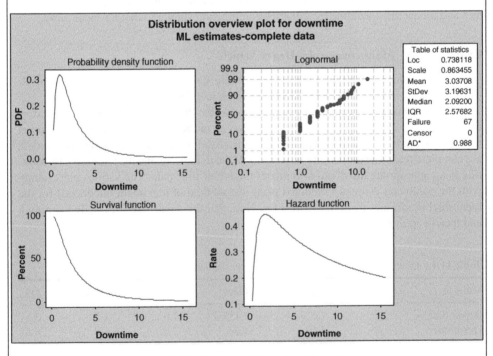

Figure 11.10 Lognormal fit to downtime data.

For the downtime, the probability plots using Minitab indicate that the lognormal distribution is also an appropriate distribution to model the data set. The various plots (along with the parameter estimates obtained using the method of maximum likelihood) for the lognormal model are given in Figure 11.10.   ■

11.5 Maintenance Costs

11.5.1 Classification of Costs

11.5.1.1 Direct Costs

The direct costs are the costs incurred in carrying out a PM or CM action. These costs can be broken down into (i) material costs and (ii) labor costs. The labor costs depend on the time duration needed for carrying out a PM or CM action. For PM actions, this depends on the set of specified tasks to be carried out. If the variability in the time needed to execute a given set of tasks is small then it can be ignored. In contrast, the variability in CM actions is large, as it depends on the severity of failure and the tasks that need to be executed to restore a system to the operational state. The same is true for the material cost in the case of CM.

11.5.1.2 Indirect Costs

The indirect costs include (i) the costs associated with the setting up (the equipment needed to carry out the CM and PM actions) and running of the workshop, (ii) the costs associated with spare part inventory, and (iii) the administration costs (which include salaries of administrative personnel, costs of developing and maintaining the maintenance management system, etc.).

11.5.1.3 Consequential Costs

Consequential costs include the following:[10]

- Losses resulting from failures on the production of goods and services;
- Loss of customer goodwill;
- Penalty costs resulting from delays in delivery of goods, completion of projects, and so on, due to failures of equipment.

11.5.2 Modeling PM and CM Costs

Maintenance cost models can be grouped broadly into the following two categories:

- *Category 1:* Modeling the cost of each action (PM and CM) separately by a parameter (if the variability is insignificant) or by a distribution function (if the variability is significant).
- *Category 2:* Modeling cost per unit time with the unit being a week, month, or year.

In this section we focus our attention on the first category models, while the second category models are discussed in Chapter 21.

Some of the PM and CM costs can be predicted precisely whilst others (for example, the time to carry out the actual repair) can be highly variable, depending on the item and the type of failure. The easiest approach is to aggregate all the above-mentioned costs into a single cost

---

[10] Indirect and consequential costs are discussed further in Chapter 21.

Z modeled as a random variable with a distribution function $G_2(z) = P\{Z \leq z\}$. We assume that $G_2(z)$ is differentiable and let $g_2(z) = dG_2(z)/dz$ denote the density function and $\bar{G}_2(z) = 1 - G_2(z)$ the probability that the aggregated cost will exceed z. If the variability is small (relative to the average) then it can be ignored and in this case the cost is modeled as a deterministic quantity. Obviously, deterministic modeling is much easier than probabilistic modeling.

For the analysis of some maintenance policies only average costs are needed. In such cases, one models the uncertain costs by their average values and in this case the modeling is simpler.

11.5.2.1   PM Costs

- *Replace by new:* This is the cost associated with replacement by a new unit and can be treated as being deterministic. We denote the cost by C_p.
- *Imperfect PM:* The cost of an imperfect PM depends on the level of PM actions (modeled by a reduction in the virtual age or in the rate of occurrence of failures (ROCOFs) intensity function). The reduction is characterized by variable τ or δ, as discussed in Section 11.2. The cost can also depend on the age of the item (a) at the time of the PM action and/or the number of times (j) that the item has been subjected to earlier imperfect PM actions.

    If the cost depends only on the age of the item and the level of imperfect PM, then it is modeled by a function $C_p(\tau, a)$ or $C_p(\delta, a)$ with the various first partial derivatives >0 implying that the cost increases as the item ages and/or the level of PM action increases. If age has no significant effect, then the cost is modeled using a simpler function $C_p(\tau)$ or $C_p(\delta)$.
- *Overhaul:* The cost of an overhaul increases with the number of times the item has been overhauled. If there is high uncertainty (due to the parts that need to be replaced) then it needs to be modeled probabilistically. In this case, the cost of the kth overhaul is denoted by $\tilde{C}_O(k)$ with $C_O(k) = E[\tilde{C}_O(k)]$, $k \geq 1$. If the variability is not significant, then one only needs to specify the function $C_O(k)$, $k \geq 1$.

11.5.2.2   CM Costs

- *Replace by new:* This is the cost of replacing a failed item by a new item and can be treated as being deterministic. We denote this cost by C_f with $C_f \geq C_p$.
- *Repair:* If the variability in the cost is large, we need to model it probabilistically. In this case, we denote it by the random variable $\tilde{C}_r$ with a given distribution function. Let C_r denote the expected value of $\tilde{C}_r$ ($\equiv E[\tilde{C}_r]$). If the variability is insignificant, then we denote the repair cost by C_r.

Comment: If the repair cost depends on the age of the component (a) then the expected cost $C_r(a)$ is an increasing function of a.

Example 11.4   Bus Maintenance

Table E.8 gives the repair time and repair cost data for 22 buses over a three-year period. A dot plot of costs for the 22 buses is shown in Figure 11.11.

Probability plots (based on the costs for failures of 22 buses) using Minitab indicated that several different distributions could be used to model the data set, as can be seen from Figure 11.12.

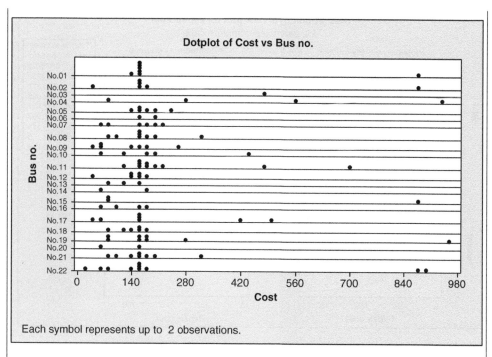

Figure 11.11   Dot plot of repair costs for buses.

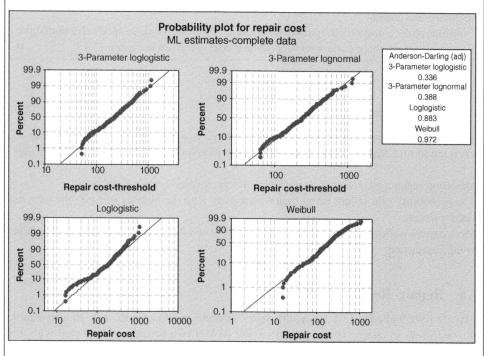

Figure 11.12   Probability plots for repair cost data.

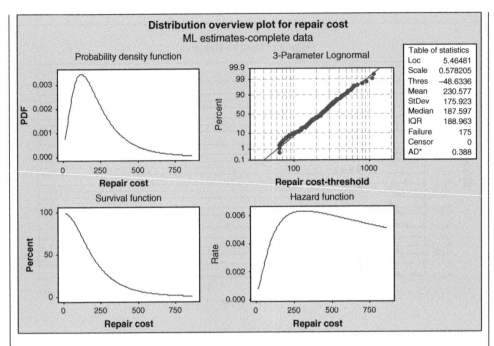

Figure 11.13   Three-parameter lognormal fit to repair cost data.

The three-parameter lognormal distribution indicates the best fit (based on visual inspection) and we discuss this further.

Various plots (along with the parameter estimates obtained using the method of maximum likelihood) for the three-parameter lognormal model are given in Figure 11.13.   ∎

11.5.3   More Complex Models

We have confined our attention to modeling repair costs using distribution functions as the underlying formulation. More complex modeling will involve one or more of the following:

- Modeling repair cost as a random variable with the mean increasing with age;
- The probability of a failure being major increasing with the number of times the item has been repaired;
- Modeling the various components of costs, repair time, and downtime separately rather than being aggregated.

11.6   Repair–Replace Decisions

So far we have looked at CM actions for repair or replace. Often, in the case of a repairable item, one needs to choose between the two. We look at two scenarios which deal with this issue.

11.6.1 Based on Number of Repair Times

In general, the efficacy of repair decreases with the number of times an item is subjected to repair after failure. This suggests a policy where an item is scrapped and replaced by a new item after it has been repaired a certain number of times. Policy 4.7 is an example of such a policy.

11.6.2 Based on Repair Cost

As seen from examples, the cost of repair is a random variable. Before repair is commenced, a reasonable (and accurate) estimate of the repair cost is made. The decision to repair or replace is based on this estimate. If the estimate is below $\bar{c}$, the failed item is repaired and replaced by a new one; if not, Policy 4.5 is one such policy.

11.7 Modeling Fleet and Infrastructure Maintenance

The modeling of PM and CM actions for fleets and infrastructures poses additional challenges. We discuss these briefly.

11.7.1 Fleet Maintenance

Consider a fleet of items (locomotives, buses, ships, etc.) used in the transport sector. In the case of buses, PM actions for each unit have a cyclical pattern that repeats over time. Over a cycle, the maintenance level for PM (at discrete time instants) can vary. The scheduling of PM actions for different units needs to be synchronized properly so as not to violate the capacity constraint of the maintenance workshop and to minimize the total maintenance cost. In the case of ships, the modeling becomes still more complex. Different maintenance actions require different resources (equipment, manpower, dry docking, etc.). Ports differ in their resource capabilities. The characterization and modeling of maintenance actions requires a more detailed description of the various PM and CM processes. Policies 4.12–4.15 are an illustrative sample of the many different policies that have been proposed for the maintenance of a fleet.

11.7.2 Infrastructure Maintenance

The characterization of different PM and CM actions in the case of infrastructures is still more complex due to the spatial nature of the asset. For example, in the case of roads, the distinction between PM and CM gets slightly blurred. Is filling a small pothole (minimal repair) a CM or PM activity? Is the relaying of the surface for a short length (imperfect repair) a CM or PM action? An upgrade starting from the foundation (major overhaul) may also be viewed as either CM or PM.

11.8 Summary

This chapter dealt with the impact of PM actions on first failure and the effect of CM actions after first failure. The focus has been on the characterization of these impacts as well as the cost implications of maintenance actions.

In modeling CM actions, we need to differentiate items (product, component, or anything in between) which are repairable from those that are non-repairable. For non-repairable items, there are two options: replace a failed item by a new one or by a used item. For a repairable item, we consider the following three types of repair: (i) *minimal repair* – the reliability of the item is unaffected by the repair action; (ii) *perfect repair* – this repair makes the failed item as good as new (in this case, the reliability function and the hazard function after repair are the same as those for a new item); and (iii) *imperfect repair* – the reliability characteristic of a repaired item is better than that under minimal repair but not as good as that for a new item. In this case, the effect of CM action is modeled by a reduction in the age or a reduction in the hazard function.

There are several different ways of modeling PM actions, depending on the kind of formulations used. In the static formulation, PM effort is modeled as a parameter which captures the different actions (or level of PM action) in an aggregated manner. In the dynamic formulation, PM actions are carried out at discrete time instants so that they are points along the time axis. The time instants can be either deterministic (for example, based on a calendar or age clock) or uncertain (also referred to as opportunistic PM actions). In the non-opportunistic case, PM actions may be one of the following three types: (i) replace by new; (ii) a reduction in age (virtual age) and a reduction in the hazard function; and (iii) overhaul (major shutdown PM) – the level of reduction being dependent on the components replaced. For complex plants which require PM actions to be done at fairly short intervals, the PM actions are better modeled by a continuous function which changes over time.

Repair time is comprised of several time periods: investigation time (the time needed to locate the fault), the time needed to carry out the actual repair, and testing time after the repair. Some of these times may be predicted precisely whilst others may be highly variable, depending on the item and the type of failure. The easiest approach is to aggregate all the above-mentioned times into a single repair time X, modeled as a random variable with a distribution function. A similar approach may be used to model downtimes.

Maintenance costs may be classified into three categories: (i) direct costs, costs incurred in carrying out a PM or CM action (materials and labor costs); (ii) indirect costs (setting up and running of the workshop, costs associated with spare part inventory, administration costs); and (iii) consequential costs (lost production, loss of customer goodwill, penalty for delays, etc.). Maintenance cost models may be grouped broadly into the following two categories: (i) modeling the cost of each action (PM and CM) separately by a parameter (if the variability is insignificant) or by a distribution function (if the variability is significant) and (ii) modeling cost per unit time with the unit being a week, month, or year.

Repair–replace decisions for repairable items may be based on the number of repair times or the repair cost considered as a random variable.

The modeling of PM and CM actions for fleet and infrastructures poses additional challenges, and some of these have been discussed in this chapter.

Review Questions

11.1 What are the different types of repairs?

11.2 Explain the differences in modeling CM actions for repairable and non-repairable items.

11.3   Define minimal repair, perfect repair, and imperfect repair.

11.4   Describe two methods for modeling imperfect repair.

11.5   What are the different ways of modeling PM actions?

11.6   What are the components of downtime?

11.7   Describe methods for modeling repair times and downtimes.

11.8   What are the main elements of maintenance cost?

11.9   Provide two categories of maintenance cost models. What is the difference between them?

11.10   Describe two approaches for repair–replace decisions.

11.11   Describe the additional complexity posed in modeling PM and CM actions for fleets and infrastructures.

Exercises

11.1 The data given below are the times (in hours) that failed buses spent in a workshop (from the time they arrived to the time they left after repair) for a sample of 50 buses.

2.26	0.41	0.45	6.03	0.22	0.23	326.53	0.32	0.03	0.73
33.78	1.15	0.21	0.06	3.77	0.10	1.60	5.04	1.46	9.54
0.58	0.47	0.92	16.36	7.71	0.28	0.08	9.42	0.65	65.11
1.67	14.09	1.30	16.69	0.45	6.05	2.94	0.19	8.27	2.19
5.47	0.62	0.05	14.02	2.31	0.50	1.75	0.07	0.06	0.83

Assume that a repair starts soon after a failed bus arrives (there is no waiting time) so that the given times are for CM action. Carry out an analysis of the data using Minitab to obtain the summary statistics (descriptive statistics). Interpret the results of the analysis.

11.2 Plot the EDF and empirical repair rate for the data set in Exercise 11.1. What can you infer from the plots?

11.3 Construct a parametric plot to see which of the distributions in Minitab are appropriate for modeling the repair times.

11.4 Assume that repair time may be modeled by a lognormal distribution. Estimate the model parameters using the method of maximum likelihood.

11.5 The data in Exercise 11.1 contain three entries (326.53, 65.11, and 33.78) which are large relative to the remaining data points. Can you give possible reasons for this?

11.6 Assume that the largest three repair times in Exercise 11.5 are treated as outliers and rejected. Is the lognormal distribution still an appropriate model for the repair times?

11.7   The mean time (τ) to carry out a maintenance action decreases due to the learning effect. One can model this as follows. Let τ_k be the mean time for the kth maintenance action, $k = 1, 2, \ldots$, then $\tau_k = \tau_0 \gamma^k$ with $0 < \gamma < 1$. The actual times for carrying out the first n maintenance actions are given by $y_1, y_2, \ldots, y_n$. If the repair times are exponentially distributed, derive the likelihood function for estimating the model parameters.

11.8   If a repairperson is not diligent then, in the process of rectifying a failed component, he/she may induce a failure of the same component within a very short period after the item is put back into operation with probability $p(0 < p < 1)$. Assume that this time delay is negligible. This implies that a failed item may require a random number of attempts before it is fixed properly. If the mean time for each repair attempt is τ, what is the expected time to fix a failed item correctly?

11.9   Show that a mixture of two exponential distribution functions has a decreasing failure rate. If repair times may be modeled by such a distribution function, derive the expression for the log-likelihood function to estimate the model parameters.

11.10   Repairing a failed multi-component system involves the following three stages:
   - *Stage 1:* Identifying the failed component(s) of the system.
   - *Stage 2:* Repairing/replacing failed component(s).
   - *Stage 3:* Testing to ensure that the system is functioning properly.

   What external factors affect the time needed to carry out the activities in each of the three stages? Which of the times are uncertain and need to be modeled by a distribution function?

References

Barlow, R.E. and Hunter, L. (1961) Optimum preventive maintenance policies, *Operations Research*, **8**: 90–100.

Doyen, L. and Gaudion, O. (2004) Classes of imperfect repair models based on reduction of failure intensity or virtual age, *Reliability Engineering & System Safety*, **84**: 45–56.

Kijima, M. (1989) Some results for repairable systems with general maintenance, *Journal of Applied Probability*, **26**: 89–102.

Kline, M.B. (1984) Suitability of the lognormal distribution for corrective maintenance repair times, *Reliability Engineering*, **9**: 65–80.

Mahon, B.H. and Bailey, R.J.M. (1975) A proposed improved replacement policy for army vehicles, *Operational Research Quarterly*, **26**: 477–494.

Pham, H. and Wang, H. (1996) Imperfect maintenance, *European Journal of Operational Research*, **94**: 425–438.

12

Modeling Subsequent Failures

Learning Outcomes

After reading this chapter, you should be able to:

- Distinguish between one-dimensional (1-D) modeling and two-dimensional (2-D) modeling and describe the variables used in 2-D modeling;
- Describe various mathematical formulations used for modeling subsequent failures;
- Describe different CM policies for non-repairable items in the context of modeling subsequent failures;
- Identify the mathematical formulation used and describe the form of the model for each CM policy including replace by new, replace by used, minimal repair, imperfect repair, and repair versus replace policies;
- Describe different subsequent failure models involving both CM and PM actions;
- Describe the data-based modeling approach to modeling subsequent failures.

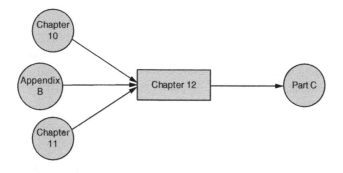

Introduction to Maintenance Engineering: Modeling, Optimization, and Management, First Edition.
Mohammed Ben-Daya, Uday Kumar, and D.N. Prabhakar Murthy.
© 2016 John Wiley & Sons, Ltd. Published 2016 by John Wiley & Sons, Ltd.

12.1   Introduction

In Chapter 10 we discussed the modeling of first failure of an item (based on the black-box approach) with no preventive maintenance (PM). In Chapter 11 we examined the modeling of the effect of PM on the reliability and first failure of an item and the modeling of corrective maintenance (CM) on item reliability subsequent to the first failure. In this chapter we look at the modeling of subsequent failures. We consider both the one- and two-dimensional cases. In the former case, reliability (and failure occurrence) depends only on age, whereas in the latter case, both age and usage have an impact.

The outline of the chapter is as follows. Section 12.2 deals with system characterization (descriptive modeling), and the focus is on both PM and CM actions as points (events) along the time axis in the 1-D case and points in a 2-D plane in the 2-D case. The mathematical formulations needed to build mathematical models are stochastic point process formulations. These are discussed in Section 12.3. Section 12.4 looks at modeling subsequent failures with only CM actions, whilst in Section 12.5 we look at subsequent failures with both CM and PM actions. Section 12.6 deals with data-based modeling, where the approach used is that discussed in Chapter 8. We conclude with a summary in Section 12.7.

12.2   System Characterization for Modeling

12.2.1   One-Dimensional Case

CM actions are random (as failures occur in an uncertain manner) points along the time axis. PM actions are additional points along the time axis. The time instants of PM actions can be either random (in the case of the age policy) or deterministic (in the case of the block policy where PM actions occur at predefined time instants). As a result, the characterization is made in terms of events along the time axis, as shown in Figure 12.1.

12.2.2   Two-Dimensional Case

In Section 10.3, we discussed three approaches to 2-D modeling of first failure. Approach 3 involves: (i) the usage rate varying across the user population and being modeled as a random variable Z with distribution function $G(z) = P\{Z \leq z\}$; and (ii) a failure distribution conditional on the usage rate. This approach can be used for the 2-D case. Conditional on the usage rate, subsequent events (failures, CM and PM actions) are characterized as random points along a line in the 2-D plane, with the slope given by the usage rate,

Figure 12.1   Events in one dimension.

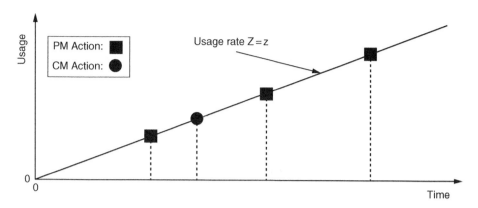

Figure 12.2   Events in two dimensions.

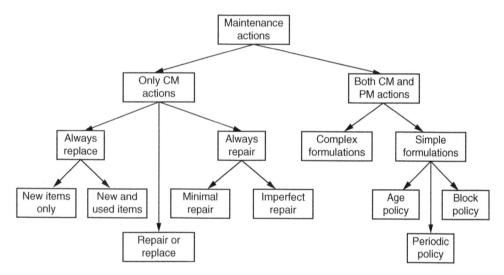

Figure 12.3   Different maintenance action scenarios.

as shown in Figure 12.2. As a result, this approach has been used in modeling failures in two dimensions.

12.2.3   Different Scenarios

The characterization of events for both the one- and two-dimensional cases depends on the type of PM and CM actions used. As a result, there are several different scenarios, as indicated in Figure 12.3. The type of formulation (stochastic point process) needed is different for each scenario, and we discuss this in the remainder of the chapter for the 1-D case.

12.2.4   Assumptions

As discussed in Chapter 8, a model is a simplified representation that is adequate for the goal or purpose in the mind of the modeler. In the modeling of subsequent failures, we assume the following:

1. Failures are detected immediately the instant they occur.
2. The time taken to carry out a PM or CM action is negligible and is assumed to be zero.[1]

12.3   Mathematical Formulations for Modeling

12.3.1   Stochastic Point Processes

A one-dimensional stochastic point process is characterized by events that occur randomly along the time continuum. Events can be failures, PM or CM actions, and so on, occurring in an uncertain manner over time. A two-dimensional stochastic point process is an extension with events occurring randomly in a two-dimensional plane, with time and usage as the two dimensions. Both types of point process belong to the theory of stochastic processes; this is discussed briefly in Appendix B, which includes a non-rigorous treatment of stochastic point processes. These processes play an important role in modeling subsequent failures in the maintenance of systems.

12.3.2   One-Dimensional Point Processes

In a one-dimensional point process formulation the time between events, the number of events in an interval, and the probability of an event occurring in a short interval are all random variables. As a result, the characterization of a point process can be made in three different (equivalent) ways, as indicated below:

C-1: *Time between events:* This is characterized by X_i, $i = 1,2,...$, the time between event i and $(i-1)$, with X_1 being the time instant for the first event measured from time $t = 0$.

C-2: *Count of events over an interval:* This is characterized by $N(t)$, the number of events over the interval $[0, t)$ and $N(t_2, t_1) = N(t_2) - N(t_1)$ denoting the number of events over $[t_1, t_2)$.

C-3: *Intensity function:* Here, the probability of an event occurring in a short interval $[t, t + \delta t)$ is given by $\mu(t)\delta t$ and the probability of two or more events occurring is $o(\delta t^2)$. $\mu(t)$ is also referred to as the *intensity function*.

Depending on the context, one form of characterization is often much simpler than the others. In the context of modeling subsequent failures, two types of point process are of particular importance, and they are discussed below.

---

[1] This is justified as, in general, the time for a repair/replacement $\ll$ the time between events (CM or PM actions). However, if downtime is needed for determining penalty costs, then this needs to be modeled. On the other hand, it can be ignored for modeling subsequent failures, as its impact is, in general, negligible. One can relax these assumptions, but the model formulation and analysis become more complex.

12.3.2.1 Non-Homogeneous Poisson Process (NHPP)

This can be characterized by using either C-2 or C-3, as indicated in Section B.5.2 of Appendix B. For a non-homogeneous Poisson process (NHPP) $\{N(t), t \geq 0\}$, the probability of j events occurring in an interval of length t (characterization C-2) is given by:

$$p_n(s, s+t) = P\{N(s+t) - N(s) = j\} = \frac{e^{-\left\{\int_s^{s+t} \lambda(t';\theta) dt'\right\}} \left\{\int_s^{s+t} \lambda(t';\theta) dt'\right\}^j}{j!} \qquad (12.1)$$

for $j \geq 0$ and for all s and $t \geq 0$. $\lambda(t;\theta)$ is the intensity function (characterization C-3). Two forms of intensity function that have been used extensively in modeling subsequent failures are the following:

1. The Weibull (power-law) intensity function given by:

$$\lambda(t;\theta) = \lambda(t;\alpha,\beta) = \frac{\beta}{\alpha}\left(\frac{t}{\alpha}\right)^{\beta-1}, \alpha > 0, \beta > 0, t \geq 0 \qquad (12.2)$$

The corresponding form for the cumulative intensity function $\Lambda(t;\theta) = \int_0^t \lambda(t';\theta) dt'$ is $\Lambda(t;\alpha,\beta) = (t/\alpha)^\beta$.

2. The log-linear intensity function,[2] given by:

$$\lambda(t;\theta) = \lambda(t;\gamma_0,\gamma_1) = \exp(\gamma_0 + \gamma_1 t), -\infty < \gamma_0, \gamma_1 < \infty, t \geq 0 \qquad (12.3)$$

The corresponding form for $\Lambda(t;\theta)$ is $\Lambda(t;\gamma_0,\gamma_1) = [\exp(\gamma_0)][\exp(\gamma_1 t) - 1]/\gamma_1$.

Comments:

- If $\beta = 1 [\gamma_1 = 0]$, the power-law [log-linear] NHPP becomes the homogeneous Poisson process (HPP).
- If $\beta < 1 [\gamma_1 < 0]$, the power-law [log-linear] intensity is a strictly decreasing function of age. The result is known as a happy system, since it improves with age.
- If $\beta > 1 [\gamma_1 > 0]$, the power-law [log-linear] intensity is an increasing function of age. The system is known as a sad system, in that it degrades with age.

12.3.2.2 Renewal Process (RP)

This is characterized very easily by C-1. For an *ordinary renewal process*, the inter-event times are independent and identically distributed non-negative random variables with an arbitrary distribution function $F(x;\theta)$.

A *delayed renewal process* is similar to the ordinary renewal process with the following important difference – the time to the first event is a non-negative random variable with

---

[2] This is also known as the exponential law or the Cox–Lewis intensity function.

distribution function $F(x; \theta)$ and the time intervals between subsequent events are independent and identically distributed non-negative random variables with a distribution function $G(x; \theta)$ that is different from $F(x; \theta)$.

Renewal Function
Let $M(t)$ denote the expected value of $N(t)$. It is related to the distribution function $F(x)$ and the density function $f(x)$ by the integral equation given below, where the notation has been simplified by ignoring the dependence on the parameter set θ.[3]

$$M(t) = F(t) + \int_0^t M(t-x) f(x) dx \qquad (12.4)$$

$M(t)$ plays an important role in the analysis and optimization of block and other policies and this is discussed further in Part C of the book. It is also commonly referred to as the *renewal function*.

In general, it is not possible to derive analytical expressions for $M(t)$. As such, one needs to use a numerical method to obtain it. A very simple algorithm to obtain $M(t)$ numerically is as follows:[4]

1. Select a small step size Δ and $M(t)$ is computed for $t = \Delta j$, $j = 0, 1, 2, \ldots$.
2. $M(0) = 0$
3. $M(\Delta j) = F(\Delta j) + \sum_{i=0}^{j-1} M(\Delta\{j-i\}) f(\Delta i) \Delta$, $j = 1, 2, \ldots$

Note: The renewal function for any intermediate values of t is obtained by simple linear interpolation.

12.3.3   ROCOF

The events are failures over time, and let $N(t)$ (a point process) denote the count of failures over $[0, t)$. The mean function of $N(t)$, often referred to as the *mean cumulative function* (MCF), is denoted by $\psi(t) = E[N(t)]$. From characterization C-3 of a point process, one of three possible events can occur over the interval $[t, t + \delta t)$:

- No failure with probability $1 - \mu(t)\delta t$.
- One failure with probability $\mu(t)\delta t$.
- Two or more failures with probability $o(\delta t^2)$.

As a result, the expected number of failures over $[t, t + \delta t)$ is given by:

$$\psi(t + \delta t) - \psi(t) = 0\{1 - \mu(t)\delta t\} + 1\{\mu(t)\delta t\} + o(\delta t^2) = \mu(t)\delta t + o(\delta t^2) \qquad (12.5)$$

---

[3] The derivation of this is given in Appendix B and Equation (12.4) is the same as Equation (B.22).
[4] More refined and accurate methods for evaluating the integral on the right-hand side of Equation (12.4) can be found in most textbooks on numerical methods and analysis.

Rearranging and taking the limit as $\delta t \to 0$, we have:

$$\mu(t) = \frac{d\psi(t)}{dt}.$$ (12.6)

$\mu(t)$ is the *rate of occurrence of failures* (ROCOFs) and is the derivative of the MCF.[5]
In the case of an ordinary renewal process we have $\psi(t) = M(t)$, given by Equation (12.4) and $\mu(t) = m(t)$, given by the following integral equation:

$$m(t) = f(t) + \int_0^t m(t-x) f(x) dx$$ (12.7)

The above equation can also be obtained by taking the derivatives on both sides of Equation (12.4). $m(t)$ is often called the *renewal density function*.
In the case of the NHPP we have $\psi(t) = \Lambda(t)$, the cumulative intensity function given by:

$$\Lambda(t) = \int_0^t \lambda(t') dt'.$$ (12.8)

and $\mu(t) = \lambda(t)$, the intensity function.

12.3.4 Renewal Points and Cycles

Let t_i, $i = 0,1,\ldots,$ be an increasing sequence in time. It defines a sequence of renewal points (for a point process $\{X(t),\ t \geq 0\}$) if $P\{\Phi(X(t)), t \geq t_i\}$ is the same for all $i \geq 0$, where $\Phi(X(t))$ is function of $X(t)$. The time interval between two adjacent renewal points defines a renewal cycle and is characterized by the fact that $P\{\psi(X(t)), t_i \leq t < t_{i+1}\}$ is the same for all $i \geq 0$. In the context of maintenance modeling, $\Phi(X(t)), t_i \leq t < t_{i+1}$, defines the cycle cost CC and $(t_{i+1} - t_i)$ defines the cycle length. Both CC and CL are random variables. The expected values of these ($ECC = E[CC]$ and $ECL = E[CL]$) play an important role in obtaining the expression for the objective function needed for maintenance optimization in terms of the maintenance policy parameters. This then allows us to decide on the optimal parameters to optimize the objective function. This is discussed further in Part C of the book.

12.4 Subsequent Failures with Only CM Actions

We will give the form of the model (obtained by linking the system characterization with an appropriate point process formulation) and present some results of analysis (for example, probabilities, mean, and variance of random variables) that will be used in Part C. The details are given in Appendix B and we indicate the results citing the appropriate equations from this appendix.

---

[5] For a more detailed discussion of MCF and ROCOF, see Rigdon and Basu (2000).

12.4.1   Always Replace

For non-repairable items this is the only option. However, it is also applicable for repairable items if the repair cost is high and replacing is more economical than repairing.

12.4.1.1   Replace by New

In this case, the model involves an ordinary renewal process with every CM action being a renewal point. The time between CM actions is a random variable with distribution function $F(x)$. Let $N(t)$ denote the number of CM actions over the interval $[0, t)$. This is a discrete random variable with probability $p_n(t) = P\{N(t) = n\}$ given by:

$$p_n\left(t\right) = F^{(n)}\left(t\right) - F^{(n+1)}\left(t\right) \tag{12.9}$$

where $F^{(n)}(t)$ is the n-fold convolution of $F(t)$ with itself (see Appendix A for details). The MCF is given by $\psi(t) = M(t)$, with $M(t)$ as the solution of the integral equation given by Equation (12.4). The ROCOF is $m(t)$, given by Equation (12.7).

12.4.1.2   Replace by Used

Let the used items have a failure distribution $G(x)$ which is different from a new item. In this case, the model involves a delayed renewal process. The MCF is given by $\psi(t) = M_d(t)$ with $M_d(t)$ given by:

$$M_d\left(t\right) = F\left(t\right) + \int_0^t M_g\left(t-x\right) f\left(x\right) dx \tag{12.10}$$

where $M_g(t)$ is the renewal function associated with the distribution function $G(t)$ and is given by:

$$M_g\left(t\right) = G\left(t\right) + \int_0^t M_g\left(t-x\right) g\left(x\right) dx \tag{12.11}$$

12.4.2   Always Repair

In this case, the characterization of the number of failures over time depends on the type of repair. We discussed the following two types of repair in Section 11.2.2.

12.4.2.1   Minimal Repair

If the failures are statistically independent, then the number of CM actions, $N(t)$, is an NHPP with intensity function $\lambda(t) = h(t)$, the failure rate associated with $F(t)$.[6] The probability $p_n(t) = P\{N(t) = n\}$ is given by Equation (12.1) with $s = 0$ and the MCF $\psi(t) = \Lambda(t)$, with $\Lambda(t)$ given by Equation (12.8). The ROCOF is given by $\mu(t) = h(t)$.

---

[6] For a proof of this, see Nakagawa and Kowada (1983).

12.4.2.2 Imperfect Repair

In Section 11.2.2 we discussed two formulations: (i) reduction in failure rate and (ii) reduction in age. Since failures occur in an uncertain manner, the characterization of $N(t)$ is nearly impossible.

12.4.3 Repair versus Replace

In this case, there are two types of CM actions: repair or replace. Let p [$(1-p)$] denote the probability that a failure is rectified through minimal repair [replace by new]. Each replacement is a renewal point and, as such, the sequence of replacements may be modeled by a renewal process. Let $\tilde{F}(x)$ denote the distribution function for the time interval (X) between renewals. The failures between renewal points occur according to an NHPP process with intensity function $\lambda(t) = h(t)$, the hazard function associated with $F(t)$.

An expression for $\tilde{F}(x)$ is obtained as follows using the conditional approach, where the conditioning is done on the number of failures rectified through minimal repair. Note that $P\{X > x\} = 1 - \tilde{F}(x)$. Let K denote the number of minimal repairs performed over the interval x. Then it is easily seen that $P\{X > x \mid K = k\} = p^k$, as all failures are repaired minimally. The probability $p_k(x) = P\{K = k\}$ is given by:

$$p_k(x) = \frac{e^{-\Lambda(x)}\{\Lambda(x)\}^k}{k!} \tag{12.12}$$

with $\Lambda(x) = \int_0^x \lambda(t)\,dt$. On removing the conditioning, we have:

$$\tilde{F}(x) = 1 - \sum_{k=0}^{\infty} \frac{e^{-\Lambda(x)}\{\Lambda(x)\}^k p^k}{k!} = 1 - e^{-(1-p)\Lambda(x)} \tag{12.13}$$

When $p = 0$, every failure results in replacement by new, so that $\tilde{F}(x) = F(x)$, and when $p = 1$, every failure is rectified through minimal repair, so that $\tilde{F}(x) = 0$ for all $x \geq 0$, as is to be expected.

12.5 Subsequent Failures with Both CM and PM Actions

In this section we look at the PM and CM actions over time for some of the policies discussed in Chapter 4.

12.5.1 Simple Policies

The three simple policies – the age, the block, and the periodic policy – were discussed in Chapter 4. We now look at modeling the occurrence of events (CM and PM actions) over time. These three policies are characterized by a single parameter (T) and this is a decision variable for optimizing the maintenance costs.

12.5.1.1 Age Policy

In the age policy, every action (PM or CM) is a renewal point. The time between renewals is no longer given by the failure distribution $F(x)$ because of PM actions. The distribution function for time between renewals is given by:

$$F_a(x) = \begin{cases} F(x), & 0 \le x < T \\ 1, & x \ge T \end{cases} \tag{12.14}$$

Note that it is also the distribution function for the cycle length (CL) defined earlier.

Let $N(t)$ denote the number of events (replacements due to CM and PM actions not being differentiated). Then $p_n(t) = P\{N(t) = n\}$ is given by Equation (12.9), the MCF is given by $\psi(t) = M(t)$, where $M(t)$ is the solution of the integral equation given by Equation (12.4) using $F_a(x)$ instead of $F(x)$. The ROCOF is given by $\mu(t) = m(t) = dM(t)/dt$ with $m(t)$ given by Equation (12.7), using $F_a(x)$ instead of $F(x)$.

12.5.1.2 Block Policy

In the block policy, every PM action is a renewal point. As a result, the *cycle length CL* is T. In other words, there is no uncertainty in the cycle length. The number of failures over a cycle is a random variable. Failures occur according to a renewal process, with the distribution function for time between failures given by $F(x)$ (the failure distribution of items).

Over each cycle there is one PM action (involving the replacement of the working item by a new item) and a random number of CM actions. The expected number of failures is given by the renewal function (given by Equation (12.4)) with $t = T$.

12.5.1.3 Periodic Policy

As with the block policy, every PM action is a renewal point, so that the cycle length is T (a deterministic quantity). Since failures are minimally repaired, the number of CM actions over a cycle is a random variable. The failures occur according to an NHPP process with intensity function $\lambda(t) = h(t)$, the hazard function associated with the failure distribution function for the item.

12.5.2 Complex Policies

12.5.2.1 Overhaul

We look at the maintenance actions over the time period from $t = 0$ to the first overhaul at time T_1 with failures in between rectified by minimal repair. As a result, the characterization of CM actions is similar to that for the periodic policy discussed earlier. In other words, the failures occur according to an NHPP process with intensity function $\lambda_0(t)$. After the jth overhaul, CM actions until the next overhaul occur according to an NHPP process with intensity function $\lambda_j(t)$ based on a local clock which is reset to zero after each overhaul.

12.6   Data-Based Modeling[7]

In the data-based (empirical) approach to modeling, the selection of the mathematical formulation (for example, a distribution function in the case of always replace or an intensity function in the case of always minimal repair) is dictated by the data available. As discussed in Chapter 8, the first step is to create empirical plots based on the data. This allows one to test assumptions (for example, increasing or decreasing trends) and to choose a formulation with which to model the data. Once the model formulation has been selected, one needs to estimate the parameters of the model and then validate the model.

In this section we first look at two scenarios: always replace (the data generated from maintenance policies such as the age policy) and always do minimal repair until replacement (the data generated from block, periodic, and other policies).

12.6.1   Always Replace

In the case of "always replace", the data consist of the failure times (for items that failed and were replaced) and censored times (either due to PM actions or the item was still working at the end of the data-collection period). If one assumes that failures are statistically independent, then the formulation to model the data is the probability distribution for failure. Since an item can only have at most one failure, the plots and the process to select an appropriate distribution are identical to those discussed in Chapter 10.

12.6.2   Minimally Repair until Replacement

Let I denote the number of new items in the data available. For item i, $1 \le i \le I$, let t_{ij} denote the age when the item failed for the jth time. The data consist of failure times for items that failed (and were repaired) and supplementary data consisting of censored times for all other items. The focus is on the empirical plot of the MCF (also called the non-parametric estimate of the MCF). The procedure is as follows:[8]

1. Order the unique recurrence times t_{ij} among all of the I items. Let n denote the number of unique times. These ordered unique times are denoted $t_{(1)} < t_{(2)} < \ldots < t_{(n)}$.
2. Compute $n_{it_{(k)}}$, the total number of recurrences for item i at $t_{(k)}$.
3. Let $\delta_{it_{(k)}} = 1$ if item i is still being observed at time $t_{(k)}$ and $\delta_{it_{(k)}} = 0$ otherwise.
4. Compute

$$\hat{\psi}\left(t_{(j)}\right) = \sum_{k=1}^{j} \left[\frac{\sum_{i=1}^{I} \delta_{it_{(k)}} n_{it_{(k)}}}{\sum_{i=1}^{I} \delta_{it_{(k)}}} \right] = \sum_{k=1}^{j} \left[\frac{n_{.t_{(k)}}}{\delta_{.t_{(k)}}} \right] = \sum_{k=1}^{j} \overline{n}_{t_{(k)}} \tag{12.15}$$

---

[7] This section is based on material from Blischke, Karim, and Murthy (2011).
[8] Additional details can be found in Meeker and Escobar (1998, p. 397).

for $j = 1, 2, \ldots, n$, where $n_{.t_{(k)}} = \sum_{i=1}^{I} \delta_{it_{(k)}} n_{it_{(k)}}$, $\delta_{.t_{(k)}} = \sum_{i=1}^{I} \delta_{it_{(k)}}$, and $\bar{n}_{t_{(k)}} = n_{.t_{(k)}} / \delta_{.t_{(k)}}$.

Note that $n_{.t_{(k)}}$ is the total number of item recurrences at time $t_{(k)}$, $\delta_{.t_{(k)}}$ is the size of the risk set[9] at $t_{(k)}$, and $\bar{n}_{t_{(k)}}$ is the average number of recurrences per item at $t_{(k)}$. As in the case of nonparametric estimation of a distribution function (EDF), the estimator $\hat{\psi}(t)$ is a step function, with jumps at recurrence times but constant between recurrence times. The estimator of the variance of $\hat{\psi}(t_{(j)})$ is:[10]

$$V\left[\hat{\psi}\left(t_{(j)}\right)\right] = \sum_{i=1}^{I}\left[\sum_{k=1}^{j}\frac{\delta_{it_{(k)}}}{\delta_{.t_{(k)}}}\left(n_{it_{(k)}} - \bar{n}_{t_{(k)}}\right)\right]^2 \tag{12.16}$$

Example 12.1   Diesel Engine

We consider Data Set E.5 given in Table E.5 (valve seat replacement data for diesel engines) and assume that failed items are minimally repaired. The data include 46 instances of valve replacement. We estimate the MCF to investigate whether the replacement rate increases or decreases with age.

Figure 12.4 is an event plot of the data, showing the observation period and the reported repair times, created by Splida[11] (Meeker and Escobar, 2005). The event plot can also be

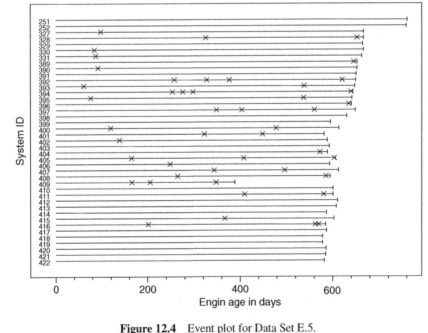

Figure 12.4   Event plot for Data Set E.5.

---

[9] The set contains counts of the items working and of the number uncensored just prior to $t_{(k)}$.
[10] Detailed derivations can be found in Meeker and Escobar (1998, p. 398).
[11] Splida (S-Plus Life Data Analysis), see www.public.iastate.edu/~splida.

Table 12.1 Estimates of the MCF, the standard errors, and 95% confidence intervals.

Age in days $(t_{(j)})$	$\hat{\mu}(t_{(j)})$	$S\hat{E}[\hat{\mu}(t_{(j)})]$	95% confidence interval	
			Lower	Upper
61	0.0244	0.0241	0.0035	0.1690
76	0.0488	0.0336	0.0126	0.1885
84	0.0732	0.0407	0.0246	0.2175
87	0.0976	0.0463	0.0385	0.2475
92	0.1220	0.0511	0.0536	0.2773
35 rows omitted				
604	1.0606	0.1852	0.7532	1.4935
621	1.1194	0.2076	0.7782	1.6102
635	1.1819	0.2070	0.8385	1.6661
640	1.2444	0.2264	0.8711	1.7777
646	1.3210	0.2286	0.9413	1.8550
653	1.6070	0.3515	1.0468	2.4670

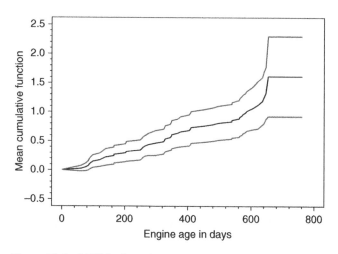

Figure 12.5 MCF for Data Set E.5 with 95% confidence intervals.

created by Minitab.[12] The plot shows when the failures occurred for each system. Each line extends to the last day of observation. Note that service times range from just under 400 days (System 409) to nearly 700 days (System 251).

Estimates of the means $\mu(t_{(j)})$ for the valve seat as a function of engine age along with point-wise ~95% confidence intervals are given in Table 12.1 and shown graphically in Figure 12.5. For example, $\hat{\mu}(92)$ is 0.122, indicating that the mean (cumulative) number of repairs per valve seat at age 92 days is 0.122. Equivalently, the mean number of repairs per hundred valve seats at age 92 days is 12.2 or 12.2%. We are 95% confident that the true mean value of the cumulative function at age 92 days is between 0.0536 and 0.2773.

---

[12] In Minitab, use Stat → Reliability/Survival → Nonparametric Growth Curve.

As can be seen from Table 12.1 and Figure 12.5, the estimate increases sharply between 620 and 650 days, but it is important to recognize that this part of the estimate is based on only a small number of systems, as indicated by the wide confidence intervals in that region. The slightly upwardly concave pattern of the plot of MCF versus time (age) indicates that the time between repairs is slightly decreasing over time, so the system reliability is deteriorating.   ∎

12.6.3 Model Selection and Parameter Estimation

The most commonly used counting process to model the failure process of a repairable system is the Poisson process (either the HPP or the NHPP).

To specify an NHPP model, we use the intensity function (also known as the rate of occurrence of failures), $\lambda(t) = \lambda(t;\theta)$, where θ is a vector of unknown parameters. The corresponding cumulative intensity function (or mean cumulative number of recurrences over (0, t]) is $\Lambda(t) = \Lambda(t;\theta)$.

For item i the data are collected over the time period $(t_{i0}, t_{a_i}]$, with $t_{i0} = 0$ the time at which the item is put into operation and t_{a_i} the time of the last failure for item i $(i = 1, 2, \ldots I)$. If we observe item i for the time interval $(0, t_{a_i}]$, with n_i failures occurring at exact times $t_{i1}, t_{i2}, \ldots, t_{in_i}$, then the likelihood is given by:

$$L_i(\theta) = \prod_{j=1}^{n_i} \lambda(t_{ij};\theta)\exp\left[-\Lambda\left(t_{a_i};\theta\right)\right], \quad i = 1, 2, \ldots, I \tag{12.17}$$

Another possible scheme for a repairable system is to observe the system until the n_ith failure. In this case, likelihood Equation (12.17) still applies, but with t_{a_i} replaced by t_{n_i}.

With multiple items, we assume that all I items are identical and so have the same failure intensity function $\lambda(t)$.[13] The NHPP complete likelihood function (based on all items) is:

$$L(\theta) = \prod_{i=1}^{I} L_i(\theta) = \prod_{i=1}^{I}\left[\left\{\prod_{j=1}^{n_i}\lambda(t_{ij};\theta)\right\}\exp\left[-\Lambda\left(t_{a_i};\theta\right)\right]\right] \tag{12.18}$$

12.6.3.1 Weibull Intensity Function

The intensity function $\lambda_1(t;\alpha,\beta)$ is given by Equation (12.2). In the case of a single item, the likelihood is given by:

$$L(\alpha,\beta) = \left(\frac{\beta}{\alpha}\right)^n \prod_{i=1}^{n} t_i^{\beta-1}\exp\left[-\Lambda\left(t_\alpha;\alpha;\beta\right)\right] \tag{12.19}$$

---

[13] The assumption that all customers (or systems) have the same intensity function is a strong assumption and might be inappropriate in some applications. If all systems are different, then each system can be modeled by its own intensity function with parameter θ_i.

where $t_1, t_2, \ldots, t_n$ are the *exact* times of the n failures. The parameter estimates are given by:

$$\hat{\beta} = \frac{n}{n \log(t_a) - \sum\limits_{i=1}^{n} \log(t_i)} \quad \text{and} \quad \hat{\alpha} = \frac{t_a}{n^{1/\hat{\beta}}} \qquad (12.20)$$

The likelihood for multiple failures for multiple items becomes:

$$L(\alpha, \beta) = \left(\frac{\beta}{\alpha^\beta}\right)^N \prod_{i=1}^{I} \prod_{j=1}^{n_i} \{t_{ij}^{\beta-1}\} \exp\left[-\sum_{i=1}^{I} \left(\frac{t_{a_i}}{\alpha}\right)^\beta\right] \qquad (12.21)$$

where $N = \sum\limits_{i=1}^{I} n_i$. For this model, the maximum likelihood estimates (MLEs) of α and β are obtained by solving the equations:

$$\hat{\alpha} = \left(\frac{\sum\limits_{i=1}^{I} t_{a_i}^{\hat{\beta}}}{N}\right) \quad \text{and} \quad \hat{\beta} = \frac{N}{\hat{\alpha}^{-\beta} \sum\limits_{i=1}^{I} t_{a_i}^{\hat{\beta}} \log(t_{a_i}) - \sum\limits_{i=1}^{I} \sum\limits_{j=1}^{n_i} \log(t_{ij})}. \qquad (12.22)$$

12.6.3.2 Log-Linear Intensity Function

The intensity function $\lambda_2(t; \gamma_0, \gamma_1)$ is given by Equation (12.3). In the case of a single item, the likelihood function is given by:

$$L(\gamma_0, \gamma_1) = \exp\left(n\gamma_0 + \gamma_1 \sum_{i=1}^{n} t_i\right) \exp\left[-\Lambda(t_a; \gamma_0; \gamma_1)\right] \qquad (12.23)$$

The parameter estimates of γ_0 and γ_1 are obtained by solving:

$$\sum_{i=1}^{n} t_i + \frac{n}{\hat{\gamma}_1} - \frac{n t_a}{1 - \exp(-\hat{\gamma}_1 t_a)} = 0 \quad \text{and} \quad \hat{\gamma}_0 = \log\left[\frac{n\hat{\gamma}_1}{\exp(\hat{\gamma}_1 t_a) - 1}\right] \qquad (12.24)$$

Similarly, the MLEs of $\hat{\gamma}_0$ and $\hat{\gamma}_1$ for multiple item data can be obtained.

Example 12.2   Diesel Engine

In Example 12.1, we calculated the nonparametric estimate of the MCF for the valve seat replacement data. This data set can be considered to contain data with multiple failures for multiple items ($I = 41$). In Table E.5, "Days observed" is the length of the observation

Table 12.2 MLEs of the parameters of two NHPP models.

Model	Parameter	Estimate	Standard error	95% confidence interval	
				Lower	Upper
Power law	β	1.39958	0.201	1.0066	1.79256
	α	553.643	57.864	440.232	667.054
Log-linear	γ_0	−6.83239	0.32748	−7.47423	−6.19054
	γ_1	0.001657	0.000801	0.000087	0.003228

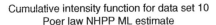

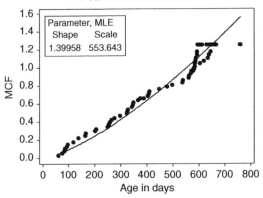

Figure 12.6 Mean cumulative intensity for power-law NHPP model.

period for the engine. Events that occur beyond this time are not observed, and the value recorded can be considered to be the censored time or retirement time of the system.[14]

Table 12.2 shows the MLEs of the parameters with 95% confidence intervals for the two intensity functions. For the power-law NHPP intensity function, $\hat{\beta} = 1.39958$ is greater than 1, indicating that the ROCOF is increasing. We are 95% confident that the interval (1.0066, 1.79256) contains the true value of the shape parameter.

For a log-linear NHPP model, $\hat{\gamma}_1 = 0.001657$, which is greater than zero, and again indicates that the ROCOF is increasing. Both models, the power law with $\hat{\beta} > 1$ and the log-linear with $\hat{\gamma}_1 > 0$, indicate that the failure intensity is an increasing function of age, that is, the system degrades with age.

Figure 12.6 is a plot of the MLE of the cumulative intensity function versus age for the power-law NHPP model, which shows a curve that is slightly upwardly concave. This plot is consistent with a shape parameter that is greater than one, implying that the system is deteriorating with time.

Figure 12.7 shows the nonparametric estimate of the MCF and the fitted cumulative intensity functions for both the power-law and log-linear NHPP models. This figure

---

[14] In order to analyze the repairable system failure data with Minitab, each system must have a retirement time, which is the largest time for that system.

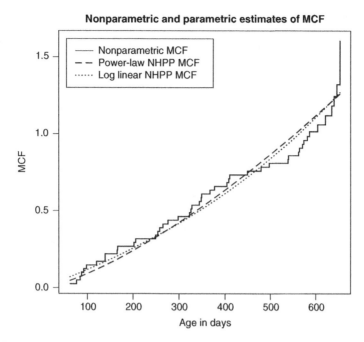

Figure 12.7 Estimates of the nonparametric MCF and the fitted cumulative intensity functions for two models.

indicates very little difference between the two fitted models. Both NHPP models seem to follow roughly the pattern in the data (as indicated by the nonparametric MCF estimate).

The slightly upwardly concave pattern of the plots of the MCFs for both models indicates that the time between repairs is slightly decreasing over time; that is, the system reliability is deteriorating with increasing age. The values of the maximum log-likelihoods are -346.5 for the power-law model and -346.8 for the log-linear model; the close agreement of these values again indicates that the two models are equally appropriate for the data. ∎

Example 12.3 Photocopier

The Data Set E.3 refers to photocopier data. Each row of the table lists a part that was replaced, giving the number of copies made at the time of replacement, the age of the machine in days, and the component replaced. The data are from a single customer (item/unit), with actual failure times for 41 failures. The two NHPP intensity functions will be examined for modeling failures (minimally repaired at the system level) using both "Days" and "Copies" as the independent variables.

Table 12.3 shows the MLEs and 95% confidence intervals for the parameters of the two models, obtained by the method discussed above. For the power-law NHPP model, the MLE of the shape parameter is $\hat{\beta} = 1.561$ for Days and $\hat{\beta} = 1.762$ for Copies. Both are greater than 1, indicating that the ROCOF is increasing for the two lifetime variables.

Table 12.3 MLEs of the parameters of two models.

Variable	Model	Parameter	Estimate	Standard error	95% confidence interval		Log-likelihood
					Lower	Upper	
Days	Power	β	1.561	0.241	1.089	2.033	−189.0
	law	α	152.919	58.358	38.54	267.298	
	Log-	γ_0	−4.720825	0.406119	−5.5168	−3.92485	−187.4
	linear	γ_1	0.001086	0.000354	0.000392	0.001779	
Copies	Power	β	1.762	0.272	1.229	2.294	−453.6
	law	α	133 528	45 164	45007.8	222 047	
	Log-	γ_0	−11.28	0.412	−12.09	−10.47	−453.5
	linear	γ_1	0.00000171	0.00000054	0.00000066	0.00000276	

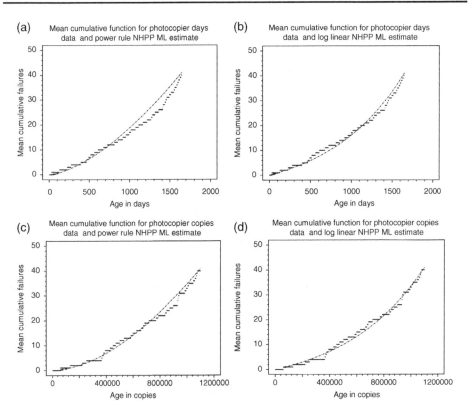

Figure 12.8 MCFs for Days with power law (a) and log-linear (b), and for Copies with power law (c) and log-linear (d).

For the log-linear NHPP model, the MLEs of γ_1 are $\hat{\gamma}_1 = 0.001086$ for Days and $\hat{\gamma}_1 = 0.00000171$ for Copies, both of which are greater than zero, indicating that the intensity functions are increasing with Days and Copies.

The intensity and cumulative intensity functions for Days and Copies are estimated for the two models by substituting the parameter estimates from Table 12.3. The MLEs of the MCFs for the two variables for both models are shown in Figure 12.8.

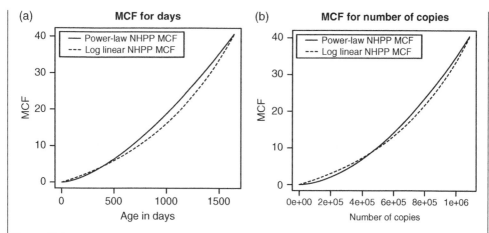

Figure 12.9 Fitted cumulative intensity functions for (a) Days and (b) Copies under the two models.

The estimated log-likelihoods in Table 12.3 and Figure 12.8 indicate that the model with the log-linear intensity function fits better than the model with the power-law intensity function for Days, and both of the models fit approximately equally for Copies.

Figure 12.9 shows plots of the MLEs of the cumulative intensity functions for Days (a) and Copies (b) under the two NHPP models. For Copies, the figure indicates a smaller difference between the two fitted models than in the case of Days. The plots in this figure may be used to predict the MCF for any specified lifetime of the photocopier machine, either in days or in number of copies, using the power-law and log-linear models. ∎

12.7 Summary

The modeling of subsequent failures can be done in one dimension (reliability and failures depend only on age) or in two dimensions (both age and usage have an impact). The characterization of events for both the one- and two-dimensional cases depends on the type of PM and CM actions used.

The mathematical formulations needed to build mathematical models are stochastic point process formulations including: (i) stochastic point processes; (ii) one-dimensional point processes (NHPP, renewal process); and (iii) ROCOF.

Several policies involving only CM actions were considered including: (i) always replace (replace by new, replace by used); (ii) always repair (minimal repair, imperfect repair); and (iii) repair versus replace.

Subsequent failures with both CM and PM actions were also considered including: (i) simple policies (age policy, block policy, and periodic policy) and (ii) complex policies (overhaul).

In the data-based (empirical) approach to modeling, the selection of the mathematical formulation (for example, a distribution function in the case of always replace or an intensity function in the case of always minimal repair) is dictated by the data available. Once the model formulation has been selected, one needs to estimate the parameters of the model and then validate the model. Two scenarios were used to illustrate this approach: (i) always replace (the data were generated from maintenance policies such as the age policy) and (ii) always do minimal repair until replacement (the data were generated from block, periodic, and other policies).

Review Questions

12.1   What is the difference between one-dimensional and two-dimensional modeling of subsequent failures?

12.2   What are the various mathematical formulations used for modeling subsequent failures? Provide a brief description of each.

12.3   State the two assumptions used in this chapter for modeling subsequent failures.

12.4   Provide a summary of the different scenarios covered in this chapter considering different CM and PM policies.

12.5   Stochastic point processes play an important role in modeling subsequent failures. Describe three different ways used for the characterization of a one-dimensional point process.

12.6   What are the renewal points in each of the following policies:
       (a) Age policy.
       (b) Block policy.
       (c) Periodic policy.

12.7   How is the mathematical formulation selected in the empirical modeling approach?

12.8   In the case of "always replace", the data consist of the failure times and censored times. What is the appropriate mathematical formulation with which to model the data?

Exercises

12.1   The failure times/CM actions (based on a usage clock using 100 hours as the unit) for a diesel engine are given below.[15]

0.860	1.258	1.317	1.442	1.897	2.011	2.122	2.439
3.203	3.238	3.902	3.910	4.000	4.247	4.411	4.456
4.517	4.899	4.910	5.676	5.755	6.137	6.221	6.311
6.6613	6.975	7.335	8.158	8.498	8.690	9.042	9.330
9.394	9.426	9.872	10.910	11.511	11.575	12.100	12.126
12.368	12.681	12.795	13.399	12.668	13.780	13.877	14.007
14.028	14.035	14.173	14.173	14.449	14.587	14.610	15.070
16.000	—	—	—	—	—	—	—

Plot the cumulative number of CM actions versus time (based on the usage clock). What can you infer from this plot?

12.2   Plot the time between CM actions for the data in Exercise 12.1. What can you infer from this plot?

---

[15] The data are from Lee (1980).

12.3 Assume that the failures over time in Exercise 12.1 may be modeled by a power-law NHPP. Estimate the model parameters of the intensity function using the method of maximum likelihood. Estimate the mean cumulative number of failures as a function of hours and compare it with the empirical plot of Exercise 12.1.

12.4 Assume that the failures over time in Exercise 12.1 can be modeled by a log-linear NHPP. Estimate the model parameters of the intensity function using the method of maximum likelihood. Estimate the mean cumulative number of failures as a function of hours and compare it with the empirical plot of Exercise 12.1.

12.5 The failure times/CM actions (based on a usage clock using 100 hours as the unit) for a diesel engine are given below.[16]

1.382	2.990	4.124	6.827	7.472	7.567	8.845	9.450
9.794	10.848	11.993	12.300	15.413	16.497	17.352	17.632
18.122	19.067	19.172	19.299	19.360	19.686	19.940	19.944
20.121	20.132	20.431	20.525	21.057	21.061	21.309	21.310
21.378	21.391	21.456	21.461	21.603	21.658	21.688	21.750
21.815	21.820	21.822	21.888	21.930	21.943	21.946	22.181
22.311	22.634	22.635	22.669	22.691	22.846	22.947	23.149
23.305	23.491	23.526	23.774	23.791	23.822	24.006	24.286
25.000	25.010	25.048	25.268	25.400	25.500	25.518	

Plot the cumulative number of CM actions versus time (based on the usage clock). What can you infer from this plot?

12.6 Plot the time between CM actions for the data in Exercise 12.5. What can you infer from this plot?

12.7 Assume that the failures over time in Exercise 12.5 may be modeled by a power-law NHPP. Estimate the model parameters of the intensity function using the method of maximum likelihood. Estimate the mean cumulative number of failures as a function of hours and compare it with the empirical plot of Exercise 12.5.

12.8 Often, maintenance data are not collected properly. An example of this is when, instead of recording the actual failure times, the data available for modeling are the pooled failures on a monthly basis. How can one estimate the parameters of the intensity function with such data?

12.9 Another case of poor data collection is when the failure times recorded may differ from the true failure times. Assume that the error in recording is uniformly distributed over the interval $[-a, a]$. How can this information be incorporated into the method of maximum likelihood to estimate the parameters of the intensity function?

---

[16] The data are from Ascher and Feingold (1984).

12.10 The failures of an object over time in a normal operating environment are given by an intensity function $\lambda(t)$, $t \geq 0$. However, when the object is used in a different environment, its failure may be affected by different factors. One such case is failure due to external shocks exceeding a threshold limit v. The shocks occur according to a Poisson process and the strength of a shock is exponentially distributed. How does this affect the failure intensity and what are the implications for CM actions needed?

References

Ascher, H. and Feingold, H. (1984) *Repairable System Reliability*, Marcel Dekker, New York.

Blischke, W.R., Karim, M.R., and Murthy, D.N.P. (2011) *Warranty Data Collection and Analysis*, Springer-Verlag, London.

Lee, L. (1980) Testing adequacy of the Weibull and loglinear rate models for a Poisson process, *Technometrics*, **22**: 195–199.

Meeker, W.Q. and Escobar, L.A. (1998) *Statistical Methods for Reliability Data*, John Wiley & Sons, Inc., New York.

Meeker, W.Q. and Escobar, L.A. (2005) SPLIDA (S-Plus Life Data Analysis), www.public.iastate.edu/~splida (accessed 5 August 2015).

Nakagawa, T. and Kowada, M. (1983) Analysis of a system with minimal repair and its application to a replacement policy, *European Journal of Operational Research*, **12**: 176–182.

Rigdon, S.E. and Basu, A.P. (2000) *Statistical Methods for Reliability of Repairable Systems*, John Wiley & Sons, Inc., New York.

Part C

Maintenance Decision Models and Optimization

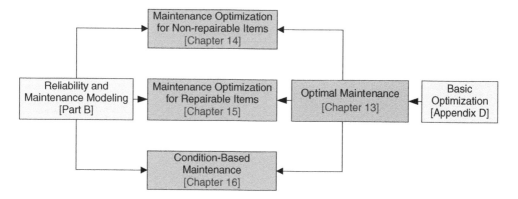

Part C

Maintenance Decision Models and Optimization

13

Optimal Maintenance

Learning Outcomes

After reading this chapter, you should be able to:

- Describe a general framework for deciding the optimal parameters of a given maintenance policy;
- Identify and describe the elements of a maintenance policy including maintenance action, maintenance effectiveness, and the triggering mechanism for maintenance action;
- Describe optimization models for determining optimal maintenance policies including the objective function, the information required, and the selection of appropriate optimization methods;
- Describe the steps required to determine the optimal decision parameters of a maintenance policy.

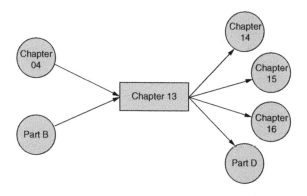

Introduction to Maintenance Engineering: Modeling, Optimization, and Management, First Edition.
Mohammed Ben-Daya, Uday Kumar, and D.N. Prabhakar Murthy.
© 2016 John Wiley & Sons, Ltd. Published 2016 by John Wiley & Sons, Ltd.

13.1   Introduction

As discussed in Chapter 2, every engineered object can be decomposed into a multi-level decomposition with the object at the top, components at the bottom, and one or more intermediate levels. Maintenance of an item (object, component, or some intermediate level) can either involve no preventive maintenance (PM) actions (so that the item is allowed to operate until failure) or some form of PM action to minimize the occurrence of failures. In Chapter 4 we discussed a variety of maintenance policies characterized by one or more parameters. The decision-maker needs to select these optimally. This requires the specification of an objective function (based on cost, operational performance, or both). This requires a proper framework and a model to determine the optimal parameter values. In this chapter we focus on the framework and the building of the model.

The outline of the chapter is as follows. Section 13.2 deals with the framework for deciding the optimal maintenance decisions for an item. This involves several elements. Some of these have been discussed in detail in earlier chapters. We review them briefly in the later sections of the chapter. Some details of the other elements not discussed earlier are also given, so that it provides a good basis for building decision models. Section 13.3 reviews various issues relating to maintenance policies and Section 13.4 discusses the nature of the decision parameters for the various policies. Section 13.5 deals with the specification of the objective function. The optimization model involves linking many different models discussed in earlier chapters, and this is discussed in Section 13.6. The information available plays a critical role in the optimal decision, and this issue is discussed in Section 13.7. The optimization of the model is discussed in Section 13.8. Section 13.9 concludes with a summary of the chapter and a brief discussion on how the results of this chapter are used in the remaining chapters of Part C to decide on the optimal parameter values.

13.2   Framework for Optimal Maintenance Decisions

There are several different options when it comes to maintaining an item. These can be grouped broadly into the following two categories:

- No PM is carried out, so that the item is allowed to operate until failure and then corrective maintenance (CM) action is initiated.
- PM is used to reduce the likelihood of failures, so the item is subjected to both PM and CM actions.

The PM actions, in turn, depend on how the state (or condition) of the item is characterized and monitored. For many items, a binary characterization of item state is used – either working or failed. For others, a multi-level characterization is used. In the former case, the first failure is modeled by a black-box approach involving a failure distribution function; subsequent failures depend on the maintenance actions carried out and these depend on the particular maintenance policy used. In Chapter 4 we discussed a variety of maintenance policies. In the latter case, the maintenance is condition-based and one needs to model the degradation process.[1]

---

[1] Chapters 14 and 15 deal with deciding on the optimal parameters based on a binary characterization, and Chapter 16 deals with deciding on the optimal parameters for CBM.

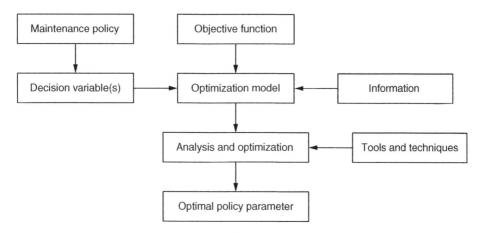

Figure 13.1 A framework for deciding on optimal maintenance decision parameters.

The first decision is to select the optimal policy from the list of potential policies under consideration. This, in turn, requires first deciding on the optimal values for each of the policies under consideration. Figure 13.1 shows the framework for deciding on optimal parameters given the maintenance policy. It consists of several elements. A proper understanding of the various issues for each of the elements is important when deciding on the optimal maintenance for an item. We discuss this further in the remaining sections of the chapter.

13.3 Maintenance Policy

A maintenance policy for an item specifies the type of action(s) for the execution of PM and CM activities and the triggering of PM activities.

13.3.1 Maintenance Actions

The different maintenance actions (discussed in detail in Chapter 4) of a maintenance policy include one or more of the following:

- Corrective maintenance;
- Preventive maintenance;
- Inspection;
- Opportunistic maintenance.

13.3.2 Maintenance Effectiveness

The effectiveness of maintenance actions characterizes the degree to which the operating condition of the item is restored after the maintenance action is performed in a planned manner. The degree of restoration specified in a maintenance policy (discussed in Chapter 4) includes the following:

- Perfect maintenance;
- Minimal repair;
- Imperfect maintenance – better or worse than before.

13.3.3   Trigger Mechanism

The different trigger mechanisms for a maintenance policy can include one or more of the following:

- Clock-based (using a calendar, age, or usage clock);
- Failure rate limit;
- Limit on repair cost or repair time;
- Number of repairs before replacement;
- Opportunity;
- Condition (for condition-based maintenance (CBM) policies).

13.4   Decision Parameters

A maintenance policy is characterized by one or more parameters which we denote by $(\wp)$. This can be either a scalar or a vector. A parameter can be either continuous or discrete valued. Table 13.1 lists the parameters for the policies defined in Chapter 4.

Table 13.1   Decision parameters for maintenance policies 4.1–4.14.

Policy	Parameters $(\wp)$	Continuous	Discrete
4.1	Replacement time (T)	√	
4.2	Replacement time (T)	√	
4.3	Replacement time (T)	√	
4.4	Failure limit (ψ)	√	
4.5	Cost limit (v)	√	
4.6	Repair time limit (T)	√	
4.7	Number of repairs before replacement (K)		√
4.8	Number of repairs before replacement (K)		√
	Replacement time (T)	√	
4.9	Age threshold level	√	
4.10	Two time control limits (t, T)	√	
4.11	Two failure rate control limits (L, U)	√	
4.12	Number of failed units (m)		√
4.13	Number of failed units (m)		√
	Inspection time (T)	√	
4.14	Number of failed units (m)		√
	Age limit (τ)	√	

Policies 4.8, 4.13, and 4.14 each have two parameters – one continuous and the other discrete. The values these parameters can assume are usually given by an interval, with the lower limit being zero and the upper limit either finite or infinite.

13.5 Objective Function

The objective of maintenance is, in general, to ensure that the engineered object delivers the output (products and/or services) and ensures the survival (and profitability) of the business. This requires ensuring adequate levels of availability (to meet desired production levels) and reliability (for safety reasons). Maintenance needs to take these into account as well as the costs associated with PM and CM actions and the indirect costs associated with each failure.

One needs to define an objective function that captures all of these factors to arrive at the optimal maintenance plan by achieving a sensible trade-off between the different factors. As such, objectives in the optimal maintenance of an item include one or more of the following:

- *Technical objectives:* These include the availability and reliability of the item, and overall equipment effectiveness (in the case of equipment used for production);
- *Economic objectives:* These include cost rate, total cost, and profits;
- *Other objectives:* Other possible objectives include safety considerations, risk, quality issues, and so on.

The objective function may be scalar, that is, a single objective such as minimizing cost or maximizing availability, or it may be a vector involving two or more objectives – for example, minimizing cost and maximizing availability are the two most commonly used.

Based on the experience and expert knowledge available, with respect to a specific application, the first step is a prioritization among the different objectives that should be included in the objective function. These may be kept separate, so that the objective function is a vector, or they may be combined (for example, adding them with different weights) to produce a scalar objective function.

An important consideration is the time interval (L) of interest, as the objective function, $J(L, \wp)$, depends on this time interval and also on the decision parameters ($\wp$). The interval may be either finite or infinite. The latter is a good approximation if L is very much larger than the mean time to first failure. The advantage of this approximation is that deriving the expression for the objective function is much simpler. When L is infinite, the objective function (technical or economic) is per unit time and is denoted by $J(\wp)$.

Since failures occur in a random manner, the objective function $\tilde{J}(L, \wp)$ is a random variable. For optimization, one needs to average over all uncertain outcomes. The first approach uses only the mean of the random variable, so that the objective function is given by:

$$J(L, \wp) = E\left[\tilde{J}(L, \wp)\right] \tag{13.1}$$

This does not take into account the variability (which is characterized by the variance). The second approach (called the mean–variance approach) takes this into account, so that:

$$J(L, \wp) = E\left[\tilde{J}(L, \wp)\right] - \varepsilon \, Var\left[\tilde{J}(L, \wp)\right] \tag{13.2}$$

The third approach is based on the Pratt–Arrow utility function used in risk analysis, and in this case the objective function is given by:

$$J(L, \wp) = E\left[\frac{1 - e^{-\gamma \tilde{J}(L, \wp)}}{\gamma}\right] \tag{13.3}$$

where γ reflects the risk-aversion factor (larger values implying greater risk aversion). In Chapters 14 and 15, we will use the expression given by Equation (13.1).

13.6 Optimization Model

The optimization model involves obtaining an expression for $J(L, \wp)$. This requires modeling PM actions and CM actions (resulting from failures) over the time interval $[0, L)$ of interest. The results of Part B play an important role in obtaining the probabilistic characterization of these quantities, as they are functions of the parameters of the maintenance policy (the PM and CM activities and the nature of their outcomes) used. From this characterization, one then obtains expressions for the various technical and economic objectives (given by $\tilde{J}(L, \wp)$). The effort required and the complexity of the expression depend on the policy. From this, one derives the expression for $J(L, \wp)$ using one of three approaches (given by Equations (13.1)–(13.3)) to yield the final objective function.

With L infinite, one can derive an expression for $J(L, \wp)$ in a simpler and more direct manner exploiting the renewal property for a renewal reward process (which defines renewal cycles and the technical and economic measures over each cycle) and the asymptotic results for such processes. These are discussed in Appendix B and their use is illustrated by looking at various maintenance policies in the next two chapters.

13.7 Information

In the context of optimal maintenance, we need to differentiate between two kinds of information: (i) that relating to item state as a function of time (working/failed in the binary characterization of state or different degrees of deterioration in the case of a multi-state characterization) and (ii) that relating to model formulation for the failure (or degradation) of the item (for example, the failure distribution function for first failure) and the parameters of the formulation.

13.7.1 Information about Item State

Information about item state can be (i) *complete,* (ii) *incomplete,* or (iii) *uncertain.*

Item state information is complete if the item is monitored continuously with no inaccuracy in the measurement. In this case, the true state of the item is known exactly at each time instant over the interval $[0, L)$. As a result, every failure is detected the instant it occurs and the item never stays in a hidden failed mode.

Item state information is incomplete if the item is not monitored continuously and the item is inspected at discrete time instants to obtain the information. The inspection times may be periodic or aperiodic (with more frequent inspections as the item ages) and may be either pre-specified (so that the time between inspections is not a decision variable) or may be treated as a decision parameter to be selected optimally. In this case, the inspection time appears explicitly in the final expression for $J(L, \wp)$.

With non-continuous inspection, the item can fail between two inspection instants and the failure is not detected until the first inspection instant after failure. The reason for non-continuous inspection is that the cost of inspections increases with increased frequency of inspection. This cost needs to be traded against the potential cost that can result from an item staying undetected in a failed state. A good example of this is a safety alarm – in the failed state it cannot warn of a plant failure (with catastrophic consequences). Maintenance with inspection time instants as decision variables is discussed in Chapter 15.

Finally, the information about item state is uncertain if the true values cannot be ascertained with certainty and there is some uncertainty in the observed (or estimated) values. There are several reasons why this may occur – the most common one being the random noise in the sensor that performs measurements, which introduces uncertainty.

13.7.2   Information Regarding Model Formulation

Information about the model (for example, the failure distribution function or intensity function) may be either (i) *complete* or (ii) *incomplete.*

- The information is complete if the structure of the formulation (type of distribution) used in the modeling and the values for model parameters are known.
- The information is incomplete if:
  - The structure of the model formulation is known but its parameters are not known, or
  - Both the structure and the parameters are unknown.

In the first case above, one uses an adaptive approach to decide on the optimal maintenance. The uncertainty in the parameters is modeled by a suitable distribution function (called the *prior distribution function*) and the mean value of the parameters is obtained. This is used in obtaining an expression for $J(L, \wp)$ from which the optimal values for the policy parameters are found. The information obtained from the next maintenance action (CM or PM) is used to update the prior in order to yield a posterior distribution function for the model parameters. The mean of this new distribution is used to recalculate the optimal values for the maintenance policy, and the posterior is then treated as the prior for the next iteration. This is called the *Bayesian approach.*[2]

13.8   Optimization

Once the objective function has been obtained as a function of the decision parameters, one can then use techniques from the theory of optimization. It is seldom possible to obtain analytical results – the optimal policy parameters as an analytical function of the optimization model parameters. Often, one needs to use computational techniques to find the optimal solution.

---

[2] We will not be discussing this approach, and interested readers can find more details in Martz and Waller (1982).

In the case when $J(L, \wp)$ is a scalar function, one can have a global optimal solution and possibly none, one, or more than one local optimal solutions. When $J(L, \wp)$ is a vector function, the solution is the Pareto optimal solution, which is discussed briefly in Appendix D.

13.9   Summary

In Chapter 4 we discussed a variety of maintenance policies characterized by one or more parameters. A proper framework and a model to determine the optimal policy parameter values are required. The elements of such a framework were described in this chapter and they include the specification of an appropriate objective function that captures important operational and economic factors, the derivation of an optimization model, the required information about item state and model formulation, and appropriate analysis and optimization of the developed model to determine the optimal values of the policy parameters.

In the next two chapters, we will use the following steps to analyze each maintenance policy:

- *Step 1: Policy description*
  - Brief description;
  - Graphic illustrating the policy.
- *Step 2: Model derivation*
  - Expected cost per cycle;
  - Expected cycle length.
- *Step 3: Model analysis and optimization*
  - Characterization of optimal solutions using first-order optimality conditions;
  - Procedure for finding optimal solutions.
- *Step 4: Example*
  - A numerical example is used for illustrative purposes.

Review Questions

13.1   Describe a general framework for deciding on optimal parameters given the maintenance policy.

13.2   What are the main elements of such a framework?

13.3   What are the main elements of a maintenance policy?

13.4   What are the types of objective function usually considered for optimizing maintenance policies? Provide different classifications.

13.5   Differentiate between two kinds of information in the context of optimal maintenance. Describe each of these.

13.6   What are the steps required for finding the optimal values of the parameters of a given maintenance policy?

Exercises

13.1   Elaborate on the framework for deciding on optimal maintenance decision parameters shown in Figure 13.1.

13.2   The effectiveness of maintenance actions characterizes the degree to which the operating condition of an item is restored after the maintenance action is performed in a planned manner. The degree of restoration specified in a maintenance policy (discussed in Chapter 4) includes the following:
(a) Perfect maintenance.
(b) Minimal repair.
(c) Imperfect maintenance.
Provide real examples where each of these degrees of restoration is applicable.

13.3   The different trigger mechanisms of a policy may include one or more of the following:
(a) Clock-based (using a calendar, age, or usage clock).
(b) Failure rate limit.
(c) Limit on repair cost or repair time.
(d) Number of repairs before replacement.
(e) Opportunity.
(f) Condition (for CBM policies).
Provide real examples where each trigger mechanism is appropriate.

13.4   The objective of maintenance is, in general, to ensure that the engineered object delivers the output (products and/or services) and thus to ensure the survival (and profitability) of the business. Different objectives are appropriate for different applications. Give examples where the following objectives are appropriate:
(a) Cost.
(b) Availability.
(c) Safety.

13.5   Information about an item state is important for building an optimization model for a particular maintenance policy. Item state information may be complete or incomplete, as discussed in Section 13.7.1. Describe some practical applications where the information is complete and other applications where it is incomplete.

13.6   Describe briefly the concept of Pareto optimality. Suggest an application where a vector objective function is appropriate and specify the appropriate objective functions.

13.7   In some applications, it may be appropriate to optimize a certain objective function subject to some constraint on reliability, safety, or failure rate. Identify applications where this appropriate.

13.8   In optimizing maintenance for products, what is the appropriate objective function?

13.9   In optimizing maintenance for plants (a refinery or power plant), what is the appropriate objective function?

13.10   In optimizing maintenance for infrastructures (rail or road), what is the appropriate objective function?

Reference

Martz, H.F. and Waller, R.A. (1982) Bayesian Reliability Analysis, John Wiley & Sons, Inc., New York.

14

Maintenance Optimization for Non-Repairable Items

Learning Outcomes

After reading this chapter, you should be able to:

- Describe several PM policies involving non-repairable items;
- Derive an optimization model to determine the optimal parameters of a given policy;
- Conduct analysis of developed models to characterize optimal solutions;
- Apply developed models to a given application;
- Derive and analyze optimization models for inspection policies.

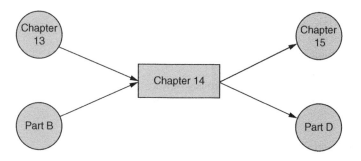

Introduction to Maintenance Engineering: Modeling, Optimization, and Management, First Edition.
Mohammed Ben-Daya, Uday Kumar, and D.N. Prabhakar Murthy.
© 2016 John Wiley & Sons, Ltd. Published 2016 by John Wiley & Sons, Ltd.

14.1   Introduction

In Chapter 4 we introduced several maintenance policies that are applicable in a variety of settings. In this chapter we focus on preventive maintenance (PM) policies that are appropriate at the component or part level for non-repairable items. Failures are modeled using the black-box approach discussed in Chapter 8. The focus is on the optimization of the timing of PM actions in order to optimize a certain measure of performance. We use material from chapters in Part B, and from Appendices B and D to build optimization models and carry out their analysis.

In this chapter, we consider a variety of replacement policies under both finite and infinite time horizons. We also discuss inspection policies in cases where failure can only be detected through inspection. The chapter is organized as follows. In the next section, we introduce common notation and assumptions used in this chapter and introduce the concept of renewal points. In Section 14.3, infinite time horizon models are presented. In Section 14.4 we take a look at group replacement, while Section 14.5 covers finite time horizon models. Inspection models are discussed in Section 14.6, and Section 14.7 concludes the chapter with a summary.

14.2   Preliminaries

Since the item is non-repairable, only first failure is relevant. We use the black-box approach to model failure time.

14.2.1   Different Scenarios

The models for obtaining the optimal parameters depend on several factors, includingthe following:

- *Policy:* Age, block;
- *Objective function:* Scalar or vector;
- *Maintenance action:* PM or no PM;
- *Time horizon:* Finite or infinite;
- *Repair time:* Negligible or not;
- *Failure detection:* Detected instantaneously or not;
- *Performance measure:* Operational and/or economic.

14.2.2   Notation

$\wp$	Parameter of the policy to be selected optimally
$\wp^*$	Optimal $\wp$
L	Time horizon for optimization
$J(L; \wp)$	Objective function (deterministic function)
$J(\wp)$	Asymptotic value of objective function per unit time $[= \lim_{L \to \infty}(J(L; \wp)/L)]$
τ_i	Renewal points $(i = 1, 2, \ldots)$
Z_i	Cycle length for cycle $i, i = 1, 2, \ldots$

ECC	Expected cost per cycle length
ECL	Expected cycle length
C_p	Cost of each PM action
C_f	Cost of each corrective maintenance (CM) action $(C_f > C_p)$
C_r	Cost of replacement
C_i	Cost of each inspection
c_d	Downtime cost per unit time
v	Ratio of CM and PM costs $[= C_f / C_p]$
$F(.)$	Distribution function for time to repair (with mean time to repair μ_f)
$G(.)$	Distribution function for time to repair (with mean time to repair μ_r)

14.2.3 Assumptions

Unless stated otherwise, we assume the following:

1. A scalar objective function.
2. The failure probability distribution is known.
3. A spare is always available when needed under CM or PM action (in other words, an unlimited number of spares is available).
4. The item state is monitored continuously, so that failures are detected the instant they occur.
5. Maintenance times (for both PM and CM actions) are negligible.
6. Replacement costs under PM and CM are deterministic quantities.

Optimization models are derived for the various replacement and inspection policies considered in this chapter in order to determine the optimal timing of PM or inspection actions that minimize an objective function of interest. The objective function considered is scalar and is based on the expected total maintenance cost per unit time. Other objective functions may be used based on reliability or availability. The computations to obtain the optimal parameters for the examples were performed using Excel.

14.2.4 Renewal Points

The time instants of a failure, the start and completion of maintenance (corrective or preventive) can be viewed as events along the time axis. In some cases, some of these might overlap, such as, for example, the case where failure is detected immediately, rectification commences with no delay, and the replacement time is negligible. Since failures (and in some instances the time to replace) are random variables, this defines a point process – $X(t)$. Some of the events are *renewal points*. These play a very important role in deriving expressions for $J(x)$, as will be indicated later.

Definition 14.1

For a point process $\{X(t), t \geq 0\}$, points $\tau_i, i = 1, 2, \ldots$, are said to be renewal points if $P\{X(\tau_i + t); t > 0\}$ is the same for all $i = 1, 2, \ldots$.

The interval between two adjacent renewal points $(\tau_{i+1} - \tau_i, i = 1, 2, \ldots)$ defines a renewal cycle. Associated with each cycle are the cycle length and cycle cost (which can be either cost or some other performance measure such as uptime)[1] and the expected values are denoted *ECL* and *ECC*, respectively.

In the case when the time horizon is infinite, the objective function $J(\wp)$ can be expressed in terms of the expected cycle cost and the expected cycle length, and is given by:

$$J(\wp) = \underset{L \to \infty}{Lim} \frac{J(\wp;L)}{L} = \frac{\text{Expected total cost per cycle}}{\text{Expected cycle length}} = \frac{ECC}{ECL} \qquad (14.1)$$

This is the very well-known Renewal Reward Theorem (see Appendix B) and is used extensively in deciding the optimal values for the policy parameter(s). Critical to using this approach is the proper identification of the underlying renewal points of the point process.

14.3   Infinite Time Horizon

Here, $L = \infty$. We use Equation (14.1) to derive the expression for the asymptotic expected total cost per unit time.

14.3.1   No PM Actions

If no preventive maintenance is used, then an item is replaced only on failure.[2] Note that there is no decision variable and the objective function is denoted simply by *J*. We look at two scenarios.

Scenario I

The objective function is the asymptotic expected total cost per unit time. Here, every failure/replacement instant is a renewal point. As a result, $ECC = C_f$ and $ECL = \mu_f$, the mean time to failure. From Equation (14.1), we have:

$$J = \frac{C_f}{\mu_f} \qquad (14.2)$$

An intuitive explanation of Equation (14.2) is that, on the average, there is a failure replacement every μ_f units of time with a cost equal to C_f. The average cost per unit time is given by the ratio of these two quantities.

---

[1] If the objective function is a performance measure such as availability, then ECC is the expected availability over a cycle. We will continue to use the term "cost" for this case also.
[2] This can be used to serve as a baseline for comparison when preventive maintenance is used to illustrate the benefit of using a PM policy.

Scenario II

Here, we drop the assumption that the time to carry out maintenance is negligible. We assume that it is a random variable with distribution function $G(t)$. The objective function is $J = A_\infty$, the asymptotic availability.

Note that the time instants when CM actions are completed are the renewal points. The cycle cost is the time to failure, so that $ECC = \mu_f$. The cycle length is the sum of life time + time to replace, resulting in $ECL = \mu_f + \mu_d$. Using these in Equation (14.1) yields:

$$A_\infty = \frac{\mu_f}{\mu_f + \mu_d}. \tag{14.3}$$

14.3.2 Age Replacement Policy (Scenario III)

This is Policy 4.1 of Chapter 4. Under this policy, an item is replaced preventively at age T or on failure under CM, whichever occurs first, as shown in Figure 14.1 where $X_i, i = 1, 2, \ldots$, are the random lifetimes of item $i = 1, 2, \ldots$. The parameter of the policy is $\wp = \{T\}$.

Example 14.1 Valve

Consider a valve used in a chemical plant. Its time to failure in years follows a Weibull distribution with shape parameter $\beta = 3$ and scale parameter $\alpha = 5$ years. Let $C_f = \$500$. From Equation (A.26), with $\gamma = 0$, we have $\mu_f = \alpha\,\Gamma[1 + 1/\beta] = 4.465$. From Equation (14.2), the asymptotic expected total cost per year for the policy of replacing on failure is given by $J = 500 / 4.465 = \$111.98$. ∎

The problem is to find the optimal replacement time T that minimizes the asymptotic expected total maintenance cost per unit time given by $J(T)$.

14.3.2.1 Model Derivation

Note that every replacement (PM or CM) is a renewal point. The cycle length is a random variable over $[0, T]$. Let $\tilde{Z}_i$ denote the length of cycle $i, i = 1, 2, \ldots$. It is easily shown that:

$$P\{\tilde{Z}_i \le t\} = \begin{cases} F(t), & \text{for } 0 < t \le T \\ R(T), & \text{for } t > T \end{cases}. \tag{14.4}$$

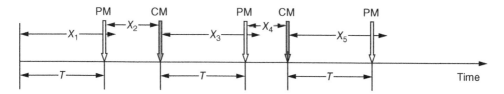

Figure 14.1 Age replacement policy.

As a result, the *ECL* is given by:

$$ECL = \int_0^T tf(t)dt + TR(T) = \int_0^T R(t)dt \tag{14.5}$$

Note that the cycle cost is also a random variable depending on whether there is a PM or a CM action. It is C_p with probability $R(T)$ and C_f with probability $F(T)$. As a result, the *ECC* is given by:

$$ECC = C_p R(T) + C_f F(T). \tag{14.6}$$

Using Equations (14.5) and (14.6) in Equation (14.1), the asymptotic expected total cost per unit time is:

$$J(T) = \frac{C_p R(T) + C_f F(T)}{\int_0^T R(t)dt} \tag{14.7}$$

14.3.2.2 Model Analysis and Optimization

T^* can be obtained from the first-order necessary condition for optimality.[3] Taking the first derivative of $J(T)$ with respect to T and equating to zero yields (after some rearrangement) the first-order condition:

$$h(T)\int_0^T R(t)dt + R(T) = \frac{\upsilon}{\upsilon-1}, \tag{14.8}$$

where $\upsilon = C_f / C_p$. T^* (if it exists) is obtained by solving this equation and then checking for the second-order conditions.

The following theorem gives the conditions for the existence and uniqueness of the optimal solution.

Figure 14.2 shows a plot of $J(T)$ versus T to illustrate this.

Theorem 14.1
If the failure rate function $h(t)$ is continuous and strictly increasing, and

1. If $h(\infty) > C_f/(\mu_f(C_f - C_p))$ then there exists a finite and unique T^* that satisfies Equation (14.8) and the optimal asymptotic expected total cost per unit time is given by Equation (14.7) with $T = T^*$, and
2. If $h(\infty) \le C_f/(\mu_f(C_f - C_p))$ then $T^* = \infty$; that is, the item is replaced only at failure and the asymptotic expected total cost per unit time is given by Equation (14.2).

---

[3] Appendix D discusses the conditions for optimality.

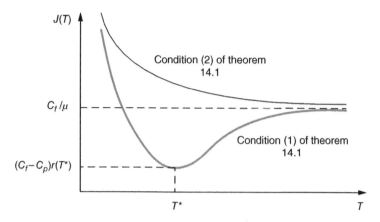

Figure 14.2 The expected total cost per unit time, $J(T)$, as a function of T.

Example 14.2 Valve

Consider the data in Example 14.1. To find the optimal replacement time instant for the valve, one needs to solve Equation (14.8) or plot $J(T)$ given in Equation (14.7). For the Weibull distribution, the integral on the left-hand side of Equation (14.8) can be written in terms of the Gamma distribution function, as follows (see Exercise 14.8).

$$\int_0^T R(t)\,dt = \int_0^T e^{-\left(\frac{t}{\theta}\right)^\beta} dt = \mu G\left(\left(\frac{T}{\theta}\right)^\beta, \frac{1}{\beta}, 1\right)$$

where $G(x, 1/\beta, 1)$ is the cumulative Gamma distribution and μ is the mean of the Weibull distribution. Using Excel, and assuming that $C_p = \$100$, the graph of $J(T)$ is shown in Figure 14.3 and we have $T^* = 2.51$. The corresponding optimal expected cost per year is $J(T^*) = \$60.63$.

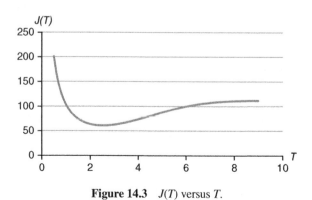

Figure 14.3 $J(T)$ versus T.

In general, it is not possible to obtain T^* analytically and one needs to use a computational scheme.

14.3.3   Age Replacement Policy (Scenario IV)

Here, the replacement times for both CM and PM are random variables with mean μ_d and μ_p, respectively. The objective function is a vector with two elements $J_1(T)$ and $J_2(T)$. $J_1(T)$ is the asymptotic availability and $J_2(T)$ is the asymptotic expected total cost per unit time. Note that time instants when replacements (under CM or PM) are completed are the renewal points. The expected cycle length is the sum of the expected time in the working state + the expected time to replace, and is given by:

$$ECL = \int_0^T R(t)\,dt + \mu_d F(T) + \mu_p R(T) \tag{14.9}$$

The asymptotic availability, $J_1(T)$, is given by:

$$J_1(T) = \frac{\displaystyle\int_0^T R(t)\,dt}{\displaystyle\int_0^T R(t)\,dt + \mu_d F(T) + \mu_p R(T)} \tag{14.10}$$

The ECC for the second objective function is given by Equation (14.6) and the ECL by Equation (14.9). Using these in Equation (14.1) yields:

$$J_2(T) = \frac{C_p R(T) + C_f F(T)}{\displaystyle\int_0^T R(t)\,dt + \mu_d F(T) + \mu_p R(T)} \tag{14.11}$$

Let T_1^* and T_2^* be the values of T that maximize $J_1(T)$ and minimize $J_2(T)$, respectively. If $T_1^* = T_2^*$ then it is the optimal value. When this is not the case, we need to find the Pareto optimal solution.[4] If $T_2^* < T_1^*$, then the Pareto optimal solution can be characterized as follows (see Figure 14.4):[5]

1. For $T < T_2^*$, as T decreases, $J_1(T)$ decreases and $J_2(T)$ increases. This means that the optimal T cannot be in this interval.
2. For $T > T_1^*$, T increases, $J_1(T)$ decreases, and $J_2(T)$ increases. This means that the optimal T cannot be in this interval.
3. For $T_2^* \leq T \leq T_1^*$, an improvement in one element of the objective function can only be obtained at the expense of the other. This defines the interval of the Pareto optimal solution (shown as a bold line in Figure 14.4).

---

[4] A brief discussion of Pareto optimality is given in Appendix D.
[5] A similar approach can be used if $T_2^* > T_1^*$.

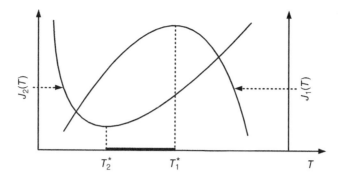

Figure 14.4 Pareto optimal interval.

Example 14.3   Valve

Consider the data in Example 14.1 and assume that $\mu_d = 0.6$ and $\mu_p = 0.3$. Using Excel, we have $T_2^* = 2.41$. The corresponding optimal asymptotic expected cost rate is $J_2(T_2^*) = 53.20$. Similarly, $T_1^* = 4.06$ and the corresponding asymptotic availability is $J_1(T_1^*) = 0.89$. Thus, the Pareto optimal solution is given by the interval [2.41, 4.06].   ∎

14.3.4   Block Replacement Policy (Scenario V)

This policy is Policy 4.2 of Chapter 4. It involves replacements under PM action at time instants $T, 2T, \ldots$, and under CM action should a failure occur between PM replacements, as shown in Figure 14.5.

Here, we also assume that replacement time, whether under PM or CM, is negligible and the cost of a preventive replacement and a failure replacement are given by C_p and C_f, respectively.

14.3.4.1   Model Derivation

In this case, every PM replacement is a renewal point. As a result, the cycle length is given by $ECC = T$, a deterministic quantity. The expected cycle cost is the expected cost of CM and PM actions over a cycle. Since failures over a cycle occur according to a renewal process, the

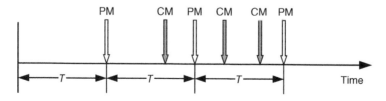

Figure 14.5 Block replacement policy.

expected number of CM actions over a cycle is given by $M(T)$ where $M(t)$ is the renewal function associated with $F(t)$ and is given by:

$$M(t) = F(t) + \int_0^t M(t-x) f(x) dx.$$ (14.12)

Since there is only one PM action during each cycle, the ECC is given by:

$$ECC = C_p + C_f M(T)$$ (14.13)

The asymptotic expected total cost per unit time is obtained using Equation (14.1) and is given by:

$$J(T) = \frac{C_p + C_f M(T)}{T}$$ (14.14)

14.3.4.2 Model Analysis and Optimization

T^* can be obtained from the first-order necessary condition for optimality. Taking the first derivative of $J(T)$ with respect to T and equating to zero yields (after some simplification):

$$Tm(T) - M(T) = \upsilon$$ (14.15)

where $m(T) = dM(T) / dT$ is the renewal density function and $\upsilon = C_f / C_p$. T^* (if it exists) is obtained by solving this equation and then checking for the second-order conditions.

In general, it is not possible to obtain T^* analytically and one needs to use a computational scheme.

Example 14.4   Valve

Consider the data in Example 14.1. The following approximation[6] is used to compute the renewal function:

$$M(T) \approx pF(T) + (1-p)H(T)$$

where $F(T)$ and $H(T)$ are the cumulative distribution function for time to failure and the cumulative hazard function, respectively. $p = W(\beta - 1, 0.9, 0.8731)$, and $W(x, \beta, \alpha)$ is the Weibull distribution function. Using Excel, the graph of $J(T)$ is shown in Figure 14.6 and we have $T^* = 2.42$. The corresponding optimal expected cost per year is $J(T^*) = \$63.62$.

---

[6] Jiang (2010).

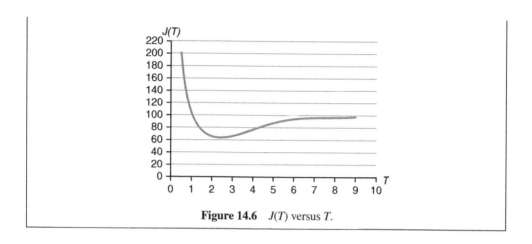

Figure 14.6 $J(T)$ versus T.

14.3.5   Block Replacement Policy (Scenario VI)

In the block policy discussed earlier, any failure between PM actions results in an immediate CM action. We look at a variation of the policy where replacements only occur at discrete time instants given by $T, 2T, \ldots$. As a result, should a failure occur between two inspections, the item stays in a failed state until the next inspection. The cost of each replacement is the same irrespective of whether the item being replaced is in a failed or working state. In other words, $C_f = C_p$, since replacements are planned and there is no unplanned replacement. However, there is a downtime cost associated with the item being in a failed state, and the cost is c_d per unit downtime. The objective function is the asymptotic expected total cost per unit time.

14.3.5.1   Model Derivation

Note that every inspection instant is a renewal point, implying that $ECL = T$. The expected downtime over a cycle is given by:

$$EDT = \int_0^T (T - t) f(t) dt = \int_0^T F(t) dt \tag{14.16}$$

and the ECC is given by:

$$ECC = C_p + c_d\, EDT = C_p + c_d \int_0^T F(t) dt \tag{14.17}$$

Using these in Equation (14.1) results in the asymptotic expected total cost per unit time, given by:

$$J(T) = \frac{C_p + c_d \int_0^T F(t) dt}{T} \tag{14.18}$$

14.3.5.2   Model Analysis and Optimization

Differentiating $J(T)$ with respect to T and equating to zero, we have, after some simplification:

$$TF(T) - \int_0^T F(t)\,dt = \frac{C_p}{c_d}, \text{ or } \int_0^T t f(t)\,dt = \frac{C_p}{c_d} \tag{14.19}$$

T^* is obtained as the solution of the above equation.

Theorem 14.2

If $\mu = \int_0^\infty t f(t)\,dt > C_p/c_d$ then there exists a unique T^* that satisfies Equation (14.19). The corresponding optimal expected cost per unit time is given by $J(T^*) = c_d F(T^*)$.

Example 14.5   Valve

Consider the data in Example 14.1 and assume that $c_d = \$25$. Using Excel to solve Equation (14.19) gives $T_2^* = 2.67$. The corresponding optimal expected cost per year is $J_2(T^*) = \$23.50$.   ∎

14.4   Group Replacement

Assume that a group of N similar items (street lamps, filters in a piping system, etc.) is managed together to accomplish economies of scale. Consider a replacement policy that calls for replacing all N items periodically at time instants $T, 2T, \ldots$. When an item fails during an interval of length T, it is replaced individually.

14.4.1   Model Derivation

In this case, every PM group replacement is a renewal point. As a result, the cycle length is given by $ECC = T$, a deterministic quantity. The expected cycle cost is the expected cost of CM and PM actions over a cycle. Since failures for each item over a cycle occur according to a renewal process, the expected number of CM actions over a cycle for the Nth item is given by $NM(T)$. $M(T)$ is the renewal function associated with $F(T)$ and is given by Equation (14.12) with $t = T$. Therefore, the ECC is given by:

$$ECC = NC_p + C_f NM(T) \tag{14.20}$$

The asymptotic expected cost per unit time is obtained using Equation (14.1) and is given by:

$$J(T) = \frac{NC_p + C_f NM(T)}{T} = N\frac{\left[C_p + C_f M(T) \right]}{T} \tag{14.21}$$

14.4.2   Model Analysis and Optimization

T^* can be obtained from the first-order necessary condition for optimality. Taking the first derivative of $J(T)$ with respect to T and equating to zero yields (after some simplification):

$$Tm(t) - M(T) = \upsilon \tag{14.22}$$

where $m(T) = dM(T)/dt$ is the renewal density function and $\upsilon = C_f / C_p$. T^* (if it exists) is obtained by solving the equation and then checking for the second-order conditions. Note that Equation (14.22) is exactly the same as Equation (14.15).

14.5   Finite Time Horizon

In this section we consider the case where the time horizon, L, is finite and we look at two scenarios. We use the conditional approach where the conditioning is based on the failure time of the first item.[7]

14.5.1   Age Replacement Policy (Scenario VII)

The problem is to find the optimal $\wp = \{T\}$, the replacement age that minimizes $J(L;T)$, the expected total maintenance cost over the time horizon.

14.5.1.1   Model Derivation

Note that every CM or PM replacement is a renewal point. Let X_1 denote the time to the first failure and $J(L;T|X_1 = x)$ denote the conditional expected cost given that $X_1 = x$. If $x \leq T$, then a CM action takes place at time x and the remaining time interval is $L - x$. Since this is a renewal point, the (unconditional) expected cost of maintenance over the remaining interval is given by $J(L-x;T)$. If $x > T$, then a PM action takes place at time T and the remaining time interval is $L - T$. Since this is also a renewal point, the (unconditional) expected cost of maintenance over the remaining interval is given by $J(L-T;T)$. This can be written in a compact form, as indicated below:

$$J\left(L;T \mid X_1 = x\right) = \begin{cases} C_p + J\left(L-T;T\right) & \text{if } x > T \\ C_f + J\left(L-x;T\right) & \text{if } x \leq T \end{cases}. \tag{14.23}$$

On removing the conditioning (by taking the expectation over X_1), we have:

$$J\left(L;T\right) = \int_0^T \left(C_f + J\left(L-x;T\right)\right) f\left(x\right) dx + \int_T^\infty \left(C_p + J\left(L-T;T\right)\right) f\left(x\right) dx \tag{14.24}$$

---

[7] The conditional approach is discussed in Appendix B.

14.5.1.2 Model Analysis and Optimization

As can be seen from Equation (14.24), $J(L;T)$ is the solution of a renewal integral-type equation and this needs to be evaluated numerically. As such, obtaining T^* requires a computational method where the above integral equation needs to be solved.

14.5.2 Block Replacement Policy (Scenario VIII)

The problem is to find the optimal $\wp = \{T\}$, the time instants to carry out PM actions that minimize $J(L;T)$, the expected total maintenance cost over the time horizon.

14.5.2.1 Model Derivation

Let n_L denote the largest integer less than L/T, that is, $n_L = \mathrm{int}(L/T)$. Therefore, the time horizon L consists of $n_L + 1$ intervals. The first n_L intervals are of length T and the last interval is of length $L - n_L T$.

At the end of each of the first n_L intervals, the item is replaced by a new one. Hence, we have n_L PM replacements. Failures within each interval result in CM actions where the failed item is replaced by a new one. As a result, the expected number of CM actions in each of the first n_L intervals is given by $M(T)$ and in the last interval this is given by $M(L - n_L T)$. These expected numbers of CM actions are obtained by solving the integral equation given by Equation (14.12). Since there is only one PM action per interval, the total expected cost is given by:

$$J\left(L;T\right) = n_L\,C_p + \left(n_L M\left(T\right) + M(L - n_L T)\right)C_f \tag{14.25}$$

14.5.2.2 Model Analysis and Optimization

Differentiating $J(L;T)$ with respect to T and equating to zero, we have:

$$\left[n_L\,m\left(T\right) - n_L\,m\left(L - n_L T\right)\right] = 0 \tag{14.26}$$

Solving for T, we have:

$$T = \frac{L}{n_L + 1} \tag{14.27}$$

In other words, T^* requires all the intervals to be of the same length. Using Equation (14.27) in Equation (14.25) yields:

$$J\left(L;n_L\right) = n_L\,C_p + \left(n_L + 1\right)M\left(\frac{L}{n_L + 1}\right)C_f \tag{14.28}$$

So now the problem is to find the optimal value of n_L (an integer variable) that minimizes $J(L;n_L)$. One can simply perform an exhaustive search by evaluating Equation (14.28) for $n_L = 1, 2, \ldots$, to find the optimal n_L and then use Equation (14.27) to obtain T^*.

14.6 Inspection Policies

In this section we deal with the case where the item state (working or failed) is not monitored continuously. Instead, the item is inspected at discrete time instants to determine its state.

A unit is inspected at times $t_k, k = 1, 2, \ldots$, where $t_0 \equiv 0$, based on an age clock, as shown in Figure 14.7. Once a failure is detected, the unit is replaced immediately by a new one. The objective function is the asymptotic expected total cost per unit time and the parameters to be selected optimally are given by the set $\wp \equiv \{t_1, t_2, \ldots\}$.

14.6.1 Model Derivation

The three costs that need to be taken into account are as follows:

1. *Inspection cost:* C_i for each inspection.
2. *Replacement cost* (under CM action): C_r for each replacement.
3. *Downtime cost:* c_d per unit downtime.

Note that each replacement is a renewal point. Let X denote the age at failure. If it is in the interval $[t_k, t_{k+1})$ then the unit remains in the failed state, and undetected, for a period $t_{k+1} - X$, as shown in Figure 14.7.

We use the conditional approach to calculate the *ECC* and *ECL* by conditioning on X. The conditional expected cycle cost is given by:

$$ECC(\wp)\Big|_{t_k \le X < t_{k+1}} = (k+1)C_i + C_r + c_d \int_{t_k}^{t_{k+1}} (t_{k+1} - x) f(x) dx \tag{14.29}$$

where the first two terms represent the inspection and replacement costs, respectively, and the last term is the expected downtime cost. Similarly, the conditional expected cycle length is given by:

$$ECL(\wp)\Big|_{t_k \le X < t_{k+1}} = t_{k+1} \tag{14.30}$$

On removing the conditioning (by taking the expectation over X), the unconditional *ECC* and *ECL* are given by:

$$ECC(\wp) = \sum_{k=0}^{\infty} \left\{ \left[(k+1)C_i + C_r \right] \left[F(t_{k+1}) - F(t_k) \right] + c_d \int_{t_k}^{t_{k+1}} (t_{k+1} - x) f(x) dx \right\} \tag{14.31}$$

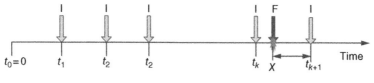

I: Inspection   F: Failure

Figure 14.7 Inspection policy.

$$ECC(\wp) = \sum_{k=0}^{\infty} (t_{k+1}) \left[F(t_{k+1}) - F(t_k) \right] \tag{14.32}$$

The asymptotic expected total cost per unit time is obtained by using Equations (14.31) and (14.32) in Equation (14.1) and is given by:

$$J(\wp) = \frac{\sum_{k=0}^{\infty} \left\{ \left[(k+1)C_i + C_r \right] \left[F(t_{k+1}) - F(t_k) \right] + c_d \int_{t_k}^{t_{k+1}} (t_{k+1} - x) f(x) dx \right\}}{\sum_{k=0}^{\infty} (t_{k+1}) \left[F(t_{k+1}) - F(t_k) \right]} \tag{14.33}$$

14.6.2   Model Analysis and Optimization

One needs to use a computational scheme to obtain the optimal parameter values (inspection times).

Comments:

1. If the failure rate is increasing, then one can intuitively expect that the sequence $t_{k+1} - t_k, k = 1, 2, \ldots$, with $t_0 = 0$, will be a decreasing sequence.
2. A suboptimal inspection scheme is the following: $t_{k+1} = \gamma^{(k)} t_1, k = 1, 2, \ldots$, with $\gamma < 0.$[8] In this case the decision variables to be selected optimally are given by $\wp \equiv \{t_1, \gamma\}$.

14.6.3   Special Case

Consider the special case where the item failure time follows an exponential distribution, that is, $F(t) = 1 - e^{-\lambda t}$ and $\mu = 1 / \lambda$. Then, from Equation (14.33), after some analysis, we have:

$$J(\wp) = \frac{C_i \sum_{k=0}^{\infty} \overline{F}(t_k) - c_d \mu + C_r}{\sum_{k=0}^{\infty} (t_{k+1} - t_k) \overline{F}(t_k)} + C_i \tag{14.34}$$

Since the item failure rate is constant, there is no aging effect. In this case, the optimal inspection times are equally spaced, with $t_{k+1} - t_k = \tau$ for all k. Using this in Equation (14.34), the objective function simplifies to:

$$J(\tau) = \frac{C_i - \left(\dfrac{c_d}{\lambda} - C_r \right) \left(1 - e^{-\lambda \tau} \right)}{\tau} + c_d \tag{14.35}$$

---

[8] One can define many other such suboptimal inspection schemes.

If $\frac{c_d}{\lambda} > C_r$, the optimal value of τ can be obtained by solving the following first-order optimality equation:

$$1-(1+\lambda\tau)e^{-\lambda\tau} = \frac{C_i}{(c_d/\lambda)-C_r} \qquad (14.36)$$

Example 14.6

Equation (14.32) can be solved using Excel Goal Seek. If $C_i = 10$, $c_d = 2$, $C_r = 10$, and $\lambda = 0.001$, solving Equation (14.32) yields $\tau^* = 103.31$, and the corresponding optimal objective function value is $J^*(103.31) = 0.29$. ∎

14.7 Summary

In this chapter we discussed PM policies that are appropriate at the component or part level for non-repairable items. Failures are modeled using the black-box approach. The focus is on the optimization of the timing of PM actions in order to optimize certain measures of performance.

We considered a variety of replacement policies under both finite and infinite time horizons. We also discussed inspection policies, in cases where failure can only be detected through inspection. For each policy, we derived an optimization model, provided model analysis to characterize the optimal solutions, and presented numerical examples for illustration purposes.

Review Questions

14.1 What is a non-repairable item? What types of PM policies are appropriate for this type of item?

14.2 What is a renewal point?

14.3 What is a renewal cycle? Why is it important to identify this in the context of the policies discussed in this chapter?

14.4 State the Renewal Reward Theorem.

14.5 What is the difference between a block replacement policy and an age-based replacement policy?

14.6 Give a few applications of group replacement policies and explain the advantages of using such a policy.

14.7 When is inspection necessary? Give a few practical examples.

14.8 What are the main steps involved in developing maintenance optimization models for the policies discussed in this chapter?

Exercises

14.1   Consider a chemical reactor used to produce toxic products. Pipes in this reactor need
       to be replaced preventively, and failure replacements are extremely costly due to air
       pollution and safety risks. PM cost is estimated as $C_p = \$1500$ and CM cost is esti-
       mated as $C_f = \$150000$. Based on past experience and previously collected data, the
       reactor's failure time distribution may be approximated well by a Weibull distribution
       with mean 1000 cycles and shape parameter $\beta = 3$.
       (a) Find the optimal replacement age T^* and the corresponding expected cost rate.
       (b) Compare the optimal cost obtained in (a) with that of the run-to-failure policy.

14.2   Consider the age replacement policy considered in Section 14.3.2. Assume that
       replacement time is not negligible and that replacement time under CM is constant and
       given by T_f and replacement time under PM is also constant and given by T_p. Develop
       a cost model to determine the optimal replacement age in this case.

14.3   Repeat Exercise 14.2 where the objective is to maximize asymptotic availability.

14.4   Consider Exercise 14.1 and assume that replacement times under CM and PM are not
       negligible and are random variables with mean $\mu_d = 40$ cycles and $\mu_p = 20$ cycles,
       respectively. Consider the multi-objective problem of minimizing asymptotic expected
       total cost per unit time and maximizing asymptotic availability. Find the Pareto optimal
       solution.

14.5   Consider Exercise 14.2 and assume that we are now interested in maximizing system
       availability. Develop the appropriate optimization model.

14.6   Consider the age replacement policy discussed in Section 14.3.2. Assume that the failure
       distribution for the item is exponential with distribution function $F(t) = 1 - e^{-\lambda t}$.
       (a) Derive the expression for the expected cost rate $J(T)$.
       (b) Derive the expression for the optimal replacement age, T^*.
       (c) What can you conclude?

14.7   Repeat Exercise 14.6 for the block replacement policy discussed in Section 14.3.4.

14.8   Computing the integral $\int_0^T R(t)dt$ is crucial for computation of the expected total cost
       per unit time for the age replacement models discussed in Section 14.3.2 and other
       models discussed in Chapter 15. In the case of the Weibull distribution for time to
       failure, show that $\int_0^T R(t)dt = \mu_f \Gamma[1/\beta, (t/\alpha)^\beta]$, where the mean time to failure μ_f is
       given by $\mu_f = \alpha \Gamma[1 + 1/\beta]$ and $\Gamma[1/\beta, z] = \int_0^z x^{1/\beta - 1} e^{-x} dx$.

14.9   Consider the data in Exercise 14.1.
       (a) Find the optimal replacement time T^* and corresponding expected cost rate using
           the block replacement policy discussed in Section 14.3.4.
       (b) Compare the optimal cost obtained in (a) with that for the run-to-failure policy.

14.10 Repeat Exercise 14.9 for the block replacement policy discussed in Section 14.3.5.
       Assume that $c_d = \$300$.

14.11   Repeat Exercise 14.1 for a finite horizon with $L = 15$.

14.12   Repeat Exercise 14.9 for a finite horizon with $L = 15$.

14.13   Show that the objective function in Equation (14.33) can be simplified to Equation (14.34) and how the objective function in Equation (14.35) leads to Equation (14.36).

14.14   Consider the inspection problem discussed in Section 14.6. Assume that the time to item failure follows an exponential distribution $F(t) = 1 - e^{-0.01t}$. Assuming that $C_i = 10$, $c_d = 2$, and $C_r = 10$, solve Equation (14.32) using Excel Goal Seek to determine the optimal inspection interval length.

14.15   Provide a proof for Theorem 14.1.

14.16   Provide a proof for Theorem 14.2

14.17   Consider the block replacement policy under a finite time horizon discussed in Section 14.5.2. Assume that the time horizon L is a random variable with probability density $\ell(t)$. Derive an expression for the expected total cost per unit time for this policy under a random time horizon.

14.18   Under the block replacement policy discussed in Section 14.3.4, it is conceivable that an almost new item is replaced at planned replacement times. To alleviate this waste, consider the following modified policy: An operating item is replaced at times $kT, k = 1, 2, \ldots$. If the item fails in the interval $((k-1)T, kT - v)$, it is replaced by a new item. If it fails in $(kT - v, kT)$, it is replaced by a used item of age T. Note that the failure distribution function for used items is given by $F_1(t) = [F(t+T) - F(T)] / [1 - F(T)]$. Derive the expression for the expected total cost per unit time for this modified block policy.

Reference

Jiang, R. (2010) A simple approximation for the renewal function with an increasing failure rate, *Reliability Engineering and System Safety*, **95**(9): 963–969.

15

Maintenance Optimization for Repairable Items

Learning Outcomes

After reading this chapter, you should be able to:

- Describe several PM policies for repairable items;
- Derive an optimization model to determine the optimal parameters of a given PM policy;
- Conduct analysis of developed models to characterize optimal solutions;
- Apply developed models to a given situation.

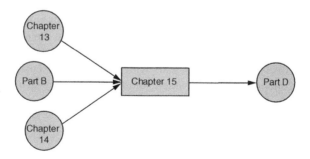

Introduction to Maintenance Engineering: Modeling, Optimization, and Management, First Edition.
Mohammed Ben-Daya, Uday Kumar, and D.N. Prabhakar Murthy.
© 2016 John Wiley & Sons, Ltd. Published 2016 by John Wiley & Sons, Ltd.

15.1   Introduction

Chapter 14 dealt with the modeling and optimization of several policies which involved replacement by new, for both corrective maintenance (CM) and preventive maintenance (PM) actions. These policies are appropriate for components and products that are either non-repairable or expensive to repair, so that replacement by new is the better option. For costly repairable components and products, policies which involve repair (in addition to replacement) are more appropriate and we discussed a few of these policies in Chapter 4. In this chapter we deal with the modeling and optimization of such policies.

There are two cases to consider, based on the characterization of the system (component, product, or plant) being maintained:

- *Case 1:* The system is viewed as a single entity – a product is viewed in the same way as a component.
- *Case 2:* The system (product or plant) is viewed as a collection of several components. The maintenance of each component is treated separately, so that the maintenance of the system is a collection of the maintenance actions for each component of the system.

We first consider Case 1 and define several maintenance scenarios where repair is always minimal but the PM actions differ. These scenarios may be categorized into the following groups:

- *Group I:* PM actions use new items.
- *Group II:* PM actions can be imperfect or uncertain.
- *Group III:* PM actions involve overhaul, so that the reliability characteristics of the item change with each overhaul.

We will look at several scenarios from each group using a format similar to that in Chapter 14. Following this we will look at Case 2.

The chapter is organized as follows. Section 15.2 deals with some preliminaries and these include assumptions, notations, and some basic results that are used in the rest of the chapter. Sections 15.3–15.5 deal with Case 1 and look at several different scenarios from Groups I–III, respectively. Section 15.6 deals with Case 2 and discusses a few scenarios. Section 15.7 is the concluding section which summarizes the key issues discussed in the chapter.

15.2   Preliminaries

15.2.1   Scenarios

As in Chapter 14, a scenario is characterized by a set comprising:

- *Maintenance actions:* PM actions (perfect, imperfect, etc.) and CM actions (overhaul, replace by new, etc.);
- *A time horizon:* Finite or infinite;

- *Time for a maintenance action:* Negligible or not;
- *An objective function:* Operational and/or economic, and so on.

15.2.2   Notation

In addition to the notation introduced in Chapter 14, we use the following:

$F(t)$	Failure distribution of a new item
$F_i(t)$	Failure distribution of the item after the ith repair $(i = 1, 2, \ldots)$
$h_i(t)$	Failure/hazard rate associated with $F_i(t)$
$H(t)$	Cumulative failure/hazard rate $[\equiv \int_0^t h(t)dt]$
c_d	Repair cost per unit time
C_a	Additional overhaul cost due to failure (Scenario VII)
C_m	Cost of a minimal repair
C_r	Cost of replacement (Scenarios VII and VIII)
$\{T_i\}_{i=0}^k$	The set $\{T_0, T_1, T_2, \ldots, T_k\}$

Additional notation will be introduced as and when needed.

15.2.3   Assumptions

Unless stated otherwise, we assume the following:

1. The item is in either a working or a failed state.
2. Failure is modeled using a failure distribution or a failure rate function.
3. The item is monitored continuously, so that a failure is detected immediately and a maintenance action is initiated after the failure is detected.
4. The failure distribution is known.
5. A spare is always available when needed under a CM or PM action (in other words, an unlimited number of spares is available).
6. There is no limit on the resources needed for maintenance actions.
7. Maintenance times are negligible, so that they can be approximated as being zero.
8. Repair, PM, and replacement costs are constant and known.
9. A scalar objective function is used.
10. The time horizon is infinite.

15.2.4   Approach to Modeling and Optimization

15.2.4.1   Objective Function

When the time horizon is infinite, the objective function is the asymptotic expected total cost (or performance) per unit time, and this is obtained using the Renewal Reward Theorem. The construction of the objective function involves properly defining the renewal points for the underlying stochastic point process and then deriving expressions for the expected cost per cycle (ECC) and the expected cycle length (ECL), as was done in Chapter 14.

15.2.4.2   Failures over Time with Minimal Repair

Let t denote the age of an item which is always minimally repaired after failure and let $N_f(t)$ denote the number of failures over the interval $(0,t]$. Since failures are independent, from Section 12.4.2 we have the following:

$$P\{N_f(t)=n\}=\frac{\left[H(t)\right]^n e^{-H(t)}}{n!}, \quad n=0,1,\dots. \tag{15.1}$$

15.2.4.3   Optimization

The objective function is a function of the decision parameters $\wp$ that need to be selected optimally to optimize the objective function. We present the optimal results in the form of theorems without proofs; the exercises at the end of the chapter involve proving these theorems based on the results in Part B and Appendix D.

15.3   Group I Scenarios

In this section we look at three scenarios.

15.3.1   Scenario I (Periodic Replacement Policy)

This is Policy 4.3 defined in Chapter 4. The PM actions are carried out at times $T, 2T, \dots$ (see Figure 14.5).

15.3.1.1   Model Derivation

In this case, each PM action is a renewal point. The ECC involves the PM cost at the end of the interval and the expected cost of minimal repairs during the interval $[0,T)$ and is given by:

$$ECC = C_p + C_m \int_0^T h(t)\,dt. \tag{15.2}$$

The ECL is T. From the Renewal Reward Theorem, the asymptotic expected total cost per unit time is given by:

$$J(T) = \frac{C_p + C_m H(T)}{T}. \tag{15.3}$$

15.3.1.2   Model Analysis and Optimization

In order to characterize the optimal value of T that minimizes the cost function given by Equation (15.3), we use the first-order necessary condition. Taking the first derivative of $J(T)$ with respect to T and equating to zero, and after simplification, we have:

$$T\,h(T) - H(T) = \int_0^T t\,dh(t) = \frac{C_p}{C_m}. \tag{15.4}$$

$T*$, the optimal T, is obtained by solving Equation (15.4), and the corresponding optimal expected cost rate is given by:

$$J(T*) = C_m h(T*) \tag{15.5}$$

Example 15.1   Valve

Consider the valve in Example 14.1 where we looked at the item being maintained under an age policy. Here, we look at the optimal periodic policy for the item. We assume that $C_m = \$40$ and the remaining model parameters are the same as in Example 14.1. The optimal replacement time is $T* = 5.39$ (obtained using Excel) and the corresponding optimal expected cost per unit time is $J(T*) = \$27.85$. Figure 15.1 shows a plot of $J(T)$ as a function of T.

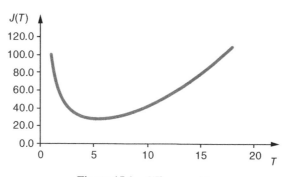

Figure 15.1 $J(T)$ versus T.

Comparing the results with that for the age policy (Example 14.2), we see that the optimal T is larger and the optimal expected cost per unit time is smaller. As C_m increases, the optimal T will decrease and the optimal expected cost per unit time will increase. ∎

15.3.2   Scenario II (Repair Count Policy)

This is Policy 4.7 defined in Chapter 4. The item is replaced at the kth failure. The first $k - 1$ failures are minimally repaired.

15.3.2.1   Model Derivation

Let X denote the time instant of the kth failure and this defines the cycle length, as every kth failure is a renewal point. Let $q_k(x)$ denote the probability density function for X. Note that:

$$P\{x < X \le x + \delta x\} = P\{(k-1) \text{ failures in } (0,x]\} P\{k\text{th failure in } (x, x+\delta x]\}$$

The first term on the right-hand side can be obtained from Equation (15.1) and the second term is $h(x)\delta x$ whilst the left-hand side is $q_k(x)\delta x$. As a result, we have:

$$q_k(x) = \frac{\left[H(x)\right]^{k-1} e^{-H(x)}}{(k-1)!} h(x) \tag{15.6}$$

The ECL is given by $ECL = \int_0^\infty t\, q_k(t)\, dt$ and the ECC is given by $ECC = C_r + (k-1)C_m$. The asymptotic expected total cost per unit time is the ratio of ECC and ECL and is given by:

$$J(k) = \frac{C_r + (k-1)C_m}{\int\limits_0^\infty t\, q_k(t)\, dt} \tag{15.7}$$

15.3.2.2 Model Analysis and Optimization

For a general $F(t)$ it is not possible to obtain k^* analytically and one needs to use some computational method. For the special case of the Weibull distribution, $F(t) = 1 - \exp\{-(t/\alpha)^\beta$, it can be shown that Equation (15.7) reduces to:

$$J(k) = \frac{C_r + (k-1)C_m}{\alpha \Gamma(k+1/\beta)} \Gamma(k) \tag{15.8}$$

The optimal value, k^*, is the smallest value of k that satisfies $J(k) < J(k+1)$ or

$$\frac{k(C_r + kC_m)}{(k+1/\beta)(C_r + kC_m)} > 1. \tag{15.9}$$

This yields:

$$k^* = \left\lfloor \frac{(C_r/C_m)-1}{\beta - 1} \right\rfloor + 1 \tag{15.10}$$

where $\lfloor x \rfloor$ is the largest integer less than or equal to x.

Example 15.2

The component has a Weibull time to failure distribution with distribution function $F(t) = 1 - \exp^{-(t/\alpha)^\beta}$ with $\alpha = 10$ and $\beta = 2.5$. The component is maintained using a repair count policy. The cost parameters are $C_r = 400$ and $C_m = 150$. Then, using Equations (15.10) and (15.8), we have the following optimal results: $k^* = 2$ and $J(k^*) = 44.28$. ∎

15.3.3   Scenario III (Repair Time Limit Policy)

This is Policy 4.6 defined in Chapter 4. Here, when an item fails, the repair starts immediately. If the repair is not completed within T units of time (called the repair time limit), the partially repaired item is discarded and replaced by a new item.[1] However, if the item is repaired, it is as good as new.

15.3.3.1   Model Derivation

Let X denote the time to repair; this is a random variable with distribution function $G(x)$. The expected cycle cost is obtained using the conditional approach conditioned on $X = x$. The cycle cost is $c_d x$ if $x \leq T$ and $c_d T + C_r$ if $x > T$. On un-conditioning, and simplifying, we have:

$$ECC = c_d \int_0^T \bar{G}(t)\,dt + C_r \bar{G}(T) \tag{15.11}$$

where $\bar{G}(t) = 1 - G(t)$. The ECL is the mean time to failure and the expected time spent on repair. This is given by:

$$ECL = \mu_f + \int_0^T \bar{G}(t)\,dt \tag{15.12}$$

The asymptotic expected cost per unit time, using Equations (15.11) and (15.12), is given by:

$$J(T) = \frac{c_d \int_0^T \bar{G}(t)\,dt + C_r \bar{G}(T)}{\mu_f + \int_0^T \bar{G}(t)\,dt} \tag{15.13}$$

15.3.3.2   Model Analysis and Optimization

The optimal repair time limit, T^*, if it exists, may be found from the first-order optimality condition. This condition is obtained by differentiating Equation (15.13) with respect to T and equating to zero, and is given by:

$$r(T)\left[\mu_f + \int_0^T \bar{G}(t)\,dt\right] + \bar{G}(T) = \frac{c_d \mu_f}{C_r} \tag{15.14}$$

where $r(t) = g(t)/\bar{G}(t)$ is the repair rate function.[2]

---

[1] Note that in this scenario the repair time is not negligible and needs to be taken into account in deriving expressions for the ECC and ECL.

[2] Similar to the failure rate function and discussed in Chapter 2.

The following theorem provides the conditions under which T^* exists and is finite.

Theorem 15.1

If $r(t)$ is monotonically decreasing, $r(0) > (c_d \mu_f - C_r)/(\mu_f C_r)$ and $r(\infty) \geq c_d \mu_f /(C_r(\mu_f + \mu_r))$, then there exists a finite and unique T^* that satisfies Equation (15.14) and the corresponding optimal expected cost per unit time is given by:

$$J(T^*) = c_d - C_r r(T^*).$$

Example 15.3

Consider an item having mean time to failure $\mu_f = 10$ and whose time to repair follows a Weibull distribution with parameters $\beta = 0.5$ and $\alpha = 4$. Assume that the replacement cost $C_r = 10$ and the repair cost per unit time $c_d = 3$.

   With the expression for $\int_0^T \bar{G}(t)dt$ derived in Exercise 14.8, Excel was used to provide values of $J(T)$ for various values of T. The optimal repair time limit was found to be $T^* = 1.25$ and the corresponding optimal expected cost rate $J(T^*) = 0.77$. It can be verified that the expression for $J(T^*)$ given in Theorem 15.1 provides the same value.   ∎

15.4   Group II Scenarios

In this section we look at scenarios which can be viewed as extensions of the periodic policy, where PM actions improve the reliability but the item is not necessarily as good as new. These correspond to imperfect PM, as discussed in Section 4.3.5. We also look at a scenario where the outcome of a PM action is uncertain. The three scenarios we discuss are as follows:[3]

- *Scenario IV:* After PM, the system (component or product) is "as bad as old" with probability p and "as good as new" with probability $(1-p)$.
- *Scenario V:* After each PM, the age of the system is reduced by a certain amount.
- *Scenario VI:* After each PM, the failure rate of the system is reduced by a certain amount.

15.4.1   Scenario IV (Uncertain Preventive Maintenance)

PM actions are carried out at time instants iT, $i = 1, 2, \cdots$, and we assume that the outcome of each PM action is uncertain. The action can lead to no improvement, so that the failure rate is unaffected (no improvement) with probability p, or the item may be restored to as good as new (back to new) with probability q $(\equiv 1 - p)$.

---

[3] More details and similar models can be found in Nakagawa (1980; 2005).

15.4.1.1 Model Derivation

In this case, every PM action which restores the item to as good as new is a renewal point and the renewal cycle is the time between two adjacent such points. Note that the number of no-improvement PM actions over a cycle is a random variable with a geometric distribution. Let $\tilde{N}$ denote the number of such actions over a cycle, then $P\{\tilde{N} = n\} = qp^n$, $n = 0,1,2,\dots$. The expected cycle costs and ECLs are obtained using the conditional approach conditional on $\tilde{N}$ and then un-conditioning. They are given by:

$$ECC = \sum_{i=1}^{\infty} qp^{i-1}\left[iC_p + C_m \int_0^{iT} h(t)dt \right] \tag{15.15}$$

and

$$ECL = \sum_{i=1}^{\infty} qp^{i-1}(iT), \tag{15.16}$$

respectively. From the Renewal Reward Theorem, the expected total cost per unit time is given by the ratio of these two functions, and on simplification yields:

$$J(T;p) = \frac{C_p + C_m q^2 \sum_{i=1}^{\infty} p^{i-1} \int_0^{iT} h(t)dt}{T}. \tag{15.17}$$

15.4.1.2 Model Analysis and Optimization

The optimal policy parameter is the value of T that minimizes the objective function given in Equation (15.17). From the usual first-order condition, the optimal T is obtained by solving the following equation:

$$\sum_{i=1}^{\infty} p^{i-1} \int_0^{iT} t\,dh(t) = \frac{C_p}{C_m q^2} \tag{15.18}$$

The following result deals with the existence of the optimal solution.

Theorem 15.2

If $h(t)$ is monotonically increasing and $\int_0^{\infty} t\,dh(t) > C_p / (C_m q^2)$, then there exists a finite and unique T^* that satisfies Equation (15.18), and the corresponding optimal expected cost per unit time is given by:

$$J(T^*;p) = C_m q^2 \sum_{i=1}^{\infty} p^{i-1} i\,h(iT^*). \tag{15.19}$$

Example 15.4

Consider a component whose time to failure follows a Weibull distribution with distribution function $F(t) = 1 - \exp^{-(t/\alpha)^{\beta}}$, $\alpha = 5$, and $\beta = 2$. In this case, we can derive the following closed-form solution for T^* using the optimality conditions in Equation (15.19), and this is given by:

$$T^* = \left[\alpha^{\beta} C_p \left/ \left(C_m q (\beta - 1) \sum_{i=1}^{\infty} i^{\beta} p^{i-1} q \right) \right. \right]^{1/\beta}.$$

Note that $\sum_{i=1}^{\infty} i^{\beta} p^{i-1} q$ is the βth moment of the geometric distribution with parameter p. The remaining model parameters are as follows: $C_p = 10, C_m = 3$, and $p = 0.3$. Then the optimal PM interval and the corresponding optimal expected cost rate, using the above formula for T^* and Equation (15.19), are as follows: $T^* = 1.795$ and $J(T^*) = 10.77$. ∎

15.4.2 Scenario V (Imperfect Preventive Maintenance with Age Reduction)

Here, PM actions are carried out at time instants jT, $j = 1, 2, \ldots$, and there are two types of PM actions. The first is replacement by new performed at time instants jkT, $j = 1, 2, \ldots$, and the second, performed at other PM time instants, reduces the age by an amount x $(0 \le x \le T)$, leading to the notion of virtual age discussed in Section 11.2.2. Figure 15.2 shows the plot of the virtual age as a function of time.

The reduction in age (x) is assumed to be constant and is specified *a priori*. All failures between PM actions are minimally repaired. The decision parameters given by the set $\wp = \{T, k\}$ are selected optimally to minimize $J(T, k)$, the asymptotic expected total maintenance cost per unit time.

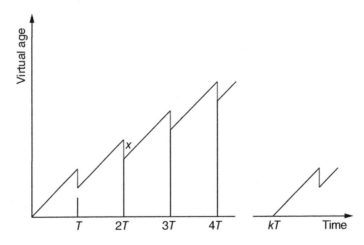

Figure 15.2 Virtual age of item over time.

15.4.2.1 Model Derivation

Note that each replacement event is a renewal point, so that the ECL is kT. Over the cycle there are $(k-1)$ PM actions (resulting in age reduction), one replacement by new, and the CM actions are minimal repairs. As a result, the ECC is given by:

$$ECC = (k-1)C_p + C_r + C_m \sum_{i=0}^{k-1} \int_{i(T-x)}^{T+i(T-x)} h(t)dt \tag{15.20}$$

and the expected total cost per unit time, using the Renewal Reward Theorem, is given by:

$$J(T,k) = \frac{(k-1)C_p + C_r + C_m \sum_{i=0}^{k-1} \int_{i(T-x)}^{T+i(T-x)} h(t)dt}{kT} \tag{15.21}$$

15.4.2.2 Model Analysis and Optimization

The optimal values for the parameters (T,k) are obtained using a two-stage approach.

In Stage (i), for a given k, one obtains the optimal $T^*(k)$ from the usual first-order condition. Differentiating Equation (15.21) with respect to T, equating to zero and simplifying, we have:

$$\sum_{i=0}^{k-1} \int_{i(T-x)}^{T+i(T-x)} t\,dh(t) = \frac{(k-1)C_p + C_r}{C_m} \tag{15.22}$$

$T^*(k)$ is obtained by solving the above equation.

In Stage (ii), the optimal k^* is obtained by minimizing $J(T^*(k),k)$ as an integer optimization problem. Since k is an integer, the value of k that minimizes Equation (15.21) should satisfy the following inequalities:

$$J(T,k+1) \geq J(T,k) \text{ and } J(T,k) < J(T,k-1), \quad k=1,2,\ldots.$$

These two inequalities imply that:

$$\Psi(k) \geq (C_r - C_p)/C_m \text{ and } \Psi(k-1) \geq (C_r - C_p)/C_m, \quad k=1,2,\ldots,$$

where

$$\Psi(k) = \begin{cases} k\int_{k(T-x)}^{T+k(T-x)} h(t)dt - \sum_{i=0}^{k-1} \int_{i(T-x)}^{T+i(T-x)} h(t)dt & k=1,2,\ldots \\ 0 & k=0. \end{cases} \tag{15.23}$$

Note that $\Psi(k+1)-\Psi(k)=(k+1)\left[\int_{(k+1)(T-x)}^{T+(k+1)(T-x)} h(t)dt - \int_{k(T-x)}^{T+k(T-x)} h(t)dt\right] \geq 0.$ Therefore, the optimal number of PM cycles until replacement is given by the smallest value of k such that $\Psi(k) \geq (C_r - C_p)/C_m$ if $\Psi(\infty) \geq (C_r - C_p)/C_m$, otherwise no replacement is made and the item is always subjected to PM actions that reduce the age.

Example 15.5

Assume that the failure distribution of an item can be modeled by a Weibull distribution with $F(t)=1-\exp[-(t/\alpha)^{\beta}], \alpha > 0, \beta > 1$ and the failure rate function is $h(t)=(\beta/\alpha)(t/\alpha)^{\beta-1}$. The expected cost rate function in Equation (15.21) can be written as:

$$J(T,k)=\frac{1}{kT}\left[(k-1)C_p+C_r+\frac{C_m}{\alpha^{\beta}}\sum_{i=0}^{k-1}\left[(T+i(T-x))^{\beta}-(i(T-x))^{\beta}\right]\right]$$

and the condition in Equation (15.22) becomes:

$$\sum_{i=0}^{k-1}\left[(T+i(T-x))^{\beta}-(i(T-x))^{\beta}\right]=\frac{\alpha^{\beta}}{(\beta-1)C_m}\left[(k-1)C_p+C_r\right]$$

∎

15.4.3   Scenario VI (Imperfect Preventive Maintenance with Reduction in Failure Rate)

This scenario is similar to Scenario V except that instead of a reduction in age, we have a reduction in the failure rate, as discussed in Section 11.2.2. As a result, the failure rate over time is as shown in Figure 15.3.

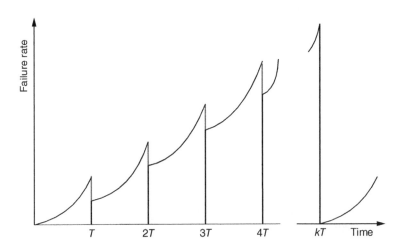

Figure 15.3   Failure rate of item over time.

Note that, here again, every replacement is a renewal point and is the start of a new cycle. Within a cycle, the failure rate after the ith PM action is given by $a^i r(t)$, $1 \le i \le k-1$, with $0 < a \le 1$, which is specified *a priori*. As before, the decision parameters are given by the set $\wp = \{T,k\}$.

15.4.3.1   Model Derivation

A renewal cycle is the time between two adjacent replacements, so $ECL = kT$. The ECC is given by:

$$ECC = (k-1)C_p + C_r + C_m \sum_{i=0}^{k-1} \int_{iT}^{(i+1)T} a^i\, h(t)dt \qquad (15.24)$$

From the Renewal Reward Theorem, the expected total cost per unit time is given by the ratio of the *ECC* to the *ECL:*

$$J(T,k) = \frac{(k-1)C_p + C_r + C_m \sum_{i=0}^{k-1} \int_{iT}^{(i+1)T} a^i\, h(t)dt}{kT} \qquad (15.25)$$

The optimal decision parameters, $\wp = \{T,k\}$, that minimize $J(T,k)$ are to be determined.

15.4.3.2   Model Analysis and Optimization

As with Scenario V, we use a two-stage approach.

In Stage (i), for a given value of k, the optimal value of the parameter $T(k)$ is obtained from the usual first-order optimality condition. By differentiating Equation (15.25) with respect to T, equating to zero, and simplifying, we obtain:

$$\sum_{i=0}^{k-1} a^i \int_{iT}^{(i+1)T} t\, dh(t) = \frac{(k-1)C_p + C_r}{C_m}. \qquad (15.26)$$

$T(k)$ is obtained as a solution of the above equation.

In Stage (ii), the optimal k^* is obtained by minimizing $J(T^*(k);k)$. Since k is an integer, the value of k minimizing Equation (15.25) should satisfy the following inequalities:

$$J(T,k+1) \ge J(T,k) \text{ and } J(T,k) < J(T,k-1), \ k = 1,2,\dots.$$

These two inequalities imply that:

$$\Psi(k) \ge (C_r - C_p)/C_m \text{ and } \Psi(k-1) \ge (C_r - C_p)/C_m, \ k = 1,2,\dots,$$

where:

$$\Psi(k) = ka^k \int_{kT}^{(k+1)T} r(t)dt - \sum_{i=0}^{k-1} a^i \int_{iT}^{(i+1)T} r(t)dt \qquad k = 1,2,\dots \tag{15.27}$$

It can be shown that $\Psi(k+1) - \Psi(k) \geq 0$. Therefore, the optimal number of PM cycles until replacement is given by the smallest value of k such that:

$$\Psi(k) \geq (C_r - C_p)/C_m \tag{15.28}$$

if $\Psi(\infty) \geq (C_r - C_p)/C_m$, otherwise no replacement is made.

Example 15.6

Assume that the failure distribution of an item can be modeled by a Weibull distribution with $F(t) = 1 - \exp[-(t/\alpha)^\beta], \alpha > 0, \beta > 1$ and the failure rate function is $h(t) = (\beta/\alpha)(t/\alpha)^{\beta-1}$. The expected cost rate function from Equation (15.25) can be written as:

$$J(T,k) = \frac{1}{kT}\left[(k-1)C_p + C_r + C_m\left(\frac{T}{\alpha}\right)^\beta \sum_{i=0}^{k-1} a^i \left[(i+1)^\beta - i^\beta\right]\right]$$

and the solution of Equation (15.26) yields:

$$T = \alpha \left[\frac{(k-1)C_p + C_r}{(\beta-1)C_m \sum_{i=0}^{k-1} a^i \left[(i+1)^\beta - i^\beta\right]}\right]^{1/\beta}$$

The optimal value of k is the smallest value that satisfies Equation (15.28). ∎

15.5 Group III Scenarios

Complex products and plants undergo one or more major overhauls before they are discarded. In Section 4.9.1 we defined two policies (Policies 4.16 and 4.17) and in this section we discuss the modeling and optimization of these two policies.

In both of these policies, a new item is subjected to K overhauls, denoted by $OH - i$, $1 \leq i \leq K$, and is then replaced (by a new one), as shown in Figure 15.4. The time to the first overhaul is $\tilde{T}_0$ and the time for which the item is in use after an overhaul ($\tilde{T}_i$, $1 \leq i \leq K$) can be either deterministic or random depending on the policy.

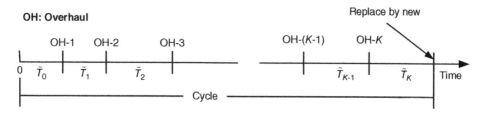

Figure 15.4 Cycle for overhaul policies.

The failure distribution of the item after the ith overhaul, $1 \le i \le K$, is given by $F_i(t; \theta_i)$. $h_i(t)$ and $R_i(t)$ are the failure rate and reliability functions associated with $F_i(t)$.[4] We denote the failure distribution function for a new item by $F_0(t)$.

The effect of an overhaul on the reliability of the item is modeled as discussed in Section 11.3.2 (see Figure 11.7) and these assumptions are restated below (for $i \ge 0$).

- A-1: $h_i(t)$ is strictly increasing and $h_i(t) \to \infty$ as $t \to \infty$.[5]
- A-2: $h_{i+1}(t) \ge h_i(t)$
- A-3: $h_{i+1}(0) = h_i(0)$

15.5.1 Scenario VII (Overhaul Policy I)

This is Policy 4.17 defined in Chapter 4. Here, $OH - 1$ occurs at time T_0 since the item was put in use and the item is then overhauled at this time or at failure, should it occur earlier. As a result, $\tilde{T}_0 = \min\{T_0, X_0\}$ where X_0 is a random variable with distribution function $F_0(\cdot)$ and is the time to first failure for a new item. Similarly, $\tilde{T}_i = \min\{T_i, X_i\}$ where X_i is a random variable with distribution function $F_i(\cdot)$ for $1 \le i \le K$. The decision parameters are given by the set $\wp \equiv \{K, \{T_i\}_{i=0}^{K}\}$ where $\{T_i\}_{i=0}^{K} \equiv \{T_0, T_1, \ldots, T_K\}$ and these are to be selected optimally to minimize the asymptotic cost per unit time.

15.5.1.1 Model Derivation

An overhaul can occur as a result of either the item not failing (resulting in a PM action) or the item failing (resulting in a CM action). For the ith overhaul, the cost is C_{oi} if it is a PM action and $C_{oi} + C_a$ if it is a CM action, where C_a is the additional cost incurred due to failure.

Every replacement by new is a renewal point and, as a result, the cycle length is given by $\sum_{i=0}^{K} \tilde{T}_i$. The cycle can be viewed as consisting of $(K+1)$ subcycles and the length and cost associated with each subcycle are random variables. Following the approach used in the analysis of the age policy (where we had only one subcycle in each cycle), we have the ECL given by:

$$ECL = \sum_{i=0}^{K} \int_{0}^{T_i} R_i(t) \, dt \tag{15.29}$$

---

[4] For notational ease we suppress the parameter θ_i.

[5] We are using a local clock which is reset to zero after each overhaul.

and the expected cycle cost by:

$$ECC = \sum_{i=1}^{K} [C_{oi}R_{i-1}(T_{i-1}) + \{C_{oi} + C_a\}F_{i-1}(T_{i-1})] + C_aF_K(T_K) + C_r$$

This expected cycle cost can be rewritten as:

$$ECC = \sum_{i=1}^{K} C_{oi} + C_a \sum_{i=0}^{K} F_i(T_i) + C_r \tag{15.30}$$

where the first expression is the cost associated with overhauls, the second is the additional costs due to CM actions as opposed to PM actions, and C_r is the cost of a replacement at the end of the cycle.

Consequently, the expected cost per unit time is given by:

$$J\left(\{T_i\}_{i=0}^{K}, K\right) = \frac{\sum_{i=1}^{K} C_{oi} + C_a \sum_{i=0}^{K} F_i(T_i) + C_r}{\sum_{i=0}^{K} \int_{0}^{T_i} R_i(t)\,dt} \tag{15.31}$$

The optimal decision parameters $\wp \equiv \{K, \{T_i\}_{i=0}^{K}\}$ are selected to minimize the objective function given by Equation (15.31).

15.5.1.2   Model Analysis and Optimization

The optimal decision parameters are obtained using a two-stage approach. In Stage (i), K is fixed and the optimal values of $\{T_i(K)\}_{i=0}^{K}$, are obtained from the usual first-order necessary condition. Differentiating Equation (15.31) with respect to T_i, equating to zero, and simplifying, we obtain:

$$h_i\left(T_i(K)\right) = \frac{\sum_{i=1}^{K} C_{oi} + C_a \sum_{i=0}^{K} F_i(T_i(K)) + C_r}{C_a \sum_{i=0}^{K} \int_{0}^{T_i} R_i(t)\,dt} = \frac{J\left(\{T_i(K)\}_{i=0}^{K}, K\right)}{C_a} \quad \text{for } i = 0,\ldots,K \tag{15.32}$$

This can be rewritten as follows:

$$h_i\left(T_i(K)\right) = h_0\left(T_0(K)\right) \quad \text{for } i = 1,\ldots,K \tag{15.33}$$

$$\sum_{i=0}^{K} \left[h_0(T_0) \int_{0}^{T_i} R_i(t)\,dt - F_i(T_i) \right] = \frac{\sum_{i=1}^{K} C_{oi} + C_r}{C_a} \tag{15.34}$$

Note that the optimal $\{T_i^*(K)\}_{i=0}^K$ are such that the item failure rates at these ages are all equal.

In Stage (ii), we compute $J(\{T_i^*\}_{i=0}^K, K)$ for $K = 1, 2, \ldots,$ and K^*, the optimal K, is the value that yields a minimum. The optimal $T_i^* = T_i^*(K^*)$, $0 \le i \le K^*$.

The following results,[6] stated without proof, can be used to develop a procedure for obtaining the optimal decision parameters $\{K^*$ and $\{T_i^*\}_{i=0}^K\}$.

Theorem 15.3

If assumptions A1, A2, and A3 are satisfied, then, for any $K \ge 1$, $\{T_i^*(K)\}_{i=0}^K$ exists. $T_i^*(K)$, $i = 0, \ldots, K$, are finite, unique, and a decreasing sequence in i.

Theorem 15.4

If assumptions A1, A2, and A3 are satisfied, then there exists a unique K^*, the optimal K, given by:

$$Min\left\{K \mid \Delta J\left(\{T_i(K)\}_{i=0}^K, K\right) \ge 0\right\} \le K^* \le Max\left\{K \mid \Delta J\left(\{T_i(K)\}_{i=0}^K, K\right) \le 0\right\}$$

where $\Delta J\left(\{T_i(K)\}_{i=0}^K, K\right) = J\left(\{T_i(K+1)\}_{i=0}^{K+1}, K+1\right) - J\left(\{T_i(K)\}_{i=0}^K, K\right)$

Solution Algorithm 15.1

The following algorithm can be used to determine the optimal K^* and $\{T_i^*\}_{i=0}^K$

Step 1: Set $K = 0$ and solve for $T_0(0)$ from Equation (15.34).[7]
Step 2: Set $K = K + 1$.
Step 3: Solve Equations (15.33) and (15.34) simultaneously for $\{T_i^*(K)\}_{i=0}^K$.
Step 4: If $T_0^*(K) < T_0^*(K-1)$, go to Step 2. If not, go to Step 5.
Step 5: Set $K^* = K - 1$, compute $J(\{T_i^*(K^*)\}_{i=0}^{K^*}, K^*)$ using Equation (15.31).

Example 15.7

Let $F_i(t) = 1 - \exp^{-(t/\alpha_i)^{\beta_i}}$ be two-parameter Weibull distribution functions with $\beta_i = 2$ and $\alpha_i = (0.85)^i$ for $i \ge 0$. In other words, the shape parameter does not change with overhaul

---

[6] For a proof, see Nguyen and Murthy (1981).
[7] When $K = 0$, there is no summation in the LHS and no overhaul costs on the RHS of Equation (15.34). This case corresponds to the age replacement policy discussed in the previous chapter.

and the scale parameter decreases geometrically. Let $C_{oi} = 10$ for $i \geq 0$, $C_r = 30$, and $C_a = 20$. Using Solution Algorithm 15.1, we have the following:

$K = 0$: $T_0(0) = 1.40$ and $J(\{T_0\}, 0) = 55.91$
$K = 1$: $T_0^*(1) = 1.18, T_1^*(1) = 0.85$ and $J(\{T_i^*\}_{i=0}^1, 1) = 47.10$
$K = 2$: $T_0^*(2) = 1.14, T_1^*(2) = 0.82, T_2^*(2) = 0.60$ and $J(\{T_i^*\}_{i=0}^2, 2) = 45.56$
$K = 3$: $T_0^*(3) = 1.15, T_1^*(3) = 0.83, T_2^*(3) = 0.60, T_2^*(3) = 0.43$ and $J(\{T_i^*\}_{i=0}^3, 3) = 45.97$

Since $T_0^*(3) > T_0^*(2)$, then $K^* = 2$ and $T_0^*(2) = 1.14, T_1^*(2) = 0.82, T_2^*(2) = 0.60$. The corresponding optimal expected cost rate is $J(\{T_i^*\}_{i=0}^2, 2) = 45.56$. ∎

15.5.2   Scenario VIII (Overhaul Policy II)

This is Policy 4.16 introduced in Chapter 4. Here, $OH - 1$ occurs at time T_0 since the item was put into use. The time interval between $OH - i$ and $OH - (i+1)$ is T_i for $i \geq 1$. All failures are rectified through minimal repair. The decision parameters are given by the set $\wp \equiv \{K, \{T_i\}_{i=0}^K\}$ where $\{T_i\}_{i=0}^K \equiv \{T_0, T_1, \cdots, T_K\}$ and these are to be selected optimally to minimize the asymptotic expected cost per unit time.

15.5.2.1   Model Derivation

Note that the cycle consists of $K + 1$ subcycles (see Figure 15.4). The results for the cycle costs and lengths are derived by summing over all the subcycles. The expected number of failures over the subcycle i, $0 \leq i \leq K$, is given by $\int_0^{T_i} h_i(t) dt$ and the cost of each minimal repair is C_m. As a result, the ECC is given by:

$$ECC = \sum_{i=0}^{K} C_m \int_0^{T_i} h_i(t) dt + \sum_{i=1}^{K} C_{oi} + C_r \tag{15.35}$$

where the first term is the expected cost of minimal repairs, the second is the cost of overhauls, and the last term is the cost of replacement at the end of the cycle. The ECL is simply $ECL = \sum_{i=0}^{K} T_i$.

Using the Renewal Reward Theorem, the expected cost per unit time is given by:

$$J\left(\{T_i\}_{i=0}^K, K\right) = \frac{C_m \sum_{i=0}^{K} \int_0^{T_i} h_i(t) dt + \sum_{i=1}^{K} C_{oi} + C_r}{\sum_{i=0}^{K} T_i} \tag{15.36}$$

So the optimal values for the decision parameters $\wp \equiv \{K, \{T_i\}_{i=0}^K\}$ minimize the objective function given by Equation (15.36).

15.5.2.2　Model Analysis and Optimization

We use the two-stage approach as in Scenario VII. In Stage (i) we use the first-order necessary condition to obtain $\{T_i^*(K)\}_{i=0}^K$ for a given K. Differentiating Equation (15.36) with respect to $\{T_i\}_{i=0}^K$, equating to zero, and simplifying, we obtain:

$$h_i(T_i) = \frac{C_m \sum_{i=0}^K \int_0^{T_i} h_i(t)dt + \sum_{i=1}^K C_{oi} + C_r}{C_m \sum_{i=0}^K T_i} = \frac{J\left(\{T_i\}_{i=0}^K, K\right)}{C_m} \quad \text{for } i = 1,\ldots,K \quad (15.37)$$

This can be rewritten as follows:

$$h_i(T_i) = h_0(T_0) \quad \text{for } i = 1,\ldots,K \quad (15.38)$$

$$\sum_{i=0}^K \left[T_0 h_0\left(T_0(K)\right) - \int_0^{T_i(K)} h_i(t)dt \right] = \frac{\sum_{i=1}^K C_{oi} + C_r}{C_m} \quad (15.39)$$

Note that the optimal $\{T_i^*(K)\}_{i=0}^K$ are such that the item failure rates at these ages are all equal.

In Stage (ii) we compute $J(\{T_i^*\}_{i=0}^K, K)$ for $K = 1, 2, \ldots$, and K^*, the optimal K, is the value that yields a minimum. The optimal $T_i^* = T_i^*(K^*)$, $0 \le i \le K^*$. Theorems 15.3 and 15.4 also apply to this scenario.

15.5.2.3　Special Case

Let $F_i(t) = 1 - \exp^{-(t/\alpha_i)^{\beta_i}}$ be two-parameter Weibull distribution functions with $\beta_i = \beta$ for $i \ge 0$. Let $C_{oi} = C_o$ for $i \ge 0$. In this case, it is possible to derive analytical expressions for $T_i(K), 0 \le i \le K$ using Equations (15.38) and (15.39) and they are given by:

$$T_1(K) = \left[\frac{\alpha_1^\beta \left[(k-1)C_o + C_r\right]}{C_m(\beta-1)\sum_{i=1}^K (\alpha_i/\alpha_1)^{\beta/(\beta-1)}} \right]^{1/\beta} \quad \text{and} \quad T_i(K) = \left(\frac{\alpha_i}{\alpha_1}\right)^{\beta/(\beta-1)} T_1(K). \quad (15.40)$$

Solution Algorithm 15.2

The algorithm to determine the optimal K^* and $\{T_k^*\}_0^K$ is the same as Solution Algorithm 15.1 with the following changes – Equations (15.31), (15.33), and (15.34) are replaced by Equations (15.36), (15.38), and (15.39), respectively.

Example 15.8

Let the lifetime distribution functions $F_i(\cdot)$, $i \geq 0$, and the distribution parameters be the same as in Example 15.7. The cost parameters are also the same and let $C_m = 3$. Using Solution Algorithm 15.2, we have the following:

$K = 0$: $T_0(0) = 3.16$ and $J(\{T_0\},0) = 18.97$
$K = 1$: $T_0^*(1) = 2.78, T_1^*(1) = 2.01$ and $J(\{T_i^*\}_{i=0}^1,1) = 16.69$
$K = 2$: $T_0^*(2) = 2.73, T_1^*(2) = 1.97, T_2^*(2) = 1.42$ and $J(\{T_i^*\}_{i=0}^2,2) = 16.35$
$K = 3$: $T_0^*(3) = 2.76, T_1^*(3) = 2.00, T_2^*(3) = 1.43, T_2^*(3) = 1.04$ and $J(\{T_i^*\}_{i=0}^3,3) = 16.57$

Since $T_0^*(3) > T_0^*(2)$, then $K^* = 2$ and $T_0^*(2) = 2.73, T_1^*(2) = 1.97, T_2^*(2) = 1.42$. The corresponding optimal expected cost rate is $J(\{T_i^*\}_{i=0}^2,2) = 16.35$. ∎

15.6   Multi-Item Policies

So far, we have considered PM policies and models for an object (product or plant) viewed as a single component based on the black-box approach and ignoring the fact that it is a multi-component system. The modeling and optimization of multi-component systems can be done by looking at each component and deriving the optimal maintenance strategies separately. The combination of the optimal policies of each component is not necessarily optimal for the whole system, as it ignores the interaction and dependencies among components. These dependencies can be divided into three broad categories:

- Economic dependence;
- Structural dependence;
- Stochastic dependence.

They should not be ignored and need to be taken into account in maintenance decision making for multi-component systems. Below, we discuss the three types of dependencies briefly.

15.6.1   Economic Dependence

The cost structure of replacement and maintenance has interdependencies among the units. This implies that economies of scale can be obtained when several components are jointly maintained instead of separately. However, in some applications, economic dependence implies that simultaneous downtime of components is undesirable and hence maintenance must be spread out over time as much as possible.

An example of economically dependent components is when the maintenance of each component of a system requires preparatory or "set-up work" that can be shared when several components are maintained simultaneously. This *set-up cost* may consist of the production loss during maintenance, or of the preparation cost associated with dismantling a machine or any other preparatory work. Set-up costs can be reduced when maintenance activities on different components are executed simultaneously and only one set-up is required.

Several opportunistic maintenance policies (see the policy descriptions in Chapter 4) address these issues of economic dependence, but we do not discuss them.[8]

15.6.2   Structural Dependence

The failure of a system is usually due to the failure of one or more components. Often, the system architecture uses a modular form, so that the system is a collection of modules which, in turn, are collections of components. When a component fails, the module to which it belongs is replaced by a working module and the module removed is sent to a workshop for repair. The modules are referred to as *line replacement units* (LRUs); this type of maintenance is discussed further in Chapter 19 and it has two advantages. First, it reduces the downtime (which is simply the time to locate the failed module and replace it by a working one) and second, it provides an opportunity to decide on the PM actions for non-failed components of the module when the failed component is being replaced or repaired.

15.6.3   Stochastic Dependence

Stochastic dependence implies that the failure of a component affects the reliability of one or more other components of the system. This dependence can be divided broadly into two categories:

- *Shock damage:* The failure of a component acts as a shock to damage one or more of the other components. An example is the failure of a suspension cable increasing the load on the non-failed cables and lowering the reliability function (or increasing the failure rate function) of the other cables.
- *Induced failures:* The failure of a component can induce the failure of one or more of the other components. One way of modeling this is that the failure of a component can either have no effect on another component (with probability p) or can result in induced failure (with probability $q = 1 - p$). An example of this is railway rolling stock. Here, the failure of a wheel (due to derailment) can either have no effect on the axle or can induce a failure of the axle.

15.6.4   Modeling and Optimization of Maintenance Decisions

In general, the modeling of interactions and dependencies among components involves complex mathematical formulations – for example, in stochastic dependence, it involves interacting point process formulations.[9]

15.7   Summary

In this chapter we discussed PM policies that are appropriate at the component level (or for products and plants treated as components) and where failures are modeled using the black-box approach. The focus was on the optimization of the timing of PM actions in order to optimize a certain measure of performance.

---

[8] For more information, see Nicolai and Dekker (2008) and Dekker, van der Duyn Schouten, and Wildeman (1996).
[9] For more details, see Murthy and Nguyen (1985a, b) and Nakagawa and Murthy (1993).

We considered a variety of PM policies including policies involving imperfect PM. For each policy, we derived an optimization model, provided model analysis to characterize optimal solutions, and presented numerical examples for illustration purposes.

Review Questions

15.1   Provide an application for which the periodic replacement policy of Section 15.3.1 is appropriate.

15.2   Provide an application for which the repair count policy of Section 15.3.2 is appropriate.

15.3   PM actions improve the reliability but the item is not necessarily as good as new. This is called imperfect PM. Provide three methods of modeling the effect of PM on the failure behavior of an item.

15.4   What is the main difference between the two PM policies discussed in Section 15.5.

15.5   For a multi-component system, what is the meaning of economic dependence between components?

15.6   For a multi-component system, what is the meaning of structural dependence between components?

15.7   For a multi-component system, what is the meaning of stochastic dependence between components?

15.8   What are the two categories of stochastic dependence?

Exercises

15.1   Consider the model developed in Section 15.3.1. The time to failure of an item has a Weibull distribution with $\alpha = 10$ and $\beta = 2.5$. Let $C_m = \$50$ and $C_p = \$140$. Determine the optimal replacement interval and the corresponding optimal expected cost per unit time.

15.2   Consider the periodic replacement policy considered in Section 15.3.1. Assume that the time to failure of an item has a two-parameter Weibull distribution with distribution function $F(t) = 1 - \exp[-(t / \alpha)^\beta]$. Derive a closed-form solution for the optimal value of T^* that solves Equation (15.4) if it exists.

15.3   Show that the T^* obtained by solving Equation (15.4) is given by Equation (15.5).

15.4   Show that the cost function given by Equation (15.7) reduces to Equation (15.8) when the time to failure follows a Weibull distribution with parameters α and β. Then show that the optimal value of k, k^* is given by Equation (15.10).

15.5   Consider the model developed in Section 15.3.3. The time to failure has a Weibull distribution with $\alpha = 20$ and $\beta = 2.5$ and the repair time distribution is also Weibull with parameters $\alpha = 3$ and $\beta = 0.6$ Let $C_r = 15$ and $c_d = 5$. Determine the optimal replacement interval and the corresponding optimal expected cost per unit time.

15.6   For the repair time limit policy discussed in Section 15.3.3, show that
   (a) The derivative of the cost function Equation (15.13) may be simplified to give Equation (15.14).
   (b) The optimal expected cost per unit time that corresponds to T^* that solves Equation (15.14), if it exists, is given by the expression in Theorem 15.1.

15.7   For the repair time limit policy discussed in Section 15.3.3, show that
   (a) The expected cost per unit time for the case of replacement only is given by $J(0) = C_r / \mu_f$.
   (b) The expected cost per unit time for the case of repair only is given by $J(\infty) = c_d \mu_f / (\mu_f + \mu_r)$.

15.8   Show that the expression for the expected cost per unit time for the model derived in Section 15.4.1 may be simplified to Equation (15.17).

15.9   Show that the optimal expected cost per unit time that corresponds to T^* that solves Equation (15.18) is given by Equation (15.19).

15.10   Show that the optimal value T^* that solves Equation (15.18) is given by the expression given in Example 15.4 when the time to failure distribution is Weibull with parameters α and β.

15.11   Consider the model discussed in Section 15.4.2 and assume that the time to failure of an item follows a two-parameter Weibull distribution with distribution function $F(t) = 1 - \exp[-(t/\alpha)^\beta]$. Show, in this case, that the optimal value T^* exists and is unique. Show that the extreme values $x = 0$ and $x = T$ provide the following lower and upper bounds for T^*:

$$\frac{1}{k}\left[\frac{\alpha^\beta\left((k-1)C_p + C_r\right)}{(\beta-1)C_m}\right]^{1/\beta} \le T^* \le \left[\frac{1}{k}\frac{\alpha^\beta\left((k-1)C_p + C_r\right)}{(\beta-1)C_m}\right]^{1/\beta}$$

15.12   Show that when the time to failure follows a Weibull distribution with parameters α and β,
   (a) The cost function given by Equation (15.25) reduces to the expression given in Example 15.6
   (b) The optimal value of T that solves Equation (15.26) for a given value of k is given by the expression given in Example 15.6.

15.13   Assume that the time to failure of an item may be approximated by a Weibull distribution with distribution function $F(t) = 1 - \exp[-(t/\alpha)^\beta], \alpha > 0, \beta > 1$ and $C_p = 10, C_r = 30, C_b = 30, \beta = 3$, and $\alpha_i = (0.85)^{i-1}$. Use Solution Algorithm 15.1 in Section 15.5.1 to determine the optimal number of overhauls and their timing and the corresponding optimal expected cost per unit time.

15.14   Assume that the time to failure of an item can be described by a two-parameter Weibull distribution with parameters $\beta = 2.5$ and $\alpha_i = (0.85)^{i-1}$. Assume that $C_p = 10, C_r = 30, C_m = 10$. Use Solution Algorithm 15.2 to determine the optimal number of overhauls and their timing and the corresponding optimal expected cost per unit time.

References

Dekker, R., van der Duyn Schouten, F.A., and Wildeman, R.E. (1996) A review of multi-component maintenance models with economic dependence. *Mathematical Methods of Operations Research,* **45**: 411–435.

Murthy, D. and Nguyen, D. (1985a) Study of a multi-component system with failure interaction. *European Journal of Operational Research,* **21**: 330–338.

Murthy, D. and Nguyen, D. (1985b) Study of a two-component system with failure interaction. *Naval Research Logistics Quarterly,* **32**: 239–247.

Nakagawa, T. (1980) A summary of imperfect maintenance policies with minimal repair. *RAIRO: Recherche Operationnelle,* **14**: 249–255.

Nakagawa, T. (2005) *Maintenance Theory of Reliability,* Springer-Verlag.

Nakagawa, T. and Murthy, D. (1993) Optimal replacement policies for a two-unit system with failure interactions. *RAIRO: Recherche Operationelle,* **27**: 427–438.

Nguyen, D.G. and Murthy, D.N.P. (1981) Optimal preventive maintenance policies for repairable systems. *Operations Research,* **29**: 1181–1194.

Nicolai, R.P. and Dekker, R. (2008) Optimal maintenance of multi-component systems: A review. In *Complex System Maintenance Handbook,* K.A.H. Kobbacy and D.N.P. Murthy (eds), Springer-Verlag, London, pp. 263–286.

16

Condition-Based Maintenance

Learning Outcomes

After reading this chapter, you should be able to:

- Describe the characterization of continuous and discontinuous degradation;
- Define diagnostics and prognostics in the context of CBM;
- Explain the elements of the approach to CBM based on diagnostics and prognostics;
- Describe different methods for assessing an object's current condition using different types of historical data;
- Classify and describe the various approaches used for fault diagnosis;
- Differentiate between diagnostics of a failed item and diagnostics for assessing the state of a non-failed item in order to predict its future state;
- Describe simple prognostic models involving continuous and discontinuous degradation;
- Describe some CBM policies and explain the approach to determine the optimal CBM policy;
- Illustrate CBM concepts through applications to different types of engineered objects.

Introduction to Maintenance Engineering: Modeling, Optimization, and Management, First Edition.
Mohammed Ben-Daya, Uday Kumar, and D.N. Prabhakar Murthy.
© 2016 John Wiley & Sons, Ltd. Published 2016 by John Wiley & Sons, Ltd.

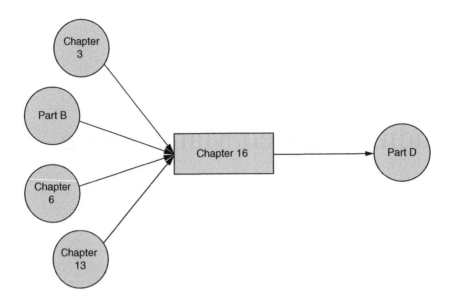

16.1 Introduction

In Part B we looked at PM (preventive maintenance) policies based on age and/or usage for an item (system, component, or some intermediate level). CBM (condition-based maintenance), defined in Chapter 4, calls for maintenance actions based on the condition or state of the item. The state of the item characterizes the level of degradation due to some degradation mechanism that depends on the physical characteristics (type of material – metal, plastic, etc.), operating environment (humidity, value of fluid, etc.), and operating load, as discussed in Chapter 3.

A CBM approach to maintenance requires four key steps, as indicated in Figure 16.1:

- *Step 1:* Collecting data using sensors and condition-monitoring technologies, as discussed in Chapter 6.
- *Step 2:* Using data to assess item state if the item has not failed or detect the fault if the item has failed.
- *Step 3:* Predicting future condition.
- *Step 4:* Taking appropriate maintenance action.

Step 4 requires formulating an appropriate CBM policy and selecting the optimal values of its parameters.

This chapter is organized as follows. Section 16.2 deals with the characterization of degradation. The diagnostic approach (to assess current condition), the prognostic approach (to predict future condition), and CBM are discussed in Section 16.3. Section 16.4 covers diagnostics, prognostics, and CBM in more detail, and includes such subjects as fault diagnostic definitions and approaches, prognostic methods, CBM policies, and illustrative examples of CBM applications for plants and infrastructures. Finally, Section 16.5 contains a summary of the chapter.

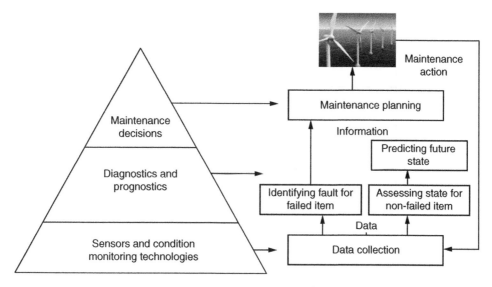

Figure 16.1 Key steps of the CBM approach.

16.2 Characterization of Degradation

Let $X(t)$ denote a measure of the state of the item at time t, where t is the age clock. There are two different scenarios:

- *Continuous degradation:* Degradation occurring continuously. For example, pipe-thickness degradation due to corrosion or bearing-lining degradation due to wear.
- *Discontinuous degradation:* Degradation occurring in a discontinuous manner due to shocks of random magnitude occurring randomly over time. For example, the wear in an aircraft's landing gear due to impact forces during landing or the degradation of a bridge due to the load of a passing train.[1]

16.2.1 Continuous Degradation

In this case, $X(t)$ is continuously changing with time. It can be increasing or decreasing with time depending on the particular application (the thickness of a pipe decreasing with time or the length of a crack in a bridge increasing with time). A higher value of change implies greater degradation.

Degradation occurs in an uncertain manner and Figure 16.2 shows a typical shape of the degradation. The degradation is usually slow in the early stages of operation up to time t_a (with the degradation being less than A). The rate of degradation starts to increase and becomes very rapid after it reaches a critical level, C (referred to as the *critical point*) by time t_c. The item fails very soon after this critical level is reached, so that the *item failure* can be defined as the instant the degradation exceeds C.

---

[1] The degradation of a bridge can be both age dependent – continuous – and also due to random loading – discontinuous.

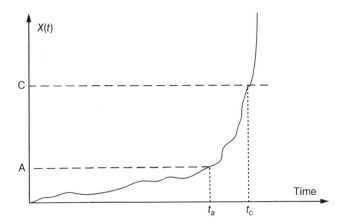

Figure 16.2   Gradual degradation of an item's state.

Example 16.1   Road Infrastructure

Cracks in a road surface are the result of dynamic traffic loads on the road and changing weather conditions. A crack usually starts small and is not easy to observe in the early stages. They are easier to observe in the coldest months of the year as the road surface contracts and are also seen more easily after light rain when the surface dries out, leaving moisture in the cracks. The light reflected from a crack is different from that reflected from a road surface with no crack. As time progresses, the cracks develop further – in length, width, and depth. The degradation of a section of a road may be characterized through the number of cracks and the length, depth, and width of each crack. As a result, the degradation is a composite of all of these measures. ∎

16.2.2   Discontinuous Degradation

In this case the changes of the state of the item occur at random discrete time instants, as shown in Figure 16.3. Again, $X(t)$ can be increasing or decreasing with time depending on the type of application and the failure mechanism involved. A lower rate of changes implies a greater level of degradation.

Similar to the continuous case, if $X(t)$ is greater than some specified value C (the *critical level*), the performance of the item is acceptable and it is unacceptable when it falls below this value.

Example 16.2   Concrete Structures

In earthquake-prone areas, structures (such as buildings, bridges, dams, etc.) are subjected to frequent earthquakes that occur randomly over time. The intensity of a quake is measured on the Richter scale, with a higher value indicating a greater quake magnitude. The quake creates high levels of stress in the structure. If the stress exceeds the strength

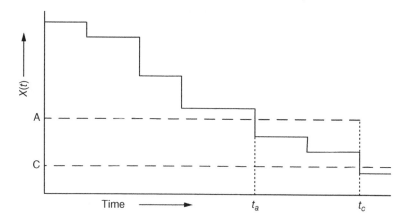

Figure 16.3 Discontinuous degradation of an item's state.

then the structure collapses. However, if the stress is below the strength, the quake does damage and weakens the structure. Many brittle elements of the structure tend to break or lose strength and may then break at loads much lower than the designed load. For example, unreinforced masonry walls that crack when overstressed in shear, and unconfined concrete elements that crush under compressive overloads generated by the quake. In addition, many ductile elements such as tension braces deform beyond their elastic strength limit and continue to carry loads with reduced strength. The magnitude of reduction in the strength is related to quake intensity. As a result, the strength of the structure decreases in a random manner, as shown in Figure 16.3. ∎

16.3 Approach to CBM

The approaches for diagnostics (to assess current condition), prognostics (to predict future condition), and CBM (based on diagnostics and prognostics) are shown in Figure 16.4.

16.3.1 Condition of a New Item

All engineered objects are expected to perform as per design specifications. Failure to deliver expected performance can be due to various reasons. Often, the material used is of substandard quality and does not conform to the design specification, leading to premature failure, or the performance may be lower due to a poor manufacturing process. For example, often, the designer does not provide enough space to reach the point for welding – this leads to poor welding strength, which, in turn, leads to failure and substandard performance.

16.3.2 Current Condition

The current object condition depends on various factors including component materials, the history of usage, the operating environment and maintenance, and the degradation mechanisms involved.

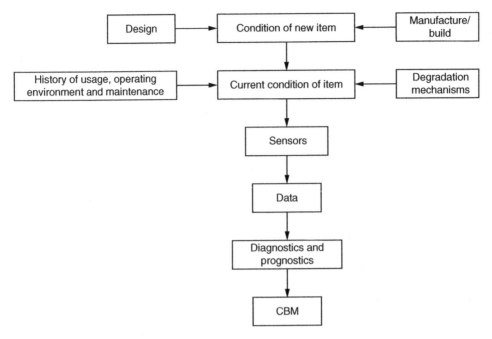

Figure 16.4   Approach to CBM.

Assessing an object's current condition requires different types of historical data from the time the object is put into operation. These data may be collected continuously or at discrete time instants and include maintenance data, data about the operating environment, usage intensity, and so on.

16.3.3   Sensors and Data

In Chapter 6 we discussed some of the commonly used sensors. In this section we discuss the types of data generated by some of these sensors and the data analysis.

16.3.3.1   Vibration Sensor Data

The vibration of an item results in displacement, velocity, and acceleration of the item.

- *Displacement:* Displacement is a measure of the actual distance an object is moving from a reference point (usually expressed in mm).
- *Velocity:* Velocity is the rate of change in position (usually expressed in mm/s). Velocity sensors operate on the principle of electromagnetic induction and are usually used for the *low to middle range of frequencies* (10–1500 Hz).
- *Acceleration:* Acceleration is the rate of change of velocity and is a measure of the force being produced (usually measured in mm/s^2). Accelerometers are used for this purpose. They work best for high frequencies where acceleration is large (>1000 Hz).

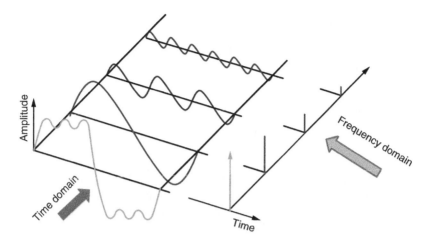

Figure 16.5   Frequency versus time domain analysis.

The signals generated by vibration may be analyzed in the time or frequency domains (see Figure 16.5):

• *Time domain:* The signal is a function of time.
• *Frequency domain:* The time domain signal is decomposed into a linear combination of a basic periodic (sine wave) signal and its harmonics. The amplitudes of different frequencies provide information that is used in assessing the condition or degradation.

16.3.3.2   Oil Debris Sensor Data

New oil used for lubrication contains no contamination. As the item (for example, a bearing) degrades (due to wear), the oil condition changes due to contamination with wear metals and this is commonly referred to as *oil debris*. The contamination may also be due to other substances (such as water).

Assessment of the oil condition is carried out using several tests including viscosity, water content (as water promotes corrosion), insolubility (provides evidence of oil filter effectiveness), total acid number (TAN) (may indicate oxidation of the lubricant), total base number (TBN) (determines alkalinity of the lubricant), flashpoint (detection of fuel in the lubricant), particle count, and elemental analysis (detects wear elements and some metals serve as catalysts for lubricant oxidation reactions).

Metallic wear debris are differentiated by their morphology (shape, texture, and color) into several classes, for example, rubbing, cutting, spherical, laminar, fatigue chunk, and severe sliding wear particles. It has been found that each type has its own generation mechanism involving a specific wear process; for instance, cutting wear particles are produced by the penetration, plowing, or cutting of mating bodies. The presence of severe sliding wear particles in a machine usually indicates a lubrication problem, caused by lubricant film breakdown.

The three main methods of wear debris analysis are:

• *Direct detection:* Involves arranging for the fluid to flow through a device sensitive to the presence of debris;

- *Particle collection:* Requires a means by which it is convenient to remove a sample of debris from the system for offline examination;
- *Fluid analysis:* Similarly requires a sample to be removed, and subsequently analyzed to determine its condition and the extent of any contamination.

Example 16.3   Bearing

The two common methods for the degradation of a bearing are (i) flaking and (ii) wear.

- *Flaking:* This occurs due to fatigue resulting from the cyclical loading experienced by the bearing. This results in a crack inside the metal which gradually extends to the surface. As the rolling elements pass over the cracks, fragments of material break away and this is known as *flaking* or *spalling*. The flaking increases with time and eventually makes the bearing unserviceable. Figure 16.6 shows the pattern of evolution of flaking

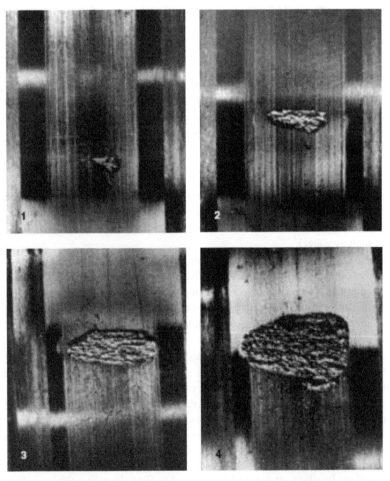

Figure 16.6   Flaking in the inner and outer races over time (SKF – product information, 1994).

---

with time through visual inspection of a ball bearing where the balls are enclosed between two circular rings (referred to as the inner and outer race).

The phenomenon of flaking is a relatively long and drawn-out process. As it evolves, noise and vibration levels increase and, as such, the bearing condition is easily detected through acoustic sensors and accelerometers.

- *Adhesive wear:* This type of wear is the progressive removal of material from the two sliding or rolling surfaces during service. Adhesive wear in a bearing occurs mostly due to a failure of the lubrication, often resulting from inappropriate seals, which results in a loss of lubricant.

Poor seals also lead to ingress of abrasive contaminant particles. These particles lead to removal of material from the raceway, cage, and rolling elements by abrasion (abrasion wear). This is an accelerating process because wear particles will further reduce the lubrication ability of the system, leading to increased wear and increased concentration of particles.

A plot of the number of particles (of a particular element, for example, iron) against time provides information regarding the degree of wear. Further analysis of the wear debris reveals the identity of the components that are wearing out, and the types of wear may be assessed from the shape and size of particles in the oil debris. In many cases, it is useful to have a composite condition variable where the concentration of particles of different elements in the lubricants is used to assess the wear. ∎

16.3.3.3   NDT Sensor Data

NDT (non-destructive testing) methods use sensors that were discussed in Chapter 6. The types of degradation measured by these methods are summarized in Table 16.1.

Table 16.1   NDT methods and defects monitored.

NDT methods	Degradation	Limitations	Remarks
Acoustic emissions	To identify propagating surface or subsurface crack	Not suitable for stable crack	Long-term data collection for decision making
Eddy current testing	Voids in thin layers, head checks	Maximum depth 2 mm, local resolution 2 mm	Commonly used in railways for head check detections
Ultrasonics	Internal voids or measuring the thickness of surface coatings	Defects below 4–5 mm below test surface are difficult to detect	Useful for rapid and automated inspection. Used for railway track inspections
Radiography	Internal voids	Hazardous in use. Maximum investigated plate thickness 70 mm	Can be used for most of the material
Color penetration testing	Surface cracks	Only surface-breaking defects can be detected	Useful for detecting small surface discontinuities

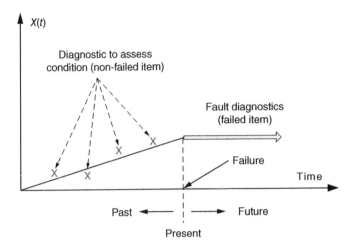

Figure 16.7   Diagnostic to assess item condition and fault diagnostics.

16.4   Diagnostics, Prognostics, and CBM

The term *diagnostic* is often used to denote *fault diagnostic*. The aim is to detect the fault when a failure occurs. Here, the item has failed and the aim is to detect the components responsible for the failure so that they can be subjected to a CM (corrective maintenance) action to restore the failed item to the operational state. However, we use the term in a broader sense, so that it includes both fault diagnostics and assessment of the condition of an item that is still operational. This information (collected over time) is needed for both prognostics (predicting the condition in the future) and for CBM. Figure 16.7 shows the two cases and these are discussed further in Sections 16.4.2 and 16.4.3. Prognostics and CBM are discussed in later sections of the chapter.

16.4.1   Diagnostics

The origin of the word "diagnostics" comes from the medical field – it refers to the identification of the nature of a health problem by examination and evaluation of signs, symptoms, and test results. In the CBM context, diagnostics deals with the examination of symptoms and sensor data to determine the nature of faults and failures of engineered objects.

In this section we present basic definitions, concepts, and brief discussions of the various approaches used for fault diagnosis. Diagnostics of a failed item is discussed in the Section 16.4.1.1 along with diagnostics for assessing the state of a non-failed item in order to predict its future state. We call the former *failure diagnostics* and the latter *fault diagnostics.*[2]

---

[2] This terminology is adopted by many authors but there is no universal agreement on these definitions.

16.4.1.1   Fault (Failure) Diagnostics

> **Definition 16.1**
>
> Fault (Failure) diagnostics refers to the process of detecting, isolating, and identifying a component that has failed (EN 13306:2010).

Fault detection deals with detecting and reporting an abnormal condition. Isolation determines which component has failed. Identification estimates the nature and extent of the failure (see Example 16.4). These tasks are carried out sequentially, as shown in Figure 16.8. Diagnosis to assess the state of non-failed items is discussed in Section 16.4.1.2.

In general, the diagnostic decision-making process may be viewed as a series of mappings on measurement data. The various transformations that measurement data go through during diagnosis are shown in Figure 16.9.

The first transformation maps the measurement space into a feature space. Developing the feature space involves feature selection and feature extraction. Feature selection involves selecting a few important measurements of the original measurement space. Feature extraction

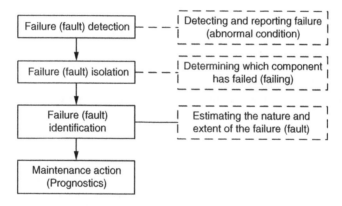

Figure 16.8   Diagnosis steps.

> **Example 16.4   Circuit Board**
>
> A circuit board in a computer consists of several elements mounted on a flat panel and connected together through wires that are embedded within the panel. The failure of a circuit board is due to failure of one or more of its elements. A fault diagnostic to locate the failed elements of a circuit board involves an automatic test system which involves inputs (voltage being applied at different locations) and outputs (currents) measured at different points in the circuit. The measured values are compared with values to be expected if there is no fault. The difference between the two is used sequentially to carry out testing to locate the faulty elements of the failed circuit board.   ∎

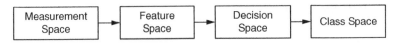

Figure 16.9   Transformations during diagnostic process.

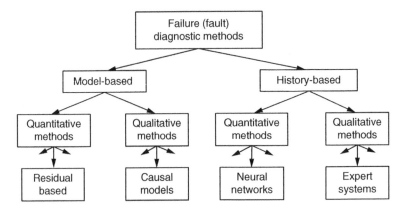

Figure 16.10   Classification of fault diagnostic methods.

uses prior knowledge of the problem to transform the measurement space into a space of smaller dimension. The points in the feature space are functions of the measurement points containing only useful information to aid diagnosis.

The transformation from the feature space to the decision space is usually designed to optimize some objective function, such as minimizing misclassification.

The class space contains the failure classes and is the final result of the diagnostic process.

Approach

Fault-diagnostic methods rely on condition-monitoring data, item-performance data, and expert opinion. Past experience plays an important role in detecting the fault quickly. Various methods have been developed to deal with failure diagnostics. Many of them are beyond the scope of this book. We present a broad classification of these methods and describe a representative set, discussing briefly their underlying concepts.

Fault-diagnostic methods can be classified broadly into two categories: model-based and history-based methods, as shown in Figure 16.10.

Model-based methods are based on fundamental understanding of the system under consideration using first principles knowledge. These can be further classified into qualitative and quantitative methods. Next, we briefly describe one method under each of these categories.

- *Residual-based methods:* These methods rely on explicit mathematical models of the monitored system based on detailed knowledge of the physical relationships and characteristics of all components in the system. Using this knowledge, a set of mathematical equations is developed to model the behavior of the system. The essence of these methods is to compare the actual system behavior against the system model for consistency. Any inconsistency expressed as residuals can be used for diagnostic purposes. The residuals should be close to zero under normal conditions but they have a significant value when a fault occurs. However, several factors, including system complexity, process nonlinearity, high dimensionality, and a lack of good data, limit the usefulness of such an approach for complex industrial systems.

- *Causal model methods:* These methods are based on cause–effect reasoning about system behavior. A good example is fault trees that use backward tracing of a chain of events until a basic event is found that represents a root cause for the problem under consideration. Fault trees have been discussed in Chapter 2.

    History-based methods rely on a large amount of system-history data and can also be subdivided into qualitative and quantitative methods. Next, we briefly describe one method under each of these categories.

- *Neural network methods:* Neural networks are composed of simple elements (nodes called *neurons*) operating in parallel. These elements are inspired by biological nervous systems. Commonly, neural networks are adjusted, or trained, so that a particular input leads to a specific target output. A neuron is a processing unit with several inputs and only one output. Each input, x_i, is weighted by a factor w_i, the neuron has a bias b, and the whole sum of the input is calculated as $a = \sum w_i x_i + b$. Then, an activation function "f" is applied to the results "a" and the neural output is taken to be $f(a)$.

    A *neural network* (NN) is organized into layers. Information is fed to the input layer – the set of input nodes for the NN (each input variable inputs directly to one node in the input layer). The output of these nodes is fed to multiple nodes in the next layer, with the final result available at the last layer called the *output layer* (each output node corresponds to an output variable). The nodes between the input and output layers are called *hidden nodes* and their layers are called *hidden layers.*

    When using neural networks for fault diagnosis, the inputs to the network can be, for example, the sensor values or alarm states of the system, and the outputs of the network represent the presence or absence of particular faults.

    Neural networks are able to learn expert knowledge by being trained using a representative set of data. In a neural network approach, the measurement space is represented by the input nodes. Hidden nodes correspond to the feature space. The network output nodes transform the feature space into the decision space. The interpretation of network outputs provides the transformation to the class space.

Example 16.5   Bearing

Neural networks are applied to motor bearing fault diagnosis. The bearing vibration frequency features and time-domain characteristics (shaft rotational frequency, fundamental cage frequency, ball pass inner raceway frequency, and ball pass outer raceway frequency) are applied to a neural network to build an automatic motor bearing fault-diagnosis machine. The neural network has three outputs, each one serving as an indicator for one of the three fault conditions (bearing looseness, defects on the inner raceway, and defects on the rolling element), producing an output of 1 for good and −1 for bad. ∎

- *Expert systems:* An expert system is computer software that attempts to act like a human expert in a particular subject area. An expert system is made up of three main parts, as shown in Figure 16.11:
  - ○ A user interface that allows a user to query the expert system;
  - ○ A knowledge base which is a collection of facts and rules created from information provided by experts in the field of interest;
  - ○ An inference engine that acts like a search engine examining the knowledge base for information that matches the user's query.

  Expert systems are intuitively attractive for process fault diagnosis, since rules can be explicitly listed linking symptoms (process measurements and alarms) to causes (specific equipment malfunctions and operational faults). The limitation of rule-based expert systems is their inability to deal with novel fault situations for which no specific rules exist. However, this shortcoming can be overcome through the inclusion of deep knowledge of the process in the knowledge base, usually in the form of models of the process structure and function. These models can be either qualitative or quantitative simulations of the process. The expert system can then reason from first principles to diagnose the novel faults. Expert systems may also be integrated with other diagnostic methods such as neural networks.

  We conclude this section by providing a set of desirable characteristics that a good diagnostic process should have. Obviously, these requirements will not usually be met by any single diagnostic approach. However, they are useful to benchmark and compare various methods (Table 16.2).

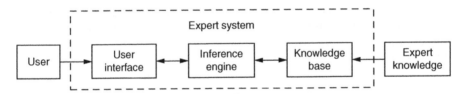

Figure 16.11   Elements of an expert system.

Table 16.2   Desirable characteristics of a failure diagnostic approach.

Desirable characteristic	Description
Quick diagnosis	Quick detection and diagnosis of system malfunctions
Isolability	Ability to distinguish between different failures
Robustness	Robust to various noises and uncertainties
Novelty identifiability	Ability to recognize the occurrence of novel (unknown) faults
Classification error estimate	Provide an *a priori* estimate of classification error to build confidence of the user in the diagnostic process
Adaptability	Adaptable to changes due to external inputs or operating conditions
Explanation facility	Provide an explanation of how the fault originated and propagated to the current situation
Modeling requirements	The modeling effort should be as minimal as possible
Storage and computational requirements	Quick real-time solutions would require a diagnostic process that balanced the competing requirements of computation and storage
Multiple fault identifiability	Ability to identify multiple faults (important but difficult requirement)

16.4.1.2 Diagnostics to Assess the Condition of a Non-Failed Item

> **Definition 16.2**
>
> Diagnostics to assess the condition of a non-failed item refers to the process of detecting, isolating, and identifying an incipient failure condition of a component of an item that is still in the operational state but in a degraded condition.

Again, the detection, isolation, and identification steps are carried out sequentially, in a manner similar to that in Figure 16.8, and they provide necessary information for the prognostics step of the CBM framework discussed earlier.

Approach

The data for assessing the condition of a non-failed (working) item may be collected either continuously or at discrete time instants from one or more sensors. We confine our attention to the case where the data are collected at discrete time instants t_i, $i = 1, 2, 3, \ldots$, with $t_0 = 0$, the time instant when the object is put into operation.

Single Sensor

The output of the sensor (signal) at time t_i is given by:

$$Z(t_i) = Y(t_i) + W(t_i) = \phi\big(X(t_i)\big) + W(t_i) \tag{16.1}$$

where $Y(t_i)$ is the true signal and $W(t_i)$ is the sensor noise (or observation error). An estimate of the condition $X(t_i)$ involves two steps:

- *Step 1:* Processing of the signal to obtain an estimate $\hat{y}(t_i | z(t_j)$, $j = 1, 2, \cdots, i)$ of $Y(t_i)$ based on the observed data $z(t_j)$, $j = 1, 2, \ldots, i$.[3]
- *Step 2:* Obtain an estimate $\hat{x}(t)$ using $\hat{y}(t_i | z(t_j))$, $j = 1, 2, \cdots, i)$ from the inverse relationship $x = \phi^{-1}(y)$ (see Equation (16.1)).

Multiple Sensors

In some cases, more than one sensor is used to assess the condition of an item (for example, vibration, oil debris, and thermal sensors for assessing the wear of bearings). Let J denote the number of sensors and the signal at time t_i from sensor j, $(1 \leq i \leq J)$, is $z_j(t_i)$. Following the same approach as for a single sensor, the estimate based on data $z_j(t_i)$ (from sensor j) is given by $\hat{x}_j(t_i)$.

Combining estimates from different sensors $[\hat{x}_j(t_i)$, $j = 1, 2, \ldots, J]$ into a single estimate $\hat{x}(t_i)$ needs to be done carefully – data from sensors with greater precision need to be given a higher rating compared with data from sensors with low precision. A simple way of doing this is a linear combination of estimates from the different sensors with different weights to yield a final estimate:

$$\hat{x}(t_i) = \sum_{j=1}^{J} a_i \hat{x}_j(t_i), \text{with } 0 < a_j < 1, 1 \leq j \leq J, \text{ and } \sum_{j=1}^{J} a_j = 1 \tag{16.2}$$

---

[3] Upper case represents random variables and lower case either observed or estimated values.

16.4.2   Prognostics

The word "prognostics" originated from the medical field, where it is used to describe the probable course of a disease. In the CBM context, prognostics deals with the estimation of the remaining useful life of the components for a given task once an impending failure condition is detected, isolated, and identified.

Definition 16.3

Prognostics is defined as the estimation of the operating time or remaining useful life before failure (ISO 13381-1:2004).

16.4.2.1   Approaches

The three approaches to building models to predict future state (and residual life) are as follows:

- *Physics-based approach:* Models based on the underlying mechanism and usually very complicated and not widely used.
- *Data-based approach:* Uses only the diagnostic data (relating to the item's condition in the past) to build models.
- *Hybrid approach:* This is a combination of the above two approaches.

We confine our attention to data-based models.

16.4.2.2   Data-Based (Empirical) Models

The models may be classified into two groups based on whether the degradation is continuous over time or at discrete time instants. A further subdivision is whether the model is deterministic or stochastic.

We look at two categories of models, indicated below:[4]

- *Continuous degradation:* Deterministic models (Models 16.1–16.3).[5]
- *Continuous degradation:* Stochastic models – discrete time formulations (Model 16.4).[6]

Deterministic Models
We consider the case where the degradation $X(t)$ is non-decreasing (as shown in Figure 16.2).

---

[4] There are many other models such as Markov chain models, neural network models, and so on.
[5] More complex models include partial differential equations. An alternative is the use of finite element modeling.
[6] Models that treat time as continuous involve complex stochastic differential equation formulations and are, in general, analytically intractable. One needs to use a simulation approach.

Model 16.1   Polynomial Equation[7]

The model is given by:

$$X(t) = a_0 + \sum_{i=1}^{k} a_i t^i \tag{16.3}$$

If k is specified, the number of parameters to be estimated is $k+1$ $(a_0, a_1, ..., a_k)$ and if not, the number of parameters to be estimated is $k+2$ (k and $a_0, a_1, ..., a_k$).

Model 16.2   Linear Differential Equation (Jantunen, 2003)

The model involves a linear differential equation and is given by:

$$\frac{dX(t)}{dt} = a \left(\frac{t_m}{t_m - t} \right), \quad X(0) = 0 \tag{16.4}$$

or

$$X(t) = -a \ln \left(\frac{t_m - t}{t_m} \right) \tag{16.5}$$

The model parameters are $\{a, t_m\}$ which need to be estimated.

Comment: Note that by time t_m, the rate of degradation is infinite. The upper limit for degradation is given by $X_m = X(t_m) = -a \ln(0)$.

The deterministic models ignore the underlying uncertainties. One can model with time being continuous or discrete. Time series models model the item state at discrete (periodic) time instants and there are many different model formulations. A simple model formulation is the following:

Model 16.3   Autoregressive (AR) Model of Order K

An autoregressive (AR) model of order K can be expressed as follows:

$$X(t_{i+1}) = \sum_{k=0}^{K} a_k X(t_{i-k}) + V(t_i) \tag{16.6}$$

where $V(t_i)$, $i = 1, 2, \cdots$, is a sequence of random variables.

The parameters to be estimated are $\{a_i, i = 0, 1, \cdots, k\}$.

Comment: More complicated than the above would be autoregressive and moving average (ARMA) models.[8]

---

[7] A special case is $k = 1$ (linear equation) with two non-negative parameters, as proposed by Onsoyen (1991).

[8] See Box and Jenkins (1970) for more on time series modeling.

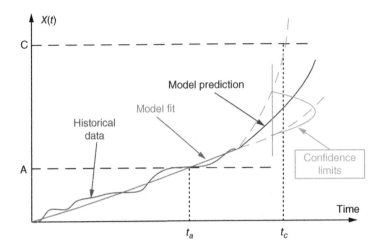

Figure 16.12   Estimating model parameters.

Stochastic Model

Model 16.4   Distribution-Based

Here, the change in degradation $\{X(t_{i+1}) - X(t_i)\}$ is modeled by a distribution function $F(x; \theta(t_i))$ where $\theta(\cdot)$ is the scale parameter. It is a function of t_i, with $\theta(t_{i+1}) < \theta(t_i)$, $i = 0,1,2,\ldots$.

Parameter Estimation
The parameters are estimated by the method of least squares using the estimates $\hat{x}(t_i)$, $i = 1,2,\ldots$, as shown in Figure 16.12.

16.4.2.3   Residual Life Prediction

The residual life is obtained by extrapolating the prediction (or the mean in the case of stochastic models) of the condition of the item (based on the model), as indicated in Figure 16.13. One can give confidence intervals for the prediction in the case of stochastic models.

16.4.3   CBM Policies

The basic concept is based on two limits: (i) the safe limit (given by A) and (ii) the critical limit (given by C) shown in Figures 16.2 and 16.3.[9] The limits define a time interval $(t_c - t_a)$ for carrying out a PM action. If this is not done, then $X(t)$ exceeds the critical limit, requiring a costlier CM action.

---

[9] The safe limit is also referred to as the *alarm limit* and the critical limit as the *failure limit*.

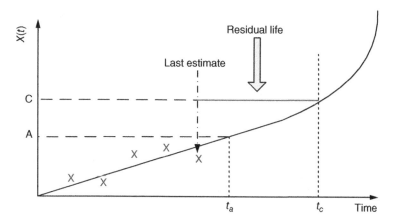

Figure 16.13 Predicting future state and residual life.

The condition variable $X(t)$ is a non-negative and non-decreasing function (as shown in Figure 16.2). The item is inspected at discrete time instants $t = t_j, j = 1, 2, 3, \ldots,$ to obtain the $z(t_j)$. An estimate, $\hat{x}(t_j)$, of the condition variable is obtained based on all the data observed until t_j. The critical and safe limits are C and A, respectively, with A < C.

Policy 16.1

The inspections are periodic, with $t_{j+1} - t_j = \tau, j = 0, 1, 2, 3, \ldots,$ with $t_0 = 0$. After each inspection, one of the following three actions takes place:

- If $\hat{x}(t_j) < A$, do nothing until the next inspection time, t_{j+1}.
- If $A \le \hat{x}(t_j) < C$, carry out a replacement with a new item (PM action).
- If $\hat{x}(t_j) \ge C$, replace the failed item by a new one (CM action).[10]

The parameters of the policy are $\{\tau, A, \text{and } C\}$.

Comment: Note that the inspections are periodic. Since degradation is slow until t_a, the inspections may be performed less frequently and then more frequently beyond t_a as the degradation is faster. The next policy allows for this.

Policy 16.2

The interval between inspections $\tau_{j+1} = t_{j+1} - t_j, j = 0, 1, 2, 3, \ldots,$ is a non-increasing sequence with $t_0 = 0$. After each inspection, one of the three actions listed in Policy 16.1 takes place. The parameters of the policy are $\{\tau_j, j = 1, 2, 3, \ldots, A \text{ and } C\}$.

---

[10] A repair may be performed instead of replacement and the repair may be imperfect.

Comment: This policy has more parameters. There are two subcases, as indicated below:

1. $\tau_j, j = 1, 2, 3, \ldots$, are decided at time t_0. This is referred to as *open loop* in the control theory literature.
2. τ_{j+1} is selected at the end of inspection j and is a function of $\hat{x}(t_j)$. This is referred to as *closed loop* in the control theory literature.

16.4.3.1 Optimal CBM

The critical level C is not a decision variable and is given. The optimal CBM decision involves selecting the remaining parameters (decision parameters) of the policy. Let θ denote the set of parameters. The process involves the following three steps:

1. Defining an objective function.
2. Building models to obtain an expression for the objective function $J(\theta)$ in terms of the decision parameter set θ.
3. Performing an analysis to obtain the optimal values for the parameters.

The simplest model assumes the following:

- The times needed to carry out each PM action, CM action, and inspection are negligible and hence may be ignored.
- The costs of each inspection, PM action, and CM action are C_I, C_p, and C_f, respectively, with $C_f > C_p$.
- The time horizon is infinite.
- The objective function is the asymptotic expected cost per unit time.

As mentioned earlier, the item degradation, $X(t)$, occurs in an uncertain manner and needs to be modeled by a stochastic process. The time instants t_a and t_c are random variables and their (probabilistic) characterization is referred to as a *level crossing problem* in the theory of stochastic processes. This problem is extremely difficult to solve analytically. One needs to use a simulation approach to obtain the optimal parameter values. As discussed in Chapter 8, one generates several time histories of $X(t)$ for a given set θ to obtain an estimate $\hat{J}(\theta)$. This information is then used to obtain the optimal values for the policy parameters.

16.4.4 CBM Decisions for Products and Plants

16.4.4.1 Railway Rolling Stock

The rolling stock consists of engines, wagons, and coaches. Each of these may be viewed as a product with several elements, and wheels are common to all. We focus our attention on the wheels of a freight wagon, as they are the most expensive components of the wagon that need periodic replacement.

Wheel Deterioration
The lifetime of a wheel is limited by several failure modes – the three important ones are: (i) wear, (ii) rolling contact fatigue (RCF), and (iii) out-of-round (OOR). The impact of the

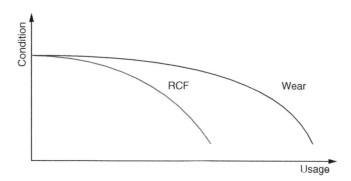

Figure 16.14 Effect of usage on RCF and wear.

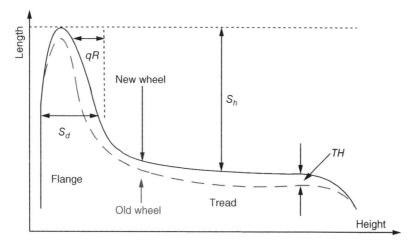

Figure 16.15 New and old wheel profiles and wheel-profile parameters.

first two modes of failure on the condition of the wheel (defined by wheel-profile parameters) is dependent on the usage (kilometers traveled), as indicated in Figure 16.14.

The wheel profiles for new and old wheels along with the wheel-profile parameters – flange height (S_h), flange thickness (S_d), flange gradient (qR), and *tread hollow* wear (TH) – are shown in Figure 16.15. Wheel-profile measurements may be done manually, using a mechanical or laser unit, or automatically, using lasers and cameras.

The interaction between wheel and rail resulting in material deterioration is a complicated process, involving vehicle-track dynamics, contact mechanics, friction wear, and lubrication. We discuss some of these mechanisms below.

- *Wheel profile wear:* Wear is the loss or displacement of material from contacting surfaces and this is related to sliding, contact stresses, and material properties.
- *Wheel flats:* Wheel flats are formed when a wheel set is locked and skids along the rail, and are the most important source of vertical loads on the rail. The cause can be poorly adjusted, frozen, or defective brakes or there may be high braking forces in relation to

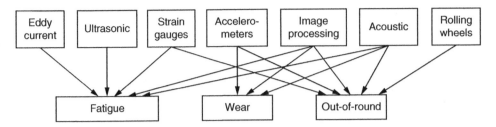

Figure 16.16   Condition-monitoring sensors and wheel-monitoring measures.

Table 16.3   Advantages and disadvantages of three methods of detecting and monitoring wheel wear and fatigue.

Method	Advantages	Disadvantages
Visual	Quick and cheap	Not all faults detected
Wayside	Objective and repeatable	CMMS (computerized maintenance management system) and decision support systems (DSSs) are needed
Workshop	Early decision	Expensive and time consuming

available adhesion. The friction between wheel and rail causes wear of the wheel surface, thus causing the wheel to become flat instead of round.

- *Out-of-round wheels:* Wheels with flats and tread damage or build-up are referred to as *out-of-round*, and it is generally accepted that these wheels cause large dynamic forces.
- *Rolling contact fatigue:* Wheel damage occurs as fatigue cracks, initiated at or below the surface, resulting in material fall-out, such as shelling or spalling.

Condition-Monitoring Methods
There are several possible sensor set-ups and monitoring techniques to detect condition, faults, and failures of a wheel, as shown in Figure 16.16.

Wheel Maintenance
The three methods to detect and monitor wheel wear and fatigue are as follows:

1. Visual inspection of the wheels at the railway yard.
2. Wayside monitoring stations (involving sensors) to detect faults or failures.
3. Opportunities during general wagon maintenance in the workshop.

The advantages and disadvantages of these three methods are indicated in Table 16.3.

When a faulty wheel is detected, a maintenance action is needed and a work order is created. Figure 16.17 shows a typical maintenance process.

16.4.4.2   Wind Turbine/Generator

Condition- Based Maintenance for a Wind Turbine
A typical layout and design of a wind turbine is shown in Figure 6.11 and is briefly described here. The reader is referred to this figure for the names of the different components. Gearbox and generator failures account for 17% of all failures reported in a study of Danish turbines in 2001.

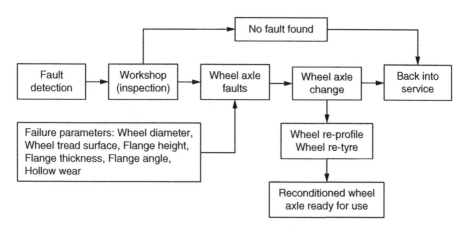

Figure 16.17   Typical wheel-maintenance process.

Historical Approach to Maintenance of Wind Turbines

Initially, "run to failure" was the prevalent strategy, with limited periodic replacement of items such as oil and filters. However, as wind turbines grew in capacity, many operators started using periodic inspections by skilled technicians who used their experience and portable analysis tools to assess the condition of the turbine during scheduled maintenance. These offline condition-monitoring practices were later replaced by online condition-monitoring schemes.

Motivation for CBM

The high cost of downtime, especially for offshore turbines, drives the need for CBM. For example, failure of a $1500 bearing could result in a $100000 gearbox replacement, a $50000 generator rewind, and $70000 to replace the failed component. In addition, the cost of lost production can be huge for large machines (more than $10000 per day for a 12 MW turbine running at 35% capacity). Furthermore, access to offshore turbines is restricted by weather conditions for much of the year. An outage may exceed 90 days for a two-day repair at a cost exceeding $900000. Clearly, there is a great need for a CBM approach to wind turbine maintenance.

CBM Approach for the Main Bearing

The main bearing is responsible for supporting the rotor shaft, on which the turbine blades are mounted, and for transmitting torque to the turbine gearbox. Continuous changes in operating loads and environmental conditions may eventually lead to damage to the main bearing, in the form of fretting on the inner and outer races and spalls and cracks in the rolling elements. We briefly describe a data-driven model which describes the fault-free behavior of the main bearing temperature signal.

The basic principle of the model-based approach is to develop a model that describes the fault-free behavior of the main bearing temperature. At each iteration of the algorithm, the estimated main bearing temperature generated by the fault-free model is then compared with the actual main bearing temperature. The difference between the estimated and actual main bearing temperatures, known as the *residual*, is then evaluated. When the turbine remains fault-free, the residual signal should generate a Gaussian distributed signal with a mean of zero and a small variance. If a fault develops, the residual signal will no longer have

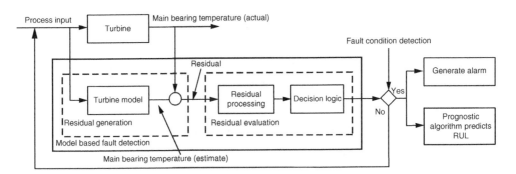

Figure 16.18   Main bearing fault diagnosis method.

a zero-mean. Analysis of the residual signal is performed by the residual processing and decision logic stages (see Figure 16.18). Assuming no fault condition is detected, the algorithm continues to iterate at ten-minute intervals as data are recorded. If a fault condition is detected, an alarm is generated and the prognostic stage is initialized to estimate the remaining useful life (RUL) of the main bearing. The algorithm then continues to iterate and the RUL predictions are recursively updated.

16.4.5   CBM Decisions for Infrastructures

16.4.5.1   Rail Track[11]

Rail infrastructure consists of several elements, and we focus on rail track. The condition of a rail track (commonly referred to as *track quality*) is characterized by the following five track geometry parameters: (i) longitudinal level, (ii) alignment, (iii) cant, (iv) twist, and (v) gauge. The degradation of track geometry is a complex phenomenon affected by dynamic loads. The rate of degradation is a function of time and/or usage intensity. The initial track quality, the initial settlement, and the deterioration rate are the major parameters of track quality deterioration. Track geometry quality is measured by the deviations from the designed track geometry. Poor track quality has serious implications: (i) slower speeds to avoid derailment and (ii) poor ride quality (in the case of passenger trains).

Track geometry maintenance activities are designed to reinstate the track quality to the original condition. This is achieved by assessing the condition of track quality by running measurement trains on the track and recording several track-quality parameters to obtain aggregate information about the condition of the track. In the track-geometry monitoring used by Swedish Rail (Trafikverket), the longitudinal level, alignment, track gauge, rail elevation, twist (over 3 or 6 m), and curvature are measured for every 25 cm of the track.

Longitudinal Level
The deviations (from the reference line) in this parameter along the track length are also referred to as *rail head corrugations* (or simply *corrugations*) along the track. For this, the vertical positions of the right and left rails are measured by a position sensor and

---

[11] This section is based on Khouy (2013) and Palo (2014).

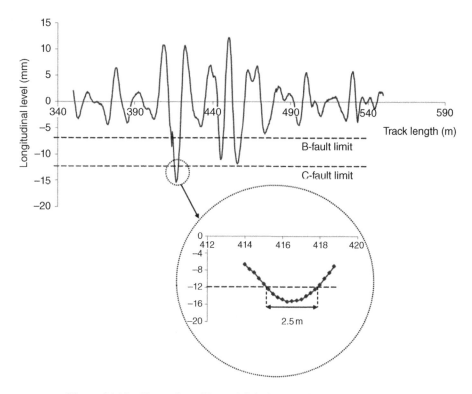

Figure 16.19   Illustration of B- and C-fault limits for longitudinal level.

an accelerometer. The accelerometer measures the vertical acceleration of the wagon. This acceleration measurement is integrated twice to determine the position. Then, the result is added to the value, which was measured by a position sensor to identify the vertical position of each rail.

Figure 16.19 shows the plot of longitudinal-level deviations (or corrugations) along a 200-meter track length and two limits (B-fault and C-fault limits) based on deviations from the reference line (corresponding to zero deviation). These measurements are classified into three classes – A–C – based on these two limits. Class A quality corresponds to deviations lower than the B-limit; class B quality corresponds to deviations between the two limits; and class C quality implies deviations greater than the C-limit.

Corrugations may be classified into long wave and short wave, and the maintenance action to improve the track quality is different for each type. In the case of long-wave corrugation, one uses tamping and for short-wave corrugation one uses grinding.

Rail Grinding[12]

This is used for removing defects caused by RCF and rail head corrugations. Maintaining the rail profile is necessary to give a contact surface for the right wheel and the rail. In general, the grinding interval is tonnage-based, and a typical maintenance interval for grinding is between

---

[12] The second case study in Chapter 23 looks at maintenance involving tamping.

20 and 30 MGT, which corresponds roughly to around one to three years depending on the tonnage carried on the track. Grinding may involve one or more passes.[13]

One-Pass Grinding (OPG)
This is a form of preventive grinding strategy where the grinding machine re-profiles the rail head in one pass by removing approximately up to 0.3 mm of material from the rail. Grinding machines are able to grind 7–8 km of rail per hour.

16.4.5.2   Rail Bridges

Most rail networks involve bridges to span rivers and valleys and to provide underpasses for other traffic (rail or road). The material used in building bridges may be reinforced concrete, masonry, or steel. The condition of a bridge degrades with age and usage. The most frequently considered aspects of bridge condition are as follows:

- *Bridge condition assessment (rating):* Evaluation of the local and/or global state (technical condition) of a bridge in the form of a numerical or linguistic rating based on a predefined scale.
- *Load capacity assessment:* Activities undertaken to determine the ability of a bridge to carry load based on the structure's technical parameters and degradation level.
- *Safety assessment:* The process of evaluation of remaining bridge safety measured in terms of partial safety index, reliability index, or probability of failure.
- *Durability assessment:* The process of evaluation of the remaining lifetime of a bridge.
- *Serviceability assessment:* Evaluation based on the criterion governing normal use of the bridge as part of the transportation system.

The degradation mechanisms can be divided into three groups:

- *Chemical:* Alkali–aggregate reaction; carbonation; corrosion, crystallization, leaching; oil and fat influences; salt and acid actions.
- *Physical:* Creep; fatigue; influence of high temperature; freeze–thaw action; modifications of foundation conditions; overloading; shrinkage; water penetration.
- *Biological:* Accumulation of dirt and rubbish; living organisms' activities.

Degradation leads to defects. A hierarchical classification for railway bridge defects involves four levels. Level I (defect type) consists of six basic defect types:

- *Contamination:* The appearance of any type of dirt, rubbish, or unplanned plant vegetation.
- *Deformation:* Geometry changes incompatible with the design, with changes in mutual distances of structure points.
- *Deterioration:* Disadvantageous changes in physical and/or chemical structural features in relation to designed values.

---

[13] Swedish Rail classifies the maintenance as PM if the number of passes required (to rectify the corrugations) is less than four and CM if it is greater than four.

- *Discontinuity:* Undesigned breaks in the structure's material continuity.
- *Displacement:* A change in structure component (components) location incompatible with the design but without deformation of the structure; also, restrictions in designed displacement capabilities.
- *Loss of material:* A decrease in the designed amount of structural material.

Each of these contains one or more Level II (defect kind) elements and each element is comprised of several Level III (defect category) elements, which, in turn, are comprised of one or more Level IV (defect class) elements. Table 16.4 shows the hierarchical classification of defects of steel railway bridges.[14]

Several NDTs (discussed in Chapter 6) are used for detection of defects in steel railway bridges. They are for detecting defects resulting from corrosion, fatigue cracking, loose connections, coating defects, brittle fracture, and so on.

The maintenance of a bridge is based on CBM. Figure 16.20[15] shows the different phases involved in the maintenance of railway bridges. Phases 1 and 2 (visual inspections using NDT methods) are performed every three and six years, respectively. The outcome of Phase 3 (using

Table 16.4 Hierarchical classification of defects in steel railway bridges.

Level I	Level II	Level III	Level IV
Contamination	Steel construction	Inorganic	Aggressive Neutral
		Organic	Aggressive Neutral
Deformation	Basic component	Deflection Distortion Torsion	
	Bolted/riveted connector	Deflection Torsion	
	Welded connector	Deflection Torsion	
Deterioration	Basic component	Modification of physical features	Hardness Impact resistance reduction Strength reduction
	Bolted/riveted connector	Modification of physical features	Loosening Strength reduction
	Protection	Modification of physical features	Adhesion reduction Embrittlement increasing Fading Thickness reduction
	Welded connector	Modification of physical features	Strength reduction

(Continued)

---

[14] Bien *et al.* (2007) have similar tables for concrete and masonry railway bridges.
[15] Adapted from Helmerich, Bien, and Cruz (2007).

Table 16.4 (*Continued*)

Level I	Level II	Level III	Level IV
Discontinuity	Basic component	Crack	Irregular Longitudinal Skew Transverse
		Delamination	
		Fracture	Irregular Longitudinal Skew Transverse
	Bolted/riveted connector	Crack	
		Fracture	
	Protection	Crack	
		Delamination	
		Fracture	
	Welded connector	Crack	Longitudinal Transverse
		Fracture	Longitudinal Transverse
Displacement	Excessive	Rotation	
		Translation	
	Limited	Rotation	
		Translation	
Loss of material	Basic component Bolted/riveted connector Protection Welded connector		

advanced NDT methods) determines the subsequent inspection of one or more components of the bridge. Finally, Phase 4 involves extensive NDT methods and advanced data-processing techniques.

16.5 Summary

CBM deals with maintenance actions based on the condition or state of an item. The state of the item characterizes the level of degradation due to some degradation mechanism that depends on the physical characteristics (type of material – metal, plastic, etc.), operating environment (humidity, value of fluid, etc.), and operating load.

A CBM approach to maintenance requires four key steps:

* *Step 1:* Collecting data using sensors and condition-monitoring devices.
* *Step 2:* Using data to assess the state if the item has not failed or to detect the fault if the item has failed.
* *Step 3:* Predicting future condition.
* *Step 4:* Taking appropriate maintenance action.

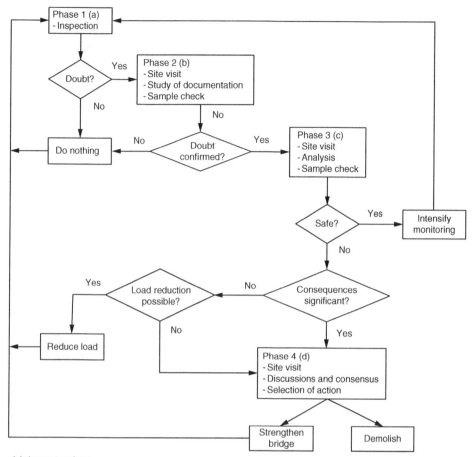

(a): Inspector alone
(b): Engineer alone
(c): Specialised laboratory specialists
(d): Engineer together with team of experts

Figure 16.20   Maintenance of bridges.

Assessing an object's current condition requires different types of historical data from the time the object is put into operation. These data can be collected continuously or at discrete time instants using various types of sensors.

The approaches for diagnostics (to assess the current condition), prognostics (to predict future condition), and CBM (based on diagnostics and prognostics) have been discussed. Fault (failure) diagnostics refers to the process of detecting, isolating, and identifying a component that has failed. Fault diagnostics rely on condition-monitoring data, item-performance data, and expert opinion. Past experience plays an important role in detecting the fault quickly. Various methods have been developed to deal with failure diagnostics and these can be classified broadly into two categories: model-based and history-based methods.

Prognostics is defined as the estimation of the operating time or remaining useful life before failure. The three approaches to building models to predict future state (and residual life) include the physics-based, the data-based, and a hybrid approach. A few data-based approaches have been presented.

A few CBM policies have been described along with the steps to determine the optimal CBM policy. The application of CBM to various types of engineered objects has been discussed for illustrative purposes.

Review Questions

16.1   What are the main steps in CBM?

16.2   What is meant by diagnostics and prognostics in the context of CBM?

16.3   What are the main elements of the CBM approach? Refer to Figure 16.4.

16.4   Describe a few methods for assessing an object's current condition.

16.5   What are the main methods used in fault diagnostics?

16.6   What is the difference between diagnostics of a failed item and diagnostics for assessing the state of a non-failed item in order to predict its future state?

16.7   What are the main approaches used in prognostics?

16.8   Describe a simple prognostic model for continuous degradation.

16.9   What is residual life? How is it used in the context of CBM?

16.10   What are the main elements of a CBM policy?

16.11   What are the main steps of optimal CBM decision making?

Exercises

16.1   Define "precision" in the context of a sensor. Data are collected from two different sensors, one with a higher precision. Should one ignore the data from the sensor with the lower precision? If not, how should one combine the two data sets?

16.2   Data from sensors are affected by sensor noise. A sequence of n observations is taken of a signal. How would you calculate the error term and obtain an estimate based on the data together with its confidence limits?
Hint: Assume that the errors due to noise are IID and normally distributed.

16.3   Bearings are critical components of any rotating machine such as the diesel engine of a truck. Following on from Exercise 3.2, consider two methods that may be used to monitor the degradation of bearings. Indicate how these methods may be used to develop a CBM strategy for bearings.

16.4   Railway transport is a complex system (see Example 2.1). The rail infrastructure is a subsystem composed of several components. Select two components of railway track and suggest variables that characterize the degradation and which may be used to assess the condition of the components. Suggest a framework for fault monitoring and diagnosis for these components.

16.5   Railway tracks are divided into of 100 m segments and the degradation of these segments is assessed using the standard deviation of the longitudinal level (measured by the distance between the rail surface and some reference line). The table below gives the measurements for two segments over the period April 2007 to May 2014.

Measurement date	Longitudinal level	
	Segment 1	Segment 2
2007-04-26	1.40	0.97
2007-10-08	1.51	1.13
2008-04-15	1.52	1.17
2008-06-10	1.53	1.18
2008-08-05	0.71	0.67
2008-09-24	0.73	0.71
2009-04-16	0.82	0.76
2009-08-05	0.78	0.60
2009-09-30	0.79	0.56
2010-05-12	0.84	0.69
2010-06-23	0.90	0.71
2010-09-01	0.89	0.70
2011-03-04	1.15	0.95
2011-06-09	1.03	0.86
2011-08-07	1.27	1.06
2011-09-21	1.05	0.81
2011-11-02	1.05	0.83
2012-03-27	1.13	0.89
2012-04-24	1.54	1.76
2012-06-05	1.15	0.92
2012-08-03	1.48	2.01
2012-09-18	1.18	0.95
2012-10-30	1.44	1.88
2013-02-05	1.46	1.12
2013-06-05	1.27	1.08
2013-08-03	1.39	1.11
2013-09-18	1.34	1.04
2013-11-05	1.37	1.08
2014-01-30	1.39	1.06
2014-04-09	1.36	1.11
2014-05-28	1.41	1.28

Assume that the data may be modeled by Model 16.1. Estimate the model parameters using the method of least squares. Corrective maintenance involves tamping, and this is done when the standard deviation reaches the value 1.8. Calculate when tamping needs to be done for the two segments.

16.6   Repeat Exercise 16.5 assuming that the degradation of track segments may be adequately modeled by Model 16.2.

16.7   Compare the results of Exercises 16.5 and 16.6. Which is the more appropriate model for the degradation of rail track segments?

16.8   Boilers are key elements of thermal power stations. A component of a boiler is a bank of pipes used for producing super-heated steam. A pipe may degrade in many different ways: (i) corrosion of the inner surface leading to a reduction in the thickness, (ii) cracks resulting from thermal stress fatigue, and so on. List the different kinds of sensors and testing that may be used to detect the two types of degradation. Discuss which of these involve destructive testing.

16.9   Degradation of aircraft wings is due to fatigue, as the wings flap constantly during flights leading to the appearance of cracks. Testing is usually done when the aircraft is not flying. What kind of sensors and testing methods are needed for this purpose?

16.10   A young maintenance engineer has been given the task of proposing a feasibility plan for the maintenance of a section of road. List the different issues that the engineer will need to address. The sensor and data-collection technologies are two important issues. List the different options that the engineer will need to discuss.

16.11   CBM is more costly to implement than the more traditional approaches such as age- or usage-based maintenance. Build an economic model that will help in deciding whether CBM is the better approach or not and discuss the limitations of the model.

16.12   In CBM implementation, the frequency of data collection from the sensor is an issue. Consider the following three data-collection scenarios for a non-repairable component with a mean time to failure of approximately 10 000 hours:
        Scenario 1: Data collected every 10 minutes throughout the life of the component.
        Scenario 2: No data collected for the first 5000 hours and then data collected every 10 minutes.
        Scenario 3: Data collection starts with a time interval of 30 minutes between each collection and then the interval decreases by a factor of 2 after every 1000 hours.

        Build a simple model to choose between the three scenarios.

References

Bien, J., Jakubowski, K., Kaminski, T., and Kmita, J. (2007) Railway bridge defects and degradation mechanisms. In *Sustainable Bridges: Assessment of Future Traffic Demands and Longer Lives*, Bien, J., Elfgren, J., and Olofsson, J. (eds) DWE Publishers, Wroclaw, pp. 105–116.

Box, G. and Jenkins, G. (1970) *Time Series Analysis: Forecasting and Control*, Holden-Day: San Francisco, CA.

Helmerich, R., Bien, J., and Cruz, P. (2007) A guideline for railway bridge inspection and condition assessment including NDT toolbox. In *Sustainable Bridges: Assessment of Future Traffic Demands and Longer Lives*, Bien, J., Elfgren, J., and Olofsson, J. (eds) DWE Publishers, Wroclaw, pp. 93–104.

Jantunen, E. (2003) Prognosis of wear progress based on regression analysis of condition monitoring parameters. Proceedings of the 16th International Congress COMADEM 2003, Vaxjo, Sweden, August 27–29, 2003, pp. 481–491.

Khouy, I.A. (2013) Cost-effective maintenance of railway track geometry. Doctoral thesis. Lulea Technical University, Lulea.

Onsoyen, E. (1991) Accelerated testing of components exposed to wear. In *Operational Reliability and Systematic Maintenance*, Holmberg, K. and Folkeson, A. (eds), Elsevier, 51–77.

Palo, M. (2014) Condition-based maintenance for effective and efficient rolling stock capacity assurance. Doctoral thesis. Lulea Technical University, Lulea.

SKF–Product Information (1994) Bearing Failures and their Causes, Product Information – 401, SKF Group, Gothenburg.

Part D

Maintenance Management

Chapter 17: Maintenance Management
Chapter 18: Maintenance Outsourcing and Leasing
Chapter 19: Maintenance Planning, Scheduling, and Control
Chapter 20: Maintenance Logistics
Chapter 21: Maintenance Economics
Chapter 22: Computerized Maintenance Management Systems and e-Maintenance

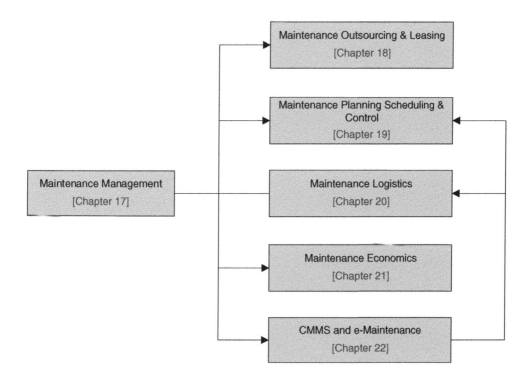

17

Maintenance Management

Learning Outcomes

After reading this chapter, you should be able to:

- Define management in general and describe the main management functions;
- Describe the three levels of management decisions;
- Explain the importance of business performance management;
- Define a performance management system and describe its main elements;
- Define maintenance management;
- Describe key strategic maintenance issues including maintenance objectives, maintenance organization, maintenance approaches, maintenance outsourcing, capital investment, and support facilities;
- Describe the different types of maintenance organization;
- Define reliability-centered maintenance (RCM);
- Describe the key features of RCM;
- Describe the RCM process and list the key implementation steps;
- Define total productive maintenance (TPM);
- Describe the key features of TPM;
- List the six major equipment losses and define overall equipment effectiveness;
- Define risk and describe a general risk management process;
- Identify key maintenance risks and define risk-based maintenance (RBM);
- Define maintenance management systems and describe their key elements.

Introduction to Maintenance Engineering: Modeling, Optimization, and Management, First Edition.
Mohammed Ben-Daya, Uday Kumar, and D.N. Prabhakar Murthy.
© 2016 John Wiley & Sons, Ltd. Published 2016 by John Wiley & Sons, Ltd.

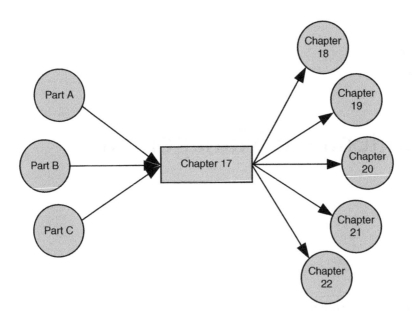

17.1 Introduction

Maintenance management deals with decision making with regards to PM and CM actions and the planning and execution of the tasks involved. It is part of the overall management of a business. An understanding of general management is essential for a proper understanding of maintenance management. The key issues in maintenance management include the following:

- In-house versus outsourcing of maintenance;
- Approaches to maintenance;
- Maintenance planning;
- Maintenance economics;
- Maintenance logistics;
- Maintenance management system (MMS).

This chapter gives an overview of maintenance management and deals with some of the issues listed above. The remaining issues are discussed in later chapters of Part D.

The structure of the chapter is as follows. We start with a brief discussion of management in general and the key issues involved in Section 17.2. Section 17.3 deals with the key issues of maintenance management and provides a brief introduction to the various chapters in Part D. Maintenance organization is discussed in Section 17.4. Section 17.5 deals with alternative approaches used in maintenance and looks at the following three methods: (i) reliability-centred maintenance (RCM), (ii) total productive maintenance (TPM), and (iii) risk-based maintenance (RBM). Section 17.6 focuses on risk and maintenance, and Section 17.7 discusses maintenance management systems. We conclude with a summary of the chapter in Section 17.8.

17.2   Management

> **Definition 17.1**
>
> Management of a business (organization) involves the coordination of the efforts of people to accomplish the business goals and objectives using the available resources in an efficient and effective manner.

Management requires decision making by managers at different levels of the business, involving many different functions, to achieve the desired goals in different time frames.

17.2.1   Management Functions

As shown in Figure 17.1, the main functions of management are planning, organizing, implementing, and controlling. We will discuss each of these briefly. The other elements of the figure are discussed later.

17.2.1.1   Planning

Planning involves establishing organizational goals and determining the means of achieving these. The three aims for planning are:

1. To establish an overall direction and goals for the future of the organization in terms of growth, profitability, social responsibility, and so on.
2. To identify the organization's resources and commit them to achieve established goals.
3. To decide what needs to be done to achieve the established goals.

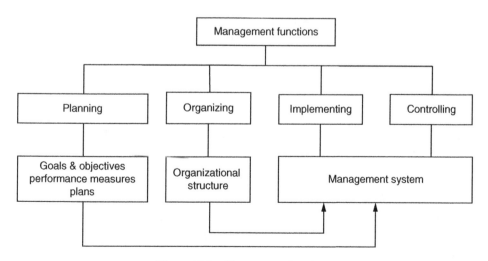

Figure 17.1   Management functions.

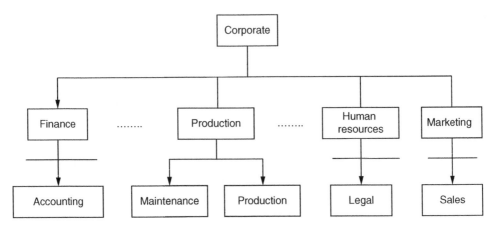

Figure 17.2   Typical organizational structure.

17.2.1.2   Organizing

Organizing involves translating plans into reality by deciding where decisions will be made, who will perform what tasks, and who will report to whom in the organization. Organizing effectively leads to better coordination of human, material, and information resources.

This requires a multi-level organizational structure. As shown in Figure 17.2, the top three levels are:

- *Level 1:* Corporate;
- *Level 2:* Functional (based on the main functions such as Production,[1] Marketing, Legal, etc.);
- *Level 3:* Departmental (based on other functions such Accounting, Sales, and so on, to support the main functions).

There are still lower levels (such as sections, subsections, etc.) which are not shown in the figure. Note that in the figure, the Maintenance Department is at Level 3 and Production is at Level 2. These basic functions continue at lower levels – for example, maintenance, as discussed in Section 17.3.

17.2.1.3   Implementing

Implementing involves the execution of the prepared plans. Proper data collection during this stage is needed to ensure that plans are executed properly. The data are vital for the controlling function discussed next.

17.2.1.4   Controlling

Controlling is the process by which performance is monitored and corrective action taken when necessary. Control is vital for continuous improvement and requires an effective

---

[1] Production can be goods and/or services and involves one or more engineered objects depending on the business.

performance-measurement system (discussed in a later section). Control requires that standards are set, performance is measured against these standards, actions are taken to correct any deviations, and adjustments are made to the standards if necessary.

17.2.2 Management Decisions

Management decisions can be categorized into three groups based on the time horizon for making decisions, as indicated in Table 17.1.

It is important that the short-term decisions (operational management) need to fit in with the medium-term decisions (tactical management) and these, in turn, need to fit in with the long-term decisions (strategic management).

17.2.2.1 Strategic Management

Strategic management deals with various tasks such as (i) diagnosing the organization's external and internal environment, (ii) deciding on a vision and mission, (iii) developing overall goals, (iv) creating strategies, and (v) allocating resources to achieve the organization's goals from a long-term (a few years to tens of years) perspective.

The strategies need to be defined for all the different levels of business. Corporate-level strategies are the major courses of action (choices) selected and implemented to achieve one or more of the business goals. Functional-level strategies refer to the actions and resource commitments established for operations, marketing, human resources, finance, and the organization's other functional areas. Functional-level plans and strategies should support business-level strategies and plans. For example, production strategies should support business-level strategies. Maintenance strategies, in turn, should support production strategies and are therefore aligned with overall business strategies.

17.2.2.2 Tactical Management

Tactical management involves making decisions regarding what to do, who will do it, and how to do it from a medium-term (months to years) perspective. This usually includes specific courses of action for implementing new initiatives or improving current operations.

17.2.2.3 Operational Management

Operational management deals with decision making from a short-term (day or week) perspective to develop specific actions that support the strategic and tactical management decisions.

Table 17.1 Categorization of management decisions.

Time horizon	Typical duration	Name
Long	5–50 years	Strategic
Medium	Months/few years	Tactical
Short	Hours/day/week	Operational

17.2.3   Classification of Managers

For management purposes, the organizational structure (see Figure 17.2) may be divided into three levels with managers at each level, as indicated in Figure 17.3. They are commonly referred to as: (i) senior-level managers; (ii) middle-level managers; and (iii) junior-level (first-line) managers.

The terms on the left side in the figure indicate the role and functions of managers at different levels. The performance targets (from the overall business perspective) and the strategic goals are closely linked and are decided by senior-level managers with inputs from middle-level managers. Planning and choosing the approach (for example, RCM or TPM in the context of maintenance) is carried out by middle-level managers and the implementation is carried out by junior-level managers. Note that the decisions made by junior- (middle) level managers need the approval of middle- (senior) level managers.

The terms on the right side in the figure deal with data, information, and measured performance. Junior-level managers use the data collected by different people within the organization and process it to produce information that middle-level managers translate into measures of actual performance at different levels (department or lower) and pass these to senior-level managers. This allows for evaluation of how the strategies (at the departmental level) are working. The aggregated information (at the functional level) is passed on to top-level managers who may evaluate (in conjunction with the CEO) the performance of individual functions or departments and of the business as a whole.

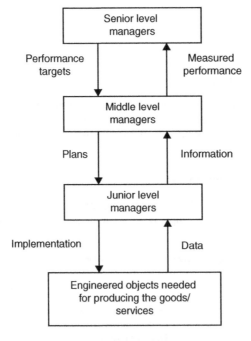

Figure 17.3   Three levels of management.

17.2.4   Management System

Definition 17.2

A management system is a framework of processes and procedures used to ensure that an organization is able to fulfill all the tasks required to achieve its objectives.

A management system is composed of several specific management systems – such as the production management system, the performance management system (PMS), the TQM (total quality management) system, and so on. The two important ones in the context of maintenance are (i) the PMS and (ii) the MMS (discussed later in the chapter).

17.2.5   Assessing Business Performance

17.2.5.1   Performance

Definition 17.3

Two dictionary definitions of *performance* are the following:

* "The action or process of performing a task or function." (Oxford Dictionary)
* "The act of performing; of doing something successfully; using knowledge as distinguished from merely possessing it." (The Free Dictionary)

Performance is the basis for businesses to evaluate how the different tasks and activities are accomplished by comparing them against predefined standards of accuracy, completeness, cost, speed, and so on, and developing strategies for improvements in performance.

Two related terms are the following:

* *Performance objective:* This is a critical success factor in achieving the business mission, vision, and strategy.
* *Performance goal:* A target level (a tangible measure), against which the actual achievement can be compared.

Objectives are long term and less specific than goals but more specific than the mission statement. However, goals provide specific, detailed target results (for example, a reduction in overtime by 10% or the implementation of a CBM (condition-based maintenance) program for a particular item, etc.).

17.2.5.2   Performance Indicators and Metrics

Definition 17.4

A *performance indicator* (PI) is a variable characterizing the performance of a task or activity. It can be at the business, engineered object, or component level.

Definition 17.5

A *key performance indicator* (KPI) is a performance indicator that is deemed to be important and is used for assessing performance.

Definition 17.6

A *performance metric* is a quantitative or qualitative characterization of a performance indicator.

Table 17.2 is an illustrative sample of a few of the performance indicators and performance metrics at the business, engineered object, and component levels. As can be seen, some are ratios of two or more variables. These variables become important in the context of data collection to measure the actual performance.

PIs can be classified broadly into two types:

- *Leading indicators:* A leading indicator (for example, trends in the capital market) warns the user about the non-achievement of goals and objectives before a possible problem (for example, the cost of borrowing jumps significantly) arises. This involves seeing the trends and patterns over time. It works as a performance driver.
- *Lagging indicators:* A lagging indicator (for example, corrective maintenance) provides information after the problem has occurred (for example, fixing failure) and is useful for initiating improvement actions.

Table 17.2 Illustrative performance indicators and performance metrics.

	Performance indicators	Performance metrics	Comments
Business level	ROI	Annual revenue (\$)/initial investment (\$)	Ratio
	Sales	Annual sales	Integer
	Market share	Fraction of total industry-wide sales	Ratio
	Reputation	Complaints/customer	Ratio
Rail transport	Efficiency	Energy consumed/ton-km	Ratio
(infrastructure)	Punctuality	Delay in departure and/or in arrival at destination	Real variable
	Cost	Cost (\$)/ton-km	Ratio
Water network	Delivery	Fraction of customers having no interruption to water supply	Ratio
(infrastructure)			
	Quality	Chemical and bacterial (g/l)	Ratio
Power plant	Efficiency	Output power/input energy	Ratio
(plant)	Pollution	Quantity of SO_2 per MW produced	Ratio
	Cost	Cost/MW	Ratio
Automobile	Fuel efficiency	km/l of fuel	Ratio
(product)	Acceleration	Time to reach specified velocity	Real variable
	Safety	Probability of survival under specified impact	Real variable
Aircraft	Fuel efficiency	Fuel consumption/passenger-km	Ratio
	Reliability	Probability of no failure during mission	Real variable
	Range	Maximum travel with full tank	Real variable

17.2.5.3 Balanced Scorecard

The balanced scorecard is a conceptual framework for translating an organization's strategic objectives into a set of performance indicators from the following four perspectives:

1. *Financial:* Encourages the identification of high-level financial measures from the shareholders' perspective.
2. *Customer:* Encourages the identification of measures that answer the question relating to how customers view the business.
3. Internal business process: Encourages the identification of measures that answer the question as to what the business must excel at.
4. Learning and growth: Encourages the identification of measures that help the business to improve, create value, and innovate.

Some of the indicators are to measure an organization's progress toward achieving its vision; others are to measure the long-term drivers of success. Through the balanced scorecard, an organization monitors its current performance – finance, customer satisfaction, and business process results – and its efforts to improve processes, motivate and educate employees, and enhance information systems – its ability to learn and improve.

The SMART test (NRC, 2005) is frequently used to provide a quick reference to determine the quality of the performance metrics and stands for the following:

- *Specific* (S)*:* Clear and focused to avoid misinterpretation. Should include measurement assumptions and definitions and should be easily interpreted.
- *Measurable* (M)*:* Can be quantified and compared to other data. It should allow meaningful statistical analysis.
- *Attainable* (A)*:* Achievable, reasonable, and credible under the conditions expected.
- *Realistic* (R)*:* Fits into the organization's constraints and is cost-effective.
- *Timely* (T)*:* Obtainable within the time frame given.

17.2.6   Performance Management

> **Definition 17.7**
>
> *Performance management* involves collecting data and information to assess actual performance (defined through PIs) and using it to produce positive change in the organizational culture and improvements in performance.

This involves (i) an iterative process similar to the PDCA (plan, do, act, check) cycle and (ii) benchmarking.

17.2.6.1 PDCA Cycle[2]

This cycle symbolizes the process for improvement and involves executing the four steps indicated below in an iterative manner.

---

[2] The PDCA cycle concept was proposed by Demming (1986) as part of TQM.

1. **P**: Plan the change by finding out what things are going wrong (that is, identify the problems faced), and develop ideas for solving these problems.
2. **D**: Do changes designed to solve the problems, on a small scale to begin with, to test whether or not the changes will work without causing undue interruptions to operations.
3. **C**: Check whether or not the small-scale changes are achieving the desired results and identify any new problems that may crop up.
4. **A**: Act to implement changes on a larger scale if the small-scale changes are successful.

17.2.6.2 Benchmarking

There are two types of benchmarking: external and internal. In external benchmarking, the performance measures of the business are compared with the measures of similar businesses, leading to what is termed *best practice*. In internal benchmarking, the performances over time are plotted to see if the operations have improved or not.

Performance Management System

Definition 17.8

A *performance management system* (PMS) is a formal procedure to assist in effective performance management throughout the organization.

A PMS comprises the methodologies, metrics, processes, and software tools that are used to manage the performance of an organization. The business objectives are translated into appropriate KPIs that are cascaded down through the organization from top management to the shop floor. Targets for these KPIs are set using benchmarking and best practices. Performance is monitored by collection of appropriate data that are transformed into useful information used to capture and resolve issues. Opportunities for improvement are identified and implementation plans are put in place. Incentives are designed to create a continuous improvement culture in the organization. These key elements of a PMS are shown in Figure 17.4.

Figure 17.4   Key elements of a PMS.

17.3   Maintenance Management

Definition 17.9

All the activities of the management that determine the maintenance objectives or priorities (defined as targets assigned and accepted by the management and maintenance department), strategies (defined as a management method in order to achieve maintenance objectives), and responsibilities and implement these by means such as maintenance planning, maintenance control and supervision, and several improving methods including economic aspects in the organization. (European Standards for Maintenance terminology).

As indicated in the above definition, maintenance objectives must be established and strategies have to be in place and should be aligned with the overall business mission objectives and strategies of the organization. This involves various activities conducted at different levels of the organization. It includes setting objectives, developing strategies, proper planning, organizing, and control through flow of information and data management, and effective performance measurement to ensure that objectives are met. Maintenance issues – strategic, tactical, and operational – are shown in Figure 17.5. In remainder of the chapter we focus on the strategic issues. We discuss some of the elements shown in the left box in

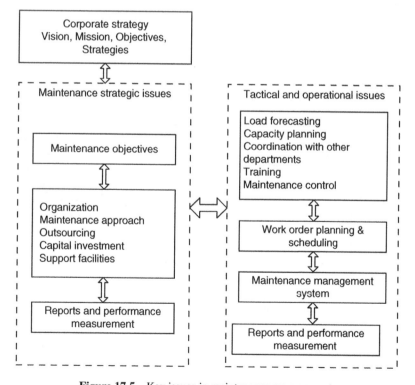

Figure 17.5   Key issues in maintenance management.

the remainder of the section and the others are discussed in later chapters.[3] The tactical and operational issues are discussed in Chapters 19 and 20.

17.3.1   Strategic Issues

17.3.1.1   Maintenance Objectives

Definition 17.10

Maintenance objectives are the targets assigned to or accepted by the management and maintenance department (CEN, 2001).

These targets may include availability, cost reduction, product/service quality, environment preservation, safety, and so on. Maintenance objectives and strategies have to be aligned with the overall business mission objectives of the organization.

The main and secondary factors (in the context of mining or manufacturing operations) that need to be taken into consideration in the formulation of maintenance objectives are shown in Figure 17.6.

Maintenance resources are used to ensure that the output (as specified under the production policy), safety standards, and plant design life are all achieved, and that the energy used and the possible adverse impact on the environment are minimized.

As can be seen in Figure 17.7, inputs from many departments need to be taken into consideration in the formulation of maintenance objectives, which need to be aligned with the corporate objectives specified in the corporate strategic plan.

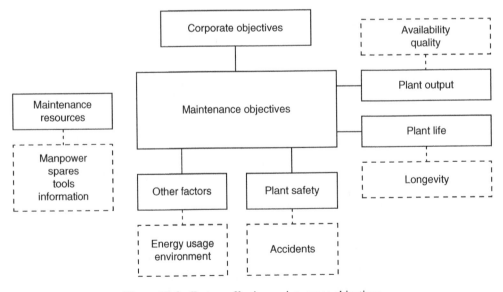

Figure 17.6   Factors affecting maintenance objectives.

---

[3] Maintenance outsourcing, support facilities, capital investment decisions, and performance measurement are discussed in Chapters 18, 20, 21, and 22 respectively.

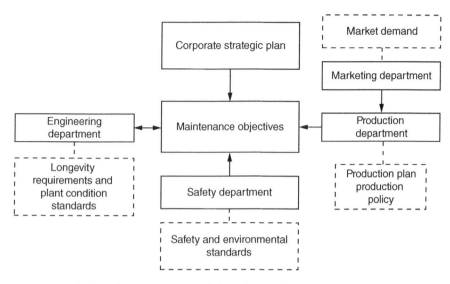

Figure 17.7   Departments influencing maintenance objectives.

   Maintenance objectives may be classified into long-term and short-term objectives or goals. The goals are derived from, and must be focused on fulfilling, the business mission. Some long-term goals and objectives are:

- Profitability or cost management;
- Productivity;
- Competitive position;
- Employee development;
- Employee relations;
- Technological leadership.

Short-term objectives must be aligned with long-term goals and established to ensure that the maintenance business is on the right track in fulfilling its mission. Some short-term goals are:

- Overtime reduction by some specified percentage;
- Implementation of a PM program for a specific area or group of equipment;
- Implementation of a training program for a specific craft or trade;
- Implementation of a work order system within a given area or for a specific trade;
- Implementation of an inventory management program.

17.4   Maintenance Organization

A good maintenance organization provides clear roles and responsibilities, an effective span of control,[4] facilitation of good supervision, effective reporting, and a continuous improvement culture.

---

[4] The number of subordinates a supervisor has.

Maintenance organizations vary across businesses and depend on several factors such as the objectives of the business, the size of the organization, whether maintenance is done using internal resources or outsourced fully or partially, and available resources (manpower, skills, tools, technology, etc.).

The location of the maintenance function in the organization chart is another issue that varies across businesses. The question is whether maintenance should report to the production/ operations departments or whether it should be located at a higher level (see Figure 17.2). The latter is a more common practice in large businesses with provision for effective coordination between the two functions. Another important issue is whether the maintenance organization (including resources such as repair shops and stores) should be centralized or decentralized.

Having qualified manpower is a pre-condition for achieving high performance in maintenance. To cope, and to keep pace with rapid advances in maintenance technologies, the knowledge and skills of the workforce are of strategic importance, and continuous upgrading through training is very important.

17.4.1   Organization Structure

Given the differences in the size of businesses, the type of engineered objects, and the technologies used, there is neither a clear guideline nor a single ideal organizational structure that is best for all businesses.

A maintenance organization structure may be considered as a hierarchy of work roles, ranked by their authority and responsibility for deciding what, when, and how maintenance work should be carried out. An example is shown in Figure 17.8, which may be seen as a continuation of Figure 17.2 with the maintenance manager at Level 3 (assisted by the maintenance planner) and supervising the staff and activities at two lower levels.

The design of a maintenance organization structure involves the following:

* Determining the responsibility, authority, and work role (the decision-making bounds) of each individual in the maintenance organization.
* Establishing the relationships, both vertical and horizontal, between individuals in the maintenance organization.

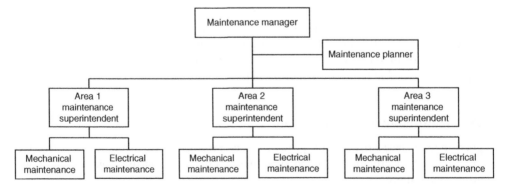

Figure 17.8   Typical organization chart of a maintenance department.

- Ensuring that the maintenance objectives are aligned with business objectives and are well understood by each individual in the maintenance organization.
- Establishing effective systems of coordination and communication vertically and horizontally in the maintenance organization.

In addition to establishing an effective organizational structure with clear roles and responsibilities, the maintenance department has to establish strong coordination, cooperation, and communication with the production/operation departments (its main customers). Such coordination should start at the top level and continue to the lower levels of the organization. This can be facilitated by making provision for joint decision making based on common goals and performance metrics.

17.4.2   Centralized versus Decentralized

Maintenance organizations may be centralized, decentralized, or a hybrid (a combination of the first two). Each of these types has its advantages and disadvantages, as described below.

17.4.2.1   Centralized Maintenance Organization

Here, the maintenance management function is centralized through the maintenance manager, who is responsible for all aspects of plant and facility maintenance and support. Almost all services are dispatched from a central location and all spares and materials are issued from the central stores. The advantages of this arrangement include:

- More flexibility and better utilization of resources such as highly skilled technicians and special tools and equipment;
- Better supervision and on-the-job training;
- Less investment in expensive modern equipment.

However, it has the following disadvantages:

- Lower productivity, as more time is spent on travel due to the remoteness of certain job locations for geographically dispersed facilities;
- Less specialization on complex production units;
- Less integration with operational units.

17.4.2.2   Decentralized Maintenance Organization

In a decentralized maintenance organization, maintenance units are assigned to specific areas or plant units. This arrangement has the following advantages:

- Better integration with operations and quicker availability of maintenance personnel when required;
- More specialization on complex plant units.

The disadvantages include:

- A high level of required resources due to duplication of personnel, special maintenance equipment, and material requirements across the organization, which add to inefficiency and cost;
- The possible requirement of additional management levels with a negative impact on communication and coordination.

17.4.2.3 Hybrid Maintenance Organization

A hybrid maintenance organization is an attempt to overcome the disadvantages of the pure centralized and decentralized arrangements and to combine the advantages of both.

A centralized part of the organization may handle specialized maintenance operations. Major equipment issues, modifications, and rebuilds that require long-term planning and scheduling may be carried out by central maintenance staff. Hiring and training specialized maintenance staff to serve the entire organization can be justified.

Smaller, less specialized crews may be decentralized in various units to provide a timely response and deal with emergencies in a timely fashion. Local groups may be in charge of routine preventive maintenance.

Traditionally, the organizational structure is hierarchical and highly functionalized, within which maintenance is organized into highly specialized trades (such as electrical, mechanical, electronic, hydraulic, etc.). This type of organization has led to many problems in terms of efficiency and effectiveness. New, process-oriented organizational structures are emerging for more effective and efficient management of business units. Within these structures, maintenance is viewed as part of a group owning the process, which may include operators.

17.4.3 Organizational Effectiveness

There are no clear criteria for selecting the most appropriate organization for a particular situation. Therefore, it may be helpful to identify the traits of an effective maintenance organization. Such an organization is characterized by the following:

- *Good control of maintenance work:* Work is planned and executed efficiently in a reasonable time frame. Jobs and skills are matched in the scheduling process. Results of maintenance work are properly recorded and periodically analyzed for opportunities to improve the system.
- *Proactive maintenance:* Compliance with the preventive maintenance program is high and corrective and emergency maintenance is reduced.
- *Effective coordination with other functions:* Good coordination and communication exist between maintenance and operations, and decisions are made in line with common objectives and goals. Spares availability and stock levels are optimized to support maintenance work in a timely fashion.

17.4.4 Training

Having a competent workforce that is continuously learning to cope with the pace of technology advancement is a key issue for an effective maintenance organization. A systematic approach that defines job requirements, evaluates personnel skill against those requirements, and defines the appropriate training needs to be in place. A combination of formal and on-the-job training may be employed depending on the organization's needs.

17.5 Approaches to Maintenance

As indicated in Chapter 1, many different approaches to maintenance have evolved over time. The three approaches that are used in many different industry sectors are: (i) RCM, (ii) TPM, and (iii) RBM. In this section we discuss the salient features of the first two, while RBM is discussed in Section 17.6.3.

17.5.1 Reliability-Centered Maintenance

The first definition is the original definition (from the aircraft industry) and the second is a later one (from the energy industry).

Definition 17.11

RCM is a disciplined logic or methodology used to identify preventive maintenance tasks to realize the inherent reliability of equipment at the least expenditures of resources. (MSG-1)

Definition 17.12

RCM analysis is a systematic evaluation approach for developing and optimizing a maintenance program. RCM utilizes a decision logic tree to identify maintenance requirements of equipment according to the safety and operational consequences of each failure and degradation mechanisms responsible for these failures. (Electric Power Research Institute)

RCM is an approach that evolved in the early 1960s. In many industry sectors, certain PM tasks were deemed to be either unnecessary, because they had little impact on plant operations, or very conservative, in the sense that PM activities were performed more frequently than actually needed. These facts were discovered in the aircraft industry when United Airlines carried out a study in the 1960s to re-evaluate the PM strategy for the Boeing 747. One of the striking findings of the study was that 89% of the components exhibited failure characteristics for which scheduled overhaul or replacement was not effective. Only 11% of the components showed a failure characteristic that justified a scheduled overhaul or replacement as part of PM actions. This was an eye opener – new thinking and a new approach were needed to deal with the maintenance of components.

The focus of RCM is on maintaining system function rather than restoring equipment to an ideal condition. The RCM approach was initially used extensively in the aircraft and nuclear industries to maintain aircraft and plants. Since then it has gained acceptance in other industry sectors such as energy (maintenance of plants), transport (maintenance of infrastructures), and others.

17.5.1.1 RCM Process[5]

The RCM process starts by asking the following seven basic questions:

1. What are the functions and associated performance standards of the asset in its present operating context?
2. In what ways can it fail to fulfill its functions?
3. What causes each functional failure?[6]
4. What happens when each failure occurs?
5. In what way does each failure matter?
6. What can be done to predict or prevent each failure?
7. What should be done if a suitable proactive task cannot be found?

Based on the answers to these questions, the RCM process suggests the appropriate maintenance requirements for the system in its operating context. The four key features of RCM are the following:

1. *Preserving system function:* This is the principal feature for understanding the RCM process, as it calls for a view of equipment maintenance that stresses functional preservation as opposed to equipment operation. This forces the analyst to systematically understand the system functions that should be preserved and how they can be lost in terms of functional failure and not in terms of equipment failure. By first addressing system function, the output of the system becomes well known and the main task is to preserve that output.
2. *Identification of failure modes that can cause functional failure:* The identification of failure modes (discussed in Chapter 2) is carried out by examining each component to delineate how it might fail and cause a specific functional failure. Identifying failure modes that may potentially defeat system functions is crucial to the RCM process.
3. *Prioritizing functional failures:* Functional failures and their related failure modes are not equally important. By prioritizing failure modes, it is possible to decide how to allocate limited budgets and resources systematically. In other words, the purpose of prioritizing is to make an efficient and cost-effective allocation of resources.
4. *Selection of applicable and effective maintenance tasks:* Each failure mode is addressed in its prioritized order to identify potential PM actions. Each PM action must be *applicable* in the sense that, if performed, it will accomplish one of the objectives of undertaking PM; that is, prevent or mitigate failure, detect the onset of failure, or discover a hidden failure. Each task must also be *effective,* in the sense that we are willing to provide the necessary resources to do that task. If several candidate PM tasks are applicable, we select the least expensive.

The RCM philosophy for PM is as follows. A preventive task is worth doing if it deals successfully with the consequences of failure which it is meant to prevent. This implies that a preventive task must be technically feasible (i.e., it can achieve the objective) and worth doing.

---

[5] For more details of the RCM process, see Moubray (1991).
[6] Failure of one or more components that leads to a failure of a particular function at the system level.

17.5.1.2   RCM Implementation

RCM implementation involves the following seven steps:

1. Select an item from the engineered object's hierarchical decomposition (see Chapter 2) for RCM implementation and collecting information.
2. Define the item boundary.
3. Describe the item and draw a functional block diagram.
4. Define the item functions and functional failures – preserve functions.
5. Perform a failure mode and effects analysis (FMEA) – identify failure modes that can defeat the function.
6. Conduct a logic decision tree analysis (LTA) – prioritize function needs via the failure modes.
7. Select from a list of maintenance tasks[7] – select only effective and applicable PM tasks.

17.5.2   Total Productive Maintenance

TPM evolved in Japan in the 1970s where it proved to be very successful in enhancing the effectiveness and profitability of several Japanese manufacturing companies.

Definition 17.13

TPM can be defined as productive maintenance with the following features:

- Equipment effectiveness is maximized.
- A thorough system of PM actions is established for the equipment's entire life span.
- It is implemented by various departments (engineering, operations, and maintenance).
- It involves every single employee, from top management to workers on the floor.
- PM is promoted through motivation and management of autonomous small group activities.

The last two features are unique to Japanese culture and are in line with the TQM approach and employee empowerment. Most companies in other countries are organized with maintenance and operations as two separate entities. Consequently, the implementation of TPM in non-Japanese companies shifts the attention from "the total involvement of every employee" to the effectiveness of equipment.

The most important feature of TPM in manufacturing is equipment management. Equipment utilization is widely used as a measure of return on assets but, in many cases, equipment utilization is very low. Consequently, a sound equipment-management program aimed at improving equipment utilization is a must for the competitiveness and profitability of any organization. Another important feature is the empowerment of employees. The organizational line between maintenance, production, and engineering is often a source of inefficiency, higher costs, and lower productivity. With TPM, operators and mechanics must realize that they both have the same goals and consequently must cooperate and have a team spirit.

---

[7] Time-based, condition-based, failure-finding, run to failure, and so on (see Chapter 4).

The goals of TPM include, but are not limited to, the following:

- Improve product quality;
- Reduce waste;
- Improve the state of maintenance;
- Empower employees.

These goals are achieved through a careful implementation of the concepts of employee empowerment and sound equipment management. The involvement of the operators in the success of TPM cannot be overemphasized. A pragmatic way of achieving this is by using a systematic approach to skills under which an operator who has been properly trained and certified can perform a mechanical task. This partnership between operations and maintenance has many benefits:

- Operators and mechanics become multi-skilled, leading to job enrichment and improved flexibility of workers;
- The involvement of operators in routine maintenance builds a sense of responsibility, pride, and ownership;
- Delay times are reduced and productivity is increased;
- Teamwork is promoted between operations and maintenance.

Equipment is the focus of TPM. The TPM process starts by identifying the major losses with regard to equipment. The following six losses limit equipment effectiveness:[8]

1. Equipment failure (breakdown);
2. Set-up and adjustment downtime;
3. Idling and minor stoppages;
4. Reduced speed;
5. Process defects;
6. Reduced yield.

These losses have to be measured and their effect on equipment effectiveness quantified. TPM defines overall equipment effectiveness (OEE) as the product of availability, performance (characterizing efficiency), and quality factors. Availability factor (A) is affected by losses 1 and 2; performance factor (P) by losses 3 and 4; and quality factor (Q) by losses 5 and 6. Figure 17.9 shows the links between the factors and the losses and the expression for calculating the OEE.

The ultimate goal of TPM, with respect to equipment, is to increase its effectiveness to its highest potential and to maintain it at that level. This can be achieved by understanding the above losses and devising means to eliminate them.

17.6   Risk and Maintenance

A sound maintenance strategy should not only increase the profitability of operations but also minimize the hazards and the resulting consequences for both humans and the environment, resulting from unexpected failures of the engineered objects. RBM (discussed in

---

[8] For more details, see Nakajima (1988).

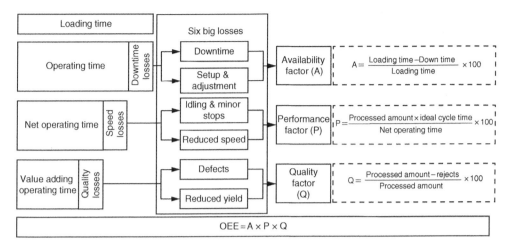

Figure 17.9   The six big losses and OEE definition.

Section 17.6.3) deals with this issue. This section provides a brief introduction to risk management and a framework for RBM.

17.6.1   Basic Concepts

Definition 17.14

Risk is the combination of the probability of an abnormal event of failure and the consequence(s) of that event or failure to a system's operator, user, or its environment.

17.6.1.1   Risk Characterization

This involves the following questions:

1. What can happen? (What can go wrong?)
2. How likely is it that that will happen?
3. If it does happen, what are the consequences?

The answers to these questions are characterized by a triplet (s_i, p_i, x_i) for each scenario, where s_i is the scenario identification description, p_i is the probability of that scenario occurring, and x_i is the consequence or evaluation measure of the scenario (measure of damage). A quantitative measure of risk (for each scenario) is given by the following expression:

$$\text{Risk} = \text{probability of failure} \times \text{consequence of the failure}$$

17.6.1.2   Risk Assessment

Risk assessment can be quantitative or qualitative. Quantitative risk assessment requires a great deal of data both for assessment of probabilities and assessment of consequences.

	Likely	Medium risk	High risk	Extreme risk
Likelihood	Unlikely	Low risk	Medium risk	High risk
	Highly unlikely	Insignificant risk	Low risk	Medium risk
		Slightly harmful	Harmful	Extremely harmful
		Consequences		

Figure 17.10   Risk matrix.

Fault trees (discussed in Chapter 2) or decision trees are often used to determine the probability that a certain sequence of events will result in a certain consequence. Qualitative risk
assessment is less rigorous and makes use of a simple risk matrix (see Figure 17.10) where
one axis of the matrix represents the probability and the other represents the consequences.
The matrix provides a prioritization of items assessed.

17.6.1.3   Risk-Management Process

Definition 17.15

Risk management is the coordinated activities to direct and control an organization with
regard to risk. (ISO 31000)

Several risk-management frameworks and standards have been proposed by different organizations. The most generic standard is ISO 31000. It provides principles and generic guidelines
on risk management and is not specific to any industry or sector. It can be applied to any type
of risk, whatever its nature and the consequences. The ISO 31000 standard proposes the
following risk-management process steps:

1. *Communication and consultation:* Communication and consultation with external and
   internal stakeholders should take place during all stages of the risk-management process.
2. *Establishing the context:* By establishing the context, the objectives, scope, and criteria for
   the remaining risk-management process are defined.
3. *Risk identification:* This step consists of identifying sources of risk, areas of impact, and
   events with their causes and consequences. The aim of the step is to create a comprehensive
   list of uncertain events, that is, risks that lack a description of likelihood and impact.
4. *Risk analysis:* The analysis of the uncertain events identified previously develops them into
   risks by quantifying both likelihood and impact, and also develops a deeper understanding
   of these risks.

5. *Risk evaluation:* During risk evaluation, based on the information gathered during risk analysis, decisions are made regarding which risks need treatment and the priorities are established among the risks.
6. *Risk treatment:* For every risk that needs treatment, different risk treatment options are analyzed regarding their cost–benefit trade off.
7. *Monitoring and review:* The monitoring and review process oversees both the risk situation of the organization and the risk-management process itself.

These steps are applicable in the context of maintenance, as we discuss in this section.

17.6.2 Risks and Maintenance

The main risks are failures of engineered objects that can have catastrophic consequences. This is true for safety-critical systems in plants (chemical, nuclear), control systems for complex products (aircraft, ships, trains, etc.), and failure of infrastructures (collapse of bridges, tall buildings, dams, etc.). Failure of these systems may result in loss of life, significant property damage, or damage to the environment, as illustrated by the real-life cases listed in Table 17.3.

Plant failure in manufacturing can result in lost production and quality problems that can also be very expensive. As with any other organization, a maintenance organization is subject to many types of risks that may adversely affect the operation and performance of the organization. RBM needs to take into account risks other than failure risks, such as:

- *Organizational risks:* Lack of qualified manpower, lack of adequate resources (for example, funding for aging public infrastructure), inadequate MMS, an ineffective PM program, inadequate coordination with other departments, and so on.

Table 17.3 Accidents due to poor maintenance.

Accident	Year	Cause	Consequences
Bhopal disaster	1984	Gas leak due to deferred maintenance	8000 deaths within two weeks and more than 0.5 million injuries
American Airline DC-10	1979	Faulty maintenance procedure	271 deaths
Railway accident in Potters Bar, UK	2002	Poor maintenance and inspections	7 deaths and 76 injuries
San Bruno, California gas pipeline explosion	2010	Gas leak due to defective welds	8 deaths
Texas City Refinery explosion	2005	Poor state of the plants and low spending on maintenance	15 deaths and 170 injuries
Highway 19 overpass bridge at Laval, Quebec, Canada	2006	Shear failure due to incorrectly placed rebar, low-quality concrete	5 deaths and 6 injuries

- *Network risks:* Unreliable suppliers, unreliable maintenance contractors, lack of coordination between multi-echelon facilities (in the case of repairable items), discontinuity of spare parts (bankrupt supplier).
- *External risks:* Weather (for example, delaying urgent maintenance for off-shore wind turbines), changes in regulations, and so on.

17.6.3   Risk-Based Maintenance

Definition 17.16

RBM is a maintenance approach that integrates risk in maintenance planning through the identification and assessment of the consequences of maintenance risks.

17.6.3.1   Key Issues

There are several issues that need to be addressed as part of RBM, and the two key ones are: (i) risk determination and (ii) risk evaluation.

Risk Determination

This consists of risk identification and estimation and involves the following steps:

1. *Failure scenario development:* A failure scenario is a description of a series of events which may lead to a system failure. A failure scenario is the basis of the risk study; it tells us what may happen so that we can devise ways and means of preventing or minimizing the possibility of its occurrence.
2. *Consequence assessment:* The objective here is to prioritize equipment and their components on the basis of their contribution to a system failure. Consequence analysis involves assessment of likely consequences if a failure scenario does materialize The total consequence assessment is a combination of three major categories: (i) system performance loss; (ii) financial loss; and (iii) human health loss.
3. *Probabilistic failure analysis:* Probabilistic failure analysis is conducted using fault tree analysis (FTA). The use of FTA, together with components' failure data and human reliability data, enables the determination of the frequency of occurrence of an accident.
4. *Risk estimation:* The results of the consequence and the probabilistic failure analyses are then used to estimate the risk that may result from the failure of each unit. In the next subsection, we will show how the estimated risk is evaluated against acceptance criteria.

Risk Evaluation

This step consists of risk aversion and risk acceptance analysis and involves the following steps:

1. Setting up acceptance criteria.
2. Risk comparison against acceptance criteria: The acceptance criteria are applied to the estimated risk for each unit in the system. Units whose estimated risk exceeds the acceptance criteria are identified. These are the units that should have an improved maintenance plan.

17.6.3.2 Maintenance Planning Based on Risk

Units (of the object) whose levels of estimated risks exceed the acceptance criteria need to be studied in detail, with the objective of reducing the level of risk through a better maintenance plan. Both the type of maintenance and the maintenance interval should be decided upon at this stage. By modifying the maintenance interval, the probability of failure changes and this will impact on the risk.

17.7 Maintenance Management System

> **Definition 17.17**
>
> A *maintenance management system* (MMS) is the framework of processes and procedures used to ensure that the maintenance function can fulfill all tasks required to achieve its objectives.

An MMS is part of the overall management system for a business. The main components of an MMS are planning, control, and performance management. Maintenance performance management (MPM) is discussed in this section and planning and control are discussed in Chapter 19.

The implementation of an MMS on computers is called a computerized maintenance management system (CMMS). CMMS-related issues are discussed in Chapter 22. A CMMS can help in the automation of many manual processes, leading to more accurate data collection, faster data analysis, and the generation of timely reports that can support effective maintenance decision making.

17.7.1 Maintenance Performance Management (MPM)

MPM deals with performance management relating to maintenance activities. Since maintenance is carried out at the component level, it impacts on the performance of the component and the object (infrastructure, plant, or product). Maintenance does not directly affect the business level – but it does so indirectly through the performance at the object level. As such, maintenance performance indicators (MPIs) and maintenance performance metrics (MPMs) are defined at the component, object, or some intermediate level.

17.7.1.1 MPIs

MPIs include the following:

1. Plant/equipment-related indicators: The MPIs measure the performance pertaining to the engineered objects used by the business.
   - Availability;
   - Performance (output delivery);

- Quality;
- Number of minor and major stoppages;
- Downtime for the number of minor and major stoppages;
- Rework.

2. *Maintenance-task-related indicators:* These MPIs pertain to the maintenance tasks carried out. These MPIs indicate the efficiency and effectiveness of the maintenance department of the organization.
   - Change-over time;
   - Planned maintenance tasks (preventive maintenance);
   - Unplanned maintenance tasks (corrective maintenance);
   - Response time for maintenance.

3. *Finance/cost-related indicators:* These MPIs relate to maintenance and production costs. Also, management of the organization may include other financial and cost-related MPIs or PIs as per their need.
   - Maintenance cost/unit;
   - Production cost per unit;
   - Total maintenance cost.

4. *Customer satisfaction:* These MPIs relate to satisfying the customers and are formulated from the organizational business strategy.
   - Number of quality complaints;
   - Low-quality returns (number/quantity);
   - Customer satisfaction (value-for-money feedback, etc.);
   - Customer retention;
   - Number of new customers added.

5. *Learning and growth:* These MPIs relate to indicators of innovation.
   - Number of new ideas generated for improvement;
   - Skills and competency development/training.

6. *Health, safety, and the environment (HSE):* Health and safety issues which form part of the societal requirements and also environmental issues are considered by the organization under this criterion.
   - Number of incidents/accidents;
   - Lost time due to HSE issues;
   - Number of legal cases;
   - Number of compensation cases/amount of compensation paid;
   - Number of HSE complaints.

7. *Employee satisfaction:* Employee satisfaction is essential to successfully implementing an MPM system and achieving the desired goals of the organization.
   - Employee absenteeism;
   - Employee complaints;
   - Employee retention.

17.7.1.2   MPMs

An MPM is involved in the process of measuring maintenance performance, to know how well the maintenance process is performing and to identify the opportunities for improvement. Performance management relies on gathering accurate data about how processes perform in order to stimulate improvement activity. The discussion in Section 17.2 about a business performance system applies to MPMs as well.

17.7.2   Maintenance Performance Management System

A maintenance performance management system (MPMS) forms part of an organization's operational system. It includes all related MPIs and their interrelationship within the whole maintenance process. Its elements are similar to those discussed in Section 17.2 and it is the backbone of the maintenance continuous improvement process.

Some of the important issues associated with the development of an MPMS are the following:

* Measuring values created by the maintenance;
* Justifying investment;
* Revising resource allocations;
* HSE issues.

17.8   Summary

The management of any business involves the coordination of the efforts of people to accomplish goals and objectives using the available resources in an efficient and effective manner. This is accomplished through the classical management functions of planning, organizing, implementing, controlling, and performance management.

Maintenance management issues arise, and need to be addressed, at three different levels: strategic, tactical, and operational. Strategic maintenance issues include maintenance organization, maintenance approaches, maintenance outsourcing, capital investment, and key supporting functions.

Maintenance organizations may be centralized, decentralized, or a hybrid (a combination of the first two). The advantages and disadvantages of each have been presented.

The three maintenance approaches that are used in many different industry sectors are: (i) RCM, (ii) TPM, and (iii) RBM. RCM is a systematic approach for developing and optimizing a maintenance program. Its key features and implementation steps have been presented. TPM can be defined as productive maintenance with a focus on maximizing equipment effectiveness using a thorough system of PM actions for the equipment's entire life span, involving every single employee, from top management to workers on the shop floor. Key features of TPM have also been described. RBM is a maintenance approach that integrates risk in maintenance planning through the identification and assessment of the consequences of different risks.

A management system is a collection of several specific systems to ensure that the tasks achieve the required business objectives. An MMS is part of this, and it is comprised of processes and procedures to ensure that the maintenance department can fulfill all the maintenance needed to ensure the business achieves its objectives.

The main components of an MMS are planning, control, and performance management. Effective MPM of all maintenance activities is vital to sound maintenance management. It involves the selection of maintenance management indicators that may be based on the balanced scorecard concept.

Review Questions

17.1   What is the definition of management and what are the four key functions of management?

17.2   What are the different levels of management decisions?

17.3   Why is business performance management important for effective business management?

17.4   What is maintenance management?

17.5   What are the main strategic maintenance issues?

17.6   What are the different types of maintenance organization?

17.7   What are the advantages and disadvantages of the various types of maintenance organizations?

17.8   What is RCM? What are its key features?

17.9   What is TPM? What are its key features?

17.10 What is risk? What are the steps of a risk-management process?

17.11 What is RBM? What are its key features?

17.12 What is a management system and what are its key elements?

Exercises

17.1   Discuss the advantages and disadvantages of having maintenance under production or operations as opposed to reporting to a higher level in the organization.

17.2   Maintenance organizations may be centralized, decentralized, or a hybrid of the two, as discussed in Section 17.4.2. Identify organizations for which each of these types of organization is appropriate.

17.3   Maintenance objectives and strategies have to be aligned with the overall business mission objectives of the organization. For a refinery, determine:
       (a) Two appropriate objectives of the maintenance function.
       (b) Measures of performance at the strategic, tactical, and operational levels and covering the four dimensions of the balanced scorecard.

17.4   Repeat Exercise 17.3 for the maintenance of the rolling stock of a railway organization.

17.5   Repeat Exercise 17.3 for road maintenance.

17.6   Consider the RCM implementation steps presented in Section 17.5.1.
       (a) Identify three criteria for selecting items from a plant to be subject to RCM implementation.
       (b) Give examples of the type of information that needs to be collected to carry out an RCM analysis.
       (c) Who should be included in an RCM team?

17.7   Discuss the two important features of TPM.

17.8   Discuss the possible relationship between TPM and RCM and whether it is possible for the two approaches to complement each other.

17.9   Consider the following data collected in a manufacturing process from one of the machines. Loading time = 800 minutes, downtime = 50 minutes, theoretical cycle time = 1.5 minutes, processed amount = 290 parts, and the number of parts rejected = 6. Calculate the following:
(a) Availability.
(b) Performance efficiency.
(c) Rate of quality.
(d) Overall equipment effectiveness (OEE).

17.10   Identify areas where RBM is recommended and justify your choice.

References

CEN (2001) *Maintenance Terminology*. European Standard, EN 13306:2001, European Committee for Standardization, Brussels.

Deming, W.E. (1986) *Out of the Crisis*, MIT Center for Advanced Engineering Study.

Moubray, J. (1991) *Reliability-Centered Maintenance*, Butterworth-Heinemann Ltd.

Nakajima, S. (1988) *Introduction to TPM: Total Productive Maintenance*, Productivity Press.

NRC (National Research Council) (2005) *Measuring Performance and Benchmarking Project Management at the Department of Energy*. The National Academies Press, Washington D.C.

18

Maintenance Outsourcing and Leasing

Learning Outcomes

After reading this chapter, you should be able to:

- Define outsourcing and explain the main reasons for outsourcing;
- Discuss the advantages and disadvantages of outsourcing;
- Explain the advantages and disadvantages of maintenance outsourcing;
- Describe the three levels of maintenance outsourcing;
- Explain the key decision problems in maintenance outsourcing from the perspectives of the two parties involved – (i) customer (recipient of maintenance service) and (ii) service agent (provider ofc maintenance service);
- Describe the key elements of the maintenance outsourcing decision-making framework;
- Define a maintenance service contract (MSC), describe its key elements, and provide a classification of MSCs;
- Describe models for optimal decision making related to MSCs including game theory models;
- Define leasing, the terms of a lease contract, and related decision problems from the perspectives of both the lessor and lessee;
- Define functional products.

Introduction to Maintenance Engineering: Modeling, Optimization, and Management, First Edition.
Mohammed Ben-Daya, Uday Kumar, and D.N. Prabhakar Murthy.
© 2016 John Wiley & Sons, Ltd. Published 2016 by John Wiley & Sons, Ltd.

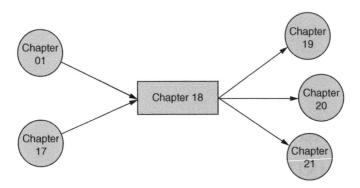

18.1   Introduction

Outsourcing can be defined as a managed process of acquiring goods and/or services from an external agent under a contract rather than doing it in-house. If the maintenance of an engineered object (product, plant, or infrastructure) is carried out by the owner, it is referred to as *in-house maintenance*. If some or all of the maintenance is carried out by an external service agent under a service contract, this is referred to as *maintenance outsourcing*.

Outsourcing involves two parties: (i) *customers* – recipients of the goods and/or services and (ii) *service agents* – providers of the goods and/or services. The *contract* is a legal document involving the two parties and deals with a range of issues (technical, financial, legal, etc.). It can be fairly simple or very complicated depending on the goods and/or services involved. The two parties have different objectives (goals) and the outcomes depend on the actions (decision variables) of both parties. As such, any rational decisions need to take into account this interaction. Game theory provides the framework for optimal decision making by both parties.

This chapter deals with maintenance outsourcing and looks at the different issues involved. The outline of the chapter is as follows. Section 18.2 deals with outsourcing in general whilst Section 18.3 deals with maintenance outsourcing and discusses some of the issues and the decision problems from both the customer and the service agent perspectives. Section 18.4 deals with the framework needed to obtain solutions to these decision problems. Section 18.5 looks at optimal decision making (from both owner and service agent perspectives) using the game-theoretic approach. Also, this section deals with agency theory, which is important in the context of contract formulation in maintenance outsourcing. Often, instead of owning the object (with maintenance either done in-house or outsourced), there is a trend toward leasing it, where the user (referred to as the *lessee*) leases the object from the owner (referred to as the *lessor*). The maintenance of the object can be the responsibility of the lessor, the lessee, or both, depending on the lease contract (LC). Leasing is discussed briefly in Section 18.6 Sections 18.7–18.8 deal with maintenance service contracts (MSCs) in the context of products/plants and infrastructures, respectively. We conclude with a summary of the chapter in Section 18.9.

18.2   Outsourcing

Businesses producing goods (products and/or services) need to come up with new solutions and strategies to develop and increase their competitive advantage. Outsourcing is one of these strategies that can lead to greater competitiveness. As mentioned earlier, it is a process of acquiring goods and/or services from an external agent under a contract. The agent charges a fee and, in exchange, the business (henceforth called the *customer* and recipient of the goods) is provided with the goods at a guaranteed quality or service level.

Most contracts stipulate specific, measurable metrics called *service level agreements* (SLAs). These depend on the goods and/or services involved. Often, SLAs also have penalties associated with not meeting the specified metrics, and sometimes rewards as incentives for exceeding the metrics. Needless to say, there is a multitude of ways of constructing outsourcing agreements.

18.2.1   Reasons for Outsourcing

The conceptual basis for outsourcing is as follows:

1. Domestic (in-house) resources should be used mainly for the core competencies of the business.
2. All other (support) activities that are not considered strategic necessities and/or whenever the business does not possess the adequate competencies and skills should be outsourced (provided there is an external agent who can carry out these activities in a more efficient manner).

There are a number of reasons that drive businesses to outsource. The list of reasons includes:

1. *To reduce costs:* Sometimes achieved through lower wage costs, but also achieved through economies of scale when the external agent provides the goods to multiple businesses.
2. *To improve service:* This often requires better-educated or skilled people who are either not available in-house or are not economical to employ.
3. *To obtain expert skills:* An external agent is often a business that is allegedly an expert in the delivery of the goods under consideration and thus should be able to do this better than the customer.
4. *To improve processes:* For complex processes, external sources often have the expertise that is needed for process improvement.
5. *To improve the focus on core activities:* Outsourcing frees management from having to worry about the inner workings of a non-core activity. The customer focuses on the internal core competencies, whilst the others are outsourced.

Unfortunately, many businesses do not look at all of these factors and often the primary reason for outsourcing is to reduce their costs.

18.2.2   Problems with Outsourcing

Outsourcing may not be appropriate for some businesses. Some of the reasons for this are the following:

1. The business may be too small to outsource effectively.
2. The culture within the business may not be appropriate for outsourcing.

3. Other reasons (such as confidentiality) may limit or prevent the business's ability to outsource.
4. The changes needed to the organizational structure make it difficult to outsource.

18.2.3 Issues in Outsourcing

Issues that need to be addressed before deciding on outsourcing are the following:

1. Is there a well-defined set of achievable business objectives?
2. Does outsourcing make sense?
3. Is the organization ready?
4. What are the outsourcing alternatives?
5. What activities should be outsourced?
6. How should the best external agents be selected?
7. What are the negotiating tactics for contract formation?
8. How is the fee to be decided?[1]
9. How are the incentives and/or penalties in the contract to be decided?
10. What systems are needed for effective monitoring?
11. What are the potential risks?

18.3 Maintenance Outsourcing

Most businesses tend not to view maintenance as a core activity and have moved toward outsourcing it.

For these businesses, it is no longer economical to carry out the maintenance in-house. There is a variety of reasons for the move to maintenance outsourcing, including the need for a specialist workforce and diagnostic tools that often require constant upgrading. In these situations, it is more economical to outsource the maintenance (in part or in total) to an external agent through an MSC.

An extended warranty (EW) is a term that is often used instead of an MSC in the context of products. EWs are extensions of product warranties (discussed in Section 4.7). Originally they were offered solely by original equipment manufacturers (OEMs), with the customer paying an additional amount.[2] Currently they are offered not only by OEMs but by many others such as retailers, insurance companies, and so on. We will use the term MSC to include EWs in the remainder of the chapter.

18.3.1 Advantages and Disadvantages

The advantages of outsourcing maintenance are as follows:

1. Better maintenance due to the expertise of the service agent.
2. Access to high-level specialists on an "as and when needed" basis.

---

[1] The fee can take many forms – based on the transaction, labor hours, cost per unit, cost per project, annual cost, cost by service level, and so on.
[2] A product warranty – also referred to as a *base warranty* (BW) – is integral to the sale and its cost is factored into the sale price whereas an EW is an optional purchase by a customer.

3. A fixed-cost service contract removes the risk of high replacement/repair costs.
4. Service providers respond to changing customer needs.
5. Access to the latest maintenance technology.
6. Less capital investment for the customer.
7. Managers can devote more resources to other facets of the business by reducing the time and effort involved in maintenance management.

However, there are some disadvantages, as indicated below.

1. Dependency on the service provider.
2. The cost of outsourcing.
3. Loss of maintenance knowledge (and personnel).
4. Being locked in to a single service provider.

For very specialized (and custom-built) products, the knowledge to carry out the maintenance and the spares needed for replacement need to be obtained from the OEM. In this case, the customer is forced into having an MSC with the OEM.

When the maintenance service is provided by an agent other than the OEM, the cost of switching often prevents customers from changing their service agent. In other words, customers get "locked in" and are unable to do anything about it without a major financial consequence.

Thus, it is very important for businesses to carry out a proper evaluation of the implications of outsourcing their maintenance. If done properly, outsourcing can be cheaper than in-house maintenance and can lead to greater business profitability.

18.3.2   Different Scenarios for Maintenance Outsourcing

Maintenance of an object involves carrying out three sequentially linked activities, as indicated in Figure 18.1. The activities are:

- *Work planning (D-1):* What (components) need to be maintained?
- *Work scheduling (D-2):* When should the maintenance be carried out?
- *Work execution (D-3):* How should the maintenance be carried out?

There are three different scenarios (S-1, S-2, and S-3) depending on which of these activities are outsourced, and they are listed in Table 18.1.

In scenario S-1, the service agent is only providing the resources (workforce and material) to execute the work. This corresponds to the minimalist approach to outsourcing. In scenario S-2, the service agent decides on *how* and *when*, while *what* is to be done is decided by the customer. Finally, in scenario S-3, the service agent makes all three decisions.

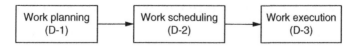

Figure 18.1   Maintenance activities.

Table 18.1   Different maintenance outsourcing scenarios.

Scenarios	Decisions	
S-1	D-1, D-2	D-3
S-2	D-1	D-2, D-3
S-3	—	D-1, D-2, D-3

18.3.3   Decision Problems

The decision problems for the customer and the service agent are different, as they have different goals or objectives.

18.3.3.1   Customer Perspective

In the case of products, some of the decision problems are:

1. Whether to buy an MSC or not.
2. How to evaluate whether the MSC price is reasonable or not.
3. How to decide on the best MSC if there is more than one option.

In the case of plants and infrastructures, some of the decision problems are:

1. Should some or all of the maintenance be outsourced?
2. What should be the terms of the MSC?
3. How should the best service provider be selected when there is more than one?

18.3.3.2   Service Provider Perspective

Some of the decision problems are listed below:

1. Should the service provider offer one or several MSCs?
2. What should be the terms of the MSCs?
3. What are the costs of servicing the different MSCs?
4. What should be the pricing of the MSCs?
5. How to deal with competition in the MSC market.
6. How to plan the servicing logistics.
7. How to tender for a MSC.

18.4   Framework for Maintenance Outsourcing Decision Making

A framework for effective decision making by both parties (customer and service agent) needs to take into account several elements and the interactions between them. Figure 18.2 shows the key elements.

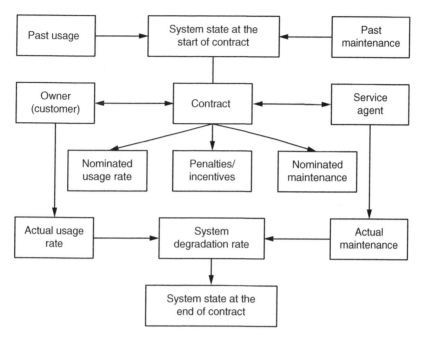

Figure 18.2   Key elements of a framework.

18.4.1   Issues Involved

18.4.1.1   Requirements

Both parties might need to meet some stated requirement. For example, the customer needs to ensure that the usage intensity and operating loads of the asset do not exceed the levels specified in the contract. These can lead to greater degradation (due to higher stresses on the components) and higher servicing costs to the service agent. Similarly, the service agent needs to ensure proper data recording.

18.4.1.2   Economic Issues

There are a number of alternative contract payment structures, as indicated below:

- Fixed or firm price;
- Variable price;
- Price ceiling incentive;
- Cost plus incentive fee;
- Cost plus award fee;
- Cost plus fixed fee;
- Cost plus margin;
- Other issues are cost deductibles and cost limits (for individual and total claims).

Each of these price structures represents a different level of risk sharing between the business (customer) and the service agent.

18.4.1.3 Legal Issues

- *Contract duration:* This is usually fixed with options for renewal at the end of the contract.
- *Moral hazard (cheating):* In maintenance outsourcing, cheating by both owner and service agent is an issue that needs to be addressed. Cheating by the owner occurs when the nominated usage is lower than the actual usage and the service agent is not able to observe this. Similarly, cheating by the service agent occurs when the actual maintenance level is below the nominated maintenance level and the owner cannot observe this. Information, monitoring, and penalties/incentives can reduce and eliminate the potential for cheating.
- *Dispute resolution:* This specifies the avenues to follow when there is a dispute. The dispute can be resolved by going to a third party (for example, an arbitration tribunal or a court).

Unless the contract is written properly and relevant data (relating to the equipment and collected by the service agent) are analyzed properly by the customer, the long-term costs and risks will escalate.

18.4.2 Maintenance Service Contracts (MSCs)

An MSC document contains some or all of the elements listed below:

- Parties involved: Service agent (SA) supplier of service and customer (recipient of the service), their names and addresses, and so on.
- Definitions: Glossary of frequently occurring words in the document.
- Description of the service (maintenance actions, materials, labor, etc.).
- Performance levels.
- Delivery of the service (single or multiple locations).
- Term: Start date and period of agreement.
- Pricing details (these can vary considerably from contract to contract).
- Pricing adjustment (for example, annual increases linked to inflation or some other index).
- Payment details: Annual, monthly, after each service, and so on.
- Responsibilities of the SA: Details of services to be performed and SLAs, if applicable.
- Responsibilities of the customer: Usage of the object.
- Indemnification and insurance.
- Bankruptcy.
- Confidentiality.
- Force majeure.
- Dispute and arbitration process.
- Termination.
- Renegotiation/renewal.

18.4.2.1   Classification of MSCs

Maintenance requires materials, parts, and labor to carry out the various activities discussed in Section 18.3. As a result, there are several different kinds of MSCs. These can be grouped broadly into three types, as indicated below:[3]

- *Type I:* SA only responsible for supply of material and parts (includes reconditioned parts).
- *Type II:* SA responsible for material and parts and carrying out some or all maintenance.
- *Type III:* SA is responsible for complete maintenance and operations.

18.5   Optimal Decisions

Determining the optimal decisions (for both customer and service agent) requires building models and proper management of information flows.

18.5.1   Information Flow Management

Many different kinds of information are needed for effective decision making by both parties and this can be grouped broadly into the following categories:

- Technical (relating to object and the services needed for maintenance);
- Operations (servicing-related data);
- Financial (relating to various types of costs);
- Legal (relating to contract details);
- Commercial (relating to customers, marketing, etc.).

MSCs for plants and infrastructures involve both parties negotiating to formulate the contract, and information flow plays a critical role in this process. We use a model which is a slight modification of the one proposed by McFarlane and Cuthbert (2012) to characterize the information flow for MSCs, and this is shown in Figure 18.3.[4] It involves six elements (shown as six boxes in the figure) and there are four different information flows, as listed below.

- Design information (from customer to provider);
- Delivery information (from provider to customer);
- Assessment/evaluation information (from customer to provider);
- Information flow between provider and external suppliers such as vendors supplying materials and spares, equipment manufacturers providing the equipment needed for carrying out the maintenance service, specialists (for example, for oil analysis), and so on.

We will discuss each of these briefly.

---

[3] Martin (1997) uses a different way of classifying MSCs. It also involves three types:

1. *Work package contract:* The customer performs all planning and scheduling and the SA carries out the execution. This corresponds to scenario S-1 and Type II in our classification.
2. *Performance contract:* This corresponds to Type III in our classification.
3. *Facilitator contract:* This corresponds to a lease contract in our definition and is discussed in Section 18.6.

[4] The information flow model of McFarlane and Cuthbert (2012) contains only the middle four boxes.

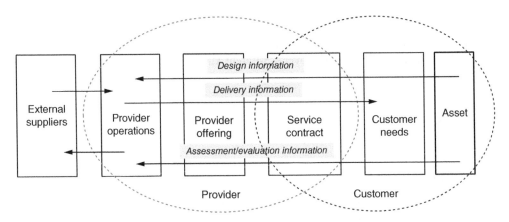

Figure 18.3   Information flows for MSC decision making.

18.5.1.1   Design Information

- *Asset:* Asset condition at the start of the contract (past history of operation and maintenance).[5]
- *Customer needs:* Conceptual information about the customer's requirements for asset performance (reliability and financial related).
- *Service contract:* Information to formalize the MSC.
- *Provider offering:* Alternative MSCs, asset performance implications, and so on.
- *Provider operations:* Technical information to plan and develop the delivery of the MSC offered, cost information, resources (organization, equipment needed, skill base), and so on.
- *External suppliers:* Technical information on equipment and materials needed by the provider for delivery of the maintenance service, cost information, and so on.

18.5.1.2   Delivery Information

- *Customer needs:* Information from the provider to enable the customer to achieve better coordination between maintenance service and asset operation.
- *Service contract:* Information regarding the details of maintenance services to be delivered by the provider.
- *Provider offering:* Information regarding the delivery details of alternative maintenance service offerings.
- *Provider operations:* Technical information on how the provider can deliver the agreed maintenance services (relating to logistics of maintenance service delivery).
- *External suppliers:* Delivery logistics for spares, materials, and so on.

---

[5] An asset can be a product, plant, or infrastructure.

18.5.1.3   Assessment and Evaluation Information

- *Asset:* Asset condition over the contract period.
- *Customer needs:* Information to determine fulfillment of customer needs.
- *Service contract:* Performance requirements defined through various metrics (reliability, financial, operations, penalties, etc.), customers' responsibilities, and so on.
- *Provider offering:* Information relating to the effectiveness of the maintenance service offerings defined through suitable performance measures.
- *Provider operations:* Operational information on performance of service infrastructure and operations.
- *External suppliers:* Component suppliers, transport services, and so on.

18.5.2   Customer's Optimal Decisions

The framework for the purchase of an MSC is as shown in Figure 18.4.

18.5.2.1   Criteria for Rating and Selection of Service Agents

A business is often faced with the strategic decision as to whether to develop its own resources to perform maintenance or purchase the required skills and performance from external service agents. To make this decision the business needs to analyze whether maintenance forms a part of its core competencies or whether it only makes a minor contribution to the value chain. Once the business has decided to outsource, it also needs to decide on the criteria to select the best service agents.

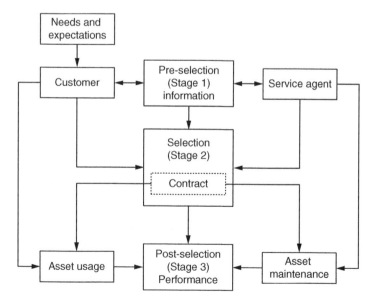

Figure 18.4   Framework for the customer's decision-making process.

The selection criteria need to be governed by the strategic intent of the business and the use of the outsourcing process to meet its goal. Therefore, the selection of the service agent is influenced by the reasons for outsourcing. These reasons can be one or more of the following:

- Concentrating on core activities;
- Reducing the maintenance costs;
- Spreading the business risk;
- Downsizing the organization;
- Supplementing knowledge to achieve the business goals;
- Bringing in strategic knowledge to meet its requirements;.
- Facilitating the building up of competence outside the organization.

In many contract situations with a large number of service agents participating, the selection of contractors is usually made in two phases: (i) the pre-selection phase and (ii) the final selection phase. We discuss these briefly below.

Pre-Selection Phase
In the pre-selection phase of a service contract process, the selection criteria are based on the following:

- *Technical capabilities:* The service agent must have the knowledge, the organizational structure, and the resource capabilities to meet the contractual agreements. That is, the service agent must have the correct organization (number of people and their competencies) and equipment, and so on, to carry out the maintenance as stated in the contract on time and correctly. Often, service agents enter a contract but lack the organizational capability to deliver the agreed performance, and this creates bottle-neck problems for the owner of the asset.
- *Experience with similar equipment:* Although the service agent might have the required manpower and competence, the agent may have had no experience in maintaining the asset under consideration. This can result in problems with the delivery and quality of service. Often, it takes some time for the service agent to understand all the factors that can cause equipment downtime and this causes bottle-necks when the agent is dealing with a specific asset for the first time.
- *Financial health of the service agent:* Often, owners are influenced by the reputation and capabilities of the service agent and fail to do a thorough analysis of the service agent's financial health. If the service agent is financially weak, there is a risk that the agent might not be able to fulfill the contract or even go bankrupt due to cash-flow problems.
- *Innovative capability of the service agent:* In recent times, the innovative capability of the service agent has become a dominant factor in an agent being awarded an MSC. If the agent has a reputation for being innovative, it provides assurance to the owners of the assets that new and innovative maintenance solutions will ensure better performance, higher quality, and/or reduced costs.
- *Demonstrated good governance/moral integrity of the service agent:* Good governance is reflected in factors such as transparency in action and moral integrity. Service agents who exhibit these characteristics are preferred to those who lack them.

Selection Phase

The selection procedure involves a detailed and in-depth analysis of the criteria used in the pre-selection phase. Some of these are listed below:

- Business plan, vision for implementation of new and proven technology: The owners of assets should demand and examine the business plan of the service agents and assess these plans in terms of the implementation of new technologies, training of personnel, and other actions to facilitate innovations.
- Special focus should be given to evaluating the service agent's quality assurance process and its implementation.
- Past experience and performance of the service agent should be assessed by talking to the agent's previous customers.
- Once short-listed, the owner of the asset must evaluate the team members that will be involved in carrying out the maintenance activities. This assessment is based on the qualification and experience of each member with respect to the maintenance of similar assets.
- Defining key performance measures.
- Proper data-collection system for monitoring and reporting.

Example 18.1   Swedish Railway Transport Administration

In order to increase the effectiveness and efficiency of the maintenance process, the Swedish Railway Transport Administration (Trafikverket) started to open up its maintenance contracts for market competition. That is, anyone with the capability to deliver a contract could participate in the contract-tendering process. Since railway maintenance is specialized and needs special tools and skills, there were only a few service agents in Sweden who could perform the service. This provided an opportunity for service agents from other European countries to also bid for the contracts. Today, at least four service agents have been awarded contracts, based on their competence, capability, and price, for carrying out maintenance in different regions. The selection of service agents at Trafikverket, in general, involves the following steps:

1. *Pre-qualification of contractors:* This is performed at head office level and all the contractors or service agents planning to bid for a contract must register and be approved by the committee based on their capability, past performance, ethics, and so on.
2. *Announcement of contract:* The contract is advertised in most of the listed major newspapers with a short description of the job and the contact details of the persons responsible for the contract.
3. *Contract procurement process:* During this step, potential contractors are informed about the type, scope, duration, and other relevant descriptions of the contract. Based on this information, interested contractors submit contract bids.
4. *Pre-selection:* Based on the details of the submitted bids and other relevant information about contractors, the client (infrastructure manager) selects two or three contractors to initiate the contract-negotiation process.

5. *Contract negotiations:* During this step, the contract together with the scope of the work and the related price tags, and so on, are discussed in detail with the selected potential service agents. This step also leads to the final selection of the service agent most suitable for the contract.
6. *Study and analysis of contract:* After selecting the service agent, the client and service agent both study and analyze the contract and enter into an agreement whereby the contract is defined at a detailed level.
7. Signing of the contract and its implementation as per the time and delivery plan.   ∎

Post-Selection Phase
- Use of data collected in evaluating the key performance measures and devising improvement strategies for continuous improvement.

18.5.2.2 Models for Decision Making

Models play an important role in the decision-making process. The models can be grouped into two categories, as indicated below:

- *Cost-based models:* Here, the objective function is the total discounted cost over the period of the contract. The expected costs with (i) in-house maintenance and (ii) outsourced maintenance are computed using standard discounting methods used by accountants to decide if maintenance should be outsourced or not and in the selection of the service agent when there are two or more available.[6]
- *AHP models:* If the objective function involves multiple criteria (a vector consisting of two or more elements), then the analytic hierarchy process (AHP) has been used in deciding whether to outsource or not and in the selection of the service agent when there are two or more available.[7]

18.5.3   Service Agent's Optimal Decisions

The service agent needs to take into account the decision-making process of the customer in order to make the best decision to achieve his/her objective. The game-theoretic approach is the most appropriate one, as this takes into account the decisions of both customer and service agent and their interactions.[8] The particular approach which has received a lot of attention in the maintenance outsourcing literature is the Stackelberg game with the service agent as the "leader." This type of game is closely linked to agency theory, and a brief description of the two is given below for the case with a single service agent and a single customer. Two other scenarios are: (i) the customer is the leader and the service agent is the follower and (ii) neither is the leader nor the follower. The former is a Stackelberg game (with the roles reversed) and the latter is a Nash game.[9]

---

[6] These kinds of models are discussed in Chapter 21.
[7] A brief introduction to the AHP is given in Appendix D.
[8] A brief introduction to game theory is given in Appendix D.
[9] Murthy and Jack (2014) deal with the game-theoretic approach to maintenance outsourcing and give a comprehensive review of the different models that can be found in the literature.

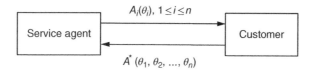

Figure 18.5   Stackelberg game for maintenance outsourcing.

18.5.3.1   Stackelberg Game

Here, the service agent offers a set of options $\{A_1, A_2, ..., A_n\}$ to a customer. θ_i is the decision variable for option i which the service agent needs to select optimally to maximize an objective function. Given the set of options, the customer chooses the option that optimizes the customer's objective function. This generates the customer's best response function $A*(\theta_1, \theta_2, ..., \theta_n)$, as shown in Figure 18.5. Using this response function, the service agent then optimally selects the values of the decision variables $\theta_1, \theta_2, ..., \theta_n$ to optimize his/her objective function.

18.5.3.2   Agency Theory

Agency theory (AT) attempts to explain the relationship that exists between two parties (a principal and an agent) where the principal delegates work to the agent who performs that work under a contract. AT is concerned with resolving two problems that can occur in principal–agent relationships. The first is the agency problem that arises when the two parties have conflicting objectives and it is difficult or expensive for the principal to verify what the agent is actually doing and whether the agent has behaved appropriately or not. The second is the problem involving the risk sharing that takes place when the principal and the agent have different attitudes to risk (due to various uncertainties). Each party may prefer different actions because of their different risk preferences.

The focus of the theory is on determining the optimal contract – behavior versus outcome – between the principal and the agent. Many different cases have been studied in depth in the principal–agent literature and these deal with the range of issues indicated in Figure 18.6. A brief discussion of the issues is given below.

- *Moral hazard:* This refers to the agent's possible lack of effort in carrying out the delegated tasks and the fact that it is difficult for the principal to assess the effort level that the agent has actually used.
- *Adverse selection:* This refers to the agent misrepresenting their skills to carry out the tasks and the principal being unable to verify this before deciding to hire them. One way of avoiding this is for the principal to contact people for whom the agent has previously provided a service.
- *Monitoring:* The principal can counteract the moral hazard problem by closely monitoring the agent's actions.
- *Information asymmetry:* The overall outcome of the relationship is affected by several uncertainties and the two parties will generally have different information with which to make an assessment of these uncertainties.

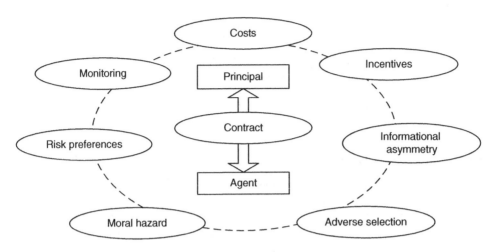

Figure 18.6   Issues in agency theory.

- *Risk:* This results from the different uncertainties that affect the outcome of the relationship. The risk attitudes of the two parties may differ and a problem occurs when they disagree over the allocation of the risk.
- *Costs:* Both parties incur various kinds of costs. These will depend on the outcome of the relationship (which is influenced by various types of uncertainty), acquiring information, monitoring, and on the administration of the contract.
- *Contract:* The key factor in the relationship between the principal and the agent is the contract, which specifies what, when, and how the work is to be carried out and also includes incentives and penalties for the agent. This contract needs to be designed taking account of all the issues involved.

18.6   Leasing

There is a growing trend for customers to lease systems rather than purchase them. A lease is a contractual agreement under which one party (the owner, who is also referred to as the *lessor*) leases to another party (also referred to as the *lessee*) an engineered object for use as per the terms of the lease contract (LC).[10] There are many types of LC. The maintenance can be the responsibility of either the lessor or the lessee depending on the type of LC. The terms of the LC are usually decided by the lessor for products (consumer, commercial, and industrial) or jointly with the lessee for plants and infrastructures. The terms can include none, one, or more of the following:

- Guarantees on asset and/or service performance;
- Incentives (penalties) if the performance exceeds (falls below) some specified level;

---

[10] The term *asset* is often used in the lease literature instead of engineered object and can include real estate such as houses, apartments, buildings, and so on.

- Compensation for some or all of the consequential losses incurred by the owner due to the unavailability of the asset.

These terms have implications for the maintenance actions of the lessee and/or lessor.

18.6.1   Terms of a Lease Contract

The LC needs to take into account the interests of both the lessor and the lessee. The contract spells out the precise provisions of the agreement. Agreements may differ, but most would include the following items:[11]

- *Object:* Description, model, serial number, date of manufacture, and so on.
- *Lease term:* The start and end dates.
- *Renewal options:* If applicable.
- The specific nature of the financing agreement.
- *Lease payments:* Amount to be paid, frequency of payment (monthly, quarterly, etc.), and due date.
- *Late charges:* If lease payments are not made by the due date.
- *Security deposit:* The lessor can use this amount to repair any damage to the asset caused by the lessee. Should the lessee breach any terms of the contract, the deposit is forfeited (subject to it not violating any law of the land).
- *Delivery:* The costs of delivery – borne by one party (lessor or lessee) or shared by both.
- *Default:* This occurs when the lessee fails to meet the obligations under the contract. The contract defines the options available to the lessor and this can include repossession of the asset.
- *Possession and surrender of the object:* Obligation of the lessee to return the object in good working condition accounting for normal wear and tear.
- *Use of the asset:* Rules and regulations with which the lessee needs to conform.
- *Maintenance:* Defines the maintenance to be carried out and the party who is responsible.
- *Insurance:* Defines the party who is responsible for covering various kinds of risks (fire, theft, collision, damage, etc.).
- Schedule of the value of the asset for insurance and settlement purposes in case of damage or destruction as a function of age and/or usage.
- *Additional terms and conditions:* Relating to penalties and/or incentives based on performance of the asset.

18.6.2   Decision Problems

18.6.2.1   Lessee's Perspective

In the case of products, some of the decision problems are:

1. How to evaluate alternative LCs.
2. How to decide on the best LC.

---

[11] It can include other items (mostly legal terms), such as encumbrances, lessor representation, severability, assignment, binding effect, governing law, entire agreement, cumulative rights, waivers, indemnification, and so on.

In the case of plants and infrastructures, some of the decision problems are:

1. What should be the terms of the LC?
2. How to select the best lessor when there is more than one.

18.6.2.2   Lessor's Perspective

In the case of products, some of the decision problems are:

1. Should the number of LCs be one or more?
2. What should be the terms of each LC?
3. What are the costs of servicing different LCs?
4. What should be the pricing of different LCs?
5. How to deal with competition in the lease market.
6. How to plan the servicing logistics.

In the case of plants and infrastructures, some of the decision problems are:

1. Should one or more LCs be offered?
2. What should be the terms of the different LCs?
3. What are the costs of servicing the different LCs?
4. What should be the pricing of the different LCs?
5. How to tender for an LC.
6. How to deal with competition in the lease market.
7. What is the optimal number of lessees to have?
8. How to plan the servicing logistics.

18.6.3   Functional Products

There is a growing trend toward *functional guarantee contracts.* Here, the contract specifies a level for the output generated from equipment, for example, the amount of electricity produced by a power plant, or the total length of flights and number of landings and take-offs per year for an aircraft. The service agent has the freedom to decide on the maintenance needed (subject to operational constraints) with incentives and/or penalties if the target levels are exceeded or are not met. However, these contracts need to take into account restrictions such as usage intensity, operating conditions, and so on.

18.7   MSCs for Products and Plants

As mentioned earlier, the terms EW and MSC are used as synonyms in the context of products. Some real-world MSCs/EWs for products are given below.

18.7.1 Consumer Products

18.7.1.1 EW/MSC

Hewlett-Packard Company (commonly referred to as HP)

HP is a multinational information technology corporation in the US that provides products, technologies, software, solutions and services to consumers, small- and medium-sized businesses (SMBs), and large enterprises. The HP service contract depends on the product (server, computer, etc.) and in its most generic form contains the following 19 elements:

1. Support Services	11. Off-Site Support and Exchange Services
2. Customer	12. On-Site Support for HP Network Connectivity Products
3. Charges	13. Maximum Use Limitations
4. Eligible Products	14. Transfer of Service
5. Limitations of Liability and Remedies	15. Post Warranty Agreement Services
6. Timeliness of Action	16. Term
7. Limitations of Service	17. Termination
8. Supported Software Versions	18. Governing Laws
9. Non-HP Products	19. Entire Agreement
10. Customer Responsibilities	

Support services include the following:

- Constant monitoring and alerting on network components;
- Full remote control and diagnostics of server equipment;
- Immediate alert and response to all events;
- Remote diagnostic and repair for all incidents and failures;
- On-site hot swap exchange of failure devices;
- Remote repair and fix, even of "hung servers;"
- Full incident logging and reporting;
- 100% cover for network server, clients, and users.

An interesting feature is the guarantee on service response time. The cost of the EW depends on the level of service offered, as illustrated by the two EW options for the HP ProLiant ML 150 servers: "4 years, 4 hours, 13×5, hardware support at an additional cost of $434.00" and "4 years, 4 hours, 24×7, hardware support at an additional cost of $690.00."[12]

Chrysler Extended Warranty

Chrysler is an automobile manufacturer in the US that sells automobiles and offers the extended warranties shown in Figure 18.7 for its automobiles.

---

[12] Quoted from Chu and Chintagunta (2009).

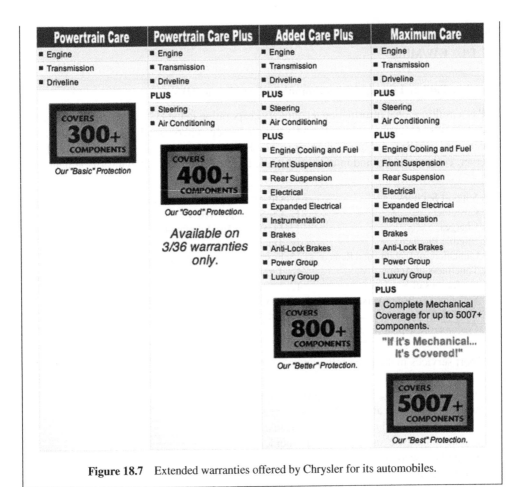

Figure 18.7   Extended warranties offered by Chrysler for its automobiles.

18.7.1.2 LC

Automobile Lease

Ford is an automobile manufacturer in the US that sells and leases its automobiles. The LC
consists of the following 35 elements.

1. Amount Due At Lease Signing or Delivery	18. Vehicle Maintenance and Operating Costs
2. Monthly Payments	19. Damage Repair
3. Other Charges	20. Vehicle Insurance
4. Total of Payments	21. Termination
5. Amounts Due At Lease Signing or Delivery:	22. Return of Vehicle
6. How the Amount Due At Lease Signing or Delivery will be Paid	23. Standards for Excess Wear and Use

7. Your Monthly Payment is Determined as
   Shown Below
8. Excess Wear and Use
9. Extra Mileage Option Credit
10. Purchase Option at End of Lease Term
11. Warranty
12. Official Fees and Taxes
13. Lessor Services
14. Late Payments
15. Life, Disability, and Other Insurance
16. Itemization of Gross Capitalized Cost
17. Vehicle Use and Subleasing

24. Odometer Statement
25. Voluntary Early Termination and Return of
    the Vehicle
26. Default
27. Loss or Destruction of Vehicle
28. Assignment and Administration
29. Taxes
30. Titling
31. Life Insurance
32. Indemnity
33. Security Deposit
34. Consumer Reports
35. General

Three elements of the contract that are important in the context of maintenance are the following:

- *Vehicle maintenance and operating costs:* Proper vehicle maintenance is Your responsibility. You must maintain and service the Vehicle at Your own expense, using materials that meet the manufacturer's specifications. This includes following the owner's manual and maintenance schedule, documenting maintenance performed, and making all needed repairs. You are also responsible for all operating costs such as gas and oil. The Lessor will provide the service(s), if any, identified in the Lessor Services section under the terms of a separate agreement. The manufacturer will invalidate warranty coverage on parts affected by a failure to maintain the Vehicle as required by the manufacturer. (See Lessor Services on the front of lease.)

- *Damage repair:* You are responsible for repairs of **All Damage** which are not a result of normal wear and use. These repairs include, but are not limited to, those necessary to return the Vehicle to its pre-accident condition, including repairs to **Exterior Sheet Metal and Plastic Components**, and to **Vehicle Safety Systems**, including air bag, seat belt, and bumper system components. Replacement of sheet metal must be made with OEM sheet metal. All other repairs must be made with OEM parts or those of equal quality. Discuss this requirement with Your insurance company prior to signing a collision repair estimate or before authorizing any collision repair work.

  If You have not had the repairs made before the Vehicle is returned at the scheduled end of this lease, You will pay the estimated costs of such repairs, even if the repairs are not made prior to Holder's sale of the Vehicle.

- *Standards for excess wear and use:* You are responsible for all repairs to the Vehicle that are not the result of normal wear and use. These repairs include, but are not limited to those necessary to repair or replace: (a) **Tires** which are unmatched, unsafe, or have less than **1/8** in. of remaining tread in any place; (b) **Electrical or Mechanical** defects or malfunctions; (c) **Glass, Paint, Body Panels, Trim, and Grill Work** that are broken, mismatched, chipped, scratched, pitted, cracked, or if applicable, dented or rusted; (d) **Interior** rips, stains, burns, or worn areas; and (e) **All Damage** which would be covered by collision or comprehensive insurance whether or not such insurance is actually in force. Replacement of sheet metal must be made with OEM sheet metal. All other repairs must be made with OEM parts or those of equal quality. Your use or repair of the Vehicle must not invalidate any warranty.

If You have not had the repairs made before the Vehicle is returned at the scheduled end of this lease, You will pay the estimated costs of such repairs, even if the repairs are not made prior to Holder's sale of the Vehicle.

18.7.2   Industrial Products

18.7.2.1   MSC

Diesel Engines

Wärtsilä is a Finnish company and a global leader in complete lifecycle power solutions for the marine and energy markets. Wärtsilä offers the following four types of service contracts for its diesel and gas engines used in power generation and marine (ships):

- *MSC-I:* Supply Agreement (Type I in the MSC classification);
- *MSC-II:* Technical Maintenance Agreement (Type II in the MSC classification);
- *MSC-III:* Maintenance Agreement (Type II in the MSC classification);
- *MSC-IV:* Asset Management Agreement (Type III in the MSC classification).

The key elements of MSC-I are as follows:[13] With supply agreement status, you get access to our global parts distribution network, and are able to order and receive spare parts 24/7, including reconditioned components, wherever your facility is located and with the shortest possible lead time. We can also guarantee the availability of a global network of trained and skilled service professionals with the right tools and onboard/on-site manpower to assist them.

- *Parts*
  - o 24/7 global logistics of spare parts;
  - o Shortening of lead time;
  - o Correct spare parts.
- *Information*
  - o Online services.
- *Manpower*
  - o Availability of a global network of trained and skilled service professionals with the right tools;
  - o Onboard/On-site manpower supply.
- *Workshop services*
  - o global component drops for reconditioning.

---

[13] Details of MSC-II, III, and IV can be found in Murthy and Jack (2014).

18.7.2.2 LC

Industrial and Commercial Equipment Lease

Wendt is a company that leases industrial and commercial equipment and the lease agreement has the following articles.

Article 1: The Parties	Article 10: Repairs and Maintenance
Article 2: The Rental Period	Article 11: Inspection
Article 3: Rent	Article 12: Insurance and Indemnification
Article 4: Overtime Rate Basis	Article 13: Title
Article 5: Terms of Payment	Article 14: Taxes
Article 6: Loading and Freight Charges	Article 15: Waivers
Article 7: Notice of Return or Recall	Article 16: Limited Liability
Article 8: Subleasing	Article 17: Indemnity
Article 9: Relocation Equipment	

Two articles of the contract that are important in the context of maintenance are the following:

- *Article 10 Repairs and Maintenance:* The Lessor is required to supply the Equipment in good operating condition. The Lessee acknowledges by signing this Lease that it has carefully examined the Equipment and accepts the Equipment as being in good operating condition. The Lessee agrees that it will pay all costs of repairs during the rental period, including labor, materials, parts and other items, except for normal wear and tear. Rent continues until the Equipment is returned to the Lessor with all necessary repairs made to the Equipment and with it in normal operating condition. "Normal wear and tear" is defined as use of the Equipment under normal work conditions, with qualified personnel providing proper operation, maintenance, and service. If repairs exceeding the normal wear and tear are necessary upon return of the Equipment, the Lessor is authorized to make such repairs and the Lessee agrees to pay the Lessor the reasonable costs of such repairs to the Equipment and rent while such repairs are being made. The Lessee agrees not to cover, alter, substitute, or remove any identifying insignia displayed on the Equipment. The Lessee will not permit the Equipment to be abused, overloaded, or used beyond its capacity. The Lessee will not alter the Equipment in any fashion and shall use and operate the Equipment in accordance with all applicable laws and the manufacturer's operating manual. The Equipment furnished is standard from the manufacturer only. Any modification or additions or optional equipment to be added to the Equipment shall be at an additional cost to the Lessee. Equipment to be used by the Lessee under normal working conditions as designed and specified by manufacturer. Unusual or abnormal working conditions, requiring work in rock, excessive mud, abrasives, and so on, or tying down, towing, demolition, adding additional, or excessive, weight will be billed to the Lessee as additional wear and tear and/or cost of repairs as provided herein.

- *Article 11: Inspection:* Before shipment is made, the Lessee may require inspection of the Equipment. If it is not in substantially the condition required by this Lease, the cost of inspection will be paid by the Lessor, and the Lessee may cancel the Lease at its option, or require the Lessor to supply Equipment in normal operating condition. The Lessor will have the right at any time to inspect Equipment and will be given free access by the Lessee to it and the necessary facilities to accomplish the inspection.

18.7.3   Plants

Plants are collections of several products. Some can be leased (from different lessors), in which case there are LCs which define the maintenance responsibilities of the lessee and/or the lessors. The maintenance of some of the products owned can be outsourced with different MSCs depending on the product.

18.8   Infrastructures

In most countries, infrastructures used to be financed by the public sector (PUS), and were constructed, maintained, and operated by agencies under the control of national, state, or local governments. Over the last few decades there has been a trend toward the involvement of the private sector (PRS) in all stages – finance (capital needed), construction, maintenance, and operation.

18.8.1   Public–Private Partnership (PPP)

In the context of infrastructures, the term *public–private partnership* (PPP) was coined to reflect the involvement of the PRS as a partner of the PUS. There are many different types of PPPs, and Hall, Motte, and Davies (2003) group them into five categories, as indicated below.

1. Outsourcing.
2. PFI (private financing initiative).
3. Concession.
4. BOT (build, operate, transfer).
5. Lease.

A PPP may be viewed as a contract, and a few of the PPPs which involve maintenance are outlined in the next few subsections.

18.8.1.1   DBFO (Design, Build, Finance, and Operate)

A contract made under the principles of the private finance initiative whereby the same supplier undertakes the design and construction of an infrastructure and thereafter maintains it for an extended period, often 25 or 30 years.

18.8.1.2   FM (Facilities Management)

This deals with the management of services relating to the operation of a commercial building. The activities involved include maintenance, security, catering, and external and internal cleaning.

18.8.1.3   O&M (Operation and Maintenance)

This involves the PRS operating a publicly-owned facility under contract with the Government.[14]

18.8.1.4   BOT (Build, Operate, Transfer)[15]

This involves a private developer financing, building, owning, and operating a facility for a specified period. At the expiration of the specified period, the facility is returned to the Government.

The maintenance depends on the terms of the PPP.

Rail System Maintenance

British Rail[16]

Prior to 1994, British Rail (BR) operated the rail system in the UK. In 1994, a new government-owned company, Railtrack, took ownership and responsibility for maintaining BR's railway infrastructure. BR's other activities were split into more than 100 companies which involved setting up "shadow" companies within BR. The ownership of railway assets was then transferred to the PRS, resulting in the structure shown in Figure 18.8.

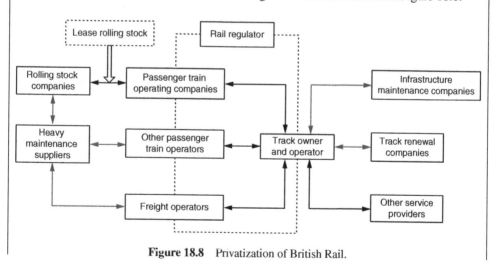

Figure 18.8   Privatization of British Rail.

---

[14] In this contract, the private sector operator assumes the risks of operating and maintaining the infrastructure, and the government retains the investment risk.

[15] This type of contract is similar to a *concession contract* or *franchise.*

[16] This section is based on material from Kain (1998) and the figure is adapted from there.

Railtrack was sold in 1996 to the PRS through flotation on the stock market. BR's infrastructure support departments were geographically and functionally divided: seven infrastructure maintenance, seven infrastructure services design, and six track renewal companies. These were then sold by tender.[17]

BR's passenger rolling stock was sold as three rolling stock leasing companies ("ROSCOs"); these companies lease vehicles to passenger and freight train operators. The ROSCOs combined to buy the company owning the vehicle spare-parts pool. Their vehicles are maintained by seven ex-BR heavy maintenance suppliers.

BR's freight train operations (including rolling stock) were split into six companies: three geographically based bulk operations, container operations, non-bulk/international freight, and a postal contractor. These were then sold by tender to the PRS.

In contrast to freight operations, passenger train operations were not sold; instead, the right to run the ex-BR passenger trains was franchised to 25 PRS train operating companies (TOCs), through the newly created (passenger) Passenger Franchising Director.

The government also set up the Office of the Rail Regulator. As a result, several different parties are now involved in the operating and maintenance of the rail system in the UK. The 25 franchises are subject to regulations overseen by the Franchising Director (of the Office of Passenger Rail Franchising – OPRAF). OPRAF's activities are centered on drawing up franchise agreements and franchise plans with TOCs, which set out TOC obligations.

A complicating factor in the maintenance of infrastructure is that it needs to take into account the interests of all the stakeholders involved.[18] The government plays a critical role in terms of providing loans to and/or acting as a guarantor for the owner, and the regulators are independent authorities responsible for ensuring public safety. The role of maintenance now becomes important in the context of safety and risk.

For PFIs, Concessions and BOT contracts, the responsibility for maintenance is with the PRS party involved. In contrast, in the case of outsourcing and leasing, it is the responsibility of the PUS parties involved. The maintenance can either be done in-house or outsourced to some third party. This results in many different scenarios for the maintenance of infrastructure. The maintenance contracts are more complex and involve performance guarantees, incentives, and penalties. An increasing issue in privatized infrastructure is the appropriate incentives needed to ensure adequate maintenance of the infrastructure as a public resource.

Swedish Rail

Some salient features of Railway Maintenance Contracts in Sweden are as follows:[19]

- Contract durations are for a period of five years and they may be extended for another period of two years, one year at a time (5 years + 1 year + 1 year).
- Tenders are open for a period of three months, to give details of the infrastructure (track areas, functional requirements, condition of the infrastructure, etc.).

---

[17] Railtrack was bought back by the UK Government in 2002 after two major rail derailments. The state-owned company now responsible for the maintenance of the railway track is called Network Rail.

[18] Depending on the infrastructure, one or more of the stakeholders might not be relevant. In some cases, two or more stakeholders might be the same – for example, the owner and the operator might be the same, or the service agent and the operator might be the same if maintenance is done in-house.

[19] Trafikverket.

- After selection, the service agent is given 9–12 months to establish his/her business to deliver the maintenance services as per the contract.
- There is a provision for an incentive for the service agent if there is a reduction in the number of faults.
- There is also provision for a penalty if the service agent fails to respond within a stipulated time period when a fault in the railway system is reported.
- Trafikverket is responsible for the supply of materials and components in the case of emergency/urgent repairs, otherwise material management is the responsibility of the service agent.

Highway Maintenance

Contract for NZTA1234 (Highway Maintenance in New Zealand)

THIS AGREEMENT is made on ("date") **BETWEEN** ("the Contractor")
AND _____.
The NZ Transport Agency, a Crown entity, established on 1 August 2008 by Section 93 of the Land Transport Management Act 2003 ("the Principal")
IT IS AGREED as follows:

1. **THE** Contractor shall carry out the obligations imposed on the Contractor by the Contract Documents.
2. **THE** Principal shall pay the Contractor the sum of $ _____ or such greater or lesser sum as shall become payable under the Contract Documents together with Goods and Services Tax at the times and in the manner provided in the Contract Documents.
3. **EACH** party shall carry out and fulfill all other obligations imposed on that party by the Contract Documents.
4. **THE** Contract Documents are this Contract Agreement and the following, which form part of this agreement:
   (a) The Conditions of Tendering;
   (b) Notices to Tenderers (give details with dates);
   (c) The Contractor's tender;
   (d) The notification of acceptance of tender;
   (e) The General Conditions of Contract, NZS 3910:2003;
   (f) The Special Conditions of Contract;
   (g) Specifications issued prior to the Date of Acceptance of Tender;
   (h) Drawings issued prior to the Date of Acceptance of Tender;
   (i) The Schedule of Prices;
   (j) The following additional documents: *(Identify any additional documents to be included, for example, agreed correspondence).*

18.9 Summary

Outsourcing is the process of acquiring goods and/or services from an external agent under a contract and is one of the strategies businesses use to enhance their competitiveness. Activities that are not considered strategic necessities and/or whenever the business does not possess the adequate competencies and skills should be outsourced. Provided there is an external agent who can carry out these activities in a more efficient manner, outsourcing can lead to improved focus on core activities, improved processes, access to better expertise, technology, financial resources, and reduced costs.

Many businesses tend not to view maintenance as a core activity and have moved toward outsourcing it through an MSC. There are several reasons for this including the need for a specialist workforce and diagnostic tools that often require constant upgrading. However, there are some disadvantages, including a dependency on the service provider and the loss of maintenance expertise.

There are different scenarios for maintenance outsourcing. In the first scenario, the service agent is only providing the resources (workforce and materials) to execute the work. This corresponds to the minimalist approach to outsourcing. In the second scenario, the service agent decides on *how* and *when*, while *what* is to be done is decided by the customer. Finally, in the third scenario, the service agent makes all three decisions.

A framework for effective decision making by both parties (customer and service agent) needs to take into account several elements, including requirements, contract payment structures, legal issues, and the interactions among them.

There are several types of MSC that depend on the type of engineered objects involved. Examples for products, industrial products, and infrastructures have been provided.

Determining the optimal decisions (for both customer and service agent) requires building models and proper management of information flows. Models that can assist the decision-making process of both the customer and service agents have been discussed.

There is also a growing trend for customers to lease systems rather than purchase them. A lease is a contractual agreement under which one party (the owner, who is also referred to as the *lessor*) leases to another party (also referred to as the *lessee*) an engineered object for use as per the terms of the LC. There are many types of LC. Maintenance may be the responsibility of either the lessor or the lessee depending on the type of LC.

The LC needs to take into account the interests of both the lessor and the lessee. The contract spells out the precise provisions of the agreement. Terms of LCs have been presented along with the related main decision problems from both the lessee's and lessor's perspectives.

Review Questions

18.1   What are the main reasons for outsourcing, in general?

18.2   What are the advantages and disadvantages of outsourcing?

18.3   What are the advantages and disadvantages of maintenance outsourcing?

18.4   What are the three levels of maintenance outsourcing?

18.5   What are the key decision problems in maintenance outsourcing from the customer's perspective?

18.6   What are the key decision problems in maintenance outsourcing from the service provider's perspective?

18.7   What is a maintenance service contract (MSC)?

18.8   What are the main elements of an MSC?

18.9   What are the different classes of MSCs?

18.10   What is leasing?

18.11   Who is responsible for maintenance in an LC?

18.12   What are the key decision problems from the perspectives of both the lessor and lessee?

18.13   What is a functional product?

Exercises

18.1   The selection of the best service agent needs to include both economic factors and non-economic factors, such as honesty, trustworthiness, efficiency, professionalism, and so on. Can you think of any other factors? Discuss each briefly and the scales for quantifying them.

18.2   List the ways in which a service agent may cheat a customer. How may the customer ensure that this does not happen?

18.3   Monitoring the actions of a service agent involves additional costs. What kind of mechanisms can be built into an LC or MSC to minimize these monitoring costs?

18.4   Compare and contrast the terms of an MSC and the LC in Section 18.7.2.

18.5   What is the difference between BOT and DBOF?

18.6   There is a lot of debate around the suggestion that PPPs have failed. Suggest one or more causes for such failures.

18.7   Information plays an important role in any game-theoretic models. One may define information as being (i) complete or incomplete; (ii) perfect or imperfect; and (iii) having different degrees of uncertainty. Discuss these terms in the context of the maintenance outsourcing of a plant.

18.8   What service-agent-related information is important for the owner of a chemical plant to help decide on the best agent to select? How can the owner obtain such information? What could be some of the problems with such information?

18.9   The relationship between a single service agent and a single customer may be characterized in terms of the dominance relationship. The figure below shows the three different types of relationship.

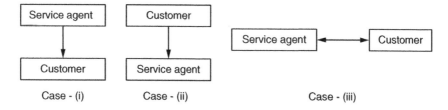

Case (i) corresponds to the scenario discussed in Section 18.5.3. It is appropriate for consumer products (for example, white goods) and commercial products (for example, an elevator in a commercial building). Cases (ii) and (iii) are more appropriate for plants and infrastructures. List the types of customer for these two cases and justify these choices.

18.10 Exercise **18.8** characterized the different dominance relationships with one customer and one service agent. What are the different dominance relationships for the following:
(a) One service agent and two customers?
(b) Two service agents and one customer?
(c) Two service agents and two customers?

18.11 Consider the case where an MSC requires a service agent to minimally repair failures and the object is not subjected to any PM actions. The service agent offers two options: (i) carry out all CM actions for a period L for a fixed price P and (ii) the customer pays for each repair and the mean cost of each repair is C_f. Carry out a system characterization to build a simple model for the owner of the object to choose between the two options.

18.12 How would you extend the model in Exercise **18.11** to assist the service agent in deciding on the pricing based on a Stackelberg game formulation?

References

Chu, J. and Chintagunta, P.K. (2009) Quantifying the economic value of warranties in the US server market, *Management Science*, **28**: 99–121.

Hall, D., Motte, R., and Davies, S. (2003) *Terminology of Private Public Partnerships (PPPs)*, Public Services International Unit, University of Greenwich, London.

Kain, P. (1998) The reform of rail transport in Great Britain, *Journal of Transport Economics and Policy*, **32**: 221–246.

Martin, H.H. (1997) Contracting out maintenance and a plan for future research, *Journal of Quality in Maintenance Engineering*, **3**: 81–90.

McFarlane, D. and Cuthbert, R. (2012) Modelling information requirements in complex engineering services, *Computers in Industry*, **63**: 349–360.

Murthy, D.N.P. and Jack, N. (2014) *Extended Warranties, Maintenance Service and Lease Contracts*, Springer-Verlag, London.

19

Maintenance Planning, Scheduling, and Control

Learning Outcomes

After reading this chapter, you should be able to:

- Identify the main maintenance-planning issues at the tactical level;
- Explain the role of maintenance load forecasting in effective tactical maintenance planning;
- Describe the two main sources of the maintenance load;
- Describe techniques used to forecast maintenance load due to CM actions;
- Define capacity planning and describe the main elements of maintenance capacity;
- Describe approaches for maintenance capacity planning;
- Identify the main maintenance-planning issues at the operational level;
- Describe maintenance-scheduling techniques;
- Define maintenance control and explain its role in the process of continuous improvement;
- Describe the main elements of maintenance control.

Introduction to Maintenance Engineering: Modeling, Optimization, and Management, First Edition.
Mohammed Ben-Daya, Uday Kumar, and D.N. Prabhakar Murthy.
© 2016 John Wiley & Sons, Ltd. Published 2016 by John Wiley & Sons, Ltd.

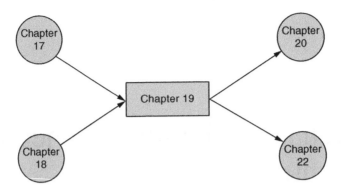

19.1 Introduction

Maintenance planning is a key element of maintenance management and needs to be done at three levels: strategic, tactical, and operational. At each level there are several issues that need to be addressed and effective decision making requires a proper framework that captures the main issues and decisions at that level. The strategic-level issues and their decision making are discussed in Chapter 17. At the tactical level, the key issues include facility capacity planning (for carrying out maintenance actions), manpower needs, and equipment and tool requirements. At the operational level, the key issue is scheduling and this depends on whether maintenance is done on site or the failed item is brought to a workshop. Maintenance control is essential to ensure that the planned maintenance and related activities are carried out properly. This involves monitoring, proper data collection, and analysis to resolve any problems and guarantee continuous improvement. This chapter deals with these topics.

The chapter is organized as follows. Section 19.2 looks at maintenance planning. Decision problems at the tactical level are addressed in Section 19.3 and at the operational level in Section 19.4. Section 19.5 looks at maintenance control. A variety of specific management systems are needed for planning, scheduling, and control, and these are elements of the overall maintenance management system discussed in Chapter 17. Section 19.6 looks at some of these systems. Sections 19.7–19.9 look at some specific issues in the context of products, plants, and infrastructures, respectively. Finally, Section 19.10 concludes the chapter with a summary.

19.2 Maintenance Planning

The issues, time frames, and managerial staff for the three levels of maintenance planning are shown in Table 19.1. The frameworks needed to address the issues and the related decision problems are different at the tactical and operational levels.

19.2.1 Tactical-Level Framework

At the tactical level, the key issues are (i) maintenance load forecasting and (ii) maintenance capacity planning. Forecasting predicts the future demand for maintenance work considering age and planned workload, and capacity planning ensures that adequate capacity is available to meet the planned and unplanned maintenance load. These issues are influenced by planned and unplanned maintenance, as shown in Figure 19.1.

19.2.2 Operational-Level Framework

Operational-level planning deals with the day-to-day preparation and execution of maintenance work. Key issues include scheduling, work order planning, and execution. Other relevant factors at this level are shown in Figure 19.2.

Table 19.1 Three levels of maintenance planning.

Level	Issues	Time frame	Managers
Strategic	Long-term maintenance aligned to strategic goals of organization	5–50 yr	Senior
Tactical	Maintenance load forecasting, maintenance capacity planning	1–5 yr	Middle
Operational	Work order planning Scheduling Execution	1 d to 1 yr	Junior

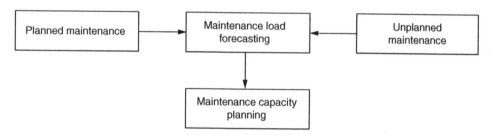

Figure 19.1 Key issues and factors in tactical-level planning.

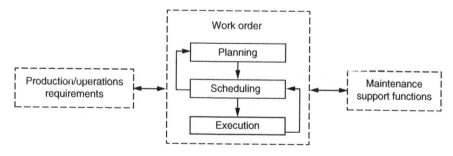

Figure 19.2 Key issues and factors in operational planning.

19.3   Tactical-Level Maintenance Planning

19.3.1   Maintenance Load Forecasting

The maintenance load denotes the volume of maintenance work anticipated over time into the future and is made up of the following two main components:

1. *Planned maintenance:* This includes all PM (preventive maintenance) work that has been planned and scheduled in advance.
2. *Unplanned maintenance:* This includes all CM (corrective maintenance) work due to unforeseen breakdowns and failures.

The load comprises the manpower, materials (including spares), and facilities (equipment and tools) needed on a periodic basis (per month, quarter, or year) for the object being maintained. For products and plants, planned maintenance is determined by the maintenance policies recommended by the OEM (original equipment manufacturer). For infrastructures, planned maintenance is decided during the design process in the building of the object, taking into account the anticipated usage and load, and needs to be revised over time based on the history of actual usage and load. The unplanned maintenance depends on the degradation and failure of the object. For products and plants, the expected number of CM actions required over a period depends on the age at the start of the period and the reliability characteristics.

19.3.1.1   Qualitative Methods for Forecasting

For a newly designed object there are often very limited data to evaluate its reliability or performance over time. In the case of products, this translates into uncertainty in the form and parameters of the ROCOF (rate of occurrence of failure).[1] In this case, qualitative (or judgmental) forecasting methods can be used. The two commonly used methods are as follows:

1. *Panel consensus:* This generates a forecast based on the average estimates of a group of experts. The idea is that a panel of people from a variety of positions is able to develop a more reliable forecast than a narrower group. Panel forecasts are developed through open meetings with free exchange of ideas from all levels of management and individuals.
2. *The Delphi method:* This is a group technique in which a panel of experts is questioned individually about their perceptions of future events. The experts do not meet as a group in order to reduce the possibility that consensus is reached because of dominant personality factors. Instead, the forecasts and accompanying arguments are summarized by an outside party and returned to the experts along with further questions. This continues until a consensus is reached.

19.3.1.2   Quantitative Methods for Forecasting

The MCF (mean cumulative function) of the object gives the expected number of CM actions as a function of the period under consideration and is given by the integration of the ROCOF

---

[1] MCF and ROCOF are discussed in Chapter 12.

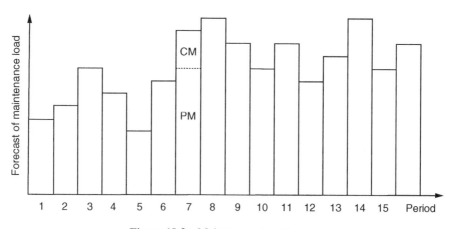

Figure 19.3   Maintenance load forecast.

over this period. One can compute the maintenance load (PM and CM) in each period, and Figure 19.3 is a typical plot of such a forecast.

For consumer products, the OEM needs to forecast the maintenance requirements during the warranty period from sales occurring over time. Time series modeling has been used for forecasting sales over different time periods.[2] The model is updated as new sales data become available. This is combined with the ROCOF to obtain the unplanned maintenance load over time.

19.3.2   Maintenance Capacity Planning

Capacity planning deals with the determination of the maintenance resources needed to meet the maintenance load on a periodic basis. The resources required can be classified into four categories: (i) human resources, (ii) spare parts and materials needed, (iii) facilities, equipment, and tools required, and (iv) information (documentation, manuals, etc.) needed to carry out the maintenance tasks. Here, we focus on human resource capacity planning, and the spare part issue is discussed in the next chapter.

Due to fluctuations in maintenance load from period to period, human resource capacity planning addresses the following issues:

- Number of maintenance workers of various trades and skills;
- Correct level of work backlog;
- Overtime capacity;
- Contract maintenance capacity.

The purpose of maintenance capacity planning is to determine how to satisfy a fluctuating maintenance load in each period. This is done by determining how much of each possible

---

[2] For more information, see Box and Jenkins (1970).

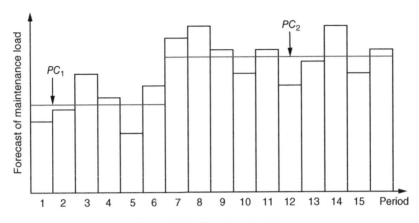

Figure 19.4 Capacity planning.

maintenance capacity (regular time, overtime, subcontracting) should be planned to meet the maintenance load. Figure 19.4 illustrates this point for the case where the demand is the human resource needed. In periods 1–6, the planned capacity is PC_1 and in the subsequent periods it is PC_2 ($> PC_1$). Note that when the demand exceeds the planned capacity, it needs to be met by either using overtime or outsourcing some of the maintenance tasks.

An important objective of capacity planning is to minimize the total cost of labor and backlog over the planning horizon. Many approaches have been proposed for determining the optimal capacity and they can be grouped broadly into (i) deterministic (when uncertainty is insignificant) and (ii) stochastic (when uncertainty is significant).

19.3.2.1 Deterministic Approaches

A common deterministic approach is to formulate the maintenance manpower capacity problem as a mathematical program[3] with:

1. *Decision variables:* The workforce size, number of workers hired (or fired), number of overtime hours, number of regular hours, number of hours subcontracted, and number of hours backlogged[4] each period.
2. *Objective function:* Minimize the total labor cost (regular, overtime, and subcontracted), the total cost of hiring and firing, and the backlog cost.
3. *Constraints:* Balance the equation for maintenance load and workforce size between adjacent periods, and limits on overtime and subcontracting that can be used.

Many different models have been proposed and we present a mixed integer programming model.

---

[3] See Appendix D for a definition and types of mathematical programs.
[4] A backlog is maintenance work that has not been carried out as planned and is deferred to a later period.

Model 19.1   The optimal capacity planning is carried out using a rolling time horizon of T periods.

Model parameters

$C_a(C_b)$	Cost of early (late) maintenance; that is, the cost of advancing (backlogging) 1 h of maintenance work by one time period
$C_r(C_o)$	Cost per hour for regular (overtime) maintenance
$C_h(C_f)$	Cost of hiring (firing) one worker
C_s	Cost per hour for subcontracted maintenance
N_{rt}	Number of regular work hours per worker in period t
N_{ot}	Number of maximum overtime work hours per worker in period t
N_{st}	Number of subcontracted work hours in period t
D_t	Demand (number of maintenance hours forecasted) in period t

Decision variables (θ)

W_t	Number of workers in period t
R_t	Number of regular hours in period t
$A_t(B_t)$	Number of advanced (backlogged) maintenance hours in period t
$O_t(S_t)$	Number of overtime (subcontracted) maintenance hours in period t
$H_t(F_t)$	Number of workers hired (fired) in period t

Objective function

The objective function is the sum of the maintenance cost incurred during regular time, overtime, and by subcontracting; the cost of advancing and backlogging work; and the cost of hiring and firing workers. It is given by:

$$\text{Minimize } J(\theta) = \sum_{t=1}^{T} \{C_a A_t + C_b B_t + C_r R_t + C_o O_t + C_s S_t + C_h H_t + C_f F_t\} \quad (19.1)$$

Constraints

1. The workforce size needs to be balanced between adjacent periods:

$$W_t = W_{t-1} + H_t - F_t, \quad t = 1, 2, \ldots, T \quad (19.2)$$

2. The maintenance workload needs to be balanced between adjacent periods:

$$A_t - B_t = A_{t-1} - B_{t-1} + R_t + O_t + S_t - D_t, \quad t = 1, 2, \ldots, T \quad (19.3)$$

3. Regular time and overtime should be related to the number of workers in each period:

$$R_t = N_{rt} W_t, \quad t = 1, 2, \ldots, T \quad (19.4)$$

$$O_t \le N_{ot} W_t, \quad t = 1, 2, \ldots, T \quad (19.5)$$

4. The number of subcontracted work hours is bounded by the available limit:

$$S_t \leq N_{st}, \quad t = 1, 2, \ldots, T \tag{19.6}$$

5. Integer constraints

$$A_t, B_t, F_t, H_t, O_t, R_t, S_t, W_t, \text{ are integers} \quad t = 1, 2, \ldots, T \tag{19.7}$$

The maintenance manpower planning problem is to obtain the optimal values of the decision variables that yield a minimum for the objective function in Equation (19.1) subject to the constraints in Equations (19.2)–(19.7).

19.3.2.2 Stochastic Approaches

The stochastic approach is seldom used in practice as it involves complex model formulations and simulations to carry out the analysis and optimization. An alternative approach used is the deterministic approach with safety factor – inflating the mean to reduce the demand exceeding the planned capacity due to uncertainties in the load demand and treating the problem as deterministic. The risk of demand not being met is reduced as the safety factor increases, but this is achieved at the expense of the capacity being underutilized when demand is below capacity.

19.4   Operational-Level Maintenance Planning

At the operational level, maintenance planning has the following main objectives:

1. Completion of maintenance work when it is needed, in a safe and efficient manner.
2. Minimization of lost production time due to maintenance.
3. Optimized utilization of maintenance labor and materials through effectively planned and balanced schedules.
4. Equitable resource allocation based on understood criteria and the varying business needs of the internal customers supported.
5. Minimization of labor delay and idle time through effective coordination with the concerned department, such as operations and stores.

It involves (i) work order planning and scheduling and (ii) maintenance scheduling.

19.4.1   Work Order Planning and Scheduling

A work order form (paper or electronic) serves as the vehicle for communicating information related to specific work requested for maintenance. The work order form must be designed to include two types of information:

- *Information needed for planning and scheduling:* This includes the requesting department, information about the item to be maintained (inventory number, location), information about the work requested (description, priority, etc.), information about resources needed (estimated time, types and trades, spare parts, tools, etc.), information about methods, safety procedures, and technical information (drawings and manuals).

- *Information needed for control:* This includes actual time taken and spares and materials used and also the causes and consequences of failures.

19.4.1.1 Planning

Work order planning is the advance preparation of maintenance work so that it can be executed in an efficient and effective manner at some future date. The maintenance planner conducts a detailed analysis of each job to determine and describe the work to be performed, the task sequence and methodology, plus the identification of required resources – including skills, crew size, man-hours, spare parts and materials, special tools and equipment.

An effective planner must have the following qualifications:

- Experience and familiarity with the engineered objects that need maintenance. This enables the planner to estimate maintenance time and other resources and select the best methods.
- Good communication skills, as this job requires coordination with other departments.
- Familiarity with planning tools and techniques and data analysis methods.

The job of a maintenance planner is greatly enhanced by the use of a computerized maintenance management system (CMMS).[5] Such a system provides timely access to available resources that need to be planned. It also assists in data collection and analysis and the generation of various reports, and is discussed in the next section.

19.4.1.2 Scheduling

Scheduling is the process by which required resources are allocated to specific jobs at a certain point in time when the engineered object is available or the job site is accessible. Effective scheduling requires coordination with production personnel.

Priorities are established in coordination with maintenance customers to ensure that the most urgent jobs are scheduled first. Most maintenance departments have three or four levels of priorities that are clearly defined, including time frames for starting the work (e.g., urgent, normal, scheduled).

Maintenance schedules are usually prepared for different time frames. A long-range schedule may cover a period between three months and one year (for example, a schedule for rail track maintenance). It is usually based on open work orders, PM work orders, and anticipated CM. The long-range schedule is usually broken down into weekly schedules that are, in turn, broken down into daily schedules. These schedules are continuously updated in light of any changes to original plans.

19.4.1.3 Execution

Good planning is a prerequisite for good execution. An effective planning function eliminates unnecessary waste from the work process, so that all materials, tools, support services, and technical information are ready for technicians to start the job without delay.

---

[5] Discussed in Chapter 22.

As mentioned earlier, the work order system plays a key role in administering, monitoring, approving, and collecting data about all maintenance jobs. In particular, data are collected about the actual time taken, the spare parts used, and the cause of the failure in case of CM actions. The approval process for execution of jobs ensures quality and identifies training needs. This information is crucial for a maintenance control system and is the cornerstone of continuous improvement.

19.4.2   Maintenance Scheduling

If planning specifies *how* to do maintenance-related jobs, then scheduling specifies *when* to do them. Maintenance scheduling deals with the decisions regarding when specific maintenance tasks are to be carried out (either at the service facility or on site). It needs to take into account various other issues – an important one being the interaction between maintenance and production/operations departments.

Maintenance scheduling is object (product/plant/infrastructure) specific and depends on whether the maintenance needs to be done on site or whether the object is being brought to a service facility. Many maintenance jobs are of short duration and can be handled by the work order system described in the previous section; other jobs, such as turnaround maintenance (TAM) for plants or major maintenance of infrastructures, are major projects that need extensive planning and scheduling, and a variety of techniques is used.

19.4.2.1   Scheduling Techniques

There are many techniques that can assist a scheduler in developing effective schedules. Some of these are graphical in nature and can be very helpful in following up the execution, especially for lengthy jobs. Other techniques are used to obtain optimal schedules in terms of cost or some other criterion, taking into account the needs of the operations department, the coordination of the maintenance of similar units, and so on. Two commonly used methods are outlined below.

Critical Path Methods
In terms of graphical methods, critical path methods (CPMs) are commonly used for large projects with complex precedence relationships between maintenance tasks, such as, TAM. CPM scheduling is a graphical technique used for illustrating activity sequences, together with each activity's expected duration, to portray project execution steps in precedence order. Several commercial software packages are available for this purpose.

Development of a CPM schedule begins by representing the project graphically by a network built up from circles (nodes) and arrows (directed arcs) which lead up to or emerge from the circles. Usually, the circles represent activities. Connecting the circles with arrows represents a sequence of activities in which each one is dependent on the previous one. In other words, the earlier activity must be completed in order to begin the next activity. Graphing out the job activities and dependencies to develop the network requires good knowledge of the constituent parts of the project.

Example 19.1   CPM Method

The following simple CPM example illustrates how the critical path is determined given a certain number of activities, their precedence relationships, and their durations. Consider the network shown in Figure 19.5, where the activities are the nodes and the duration of each activity is shown on the arc out of the node. The arc out of a node points to its successor activity. One can follow the arrows backwards to find what is required for each task and follow them forwards to see what task is next.

The critical path is found by calculating the earliest start and finish times for each node, beginning from the start point and moving forward to the end node. This is called the *forward pass*. The results are indicated in the upper part of the table above each node. The *backward pass* calculates the latest start and finish times for each activity. The results are indicated in the lower part of the table above each node, as shown in Figure 19.5.

The critical path is then identified from the difference between the earliest start times and the latest finish times. These differences are called the *slack times*. The critical path is the path where the earliest start and the latest finish time are the same and therefore there is no slack in these activities – a delay in these activities leads to a delay in the entire project. Activities that have slack time may be delayed without causing a delay in the entire project. Such activities are not on the critical path.

The critical path for this example is then N_1-N_2-N_5-N_6-N_7.

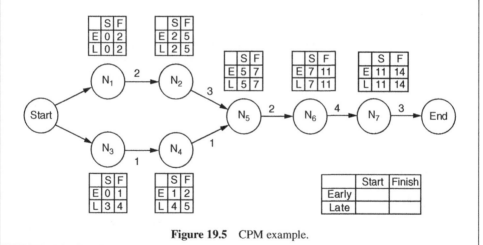

Figure 19.5   CPM example.

Mathematical Programming Techniques

Scheduling of maintenance work for a fleet of objects (buses, airplanes, locomotives, etc.), or a large number of interdependent plants, leads to complex problems with many constraints arising from the need to coordinate maintenance timing with the operational requirements of the engineered objects. Finding optimal maintenance schedules that minimize the overall maintenance cost subject to various constraints may be formulated as a mathematical program problem.

19.5 Maintenance Control

The execution of maintenance plans needs to be monitored to ensure that the actual outcomes match the expected outcomes. Monitoring requires tools and procedures and is referred to as *maintenance control*, and the process is referred to as the *maintenance control process*. It not only provides the basis for evaluation of the implementation but also valuable information for continuous improvement. The overall maintenance control process is shown in Figure 19.6.

An effective maintenance control process involves the following:

- *Established objectives:* Clear objectives aligned with overall business goals need to be established to guide maintenance activities and the maintenance control system.
- *Effective performance measures:* Maintenance objectives should be translated into effective measures of performance covering all aspects of maintenance. Examples were provided in Section 17.2.5. Targets for these performance measures should be established based on set objectives in order to be able to assess results and identify any deviations from established targets.
- *Data collection and analysis:* Measuring performance requires the collection of comprehensive and accurate data. The work order form discussed in Section 19.4.1 is the main vehicle for data collection. Data are then turned into information and analyzed to evaluate the actual outcomes. Timely availability of feedback from this analysis for effective decision making is enhanced with an effective CMMS.
- *Continuous improvement:* This is the most important step of the maintenance control process. Through a regular management review process, results are compared with objectives, performance gaps, and opportunities for improvements are identified and plans for corrective action are then implemented. Therefore, the maintenance control system should focus not only on monitoring and measurement but on action-centered management where timely decisions are made based on facts. This makes continuous improvement part of the organizational culture, leading to sustainable improvement in quality, cost, delivery, and safety.

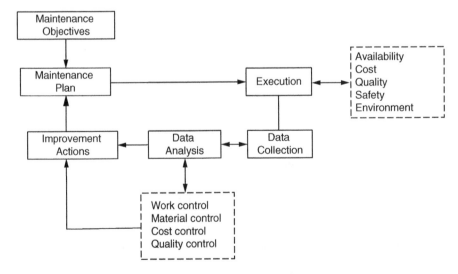

Figure 19.6   Maintenance control process.

19.5.1   Elements of Maintenance Control

The three elements of maintenance control are: (i) work control, (ii) quality control, and (iii) cost control.

19.5.1.1   Work Control

Work control deals with monitoring the efficiency of performed maintenance work. This includes manpower productivity and utilization, status of planned work such as compliance with PM schedules, emergency work as a fraction of the total maintenance load, and the amount of backlog (deferred work).

Reports on manpower productivity can reveal training needs, the need to review planning and scheduling procedures, and so on. PM and CM reports can point to the effectiveness of PM programs and the need for their review.

Maintenance load is usually fluctuating over time due to many uncertainties and unforeseen events. Therefore, keeping a certain backlog within acceptable limits can be a good practice in order to smooth the maintenance load over time. Backlog reports can help in adjusting resources to best meet the maintenance load. An excessive and increasing backlog can point to the inadequacy of manpower resources. Possible corrective action may be to use overtime, subcontracting, or both, or to increase internal manpower. Too little backlog is also not healthy and may point to excess manpower, unjustified use of overtime, or subcontracting. Possible corrective action may involve reducing subcontracting or reducing the size of the internal workforce.

19.5.1.2   Quality Control

Quality control in maintenance has two main dimensions: (i) the effect of maintenance on the quality of the business output (products and/or services) and (ii) the quality of maintenance work carried out.

Engineered objects used in the production of goods that are not well maintained fail frequently and experience (i) speed losses and (ii) loss of the precision needed. The latter leads to an increase in the defective outputs produced. The process capability[6] is affected and can lead to the process going out of control. Obviously, product quality problems can be traced to other assignable causes such as operators, raw materials, and so on. However, if the causes can be traced to maintenance, then corrective action is needed (through changes to PM actions) to ensure that the process stays in control.

The quality of the maintenance is due mainly to the actions of the maintenance technicians. A poorly trained technician carrying out maintenance might need more than one attempt to fix a failure. Through proper data collection, one can identify such situations, and through enhanced training, the skills and performance of the maintenance workforce can be improved to reduce the chances of this happening.

19.5.1.3   Cost Control

Maintenance costs may be grouped into (i) direct costs and (ii) indirect costs. Direct costs include the cost of labor, spares, materials, and tools and equipment used. Indirect costs

---

[6] Process capability compares the output of an in-control process to the specification limits by using capability ratios such as $(USL - LSL)/(6\sigma)$, where USL and LSL are the upper and lower specification limits and σ is the standard deviation of the output of the process.

include production/operation losses due unplanned shutdowns, deterioration of the object at a faster rate due to lack of proper maintenance, and quality-related costs such as rectifying defective items produced in the case of a manufacturing plant. Both types of cost need to be measured in order to control maintenance costs.

Reviewing the trends in different categories of these costs may reveal areas that need to be analyzed carefully to identify those that need attention in order to control the unjustified escalation of maintenance costs.

It is important to note that the various measures monitored should not be analyzed in isolation. Sometimes, there is a cause and effect relationship between them. For example, PM compliance (work control, which is a result of how much PM work is completed as scheduled) has a direct effect on item reliability. A higher level of compliance implies greater reliability and reduced CM costs.

19.6   Maintenance Control System

A maintenance control system is part of the overall maintenance system and is made up of several subsystems including procedures for effective execution of planned and scheduled work, and work approval procedures based on clear standards to ensure quality. The key elements of the control system are the work order system and the system for data collection and reporting. The work order system is the vehicle for administering and monitoring the work and also data collection. Performance measurement and reporting are crucial for closing the loop and creating a culture of continuous improvement.

19.6.1   Work Order System

The work order system and its procedures provide a uniform means of information flow for requesting, planning, scheduling, controlling, recording, and analyzing the performance of all the work done by the maintenance department.

Some of the many purposes of the work order system include the following:

- It serves as a single common means of transmitting requests for services by the maintenance department;
- It screens all work requested to ensure that it is needed;
- It allows maintenance work to be preplanned so that the best method is used and the job is matched with correct labor skills;
- It reduces costs and delays through the effective utilization of resources (manpower, spares, and tools) by checking availability of these resources prior to scheduling work;
- It controls the work to ensure that the most important work is performed first;
- It serves as a means of collecting data that may be used for control and continuous improvement.

19.6.2   Performance Measurement and Reporting System

An important element of the maintenance control system is data collection and analysis. It is needed to estimate measures of performance and KPIs (key performance indicators) that are reported to the various layers of management on a regular basis. This information is then used by management to evaluate various aspects of the overall maintenance for the purpose of identifying problems and areas for potential improvement. Corrective action should result in plans that are implemented and monitored. A maintenance control system is only successful when it creates a culture of continuous improvement.

19.7 Maintenance of Products

For some products (such as cell phones, computers, small appliances, etc.), failed items are usually brought to a service facility for repair. For others (such as washing machines, refrigerators, air-conditioners, etc.), the maintenance actions (CM or PM) need to be carried out on site, with these being dispersed over a wide geographical area. The maintenance may be carried out by the OEM, a dealer, or a third party acting independently or on behalf of the OEM.

19.7.1 Maintenance at a Service Facility

A service facility is a workshop that is owned either by an independent service agent servicing products under a maintenance contract or by the owner for in-house maintenance. Jobs arrive at the workshop for servicing involving either PM or CM actions and depart after being serviced. The overall process is shown in Figure 19.7 and we will discuss briefly some of its elements.

- *Servicing of items:* Some products need both PM and CM (for example, cars, locomotives, aircraft) and others only CM (for example, computers and electronic appliances).

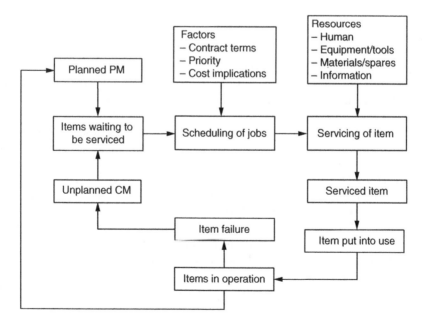

Figure 19.7 Servicing at a facility.

- *Resources:* The number of service channels (or bays) in a facility can be one or more than one. In the latter case, the number of jobs that may be serviced simultaneously equals the number of channels. Also, each channel may have one or more stages, so that the servicing is a pipeline activity with items in different stages of servicing.
- *Scheduling of jobs:* Jobs may be classified into various categories of priority depending on (i) the terms of the contract (when an item is maintained under a contract); (ii) the pricing structure of the independent service agent for items not covered by a maintenance contract; and (iii) by the production department in the case of in-house maintenance. Jobs with higher priority take precedence over jobs with lower priority.
- *Information:* Information may be grouped into two categories: (i) information (such as manuals, procedures, etc.) needed to carry out different maintenance tasks and (ii) data (such as repair times, waiting times, time in the system, etc.) collected for evaluating various measures (such as utilization of equipment and resources, etc.) for monitoring, controlling, and improving the service facility performance.

19.7.1.1   Decision Problems

A variety of decision problems arise in the context of maintenance at a service facility and include the following:

- The level of human resource utilization (expected fraction of the time a repair person is busy);
- Optimal scheduling of jobs on hand at the start of each week (or other nominated time period) based on different criteria (maximizing the number of jobs completed, minimizing the average waiting time for jobs, etc.)

Models play an important role in determining the optimal decisions. The system characterization and the type of mathematical of formulations needed depend on the problem.

Modeling of Human Resource Utilization

We will illustrate by considering an in-house workshop with J maintenance technicians responsible for maintaining K machines of a manufacturing business. We focus only on CM actions. At any given time, each technician is either busy or idle. The variable of interest is the expected fraction of time a technician is idle (over a long time interval). This is important in deciding on the number of technicians that the business should have. If it is high, the business is losing money through higher labor costs and if it is too low, then the waiting time for machines to get fixed becomes too large. In each case the business loses money.

The building of a model to make the optimal decision requires modeling the arrival and servicing of jobs. The arrival of jobs depends on how many machines are operational. Let $\tilde{K}(t)$ denote the number of machines in operation at time t, so that $K - \tilde{K}(t)$ are either getting serviced or waiting to be serviced. We assume that the repairs are minimal. The arrival of jobs occurs according to a point process and the arrival rate is modeled through an intensity function given by:

$$\psi(t) = \sum_{k \in \tilde{K}(t)} h_k\big(A_k(t)\big) \tag{19.8}$$

$A_k(t)$ is the age of machine k and the hazard function of the machine is $h_k(t)$.

The jobs are processed on a first-come-first-served basis and the time to repair machine k is a random variable with a distribution function $G_k(\cdot), 1 \leq k \leq K$. The analysis of the flow of jobs may be performed using techniques from queuing theory. At time t if $K - \tilde{K}(t) < J$, then $J - (K - \tilde{K}(t))$ technicians are idle. As a result, the number of technicians idle may be modeled as a discrete-state, continuous-time stochastic process $\{Z(t), t \geq 0\}$, with $Z(t) = j$ implying j technicians are idle. Through an analysis of the process, one can obtain the fraction of time j technicians $(0 \leq j \leq J)$ are idle.

In general, it is not possible to obtain analytical expressions for the variables of interest. However, in some special cases (such as all machines are identical with a hazard function that is not changing with time and the repair times are exponentially distributed) it is possible to derive analytical expressions for these variables.

19.7.2 Maintenance on Site

Maintenance on site is an issue for both in-house maintenance (for example, a traffic department maintaining the traffic lights in a city) and outsourced maintenance (for example, a service agent maintaining a number of elevators across a city). In both cases the maintenance (CM and PM) of objects needs to be done on site. Job requests arrive at a call center and then a repair crew, from one of the bases, is dispatched to carry out the service on site. The overall process is shown in Figure 19.8, and many of the elements are the same as for the in-house maintenance discussed earlier. We will discuss the elements that are different from before.

- *Call center:* The operating times of the call center may be 24/7 (24 hours a day and seven days a week), 12/5 (8:00 a.m.–8:00 p.m. during weekdays only), or even less depending on the type of product and the terms of the maintenance service contract.
- *Travel to site:* This requires vans to carry a range of tools, spares, manuals, and so on.
- *Dispatch rule:* This depends on the penalty terms of the contract for the failed items. Items covered by high penalties get precedence over items with low or no penalties.

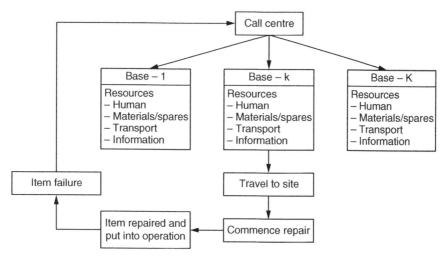

Figure 19.8 Servicing on site.

19.7.2.1 Decision Problems

A variety of decision problems arise in the context of maintenance on site, and these include the following:[7]

- Dispatching schedule for different bases at the start of each day to maximize the number of jobs serviced per shift;
- Traveling repairperson problem (sequencing of jobs to minimize the travel time between jobs);
- Knapsack problem (deciding on the spares of different components to carry on board to minimize the need for a second visit).

19.8   Maintenance of Plants

The maintenance of plants such as petrochemical refineries and power plants is dominated by TAM events. These events are large projects that last for several weeks and need to be planned carefully to control their duration, as each day of outage results in millions of dollars of lost production.

19.8.1   Turnaround Maintenance

TAM is periodic maintenance for process plants during which plants are shut down to carry out inspections, repairs, replacements, and overhauls. All major process industries (refineries, power plants, steel plants, petrochemical plants, etc.) schedule TAM regularly every two to four years to increase plant reliability, safety, and reduce the risk of unscheduled costly outages or catastrophic failures. TAM activities are very expensive in terms of direct costs and lost production, so they need to be planned and executed carefully. Maintenance planning and scheduling play a key role in managing these complex events.

19.8.1.1   TAM Initiation

TAM is usually initiated long enough before the actual commencement of maintenance (up to one year) to allow for detailed planning and preparation of all aspects of the project including work scope, job packages, procurement of materials and items, contracting, site logistics, safety and quality programs, communication and work control processes, plant start-up, and the TAM closing process.

19.8.1.2   Work Scope Determination

The work scope is the list of activities that need to be carried out during TAM. This forms the basis for all other aspects of the event. The work scope is based on input from the maintenance, production, engineering, and safety departments. It comprises the following categories of maintenance tasks and projects:

- Major maintenance tasks such as the overhaul of a large turbine in a power plant;
- Bulk work such as the overhaul of a large number of small items such as small pumps and valves;
- Small maintenance tasks such as the inspection and cleaning of various items.

---

[7] There is a vast literature on each of the three topics listed below, and readers should refer to books and journals on Operations Research.

19.8.1.3   Long Lead Time Resources

Many items required for TAM are engineered to order and have long lead times. For example, the lead time for a compressor rotor may be as much as 16 months. The work scope needs to be analyzed to identify such items, with special attention paid to prefabricated work, special technologies, vendors' representatives and services, and utilities.

19.8.1.4   TAM Planning

The TAM event involves many tasks executed by a large number of people working under time constraints. Proper planning must ensure that the right job is carried out at the right time by the right people. TAM planning also covers the shutdown/start-up logic, the shutdown network, the start-up network, and the critical path program, as work schedules are usually generated using project management techniques and software (see Section 19.4.2).

Another aspect of TAM that needs good planning is the use of contractors because of the size and complexity of the projects. A well-defined work scope is vital for determining the type and amount of contracting required. A good process for contract selection (discussed in Chapter 18) is essential for ensuring quality work.

19.8.1.5   TAM Organization

The most suitable personnel should be selected to plan and manage the TAM project. TAM organization should be a blend of the required knowledge and experience that includes the following:

- A dedicated TAM manager;
- Plant personnel who possess local knowledge and can supervise contract workers;
- TAM personnel who are skilled in planning, coordination, and work management;
- Technical personnel who possess engineering design and project management skills;
- Contractors who have the skills and knowledge to execute the work.

Site Logistics
Site logistics organizes the TAM operation and shows locations for the storage of material, equipment, and accommodation for contract personnel. An important element of the site logistics is the plot plan, which is like a map that provides all information about the plant site.

Quality and Safety Plans
The quality plan ensures that the TAM tasks are performed according to set standards. It also ensures that the quality of jobs is planned, executed, and controlled. The quality plan includes:

- A quality policy that guides the practices and behaviors needed for high-quality work execution;
- A quality control system to ensure the implementation of the quality policy;
- A quality plan for critical jobs that may affect the reliability and safety of the plant.

The quality plan should provide a coherent auditable quality trail from initial work request to the final acceptance of the completed work.

19.8.1.6   TAM Execution

The focus of TAM execution is on carrying out the tasks as planned through effective monitoring and control, as discussed earlier in this chapter. Execution can benefit from the availability of clear guidelines for plant shutdown, work control guidelines, procedures for handling unexpected work, and plant start-up guidelines. Communication plays a key role in reducing delays, conflicts, and accidents.

19.8.1.7   TAM Closing

The plant start-up procedure follows immediately from the completion of all TAM tasks. The whole event is reviewed to gather and document the event and the lessons learned. This is usually part of a comprehensive report about the event that can be very useful for the preparation of future TAM events.

The organization must measure TAM performance using a balanced set of KPIs that provide a comprehensive measure of performance. This is essential for continuous improvement.

19.9   Maintenance of Infrastructures

19.9.1   Railway Maintenance Planning

An ongoing challenge for railway companies is to balance the conflict between providing fast and reliable train services and allowing enough time for adequate maintenance activities. Railway infrastructure maintenance (for example, rail, ballast, sleepers, switches, and fasteners) planning is a complex problem that involves high costs and needs to account for train operations amongst other constraints. In particular, the train operations restrict the length and frequency of maintenance activities. The complexity of the problem varies depending on whether we have a single-line track or a network with duplicate tracks. The other factor is the intensity of train operations.

Railway preventive maintenance is made up of short-duration routine activities (for example, switch inspections, switch lubrication, maintenance at level crossings, rectifying track gauge, tamping, etc.) and long-duration project-type activities (for example, grinding, rail renewal, etc.). Operations Research (OR) tools may be used to assist planners in developing optimal maintenance plans.

19.9.1.1   Decision Problems

An important decision problem is to determine a schedule for the maintenance activities (routine and projects) for one link of rail track such that possession[8] and maintenance costs are minimized. The possession costs are mainly determined by the possession time. During a possession time required for maintenance, the track cannot be used for railway traffic. The focus is on medium-term planning and the problem is to determine which preventive maintenance activities will be performed in what time periods. Such a problem may be formulated using mathematical programming techniques to find the optimal PM schedule.

---

[8]A section of track blocked from train traffic to conduct maintenance activities is handed over to maintenance engineers who take "possession" of the track.

19.9.2   Road Maintenance Planning

Road maintenance and rehabilitation (M&R) involves major maintenance carried out to enhance the structural capacity of pavements. Typical rehabilitation actions include resurfacing (overlay), resurfacing with partial reconstruction (localized reconstruction), and complete reconstruction. The stringent yearly M&R budgets available to state departments of transportation usually cannot support every M&R need. Planning which road network sections to include in the yearly M&R projects list for a planning horizon of several years is a major undertaking.

Typical decision problems regarding the planning of M&R activities include:

- Estimating the current status (by surveying pavement conditions) and the prediction of future pavement condition as it deteriorates due to external factors such as usage, climate, and aging (using various models for pavement deterioration).
- Planning at the road network level: This considers the whole road network and develops:
  - An M&R budget plan.
  - A prioritization program to identify which projects should be carried out in each year of the planning horizon. The project selection is prioritized according to four elements, including average daily traffic, performance serviceability index, structure strength index, and required maintenance.
  - A schedule of maintenance work over the planning horizon.
- Planning at the project level deals with engineering concerns for the implementation for an individual project.

Effective M&R planning may significantly lower the total life cycle cost of the facility and provide a consistent level of service for the network users. Several objectives may be used in determining optimal solutions to these decision problems, including: maximizing network performance, maximizing the cost-effectiveness of maintenance activities, minimizing road user cost, minimizing the present worth of the total maintenance cost, and so on, with a certain set of constraints (for example, budget, pavement standard, manpower, equipment, safety, etc.).

19.10   Summary

Planning is a key element of maintenance management. In this chapter we focused on tactical and operational maintenance planning. At the tactical level, the key issues include maintenance load forecasting and capacity planning.

Maintenance load denotes the volume of maintenance work anticipated over time into the future and is made up of the following two main components: (i) planned maintenance and (ii) unplanned maintenance. Qualitative and quantitative forecasting techniques have been introduced.

Capacity planning deals with the resources required to meet the maintenance load including manpower, materials (including spares), and facilities (equipment and tools) needed over periodic intervals (per month, quarter, or year) for the object being maintained. Capacity planning approaches have been briefly described.

At the operational level, work order planning is advance preparation of maintenance work so that it can be executed in an efficient and effective manner at some future date. The work order plays a key role in administering, monitoring, and collecting data about maintenance jobs.

Scheduling is the process by which required resources are allocated to specific jobs at a certain point in time when the engineered object is available or the job site is accessible. Various scheduling techniques may be used for maintenance scheduling for products, plants, and infrastructures.

Proper maintenance control is essential for ensuring that the planned maintenance and related activities are carried out properly. This involves monitoring, proper data collection, and analysis to resolve any problems and ensure continuous improvement.

Key maintenance planning and scheduling issues for products, plants, and infrastructures have also been presented.

Review Questions

19.1   What are the main maintenance planning issues at the tactical level?

19.2   What is the maintenance load and what are its components?

19.3   Which parts of the maintenance load need to be forecast?

19.4   What are the main forecasting techniques used to predict maintenance load?

19.5   What is capacity planning and what are the main elements of maintenance capacity?

19.6   What are the key issues of maintenance planning at the operational level?

19.7   What is maintenance scheduling?

19.8   What are the techniques commonly used for maintenance scheduling?

19.9   What are the main elements of sound maintenance scheduling?

19.10   What is maintenance control?

19.11   What are the main elements of maintenance control?

19.12   What are the main elements of a maintenance control system?

19.13   What is the relationship between maintenance control and continuous improvement?

19.14   What are the key maintenance planning issues for products and plants?

19.15   What are the key maintenance planning issues for infrastructures?

Exercises

19.1   The model developed in Section 19.3.2 for manpower planning did not take into account the fact that jobs may require different skills and that we have a certain number of skills in the workforce. Develop a similar model taking this into consideration.

19.2   Railways are an important mode of transportation. Proper maintenance of the existing railway track with repairs and replacements being carried out on time are all important to ensure efficient operation and also the safety of passengers. The main question is when to carry out the maintenance such that the inconvenience for the train operators, the disruption to the scheduled trains, and the infrastructure possession time for maintenance are minimized with the lowest possible maintenance cost.

(a) What are the main issues involved in such a decision problem?

(b) Suggest an approach for addressing this problem.

19.3   Repeat Exercise 19.2 for road maintenance.

19.4   Maintenance of power systems is costly because it is impossible to store generated electrical energy. Moreover, the continuity of supply is very important for customers. Thus, maintenance scheduling of power-generation units has to be coordinated to ensure continuity of supply.

(a) What are the main issues involved in such a decision problem?

(b) Suggest an approach for addressing this problem.

19.5   The work order system is the backbone of maintenance planning, scheduling, and control. Design a work order form that satisfies the requirements outlined in Section 19.4.1. Note that the designed form has to be tailored to a particular application.

19.6   Consider a manufacturing plant with an in-house maintenance department. There are K machines to be maintained by J technicians. Build a model that will predict the expected time a technician is busy or idle, assuming that the time horizon is long so that one can use asymptotic results.

19.7   Identify capacity planning decision problems for aircraft maintenance and suggest approaches for addressing these problems.

19.8   Repeat Exercise 19.7 for a fleet of buses.

19.9   Maintenance control is important for the continuous improvement of maintenance activities. Selecting the appropriate maintenance measures is important in this regard. Which performance measures would you include for maintenance control in a plant such as a refinery or petrochemical plant?

19.10   Repeat Exercise 19.9 for a manufacturing plant.

19.11   Repeat Exercise 19.9 for rail maintenance.

19.12   Repeat Exercise 19.9 for building maintenance

Reference

Box, G.E.P. and Jenkins, G. (1970) *Time Series Analysis, Forecasting and Control*, Holden Day, San Francisco, CA.

20

Maintenance Logistics

Learning Outcomes

After reading this chapter, you should be able to:

- Provide a general classification of logistics;
- Define maintenance logistics;
- Describe the main elements of maintenance logistics;
- Identify the different kinds of inventories supporting maintenance activities;
- Describe the characteristics of spare parts;
- Describe methods for spare parts forecasting;
- Identify the factors influencing the demand for spare parts;
- Describe a general framework for the management of consumable spare parts and discuss its main elements;
- Develop models for determining optimal parameters for various inventory ordering policies for consumable spares;
- Describe a general framework for managing the inventory of repairable spare parts;
- Describe conceptually single-echelon and multi-echelon models for the inventory management of repairable spares.

Introduction to Maintenance Engineering: Modeling, Optimization, and Management, First Edition.
Mohammed Ben-Daya, Uday Kumar, and D.N. Prabhakar Murthy.
© 2016 John Wiley & Sons, Ltd. Published 2016 by John Wiley & Sons, Ltd.

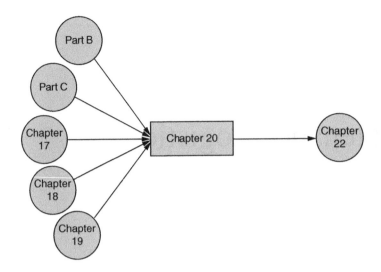

20.1   Introduction

Effective execution of maintenance requires resources, such as spare parts, repair facilities, trained personnel, and relevant information to carry out the various tasks involved. Maintenance logistics deals with these issues and is the focus of this chapter. The outline of the chapter is as follows. We start with a brief discussion of logistics in general in Section 20.2, and this sets the scene for the later sections of the chapter. Section 20.3 outlines the framework which links the different issues in maintenance logistics. Carrying out maintenance activities requires service facilities, human resources, and inventories of material and spare parts. These three issues are discussed in Sections 20.4–20.6, respectively. Inventory management of consumable parts is discussed in Section 20.7, and that of repairable parts in Section 20.8. Key issues related to maintenance logistics for products, plants, and infrastructures are presented in Sections 20.9–20.11, respectively. Finally, we conclude with a brief summary in Section 20.12.

20.2   Logistics

Definition 20.1

Logistics is the integrated design, management, and operation of human, physical, financial, and information resources, during product, system, or service life time. (The Society of Logistic Engineers, SOLE)

20.2.1   Classification of Logistics

Logistics may be divided broadly into the following three categories:

- *Supply chain logistics:* Supply chain logistics deals with the delivery of inputs from suppliers to the manufacturing plant and the delivery of finished goods to various demand centers. It deals with raw materials and components on the input side and finished products on the output side.

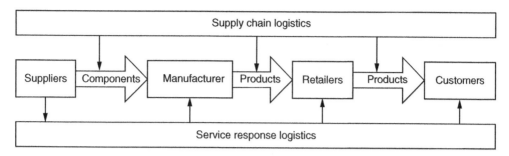

Figure 20.1   Supply chain logistics and service response logistics.

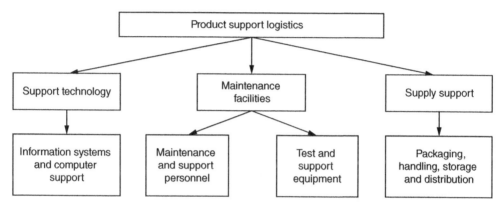

Figure 20.2   Elements of product support logistics.

- *Service response logistics:* Service response logistics is the process of coordinating non-material activities necessary for the fulfillment of the service in an effective way. Service response logistics has a different focus from supply chain logistics, in that supply chain logistics focuses on physical supply and distribution of products, whilst service response logistics emphasizes building responsive organizations, which can respond to customer requests. This difference in emphasis is illustrated in Figure 20.1.
- *Product support logistics:* Product support logistics deals with the provisioning, procurement, materials handling, transportation and distribution, and warehousing of the items and the support infrastructure needed for carrying out these activities over the life of the product. Figure 20.2 shows the main elements of product support logistics.

20.2.2   Logistics Management

Logistics management deals with decision making and this is done at three different levels. In the context of manufacturing logistics, the three levels are as follows:

- *The strategic level* deals with decisions that have a long-lasting effect on the firm. This includes decisions regarding the number, location, and capacities of warehouses and manufacturing plants.

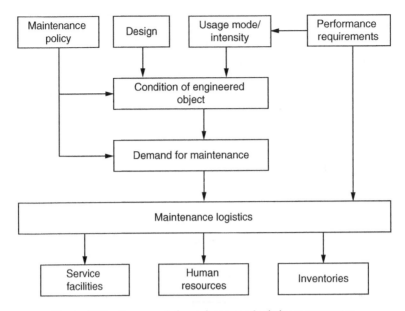

Figure 20.3 Framework for maintenance logistics management.

- *The tactical level* typically includes decisions that are updated anywhere between once every quarter and once every year. This embraces purchasing decisions, inventory policies, and transportation strategies.
- *The operational level* refers to day-to-day decisions such as scheduling, routing trucks, and measuring performance.

When some or all of the maintenance activities are outsourced, the responsibility for logistics depends on the maintenance contract.

20.3   Key Elements of Maintenance Logistics

Maintenance logistics overlaps with product support logistics when one is dealing with products. Since our focus is not only on products, but also on other engineered objects such as plants and infrastructures, maintenance logistics needs to be looked at from the provider perspective – this may be the maintenance department (for in-house maintenance) and/or the external service provider (for outsourced maintenance). The key elements of maintenance logistics for engineered objects are shown in Figure 20.3. As can be seen, several elements are involved. We discuss each of the three elements in the bottom row of Figure 20.3 in the next three sections. The remaining elements have been discussed in earlier chapters.[1]

---

[1] Design and usage mode/intensity in Chapter 2; maintenance policies in Chapter 4; condition of an engineered object in Chapter 16; and performance requirements in Chapter 17.

20.4   Service Facilities

Carrying out maintenance activities requires several service facilities including workshops to repair failed items, warehouses and other storage facilities to store materials and spare parts, and so on. Having the tools and equipment to carry out maintenance is an important issue that needs to be addressed.

20.4.1   Location

Maintenance service facilities may be located in one place, as in the case of plants, or may be distributed over a wide geographical area to be close to customers in the case of consumer products. In the case of infrastructures, these facilities have to be distributed due to the nature of the engineered object itself. The facilities may be owned by a single or several different service providers. In some cases, such as air forces or airlines, these facilities may have a multi-echelon structure where different types of maintenance are carried out at different levels.

20.4.2   Tools and Equipment for Maintenance

Different types of engineered objects require different types of tools and equipment. These also may be centralized, mobile, or distributed depending on the object. Example 20.1 provides examples of tools and equipment required by products, plants, and infrastructures.

Example 20.1   Tools and Equipment Needed by Products, Plants, and Infrastructures

Table 20.1 shows a variety of tools and equipment needed to maintain a number of engineered objects.

Table 20.1   Tools and equipment needed to maintain various engineered objects.

Engineered object		Tools and equipment
Products	Computers	Diagnostics software, soldering tools, tool kits, power supply testers, memory testing machines
	Cars	Diagnostic code reader, jacks, wrenches, air compressor
	Appliances	Refrigerant-charging systems, electronic leak detectors
Plants	Plants	Cranes, machine tools, bearing puller, welding equipment
	Fleet of buses	Mobile truck repair, jacks, machine tools
	Rolling stock	Cranes, machine tools, wheel test equipment, wheel mounting and dismounting machines
Infrastructures	Road	Asphalt recycler, pothole patcher, compacters, paving machine, wheel loaders, excavators, motor graders
	Rail	Ballast-regulator machine, rail-lubrication machines, track-inspection machines, rail and sleeper drilling machines, rail-grinding machines
	Pipeline	Automated welding machines, programmable welding arc controller

20.4.3   Management of Facilities

At the strategic level, decisions have to be made on the number and location of facilities such as workshops, storage facilities, and major equipment. This will depend on the amount and nature of the maintenance load. The estimation of maintenance load has been discussed in Chapter 19.

At the operational level, the effective allocation of these facilities to various maintenance activities during maintenance planning and scheduling (see Chapter 19) is vital for sound maintenance management.

20.5   Human Resources

The maintenance of most objects (products, plants, or infrastructures) is labor intensive. Having the right mix of skills and the right workforce size are key to ensuring effective maintenance. Maintenance personnel may be specialized in certain areas (for example, mechanic, electrician, welder, etc.) or multi-skilled to deal with a particular item (car, aircraft, ship, air conditioner, turbine, rail, etc.).

Key issues in terms of maintenance human resources include having the right mix of skills supported by adequate training programs in order to provide the required level of service. The other important decision involves having the right number of craft, as discussed in Chapter 19.

20.6   Inventories

Maintenance of an object requires various kinds of physical goods and they can be grouped broadly into two categories: (i) *consumables* – such as oil and grease in plants, paint in infrastructures, and so on, and (ii) *spare parts* – items (from components to objects and anything in between) that may be bought new from external suppliers or repaired/reconditioned either in-house or by an external agent.

Inventory management (of materials and spares) is important, as holding inventories implies capital being tied up. Not having enough inventories of spare parts and materials may affect the functioning/operation of the object, with serious consequences in terms of availability and cost. In this section we focus on spare parts from the point of view of the maintenance service provider.

20.6.1   Characterization of Spare Parts

Spare parts and maintenance are a significant part of most industrial world economies, as illustrated by the statement given below.

Spare parts and services account for 8% of the annual gross domestic product in the United States. Consumers and businesses spend more than $700 billion each year on spare parts and services for previously purchased assets, such as automobiles, aircraft, and industrial machinery. On a global basis, the annual spending on such aftermarket parts and services totals more than $1.5 trillion.

There are many different ways of characterizing the spare parts used in maintenance, and these include the following.

20.6.1.1   Repairable versus Non-Repairable Items

An engineered object can be viewed as a multi-level system (see Chapter 2). The number of appropriate levels[2] depends on the object under consideration. Deciding whether an item is repairable is not a straightforward decision.

We distinguish between two types of spare parts:

1. *Repairable parts:* Parts that are repaired rather than procured; that is, parts that are technically and economically repairable. After repair, the part becomes ready for use again.
2. *Non-repairable parts or consumables:* Parts which are scrapped after replacement.

Non-Repairable Spares

The primary question that service providers encounter for spare parts planning is how to place the spare part inventories throughout their service network. Possible options include delivering parts to the field where they are required, channeling the parts through a central warehouse – a two-echelon solution, or a three-echelon solution with a central distribution center and regional warehouses close to customers. Once the distribution network is in place, the next issue is the ordering and inventory policies for the different echelons.

Repairable Spares

Given an item design and a repair network, a level of repair analysis (LORA) determines, for each component in the item, (i) whether it should be discarded or repaired upon failure and (ii) at which echelon in the repair network this should be done. The objective of the LORA is to minimize the total (variable and fixed) costs.

A typical structure of the repair network is to have a single- or multi-echelon system. The details of these types of network and their operation are discussed later in this chapter.

20.6.1.2   Other Issues

Other issues include the following:

- *Criticality:* This is based on the consequences caused by the failure of the part. The unavailability of some parts may shut down a whole unit or plant, resulting in high losses.
- *Specificity:* Some parts are custom-made whilst others are generic and common to many objects.
- *Lead time:* Many spare parts have a long lead time, especially for custom built items or repairable items that have to queue for service at a repair facility.

20.6.2   *Framework for Spare Parts Inventory Management*

Figure 20.4 is a system characterization of the real world of spare parts management which shows the key elements and the interactions among them.

---

[2] In the spare parts literature, a level is called an *indenture.*

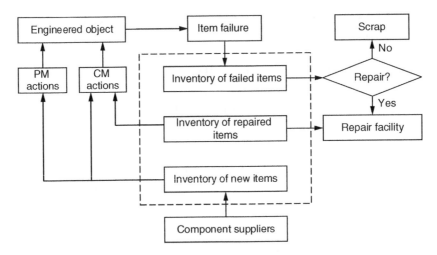

Figure 20.4   Framework for spare parts inventory management.

The key issues we will discuss include:

- Inflow and outflow of items from the inventory;
- Forecasting of the demand of spare parts;
- Inventory control to manage the flow.

20.6.3   Forecasting of Demand for Spares

There are two main approaches for forecasting the demand for spare parts. The first is the reliability-based approach and the second is the black-box approach based on historical data on spare parts consumption. In some cases, spare parts demand exhibits patterns that cannot be predicted well using traditional forecasting methods. We focus on reliability-based forecasting.[3]

Demand for non-repairable items depends on several factors, and the factors influencing the demand for spares are shown in Figure 20.5.

Replacements of components are points (some random and others non-random) along the time axis. Let $N(t)$ denote the count of failures and replacements over $[0,t)$ and this is the demand for replacement items. The demand is uncertain and can be characterized in terms of the mean and variance, as shown in Figure 20.6. The mean demand for an item is given by $E[N(t)]$ (often referred to as the *mean cumulative function* (MCF)) for the item.[4]

One can divide items into three groups, as shown in Figure 20.7: (A) fast-moving, (B) medium-moving, and (C) slow-moving items based on their reliability (low, intermediate, and high).

---

[3] There is a vast literature dealing with the black-box approach and we will not discuss this in this chapter.

[4] Expressions for the expected value $E[N(t)]$ and the variance $\text{var}[N(t)]$ may be obtained from results given in Part B.

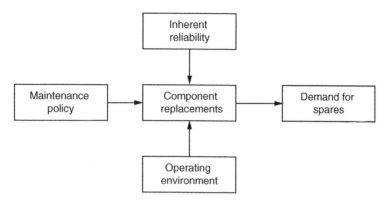

Figure 20.5 Factors influencing the demand for spares for non-repairable items.

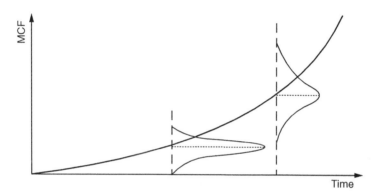

Figure 20.6 Demand for spares.

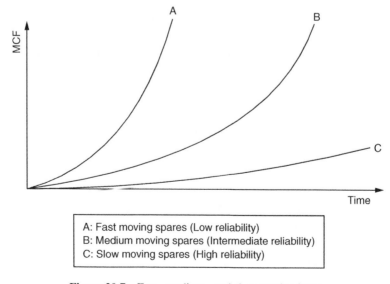

A: Fast moving spares (Low reliability)
B: Medium moving spares (Intermediate reliability)
C: Slow moving spares (High reliability)

Figure 20.7 Fast-, medium-, and slow-moving items.

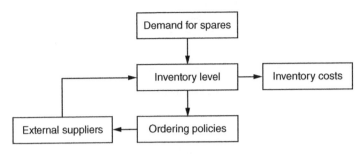

Figure 20.8   Framework for ordering policy decisions.

20.7   New Item Inventory Management

Inventory management involves the selection of suppliers (a tactical decision) and ordering policies. This section deals with ordering policies and inventory costs.[5]

20.7.1   Framework for Ordering Policy Decisions

The framework for ordering policies is given in Figure 20.8. We discuss briefly the various elements of the framework.

20.7.1.1   Inventory Level

The inventory level of new parts changes dynamically in an uncertain manner. It decreases when an item is issued for maintenance activities and increases when an order is received from suppliers.

20.7.1.2   Ordering Policies

There are three key issues related to ordering policies. For the single-item case, the first two issues are important and they deal with (i) when to order and (ii) how much to order, with one or both being the decision variables. For the multi-item case, the coordination of times to order is issue (iii), and this is commonly referred to as *joint replenishment*.

Several different inventory policies dealing with issues (i) and (ii) have been studied, and the two most commonly used are as follows:

- *Fixed ordering time policy:* The inventory is reviewed at ordering times jT, $j = 1, 2, \ldots$, and the quantity ordered may change with time.
- *Fixed ordering quantity policy:* Here, the quantity ordered each time is the same and the time between orders changes with time.

If the demand for spares is based on the MCF, then the ordering times and quantities for the two policies are as shown in Figure 20.9.

---

[5] The selection of suppliers is a multi-criteria decision problem and AHP models have been used for making the optimal decision.

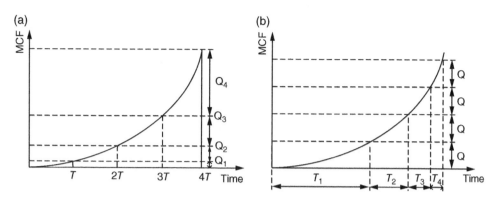

Figure 20.9   (a) Fixed ordering time policy and (b) fixed ordering quantity policy.

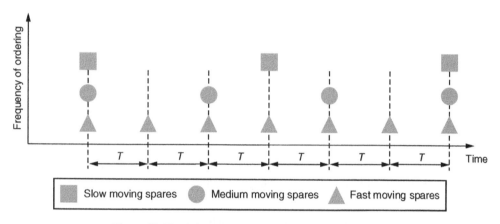

Figure 20.10   Joint replenishment for the multi-item case.

A commonly used policy that deals with issue (iii) is a joint replenishment of two or more items by synchronizing the frequency of ordering using a common base cycle, T. For example, fast-moving items are ordered every cycle, whilst medium-moving items are ordered less frequently and slow-moving items are ordered even less frequently than medium-moving items, as shown in Figure 20.10.

20.7.2   Decision Problems

The fixed ordering time [quantity] policy has a single decision variable T [Q] and the optimal selection of this requires a suitable objective function. One that is commonly used is the asymptotic expected total inventory cost per unit time (over an infinite time horizon). The total cost consists of the following elements:

- *Ordering cost:* This depends on the quantity ordered, Q + administration cost, and is given by $a + bQ$ where b is the sale price of a spare item.
- *Inventory holding cost:* This is given by hy where h is the holding cost per item per unit time and y is the duration the spare item stays in inventory.

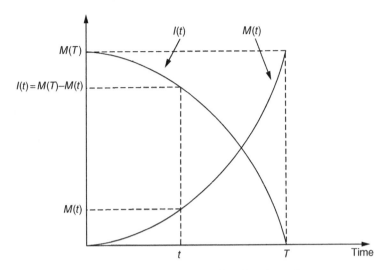

Figure 20.11   Inventory profile of cycle stock.

- *Shortage cost:* This cost may include losses resulting from the downtime due to unavailability of spares.
- *Emergency ordering cost:* This cost is incurred when the last spare is used before the next regular ordering time instant.

It is difficult to obtain analytical expressions for these cost elements as failures occur randomly and, as such, the inventory levels change in an uncertain manner over time.

20.7.3   Integrated Maintenance Inventory Model

In this section we consider a periodic inventory policy for an item maintained using the block policy discussed in Chapter 4.[6] Note that the expected number of spares needed between two preventive maintenance (PM) actions is given by $M(T)$, where $M(t)$ is the renewal function associated with $F(t)$, the failure distribution for the item. The quantity ordered is $1 + M(T)$ at time instants which correspond to PM actions, as shown in Figure 20.11 for the first cycle. Since one item is used in the PM replacement, the inventory level at the start is $M(T)$. The inventory reduces by one each time a spare is used and this occurs in an uncertain manner. However, the mean value of the inventory is given by $I(t) = M(T) - M(t)$ where $t = 0$ is the time instant of PM action. The mean inventory profile of the cycle stock is shown in the figure.

Since the inventory level changes in an uncertain manner, the total demand for spares over a cycle is a random variable which can be either less than $M(T)$ or greater than $M(T)$. In the former case, the order quantity needed is less than $M(T)$. However, in the latter case there is a shortfall.

---

[6] The block replacement policy is used for streetlights where a large number of items are replaced at the same time as part of PM action. In this case, the inventory order quantity is given by $K[1 + M(T)]$ where K is the number of streetlights that get replaced.

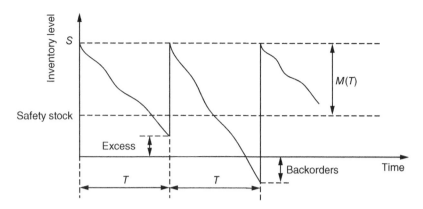

Figure 20.12 Inventory profile for the fixed-interval inventory policy.

A periodic inventory policy with maximal level S and safety stock level s is shown in Figure 20.12. Note that even in this case, the demand over a cycle can exceed S, so that the inventory is depleted and additional spares need to be ordered as emergency orders.

20.8 Repairable Items Inventory Management

20.8.1 Single-Echelon Inventory Model

The single-echelon inventory system for a repairable item is shown in Figure 20.4. When an item fails, it is replaced by a working item from the new or repaired item inventory, if available. Otherwise, the system (for example, an aircraft) waits until a working item becomes available. The failed item is removed and joins the repair queue. At a later time it is either scrapped or repaired (and then joins the inventory for repaired items).

The key issues in the single-echelon repairable item problem include:

- *The distribution of the arrival of the failed items to the repair facility:* This depends on the number of engineered objects involved (if the item is an aircraft engine then the arrival distribution depends on the number of aircraft in the fleet), their intensity of usage, the maintenance policy adopted, and so on.
- *The capacity of the repair facility:* This capacity determines the service rate of repair, which is an important parameter of the problem.
- *The appropriate measures of performance for the system:* The common measures include:
  - Average fill rate: This is the percentage of parts required for repair that are available from on-the-shelf inventory.
  - Total (system) backorders: The system-wide backorders simply represents the sum of expected backorders of all parts that are used to support the system.
  - System availability: This is a measure that is both intuitive and directly reflects the customer goal of generating value through the use of the system.
- *The optimal number of spares in the system:* This is usually the key decision variable of the problem and is determined to optimize one of the above measures of system performance.

20.8.2   Multi-Echelon Inventory Model

Consider a two-echelon inventory system for a repairable item where the system consists of a repair depot and N operating sites. Each site requires a set of working items and maintains an inventory of spare items. All failed items are repaired at the repair depot, which also maintains an inventory of spare items. We consider a one-for-one replenishment policy, which is appropriate when the item has high value and is subject to infrequent failures. When an item fails at a site, three events occur simultaneously:

1. The failed item is replaced with a spare item from the site's inventory, if one is available; otherwise, there is a shortage at the site that will last until a replacement arrives from the repair depot.
2. The failed item is sent to the depot for repair.
3. The depot ships a replacement item if it has available inventory; otherwise, the depot places the replacement request on backorder and will fill it when stock is available. When the failed item arrives at the repair depot, it enters the repair process; upon completion of the repair process, the item goes into the depot inventory or fills a backorder if any exist.

The decision problem is the quantity of spare items to be stocked and their locations.

20.8.2.1   The METRIC Model

The Multi-Echelon Technique for Recoverable Item Control (METRIC) was developed in the late 1960s for the US Air Force by the RAND Corporation.[7] The METRIC model determines, for every item of a system, the optimal stock level at each of several different bases, which may be different in terms of item demand rates and other characteristics, and the supporting depot. The objective function is the sum of backorders across all bases.

20.9   Maintenance Logistics for Products

20.9.1   Characteristics of Service Part Logistics for Products

Delivering after-sales services is more complex than manufacturing products. When delivering after-sales services, firms have to deploy parts, people, and equipment at more locations than they do to make the products. An after-sales network has to support all the products a company has made in the past as well as those it currently makes. As a result, the service network often has to cope with 20 times the number of stock-keeping units that the manufacturing function deals with. Businesses also have to train service personnel, who are dispersed all over the world, in a variety of technical skills. Moreover, after-sales networks operate in an unpredictable and inconsistent marketplace because of the unpredictable nature of the demand for product repair.

In addition, companies must design a portfolio of service products, since different customers have different service needs even though they may own the same product. Those needs

---

[7] Sherbrooke (1968).

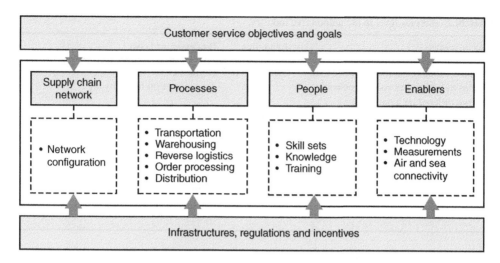

Figure 20.13   Framework for service parts logistics.

also change with time. For example, the failure of a computer in a nuclear power plant will have a more severe impact than when a computer in a library goes down. Also, a grounded aircraft means more to an air force during a war than it does during the course of a training exercise.

The management of service part logistics encompasses planning, fulfillment, and execution of service parts through activities like demand forecasting, parts distribution, warehouse management, repair of parts, and collaboration processes with all the relevant parties in the after-sales service supply chain. In the next section we present a framework that captures the main elements of service part logistics in the automotive and aerospace industries.

20.9.2   Service Part Logistics in the Automotive and Aerospace Industries[8]

A framework encompassing the main elements of service part logistics in the automotive and aerospace industries is shown in Figure 20.13. We focus on the network configuration for delivering spare parts.

20.9.2.1   Customer Service Objectives and Goals

The amount of time it takes to restore a failed item is often seen as a key performance indicator, especially in the aerospace industry, where any part unavailability translates into huge losses. Companies need to design a portfolio of service products, as each customer segment demands a different level of service.

In general, both automotive and aircraft companies offer three different levels of service, as indicated in Tables 20.2 and 20.3.

---

[8] The Logistics Institute – Asia Pacific white papers series.

Table 20.2 Service level in the automotive industry.

Type of service	Lead time	Processes and cost
Vehicle off road (customers require their vehicles to be fixed as soon as possible. Parts are ordered on a daily basis)	Within 24 h	Highest cost Transportation mode by air
Emergency (customers require a part soon but can afford a short waiting time)	2–4 d	To decrease shipment cost, parts requisitions from the same warehouse are consolidated at the regional logistics center (RLC) before being shipped Transportation by air
Regular	7 d	Lease cost Orders are consolidated and shipped out by sea

Table 20.3 Service level in the aerospace industry.

Type of service	Lead time	Processes and cost
Aircraft on ground (serious problem that prevents the aircraft from flying. Special arrangements will be made to deliver parts; often the RLC has a private fleet to cater for this)	12 h	Highest cost Companies recommend that airlines keep some spare parts on hand to minimize AOG delay. However, some airlines prefer not to invest in expensive parts
Critical (airlines need the parts soon)	2–4 d	Less costly
Ordinary (parts that are normally used for servicing)	1 wk or as promised	Least cost

20.9.2.2 Supply Chain Network

A typical after-sales supply chain network consists of four entities, namely the parts supplier or original equipment manufacturer (OEM), the regional logistics center (RLC), the importers or country warehouses, and the dealers. Three typical configurations that are commonly used are as follows:

- A *centralized configuration* where parts from suppliers will be stored in the RLC and delivered directly to the dealers whenever demand arises.

Example 20.2   Centralized Configuration

Toyota uses this centralized configuration, as all spare parts are kept at a central warehouse. As part of the company's just-in-time (JIT) philosophy, it tends to buy these parts in small lot sizes frequently from the manufacturer (as opposed to large lot sizes). There are, however, some products that move slowly and others that are fast-moving parts, such as filters, which are purchased on a daily basis.   ∎

- A *decentralized configuration* where parts from a supplier will be forwarded to the RLC first. The RLC usually breaks up the large shipments received from suppliers/OEM and then sends the smaller shipments to various warehouses in other countries in the region. Some of these warehouses are owned by the company whilst others are outsourced to third party logistics providers.

Example 20.3   Decentralized Configuration

This configuration is adopted by many companies. Volkswagen-Singapore replenishes all of its genuine parts supplies from Kassel Parts Center in Germany. Parts are delivered regularly by ocean or airfreight to the warehouse in Singapore. Orders are usually placed with suppliers once a week unless serious shortages arise. From Singapore, parts are shipped directly to countries in Asia (for example, India, Malaysia, and Thailand). Batteries are the exception, with these being sent directly to dealers from suppliers because they require special handling.   ∎

- A *hybrid configuration* that combines the above two networks.

Example 20.4   Hybrid Configuration

Mercedes Benz in Singapore uses the hybrid approach, as the RLC receives large shipments from Germany and then redistributes smaller shipments to other countries in the region. Each country has its own warehouse – some are outsourced and others are owned by Mercedes Benz. From the country warehouse, parts are usually forwarded to a network of dealers. Mercedes feels that it is more convenient for customers to have the country warehouse as an additional part of the supply chain. Having the RLC deal directly with dealers may create too much complexity in the system.   ∎

20.10   Maintenance Logistics for Plants

In Chapter 19 we discussed turnaround maintenance (TAM) for plants and the focus was on the planning and scheduling of these events. In this section we address logistics issues related to coordination with other plants and key stakeholders that are affected by the plant interruption.

The TAM event affects and is affected by many internal and external stakeholders in a wider supply chain context. In the petrochemical industry, plants feed each other; that is, the product of one plant is the raw material of another. Also, a large number of plants in a given area (for example, Jubail in Saudi Arabia) will compete for a limited number of subcontractors. As explained in the previous section, TAM also requires the ordering of many spare parts, involves long lead times for items from suppliers, and sometimes the assistance of technology providers is needed. This supply chain view of TAM is depicted in Figure 20.14. This system view requires integrated TAM planning and coordination involving all stakeholders to secure maximal utilization of resources to benefit the entire system.

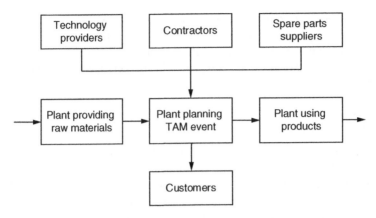

Figure 20.14   Supply chain view of TAM.

In particular, the timing of TAM for the various plants should take into consideration the interdependence between them to minimize the disturbance to the whole system. Timing coordination between plants and the sharing of experiences can benefit an entire industry if the TAM event is viewed in this wider supply chain context.

This coordination of TAM events is also an important issue in the power-generation industry. Unlike the petrochemical industry, where an inventory buffer of final products and raw materials may be built ahead of a TAM event, electricity cannot be stored. Thus, the timing of TAM is crucial to avoid an interruption to the electricity supply.

Many models have been developed in the literature to generate optimal maintenance schedules, taking into account the interdependence between plants to minimize the adverse effect of plant shutdown on all stakeholders.

20.11   Maintenance Logistics for Infrastructures

In this section we discuss some aspects of rail track maintenance logistics. Maintenance and renewal activities for rail track require several items such as rails, switches and crossings (S&C), sleepers, and ballast and also machinery such as welding machines, rail-grinding machines, and so on. Here, we focus on the logistics for rails and also provide some information on S&Cs, sleepers, and ballast.

20.11.1   Rail Logistics

Rails may be delivered directly from the plant to the renewal or maintenance site or they may be stocked for later use. The welding of rails may be done on site or in welding plants located close to rail rolling mills. The rail length may be up to 400 m and they are transported by train. This is the case for track renewal. However, for track maintenance, shorter rails are often needed and are located at many sites along the track, since defects may occur in any part of the network. In this case, the replacement short-length rails are best stocked at discrete locations such as maintenance depots. Hence, the most flexible logistical solution for the delivery of

Table 20.4 Information about rail track logistics.

Item	Lead time (mo)	Number of suppliers	Stock held % of annual use	Mode of transportation
Rail	>3	3	10	Rail
S&C	6–12	2	Not held in stock	95% rail
Sleepers	3–12	3	10–40	80% rail
Ballast	1–3	50	<10	70% rail

short-length rails (up to 27 m) is by road using flatbed trailers. Other logistics information in terms of lead time, number of suppliers, and relationships with suppliers is indicated in Table 20.4 based on a study in the European Union.[9]

20.12 Summary

Logistics deals with the resources required for effective maintenance such as spare parts, repair facilities, trained personnel, and relevant information to carry out the various tasks involved.

The key elements of the maintenance logistics for engineered objects include service facilities, human resources, and inventories of materials and spare parts.

Carrying out maintenance activities requires several service facilities, including workshops to carry out repairs, warehouses and other storage facilities to store materials and spare parts, and so on. Tools and equipment to carry out maintenance are also needed.

In most organizations, maintenance activities are labor intensive. Having the right mix of skills and the right workforce size are key to effective maintenance work. Key related issues were covered in Chapter 19.

Inventories and maintenance inventory management were the focus of this chapter. There are two kinds of inventories supporting maintenance activities: *consumables* – such as oil and grease in plants, paint in infrastructures, and so on, and *spare parts*. Spare parts are items (from components to objects and anything in between) that can be bought new from external suppliers or repaired/reconditioned either in-house or by an external agent.

The main way of characterizing the spare parts used in maintenance is whether they are repairable or non-repairable (consumables). Other characteristics include criticality, specificity, and lead time.

Effective inventory management for consumables is based on sound forecasting techniques. Although classical forecasting techniques based on time series are applicable to certain categories, we focused on reliability-based techniques, as these relate the reliability of engineered objects and their failure behavior. The MCF for various maintenance policies may assist in the prediction of future spare part requirements.

The inventory ordering policies discussed rely on item reliability and the maintenance policies adopted. Some models for determining optimal ordering policies have been presented.

---

[9] Bouch and Roberts (2010).

Inventory policies and models for repairable items have been discussed for single-echelon and multi-echelon systems. Key issues have been highlighted.

Maintenance logistics applications for products, plants, and infrastructures have been discussed.

Review Questions

20.1   What is a possible classification of logistics in general?

20.2   What is maintenance logistics?

20.3   What are the main elements of maintenance logistics?

20.4   What are the different kinds of inventories supporting maintenance activities?

20.5   What are the main characteristics of spare parts?

20.6   What are the factors that influence the demand for spare parts?

20.7   What are the different methods for spare parts forecasting?

20.8   What are the main elements of inventory management for consumable parts?

20.9   What are the benefits of relating parts demand to the reliability of the item?

20.10 What are the different types of inventory costs?

20.11 What are the different network structures for handling the inventory of repairable items?

20.12 What are the key issues involved in handling the inventory management of repairable items?

Exercises

20.1   Discuss the differences between products and service parts supply chains in terms of demand, supply chain network, and management.

20.2   Discuss the advantages and disadvantages of reliability-based and black-box-based forecasting approaches for predicting spare part needs. Identify areas where each approach may be used and discuss the data requirement for each.

20.3   Consider a replacement problem where it is possible to keep one spare item in stock in case the operating item fails. After a fixed time $T_1 \geq 0$ of operation, an identical spare is ordered. It is delivered after a constant lead time, L. From time $T_1 + L$ onwards, we have a spare unit and at some time $T_2 (T_1 + L < T_2 < \infty)$ the operating item is replaced with the spare item. If the operating item fails before time T_1, an emergency order is placed and the failed item is replaced as soon as the ordered item arrives. If the operating item fails between T_1 and $T_1 + L$, it is replaced at time $T_1 + L$ when the ordered spare arrives, otherwise it is replaced at time T_2. Assume that the inventory cost of holding a spare per unit time is h, the regular ordering and emergency ordering costs are A_r and A_e ($A_e > A_r$), respectively. There is also a penalty cost C_ℓ per unit time when the item is waiting for a spare to arrive.

Develop a model for deciding the optimal time of ordering T_1 and the optimal time of replacement T_2 that will minimize the asymptotic expected total cost per unit time.

20.4   Develop a model for the single-echelon inventory model for repairable items described in Section 20.8.1.

20.5   Service parts management for products is a challenging task. To understand the key issues relating to spares supply chain design, cooperation between different stakeholders and inventory management, you need to contact a local service provider in one of the following areas: (i) cars, (ii) washing machines, and (iii) TVs. Write a report that highlights the key issues.

20.6   How does spare parts management for plants differ from that for products?

20.7   More often than not, plants have several brands of capital equipment on their production floor performing exactly the same function, which leads to a high level of inventory in the storerooms. Standardization of equipment across plants helps in standardization of spare parts and can reduce overall inventory levels and maintenance costs for the organization.

Discuss the effect of standardization on the spare parts inventory management.

20.8   Some aspects of logistics for rail maintenance were discussed in Section 20.11. Provide a similar discussion of logistics for road maintenance.

20.9   The use of wind power as an energy source is growing worldwide. From 1995 to 2014 the installed capacity of wind turbines increased from 5000 to 370 000 MW.[10] After installing new facilities, the focus is set on securing plant availability. Maintenance and servicing become major fields of activity, which draw attention toward spare parts management.

What are the challenges of the management of spare parts in the wind energy sector? Focus on the issues related to network design, cooperation with suppliers, customers or logistics service providers, and inventory management.

20.10   Collaboration between material and parts suppliers and maintenance service providers is important in terms of reducing costs and improving service. Discuss how this collaboration may be done for:
(a) Products (take a specific example).
(b) Plants (take a specific example).
(c) Infrastructures (take a specific example).

References

Bouch, C.J. and Roberts, C. (2010) State of the art for European track maintenance and renewal logistics, *Proceedings of the MechE Part F: Journal of Rail and Rapid Transit*, **224**: 319–326.
Sherbrooke, C.C. (1968) METRIC: A multi-echelon technique for recoverable item control. *Operations Research*, **16**(1): 122–141. The Society of Logistics Engineers (SOLE) http://www.sole.org/

---

[10] Global wind energy council.

21

Maintenance Economics

Learning Outcomes

After reading this chapter, you should be able to:

- Define capital items;
- Explain the reasons for capital item investment decisions;
- Explain the role of maintenance in capital item investment decisions and the need for a life cycle perspective;
- Identify the cost elements of capital investment and the risks involved;
- Define life cycle cost and identify its elements;
- Define life cycle cost analysis (LCCA);
- Describe the steps of LCCA;
- Describe the procurement process for capital equipment;
- Explain the reasons for capital item replacement decisions;
- Apply different capital item replacement models to determine optimal replacement times;
- Explain the different approaches for making leasing versus buying decisions.

Introduction to Maintenance Engineering: Modeling, Optimization, and Management, First Edition.
Mohammed Ben-Daya, Uday Kumar, and D.N. Prabhakar Murthy.
© 2016 John Wiley & Sons, Ltd. Published 2016 by John Wiley & Sons, Ltd.

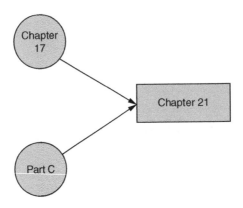

21.1   Introduction

Many objects (products, plants, or infrastructures) used for producing goods and/or services are costly (anything from a few million dollars to tens and hundreds of millions of dollars). These are called *capital items* as the acquisition and the upgrade/replacement (partial or complete) involves a lot of capital. The replacement/upgrade of capital items (for example, a fleet of expensive products such as aircraft, plants, and infrastructures) impacts significantly on the financial expenditure of a business. The expenditure (investment) on these items is termed *capital expenditure* (CAPEX) (investment) and is under the control of senior management. A proper understanding of the different issues involved is critical for proper decision making and effective life cycle management using models. Maintenance economics deals with these issues and is the focus of this chapter.

The outline of the chapter is as follows. We start with some basic concepts and terminology from economics and accounting in Section 21.2. These form the foundation for later sections of the chapter. Section 21.3 deals with capital investment and the framework needed for decision making. Section 21.4 looks at the cost elements of capital investment, and Section 21.5 deals with life cycle costing and its main elements. Section 21.6 deals with capital equipment replacement, and buy versus lease decisions are discussed in Section 21.7. Section 21.8 [Section 21.9] deals with life cycle cost analysis (LCCA) for products and plants [infrastructures]. We conclude with a summary of the chapter in Section 21.10.

21.2   Basic Concepts and Terms

21.2.1   Capital Items

Although the term capital item is widespread in theory and practice, no standard definition is available, as is illustrated by the following two definitions.

Definition 21.1

Capital items are of considerable value and durability and are used to provide a service or to make, market, keep, or transport products. (www.business dictionary.com)

> **Definition 21.2**
>
> Capital equipment present tangible and intangible goods that are procured by organizations and that present the technical prerequisites for the production of goods and services. One characteristic of capital equipment is the permanence of use with the possible inclusion of services of provision, maintenance, and repair; another characteristic is the high value of an individual object compared with the material used. (Hofmann *et al.*, 2012)

Most businesses establish criteria for designating acquired items as either capital or non-capital items. These criteria are based partly on local tax laws, but they also represent accounting policy choices by management. The criteria usually specify that capital items must have a minimum useful life (for example, one year or more), have an acquisition cost above a certain threshold (for example, $5000 or more), and contribute value to the business.

21.2.2   Economics and Accounting Terms

In this section we define various terms from economics and accounting that are needed in building models for capital investment decisions.

21.2.2.1   Time Value of Money

Engineered objects such as infrastructure and plants are items with long lives extending over many years or even decades. Financial analysis related to their acquisition, their operations, and maintenance throughout their useful life and their disposal needs to take account of the time value of money. In this section, we introduce the basics of discounted cash flow, including the concepts of interest rate, discount factor, present value, cash flow diagrams, and equivalent annual cost (EAC).

21.2.2.2   Interest Rate

Having money now is worth more than having it later. Interest is simply the cost of money. It is either the rent one pays on money borrowed from a bank, bonds issued by a corporation, or the money one receives for investing money in a bank or other financial institution. The interest rate on borrowing is higher than that on lending, the difference being the bank's margin or source of income. Interest rates are normally given on an annual basis but they may be monthly or even daily rates.

21.2.2.3   Future Value

The future value, F_n, of an amount F_0 invested at interest rate i for n years is:

$$F_n = F_0 \left(1+i\right)^n \tag{21.1}$$

21.2.2.4   Present Value

The present value, PV, of an amount F_n received in n years for an interest rate i is:

$$PV = F_n / (1+i)^n \qquad (21.2)$$

21.2.2.5   Discount Factor

The multiplier $1/(1+i)$ which occurs in calculating present values arises frequently in discounted cash flow analysis and is known as the *discount factor*.

21.2.2.6   Net Present Value

The *net present value* (NPV) of a series of amounts received or expended over a number of years is the sum of the present values of these amounts.

21.2.2.7   Annuity Factor

The annuity factor is used to convert the NPV into an equivalent annual amount and is given by:

$$A(i,d) = \frac{i(1+i)^d}{(1+i)^d - 1}$$

where d is the number of years considered.

21.2.2.8   Equivalent Annual Cost [Value]

The equivalent annual cost EAC [equivalent annual value (EAV)] is the amount of a regular annual cost [profit] which, over a given period of years, has the same NPV as any given series of costs [profits]. The EAC [EAV] converts the NPV into an equivalent annual amount by multiplying it by the annuity factor. It helps in the comparison of options, particularly where the options are dissimilar in type or duration.

21.2.2.9   Cash Flow Diagram

A cash flow diagram is a schematic representation of cash received and expended over the course of an activity.

Example 21.1   Mining

A mining company would like to buy a new load–haul–dump truck with a useful life of four years. The costs (in $1000) involved are as follows:

- Acquisition cost = $600;
- Resale value at the end of year 4 = $150;
- Maintenance costs over the four years = $70, $90, $130, $180.

The cash flow diagram is as shown in Figure 21.1.

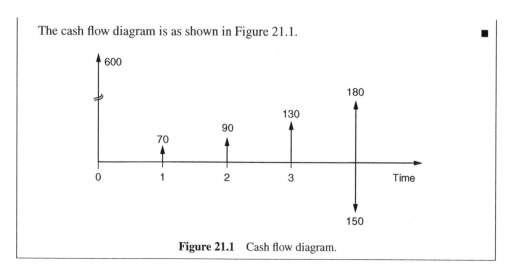

Figure 21.1   Cash flow diagram.

21.2.2.10   Inflation

Inflation means that identical goods cost more money now than they did a year ago.[1] The inflation rate, f, is not the same as the interest rate. However, the two are related, as the interest rate normally exceeds the inflation rate, otherwise there would be no attraction in saving or investing. The inflation rate usually varies from year to year.

The expression *real terms* is used when money quantities are adjusted to allow for inflation. The unadjusted figures are referred to as *nominal* dollars.

21.2.2.11   Real Rate of Interest

The "real" rate of interest, r, is the annual increase in the value of invested money in terms of its purchasing power, allowing for inflation, and is related to the bank interest rate, i, and the inflation rate, f, by the following relation:

$$1 + r = \frac{(1+i)}{(1+f)} \tag{21.3}$$

21.2.2.12   Acquisition Cost

The acquisition cost of an item is the full purchase price plus any initial costs for installation or transportation, and so on.

21.2.2.13   Effective Life

The effective or useful life is the age to which an item retains value, as specified in tax legislation as a basis for depreciation allowances. Typical effective lives (ELs) of a sample of items are shown in Table 21.1.

---

[1] *Deflation* is also possible and corresponds to a decrease in the price of goods with time.

Table 21.1 Sample of effective lives for items (in years).

Item	EL	Item	EL
Aircraft general use	20	Mining compressors	15
Cars	8	Dragline (coal mining)	20
Medical CAT scanner	7	Electricity industry storage batteries	13
Pumps	20	Cranes	20
Conveyor belts	7	Sewage treatment plant	20

21.2.2.14 Depreciation

Depreciation is an amount by which the value of a capital item is decreased to approximately reflect its decreasing value with age, for accounting purposes. Depreciation is deductible from income in determining taxable income. As an item depreciates, its reduced value is referred to as its *book value*, which may not necessarily reflect the actual value of the item.

21.2.2.15 CAPEX and OPEX

• CAPEX (*capital expenditure*) is an expense that a business incurs in acquiring new assets (tangible assets such as products, plants, and infrastructures, and intangible assets such as patent licenses) and/or upgrading its existing assets to create future benefits with a time horizon of several years. CAPEX can be financed either internally or externally. In the latter case, the investors are interested in interest payments and getting their money back in the end and/or a part ownership of the business. This expenditure is usually shown in the financial statement as cash flow or investment in an asset.

• OPEX (*operating expenditure*) is the expenditure required for the day-to-day functioning of the business and includes items such as wages, utilities, and so on, needed to produce the output (goods and/or services). It also includes depreciation of plants and machinery which are used in the production process.[2] Thus, OPEX is expenses that are necessary to maintain capital assets.

The distinction between CAPEX and OPEX has become very complicated today, especially in businesses where products and services are driven by knowledge workers.[3]

21.3 Capital Investment

CAPEX for acquisition is typically driven by operating cost control, technical obsolescence, requirements for performance and functionality improvements, and safety. Rational decision making about capital replacement must take account of engineering, economic, and safety aspects.

---

[2] Depreciation of the asset takes place every year over the effective life (as determined by the taxation office) after which it becomes zero.

[3] Business valuation often starts with the measurement of CAPEX and OPEX.

The term "upgrade" is general and many other terms have been used as synonyms. These include the following:

- *Shutdown maintenance* is a term discussed in Chapter 4.
- *Overhaul* is a term that is used in many private and public sectors (Example 21.2 is an illustrative example).
- *Turnaround maintenance* (referred to as TAM) is a term used in manufacturing and processing industries and is discussed further in Section 19.8.1.

Example 21.2   US Navy Fleet Modernization

The *Fleet Modernization Program* provides the management structure by which the characteristics of ships of the active and reserve fleets are improved. Changes to ship characteristics are accomplished by Field Changes and are developed and installed when military, survivability, or technical characteristic considerations dictate ship configuration changes.

  Ship overhauls constitute alterations at a naval shipyard or other shore-based depot to update the ship's capabilities and other large-scale maintenance that cannot be undertaken at other times.   ∎

21.3.1   Framework for Investment Decisions

A framework for capital investment decisions needs to take into account various factors and we discuss each of these briefly.

21.3.1.1   Cost of Upgrade

The cost of upgrade is comprised of several elements which can be grouped broadly into two categories which are discussed in the next two subsections: (i) direct costs and (ii) indirect costs. Some illustrative cost figures are given below.

Upgrade Costs in Two Industry Sectors

- *Transport sector:* In mass urban transportation, annual expenditure on equipment replacement for the Hong Kong underground is of the order of $50 million, and further, the Hong Kong underground network is a fraction of the size of that in London, Paris, or New York.
- *Mining sector:* A dragline undergoes shutdown maintenance every five years. The direct cost of maintenance is roughly 50% of the purchase price and the indirect costs due to the loss of revenue are around $40 million (based on a loss of $1 million per day for 40 days).

21.3.1.2 Capital

Capital is the money needed to acquire or carry out an upgrade. The alternatives available for the business are any one or a combination of the following:

- Self-financing: This reduces the capital for other projects;
- Borrowing from banks;
- Issuing new float on the stock market.

Each option has advantages as well as disadvantages and they need to be taken into account in deciding on the best option. If the business is not able to raise the capital, it might look at the following two options:

- *Leasing the object as opposed to buying:* The decision between buying and leasing is discussed in Section 21.7.
- *Private–public partnership (PPP):* Several PPPs are discussed in Chapter 18 and these are used in the context of infrastructures.

21.3.1.3 Technology

The technology for the new/replacement item may be well established or something new and innovative. In the latter case, the performance of the object is uncertain and so there are risk implications. The level of risk depends on the degree of technological innovation in the new object.

21.3.1.4 Risks

There are many different types of risk associated with capital investment, as indicated below.

- *Demand risk:* Risk associated with the demand for goods and/or services falling short of the initial expectations.
- *Financing risk:* Risk associated with variation in the financing costs from initial expectations, which includes the following:
  - ○ *Interest rate risk:* Risk that interest rates and/or their implied volatility will change;
  - ○ *Currency risk:* Risk that foreign exchange rates and/or their implied volatility will change.
- *Legal risk:* Risk that a business may incur losses due to violation of laws andregulations, breach of contract, entering into improper contracts, or other legal factors.
- *Liability risk:* Risk to a business arising from the possibility of liability for damages resulting from the new or upgraded object.
- *Operational risk:* Risk of loss resulting from performance of the new/upgraded object not performing as expected.
- *Regulatory change risk:* Risk that may result in losses due to changes in various regulations or systems, such as those related to law, taxation, and accounting.
- *Technological risk:* Risk associated with technological change that could render the new object obsolete earlier than expected or the object performance falling well below expectations.

21.3.1.5   Time Horizon

The time horizon may be (i) one period or (ii) multi-period. In the one-period case, it is the useful life of the new object – from the time it is put into operation to the time it is discarded. In the multi-period case, the time horizon includes the useful life of the new item and several subsequent replacements.

21.3.2   Objective Function

The objective function may be either a scalar or a vector and it may be based solely on economic factors or on both economic and non-economic factors.

21.4   Cost Elements of Capital Investment

There are many cost elements that need to be taken into account in making capital investment decisions. The process of upgrading/replacing an object may be viewed as a project, and a business might be looking at several projects either concurrently and/or sequentially. The various costs may be divided into two categories: (i) costs associated with the execution of the project and (ii) the *potential costs* that may be incurred once the upgraded/replaced object is put into operation. The first category costs may be further divided into *direct* and *indirect* costs, and the potential costs may be viewed as indirect costs in the context of investment decision making.

21.4.1   Direct Costs

Direct costs are the costs attributable to a specific project and are charged on an item-by-item basis. The items include:

- Labor-related costs;
- Material-related costs;
- Equipment-related costs;
- Other costs, such as insurance premiums, and so on.

21.4.2   Indirect Costs

Indirect costs are costs that are not directly attributable to a specific project and benefit more than one project. Their precise benefits to a specific project are often difficult or impossible to trace. However, they need to be distributed over several projects and over time. These costs may be categorized into two groups:

- *Fixed:* These do not vary substantially and are usually associated with the initial stage of the project (for example, building a temporary road to access a site);
- *Recurring:* These are costs incurred on a regular basis (for example, maintenance of records, payment of salaries, etc.).

Example 21.3   Rail Track Upgrade

Some of the direct and indirect costs associated with the upgrading of rail track are as follows:

- *Direct costs:*
  - *Labor:* Cost of all craftsmen (track laborers, welders, machine operators, and any helpers) and their direct supervision (the crew foremen for most organizations).
  - *Materials:* Cost of components and materials needed, including delivery to the site.
  - *Expendables:* Cost of fuel.
  - *Premiums:* Extra payments for working in tunnels or other constricted areas, at night, and so on.
- *Indirect costs:*
  - *Labor:* Costs of preparing crews, mid-level supervision, and so on.
  - *Materials:* Storage costs (material stock-pile efforts, including purchasing activities, inventory, etc.).
  - *Equipment:* Procurement and associated maintenance costs.
  - *Other:* Travel from a staging area to a site; delays (for example, due to limited track access or material not being available); training; and organization overheads.   ■

21.4.3   Potential Costs

There are several different potential costs depending on the object, and a small illustrative list is given below.

- The new object not performing as expected – for example, OPEX being significantly higher than planned.
- Quality of output (goods and/or services) falling short of expected targets and this, in turn, leading to customer dissatisfaction and possibly losing some customers.
- Unforeseen catastrophic damage to property, humans, and environment (for example, water or air contamination).
- Increased insurance costs due to unanticipated risks.

21.5   Life Cycle Cost

There are many different notions of life cycle, as discussed in Chapter 5, and one of them is the useful life, with the object being discarded (and in some cases replaced) at the end of its useful life. *Life cycle cost* (LCC) refers to the total costs associated with an object (product, plant, or infrastructure) over its useful life. It is comprised of several cost elements, shown in Figure 21.2, some of which were discussed in earlier chapters of Part D of the book.

LCC is a concept originally developed by the U.S. Department of Defense (DoD) in the early 1960s, with the main aim of increasing the effectiveness of government procurement. Two related purposes were (i) to encourage a longer planning horizon that would include operating and support costs and (ii) to achieve a reduction in total costs by spending more on design and development.

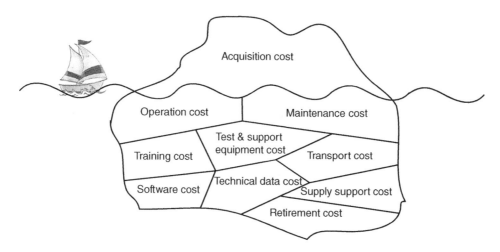

Figure 21.2 Different elements of LCC.

The LCC from the manufacturer's perspective is different from that of the customer. They differ in the cost elements (shown in Figure 21.2) considered in the LCC analysis discussed later in the section.

As can be seen from Figure 21.2, acquisition cost is only the tip of the iceberg if one takes into account all life cost elements. Therefore, making an investment decision based only on initial cost is shortsighted, as the initial investment represents only a small percentage of LCC for many engineered objects. For example, the energy cost of a pump over its useful life is many times more than its acquisition cost.

21.5.1 Elements of Life Cycle Cost

LCCs can be divided into three broad categories and several subcategories, as indicated below.

- *Acquisition costs:*
  - Research and development costs;
  - Non-recurring investment costs;
  - Recurring investment costs.
- *Sustaining costs:*
  - Maintenance (PM and CM) costs;
  - Facility usage costs.
- *Disposal costs.*

Each of these is comprised of several different costs, as indicated in Table 21.2.

It should be noted that not all of the cost categories are relevant to every project. The selection of relevant cost elements to perform a realistic LCC comparison of project alternatives is dependent on the type of the engineered object under consideration and its intended function and intensity of usage. If costs in a particular cost category are the same for all alternatives under consideration, they can be documented as such and removed from consideration in the LCC comparison.

Table 21.2 Acquisition cost categories.

Research and development costs	Non-recurring investment costs	Recurring investment costs
Program management	Spare parts and logistics	Upgrade parts
R&D	Manufacturing and operations and maintenance	Support equipment upgrades
Engineering design	Facilities and construction	System integration and improvement
Equipment development and testing	Initial training	Utility improvement costs
Engineering data	Technical data	Green and clean costs
Maintenance (PM and CM)	Facility usage costs	Disposal costs
Labor, materials and overheads	Energy costs and facility usage costs	Permits and legal costs
Replacement and renewal	Support and supply maintenance costs	Wrecking/disposal
Replacement/renewal and transportation	Operations costs	Remediation costs
System/equipment modification	Ongoing training for maintenance and operations	Write-off/asset recovery costs
Engineering documentation	Technical data management	Green and clean costs

Depending on the type of engineered object, not all cost categories are equally important. Certain costs may have a significant influence, whilst others have only a minor influence on the total cost. For example, for a nuclear power plant, environmental protection aspects and issues of disposal costs are eminently important. In other investments, energy might be a considerable cost driver, such as for pumps or aircraft. The evaluation criteria must be crafted for each engineered object, and Sections 21.6–21.8 deal with this.

21.5.2 Life Cycle Cost Analysis

Definition 21.3

Life cycle cost analysis (LCCA)[4] is a tool that combines financial, technical, and other information covering the life cycle of an engineered object to facilitate capital investment decision making.

According to IEC60300-3-3, LCCA may be used for the following:

- The evaluation and comparison of alternative designs;
- The assessment of economic viability of projects/products;
- The identification of cost drivers and cost-effective improvements;

---

[4] The term *life cycle costing* is sometimes used instead of LCCA.

- The evaluation and comparison of alternative strategies for product use, operation, test, inspection, maintenance, and so on;
- The evaluation and comparison of different approaches for replacement, rehabilitation/life extension, or disposal of aging facilities;
- The optimal allocation of available funds to activities in a process for product development/ improvement;
- Long-term financial planning.

LCCA may be implemented by following these steps:

Step 1: Determine the item's useful life.
Step 2: Estimate relevant costs (acquisition, operation, maintenance, disposal, etc.).
Step 3: Estimate the terminal (residual) value of the item.
Step 4: Discount all costs to present value.
Step 5: Obtain the LCC by adding all cost categories.
Step 6: Repeat the above steps for all items under consideration.

Usually Step 2, where all relevant costs should be estimated, requires the most time and resources. Furthermore, effective LCCA requires skill in many areas including engineering, finance and accounting, logistics, statistical analysis, maintenance and reliability, contracting, and so on. It should be conducted by a cross-functional team covering all areas relevant to the problem at hand. Furthermore, LCCA should also be combined with risk analysis to assess the degree of financial or safety risk associated with individual alternatives.

21.6   Capital Equipment Replacement

Capital equipment refers to costly machines used in the production of goods (such as a drag-line in an open-cut mine or machinery in manufacturing or processing plants) or services (such as a fleet of locomotives or aircraft). We first discuss the procurement process and then look at some models (based solely on economic considerations) for optimal capital equipment decision making. The models may be categorized into two groups: (i) one-period models and (ii) multi-period models.

21.6.1   Procurement Process

The procurement process involves the following steps.[5]

1. *Identification of capital investment needs:* Managers identify capital investment needs and opportunities from many sources such as changes in sources and quality of materials, production bottlenecks caused by old or obsolete equipment, new production or distribution methods, a requirement for adding new products to the product line or the need to expand capacity in existing product lines to capture market share, or a requirement to reduce cost by automating existing production processes.

---

[5] Based on Needles *et al.* (2010).

2. *Formal request for capital investment:* Such a request includes a complete description of the investment, justification, and specifications.
3. *Preliminary screening:* This ensures that only proposals that meet both company strategic goals and produce the minimum rate of return set by management are considered.
4. *Establishment of the acceptance–rejection standard:* The organization establishes an acceptance–rejection standard in order to attract and maintain funding for capital investments. For example, the standard may be expressed as a minimum rate of return, and if the number of acceptable requests for capital investments exceeds the funds available for such investments, the proposals must be ranked according to their rates of return.
5. *Evaluation of proposals:* Proposals are evaluated using key decision variables including (i) expected life, (ii) estimated cash flow, and (iii) investment cost. The commonly used method of evaluating proposed capital investments is the NPV method. In addition, management will take into consideration some qualitative factors, such as availability and training of employees, possible future technological improvements, and the investment's impact on other company operations.
6. *Capital investment decisions:* The acceptable proposals are usually ranked in order of the company acceptance–rejection criteria and the highest-ranking proposals are funded first, taking into account available funds. The final capital investment budget is then prepared by allocating funds to the selected proposals.

21.6.2 One-Period Models

In this section we consider a model for optimal useful life determination for a capital item used for only one period (one-time investment). This would be the case if the products produced with the acquired capital equipment would be discontinued after the end of the capital equipment's useful life.

21.6.2.1 Assumptions and Notation

It is assumed that the item generates income (in-payments) and incurs costs (out-payments) and therefore the objective function is to maximize NPV. The following notation will be used.

I_0	Acquisition out-payment
SP_t	Payment surplus at time t calculated from the difference between in-payments (revenue) and out-payments (OPEX)
R_k	Residual value after a useful life of k periods
i	Interest rate
r	Discount factor, that is, $r = 1/(1+i)$
NPV_k	NPV at a useful life of k years
NPV_k^j	The NPV for period j of duration k years calculated at the beginning of period j
$NPV(k)$	The total NPV for a finite number of periods of duration k years each

Model 21.1   NPV Approach

This is the simplest model for useful life determination. It relies on the calculation of the net present value introduced in Section 21.2. The net present value is calculated as follows:

$$NPV_k = -I_0 + \sum_{t=1}^{k} r^t SP_t + r^k R_k \qquad (21.4)$$

The optimal useful life is the period k which maximizes the net present value of the investment.

The determination of optimal useful life is carried out following the steps given below, and is illustrated in Example 21.4:

1. Payment surpluses SP_t of every year are calculated from revenue minus OPEX (columns 1 and 2 respectively, and placed in column 4).
2. The discount factor for year t is calculated as $r^t = (1+i)^{-t}$ and placed in column 5.
3. Payment surplus and residual value are multiplied by the discount factor to discount them to the present point in time and are placed in columns 6 and 7, respectively.
4. For every useful life k, the cumulated payment surplus is calculated up to that point in time and is placed in column 8.
5. Finally, the net present value of the capital equipment is calculated from the sum of the discounted residual value and the cumulative discounted payment surpluses of the corresponding year, minus the initial acquisition payment I_0 according to Equation (21.4) and is placed in column 9.

Example 21.4   Wind Turbine

Table 21.3 shows useful life calculations for a wind turbine ($I_0 = \$800\ 000$). All entries are expressed in $1000.

Table 21.3   Useful life calculations for a wind turbine.

t	(1) Revenue	(2) OPEX	(3) Resale value	(4) Net surplus	(5) Discount rate	(6) Discount surplus	(7) Discounted resale value	(8) Cumulative discounted surplus	(9) NPV
1	450	207	720	243	0.91	221	655	221	75.455
2	450	215	652	235	0.83	194	539	415	153.405
3	450	224	593	226	0.75	170	445	585	230.268
4	450	233	543	217	0.68	148	371	733	303.662
5	450	242	499	208	0.62	129	310	862	372.075
6	450	252	462	198	0.56	112	261	974	434.619
7	450	262	429	188	0.51	97	220	1071	490.855
8	450	272	401	178	0.47	83	187	1153	540.656

(Continued)

Table 21.3 (*Continued*)

t	(1) Revenue	(2) OPEX	(3) Resale value	(4) Net surplus	(5) Discount rate	(6) Discount surplus	(7) Discounted resale value	(8) Cumulative discounted surplus	(9) NPV
9	450	283	377	167	0.42	71	160	1224	584.113
10	450	295	357	155	0.39	60	137	1284	621.463
11	450	306	339	144	0.35	50	119	1334	653.041
12	450	319	324	131	0.32	42	103	1376	679.234
13	450	331	311	119	0.29	34	90	1410	700.463
14	450	345	300	105	0.26	28	79	1438	717.157
15	450	358	291	92	0.24	22	70	1460	729.744
16	450	373	283	77	0.22	17	62	1477	738.636
17	450	388	278	62	0.20	12	55	1489	744.230
18	450	403	274	47	0.18	8	49	1498	746.898
19	450	419	271	31	0.16	5	44	1503	746.989
20	450	436	270	14	0.15	2	40	1505	744.824

The optimal useful life is 19 years, as this yields the maximal net present value. ∎

21.6.3   Multi-Period Models

Let *N* be the number of periods (or future replacements) that defines the time horizon for optimal decision making and let's assume that this is specified and finite. One needs to look at two scenarios: (i) no change in technology occurs, so that all replacements are identical and (ii) changes in technology occur, so that an item used for replacement is an improvement over the one being replaced. The time between two replacements is a period. The length of period *j* is k_j (years) for $1 \le j \le N$ and is a decision variable to be selected optimally. We first look at the simple case where all the $k_j's$ are the same (so that $k_j = k$) if there is no change in technology.

Model 21.2   No Change in Technology

In this case, we consider the replacement with an identical object over *N* periods. Each period (the useful economic life of the object) is *k* (years) and is a decision variable to be selected optimally. Figure 21.3 shows the time history of replacements.

It is assumed that the trends in OPEX costs following each replacement will remain identical in every period. This assumption may be relaxed if the OPEX changes from period to period.

Note that $NPV_k^j = NPV_k^1$ $j = 2,3,\dots,N$, where NPV_k^1 is given by Equation (21.4). Therefore, the total net present value, *NPV(k)*, is calculated by discounting and adding all net present values of the *N* periods to time 0, namely:

$$NPV(k) = \sum_{i=1}^{N} NPV_k^1 r^{(i-1)k} = NPV_k^1 \sum_{i=1}^{N} r^{(i-1)k} = NPV_k^1 \frac{1 - r^{kN}}{1 - r^k} \qquad (21.5)$$

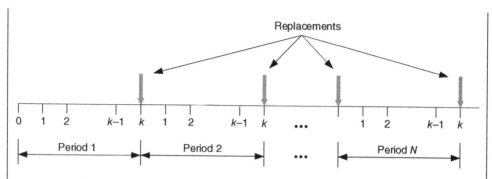

Figure 21.3 Replacement under a finite horizon with an identical item.

The last equality results from the finite geometric progression. Using the expression for NPV_k^1 (from Equation (21.4)) we have:

$$NPV(k) = \left[-I_0 + \sum_{t=1}^{k} r^t SP_t + r^k R_k \right] \frac{1 - r^{kN}}{1 - r^k} \tag{21.6}$$

The problem is to find k^* (the optimal k) which maximizes $NPV(k)$.

Example 21.5   Wind Turbine

Table 21.4 shows useful life calculations for a wind turbine ($I_0 = \$800\,000$) in the case of recurrent investment cycles over a finite time horizon of 10 periods, so that $N = 10$. We make use of the calculations of net present value for the first period from Example 21.4. All entries are expressed in \$1000.

The optimal useful life (k^*) is nine years, which yields the maximal net present value. Note that in terms of the time horizon, this corresponds to 90 years. ∎

Table 21.4   Useful life calculations for a wind turbine in the case of recurrent investment cycles over a finite time horizon of 10 periods, so that $N = 10$.

k	NPV_k^1	$NPV(k)$	k	NPV_k^1	$NPV(k)$
1	75.5	539.1	11	653.0	1005.4
2	153.4	775.3	12	679.2	996.9
3	230.3	886.1	13	700.5	986.1
4	303.7	943.5	14	717.2	973.5
5	372.1	976.3	15	729.7	959.4
6	434.6	996.1	16	738.6	944.1
7	490.9	1007.6	17	744.2	927.8
8	540.7	1013.2	18	746.9	910.7
9	584.1	1014.2	19	747.0	893.0
10	621.5	1011.4	20	744.8	874.9

Model 21.3   Changes in Technology

In this section we consider a two-period model where one can either use an item based on the current technology for replacement or one based on a new technology. The new item can be more expensive, have lower operating and maintenance costs, generate higher revenue and/or have a longer useful life. The decision choices are: (i) remain with the current technology for both periods; (ii) switch to the new technology at the end of the first period; and (iii) switch to the new technology at the beginning of the first period and use it in the second period. This leads to the three scenarios (S_1–S_3) shown in Figure 21.4. The replacement times of an item used for j periods are denoted k_o^j and k_n^j for current and new technologies, respectively. These are decision variables to be selected optimally to determine the optimal scenario that maximizes the equivalent annual value.

To build the model to obtain the optimal decisions, we need some additional notation, given below.

k_o^j	Optimal replacement of old item if used for j periods
k_n^j	Optimal replacement of technologically improved item if used for j periods
$NPV_o(j)$	Optimal net present value for the current item if used for j periods
$NPV_n(j)$	Optimal net present value for the technologically improved item if used for j periods
$A(i,d)$	Annuity factor corresponding to an interest rate i and d years
$EAV(S_j)$	Equivalent annual value corresponding to scenario S_j, $j = 1,2,3$

Note that $NPV_o(j)$ ($NPV_n(j)$) is given by Equation (21.4) for $j = 1$ and by Equation (21.6) for $j = 2$.

The equivalent annual values for these three scenarios are given by:

$$EAV(S_1) = NPV_o(2)A(i,2k_o^2) \tag{21.7}$$

$$EAV(S_2) = (NPV_o(1) + NPV_n(1))A(i,k_o^1 + k_n^1) \tag{21.8}$$

$$EAV(S_3) = NPV_n(2)A(i,2k_n^2) \tag{21.9}$$

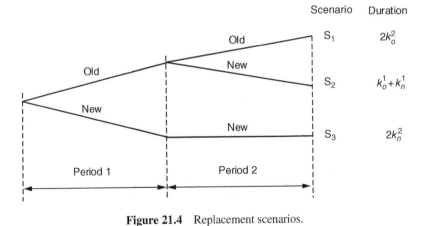

Figure 21.4   Replacement scenarios.

Example 21.6   Wind Turbine

Consider again Example 21.4 and assume that there is a technologically improved turbine that incurs the same costs and has the same acquisition cost but generates higher revenues per year ($460 000 instead of $450 000 for the old turbine). Assume that the interest rate is 10% and the useful life is 20 years for both machines.

From Example 21.4, $k_o^1 = 19$ and $NPV_o(1) = 747.0$. Using Model 21.1 for the new turbine yields $k_n^1 = 19$ and $NPV_n(1) = 830.6$. Using Model 21.2 with $N = 2$ yields $k_o^2 = 12$ and $NPV_o(2) = 906.01$ for the old turbine and $k_n^2 = 12$ and $NPV_n(2) = 999.07$ for the new turbine.

Now, using Equations (21.7)–(21.9), we have:

$$EAV(S_1) = NPV_o(2)A(i, 2k_o^2) = 906.01 \times 0.1113 = 100.84$$

$$EAV(S_2) = (NPV_o(1) + NPV_n(1))A(i, k_o^1 + k_n^1) = (747.0 + 830.6) \times 0.1027 = 90.7$$

$$EAV(S_3) = NPV_n(2)A(i, 2k_n^2) = 999.07 \times 0.1113 = 111.20$$

Thus, the optimal solution is to switch to the new turbine at the beginning of the first period and then use it for both periods.

21.7   Buy versus Lease Decisions

In this section we discuss approaches for making leasing versus buying decisions. The models are based on one of two approaches, and we discuss each of these briefly.

21.7.1   Approach 1

This approach is based solely on economic considerations. We consider a simple deterministic model based on the NPV criterion.

Model 21.4   NPV Approach

- *Notation:* We use the following notation:

N	Duration of period 1 (years)
t	Discrete time $t = 0, 1, 2, \ldots, N$ (years)
k	Marginal cost of capital for the lessee (this is the required return on new investments that will leave the market value of the lessee's equity unchanged and it depends on the financial status of the lessee to borrow capital)
r	Marginal cost of debt to the lessee (this is the rate of borrowing)
I_t	Interest payment in period t for any loan taken
A_t	Amortization of loans for purchase of asset in time period t
F_t	Maintenance and operating cost in time period t
O_t	Additional expenses that would not exist (such as administration, insurance, etc.) in the case of lease in time period t
L_t	Lease payment in time period t
I_0	Initial value of the asset before purchase or lease
R_N	Residual value of the asset at the end of the lease period

- *Assumptions:* Complete information so that the values of the variables and parameters are known to the decision maker.
- *Decision:* The customer must decide whether to buy or lease the object.
- *Objective function:* The customer makes the decision by comparing the net present value (NPV) of both options.

If the asset is purchased, the NPV of cash flows (expenses) to the customer involves the following four components:

1. *Cash expense:* $C_P^E = \sum_{t=0}^{N} \dfrac{O_t}{(1+k)^t} + \sum_{t=0}^{N} \dfrac{I_t}{(1+r)^t} + \sum_{t=0}^{N} \dfrac{A_t}{(1+r)^t}$

2. *Maintenance and operation:* $C_P^M = \sum_{t=0}^{N} \dfrac{E_t}{(1+k)^t}$
3. *Salvage value:* R_N
4. *Initial asset value:* I_0

Thus, the NPV of cash flows to the customer if the asset is purchased is given by:

$$NPV_P = I_0 + C_P^M + C_P^E - R_N \tag{21.10}$$

The NPV of cash flows (expenses) to the customer in the case of leasing is the sum of the following two components:

1. *Cash expense:* $C_L^L = \sum_{t=0}^{N} \dfrac{L_t}{(1+r)^t}$

2. *Maintenance and operation:* $C_L^M = \sum_{t=0}^{N} \dfrac{E_t}{(1+k)^t}$

Thus, the NPV of cash flows to the customer if the asset is leased is given by:

$$NPV_L = C_L^M + C_L^L. \tag{21.11}$$

The optimal decision for the customer is as follows: If $NPV_P < NPV_L$ then buy the asset; if $NPV_P > NPV_L$ then lease and if $NPV_P = NPV_L$ then the customer is indifferent between the two options.

Example 21.7   Photocopier

An office manager is planning to replace a photocopier with a new one. The manager needs to decide whether to buy[6] or lease the machine using a discount rate of 12% given the information shown in Table 21.5.

    Table 21.6 provides a summary of the NPV calculations for both options.

    Since $NPV_L < NPV_P$ the better option is to lease the photocopier. Note that there is no operating cost for the lease option.   ∎

---

[6] If the customer buys, internal funds are used and the money needed is not borrowed so there are no cash expenses.

Table 21.5 Information on buying/leasing a photocopier.

Buy option	Lease option
Purchase price $25 000	Annual lease payment $7500 (including all supplies and maintenance)
Useful life four years	Length of contract four years
Salvage value after four years $5000	
Annual operating cost $1100	

Table 21.6 Summary of the NPV calculations for the two options.

Cost item	Present value	Cost item	Present value
Purchase price	25 000	Lease	$= \sum_{t=1}^{4} 7500/1.12^t = 22\ 780$
Salvage value	$= 5000/1.12^4 = 3415$		
Operating cost	$= \sum_{t=1}^{4} 1100/1.12^t = 3341$		
NPV_P	$= 25\ 000 - 3415 + 3341 = \mathbf{\$24\ 926}$	NPV_L	**$22 780**

21.7.2 Approach 2

In many cases, especially for products, purchasing decisions are not solely made based on economics. Other intangible factors are taken into consideration, leading to a multi-objective problem. The analytic hierarchy process (AHP) method is an effective method for dealing with such multi-criteria problems, which involve criteria that can be carefully measured or roughly estimated, well-, or poorly understood – anything at all that applies to the decision at hand.[7] We illustrate by the following example.

Example 21.8 Buying versus Leasing a Fleet of Vehicles

When a firm decides on a fleet of vehicles for use by its staff, then factors such as comfort, safety, company image, and ego are important, and cost and maintenance also play an important role. The benefits accrued by the firm from leasing versus purchasing vary from firm to firm. Some of the benefits are related to:

1. The financial status of the firm (wealth of the firm);
2. The availability of working capital;
3. Centralization or decentralization of the fleet operations;
4. Tax advantages;
5. Company image;
6. Specifications;

---

[7] See Appendix B for a brief introduction to the AHP.

7. Greater rate of return and earning power of the firm;
8. Increase in the total resources of the firm;
9. Economic security and others.

The last four items play an important role in the decision if fleet financing requires large amounts of capital. Different answers would be obtained if all benefits were considered separately. The impact of buying and leasing on the centralization of the operation depends on factors such as the size of the fleet, the disposition of the firm to have the leasing company manage the operation of the fleet, availability of experienced personnel, and so on. The firm may be correct in assuming that it can do a better (i.e., cheaper) job than a leasing company. The principle of comparative advantage suggests that the firm may increase its benefits (for example, profits) by hiring a leasing company to free up manpower and time which the firm can then use in some other, more profitable activities. One must also compare the alternatives according to the tax advantages provided, and this depends on the debt rate of the firm and the depreciation rate allowed by the law.

In practice, one needs to make a trade-off amongst the factors and the AHP provides the framework for doing this. The structure that captures the relationships among the overall wealth of the firm, the alternatives at hand, and the four benefits is the *analytic hierarchy* shown in Figure 21.5.

To apply the AHP procedure one must compare all the benefits. Two benefits are compared at a time by answering the question: Given two benefits, which is the more desirable one to pursue and how strongly? We illustrate by looking at the tax advantage benefit. Leasing expenditures are tax deductible, and so are operating expenses, interest payments, and depreciation. By leasing, the firm may obtain some tax advantages due to leasing expenditure. However, leasing has an impact on the debt rate of the firm and possibly leads to the loss of some other tax advantages due to the decreasing borrowing capability of the firm. Thus, it is not clear that either one of the alternatives would provide better tax advantages. We omit the subsequent steps as it requires a deeper understanding of AHP, which is beyond the scope of this book. ∎

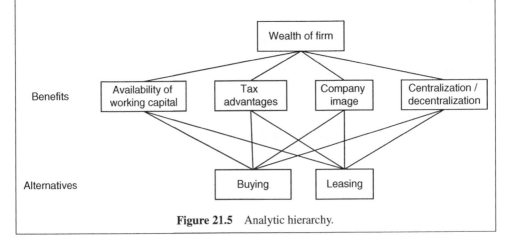

Figure 21.5 Analytic hierarchy.

21.8 LCCA for Products and Plants

21.8.1 Pump

Pumping systems represent about 20% of the world's electrical energy demand. In certain plant operations, they consume from 25 to 50% of the energy usage.[8] Pumping systems are widespread in many sectors of the economy, ranging from domestic services, commercial and agricultural services, municipal water/wastewater services, to industrial services for chemicals, petrochemicals, pharmaceuticals, and so on.

The components of an LCCA typically include initial costs, installation and commissioning costs, energy costs, operation costs, maintenance and repair costs, downtime costs, environmental costs, and decommissioning and disposal costs. The typical LCC for a medium-sized pump is comprised of roughly 10% initial cost, 30% maintenance costs, 50% energy costs, and 10% other costs. Note that energy costs have the largest share and are many times higher than initial costs and therefore weigh heavily in the decision.

Because of the widespread use of pumps in many sectors, reliable information is available from many sources about all of these cost elements. This is an area where LCCA is able to assist rational decision making based on sound information.

21.8.2 Machine Tools

The cost categories and the LCCA for machine tools[9] are summarized in Table 21.7.[10] The results indicate, among other things, that CM is one of the major drivers of LCC (23.8% of LCC). These results prompted a systematic improvement process to deal with the high CM costs. The main cause of machine stops was uncontrolled chip removal. Subsequently, several design solutions were proposed to deal with this problem. The implementation of these design solutions lowered the corrective maintenance (CM) task costs by a factor of 2. Although preventive maintenance (PM) cost did not have the impact of CM, PM frequencies were revised and adjusted to real components' behavior. As a consequence, frequencies of spare part replacements were reduced. This allowed an increase in machine availability without causing more breakdowns. Therefore, the LCCA provided the information required for acting on reducing costs on those cost drivers that had greater influence on the total LCC.

Table 21.7 Machine tool LCC categories and their contribution to LCCs.

Acquisition cost 50%	Operation cost 2%	Maintenance cost 28%	Turnover/scrap cost (negligible)
Purchase price	Tooling cost 50%	PM labor 13%	—
Administration cost	Consumables 10%	CM 85%	
Installation	Utilities 35%	Spares 2%	
Training	Floor space 3%		
Support equipment	Inventory cost 2%		
Extended warranty			
Shipping cost			

---

[8] Hydraulic Institute (2001).
[9] Machine tools are expensive machines that include lathes, CNC machines, and so on.
[10] Based on Enparantza *et al.* (2006).

21.9   LCCA for Infrastructures

In this section we illustrate LCCA for a road infrastructure. We briefly describe a study used to evaluate alternative pavement designs.[11] The LCCA adopted for pavement design is based on the eight-step procedure described briefly below.

Step 1 Establish alternative pavement design strategies for the analysis period: A pavement design strategy is a combination of the initial pavement design and necessary supporting maintenance and rehabilitation activities. As a rule of thumb, the analysis period should be long enough to incorporate at least one rehabilitation activity. The American Federal Highway Authority's September 1996 Final LCCA Policy statement recommends an analysis period of at least 35 years for all pavement projects, including new or total reconstruction projects as well as rehabilitation, restoration, and resurfacing projects.

    Typically, each design alternative will have an expected initial design life, periodic maintenance treatments, and possibly a series of rehabilitation activities.

Step 2 Determine performance periods and activity timing: Performance life for the initial pavement design and subsequent rehabilitation activities has a major impact on LCCA results. It directly affects the frequency of agency intervention on the highway facility, which, in turn, affects agency cost as well as user costs during periods of construction and maintenance activities.

    Work zone requirements for initial construction, maintenance, and rehabilitation directly affect highway user costs and should be estimated along with pavement strategy development. The frequency, duration, severity, and year of work zone requirement are critical factors in developing user costs for the alternatives being considered.

Step 3 Estimate agency costs: Agency costs include all costs incurred directly by the agency over the life of the project. They typically include initial preliminary engineering, contract administration, construction supervision and construction costs, as well as future routine and preventive maintenance, resurfacing and rehabilitation costs, and the associated administrative cost. Agency costs also include maintenance of traffic cost and may include operating costs such as pump station energy costs, tunnel lighting, and ventilation. At times, the salvage value, the remaining value of the investment at the end of the analysis period, is included as a negative cost.

Step 4 Estimate user costs: In the simplest sense, user costs are costs incurred by the highway user over the life of the project. User costs are an aggregation of three separate cost components: vehicle operating costs (VOCs), user delay costs, and crash costs. In the LCCA of pavement design alternatives, there are user costs associated with both normal operations and work zone operations. During periods of initial construction and future maintenance and rehabilitation activities (i.e., work zone operation), VOC, user delay, and crash costs may be significantly different between alternative pavement design strategies.

    User costs are calculated by multiplying the quantity of the various additional user cost components (VOC, delay, and crash) incurred by the unit cost for those cost components.

---

[11] *Pavement design* is defined as "a project-level activity where detailed engineering and economic considerations are given to alternative combinations of sub-base, base, and surface material which will provide adequate load-carrying capacity. Factors that are considered include: materials, traffic, climate, maintenance, drainage, and life cycle costs."

Step 5 Develop expenditure stream diagrams: Expenditure stream diagrams are graphical representations of expenditures over time. They are generally developed for each pavement design strategy to help visualize the extent and timing of expenditures. Figure 21.6 is a cash flow diagram of an expenditure stream for illustrative purposes.

Step 6 Compute net present value: An NPV calculation is illustrated using Example 21.9.

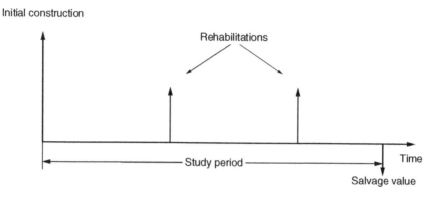

Figure 21.6   Cash flow diagram for pavement study.

Example 21.9   Road Infrastructure

Consider a pavement design alternative for a study period of 35 years that calls for two rehabilitations at years 15 and 30. The different costs and calculations are summarized in Table 21.8 for a discount rate of 4%.

This alternative can be compared to other alternatives that call for different rehabilitation timings or a different number of rehabilitations.   ■

Table 21.8   NPV calculations for a pavement design alternative.

Cost component	Years (*t*)	Costs ($1000)	Discount cost = cost/$(1.04)^t$
Initial construction	0	1000.00	1000
Initial work zone user cost[12]	0	300.00	300
Rehabilitation 1	15	325.00	180
Rehabilitation 1 work zone user cost	15	269.00	149
Rehabilitation 2	30	325.00	100
Rehabilitation 2 work zone user cost	30	361.00	111
Salvage value	35	−216.60	−55
Total NPV			1785

Step 7 Analyze results: Once completed, LCCA should, at a minimum, be subjected to a sensitivity analysis. The sensitivity analysis allows the analyst to subjectively get a feel for the impact of the variability of individual inputs on the overall LCCA results.

---

[12]User cost, such as delays due to the zone of the road under maintenance.

Example 21.10   Road Infrastructure

Consider Example 21.9 and let us conduct a sensitivity analysis with respect to the discount rate. The results are shown in Table 21.9.

Sensitivity analysis with respect to other parameters deemed uncertain may also be conducted. ■

Table 21.9   Sensitivity analysis – effect of discount rate.

	Year	Cost	NPV				
Cost component (%)			2	3	4	5	6
Initial construction	0	1000	1000	1000	1000	1000	1000
Initial work zone user cost	0	300	300	300	300	300	300
Rehabilitation 1	15	325	241	209	180	156	136
Rehabilitation 1 work zone user cost	15	269	200	173	149	129	112
Rehabilitation 2	30	325	179	134	100	75	57
Rehabilitation 2 work zone user cost	30	361	199	149	111	84	63
Salvage value	35	−217	−108	−77	−55	−39	−28
Total NPV			2011	1888	1785	1705	1640

Step 8 Reevaluate design strategies: Once the net present values have been completed for each alternative, the analyst needs to reevaluate the competing design strategies. The overall benefit of conducting a life cycle cost analysis is not necessarily the LCCA results themselves, but rather how the designer can use the information resulting from the analysis to modify the proposed alternatives and develop more cost-effective strategies.

21.10   Summary

Many objects (products, plants, or infrastructures) used for producing goods and/or services are costly (running from a few million dollars to tens and hundreds of millions of dollars). These are called *capital items*, as acquisition and upgrade/replacement (partial or complete) involves a lot of capital. Maintenance economics deals with several related issues.

There are many cost elements that need to be taken into account in making capital investment decisions. The process of upgrading/replacing an object may be viewed as a project, and there can be several projects either occurring concurrently and/or sequentially. The various associated costs that are relevant may be divided into two categories: (i) costs associated with the execution of the project (*direct* and *indirect* costs) and (ii) the *potential costs* that may be incurred once the upgraded/replaced object is put into operation.

This chapter dealt with three main issues related to capital investment decisions: (i) the LCC perspective of capital investment decisions; (ii) capital item replacement decisions; and (iii) lease or buy decisions.

LCC refers to the total costs associated with an object (product, plant, or infrastructure) over its useful life. It is comprised of several cost elements. Making investment decisions based only on initial cost is shortsighted, as the initial investment represents only a small percentage of the LCC for many engineered objects. LCCA is a tool that combines financial, technical, and other information covering the life cycle of an engineered object to facilitate capital investment decision making.

The main reason for replacement is economic – the item being no longer profitable due to increased maintenance and operation costs. In some cases, the replacement is driven by new technology – the new item being far superior to the existing one in terms of economic and operational performance. Several models for optimal replacement decision making have been presented under various conditions.

Some engineered objects may be leased instead of owned for several reasons. Different approaches for making leasing versus buying decisions have been presented.

Finally, capital investment issues for products, plants, and infrastructures have been presented in the last three sections of this chapter.

Review Questions

21.1   What is a capital item?

21.2   Why are capital investment decisions important for a business?

21.3   What are the cost elements of capital investment?

21.4   What are the risks involved in capital investment decisions?

21.5   What are the elements of life cycle cost (LCC)?

21.6   Why should capital investment decisions be based on a life cycle perspective?

21.7   What is LCCA?

21.8   What are the main steps of LCCA?

21.9   What are the reasons for capital item replacement?

21.10   What are the main elements of a capital item replacement model?

21.11   What are the reasons for leasing capital items instead of buying them?

21.12   What are the main issues involved in lease versus buy decisions?

Exercises

21.1   Switches and crossings (S&Cs) are important elements of rail track. Their renewal requires track possession and therefore minimizing the replacement time is important for avoiding train delays. The conventional renewal methods assemble S&Cs on site. However, some new technologies provide modular designs and pre-assembled units that have several advantages: shorter possession time, better quality leading to fewer failures, and lower renewal costs in terms of manpower and renewal time.

What are the factors that need to be taken into account in order to conduct an LCCA to estimate the expected savings due to the adoption of the new technologies?

21.2 The selection of relevant cost elements to perform a realistic LCC comparison of project alternatives is dependent on the type of the engineered object under consideration and its intended function.

What are the most relevant cost elements of an LCCA for hybrid cars?

21.3 Repeat Exercise 21.2 for wind turbines.

21.4 Repeat Exercise 21.2 for rail maintenance projects.

21.5 A capital item is currently in use. The estimated O&M costs and resale values for this item for the next six years are given in the following table:

Year	0	1	2	3	4	5	6
$C_{o,i}$	—	3000	4000	5000	6000	7500	8000
$R_{o,i}$	7000	6500	6000	5000	4500	4000	4000

A technologically superior capital item has appeared with the following data:

Year	0	1	2	3	4	5	6
$C_{n,i}$	—	500	600	800	1 000	1 500	2 500
$R_{n,i}$	10 000	8 000	7 000	7 500	7 000	6 000	5 000

(a) What is the optimal replacement time for the current item if it is used for only one period?
(b) Repeat (a) for the technologically superior item.
(c) For a two-period planning horizon, when should we switch from the current item to the new one?

21.6 Consider Example 21.6.
(a) Extend Model 21.3 to the case of three periods.
(b) Determine the switching period to the new equipment based on a three-period planning horizon.

21.7 A farmer is about to make a lease or buy decision about a tractor. The tractor is financed at an interest rate of 8% for seven years, with an annual payment of $10 564. The cash flows of both options for seven years and the factors taken into consideration are summarized in the table below. Also, it is assumed that there are no repair costs associated with leasing the equipment.

Year	1	2	3	4	5	6	7
Buy option							
Down payment	5 000	—	—	—	—	—	—
Repairs	1 000	1 500	2 500	3 000	3 500	4 000	4 500
Loan payment	10 564	10 564	10 564	10 564	10 564	10 564	10 564
Tax reduction	3 912	3 914	4 045	4 024	3 990	3 942	3 879
Salvage value	—	—	—	—	—	—	31 000
Lease option							
Up-front charges	3 000	—	—	3 000	—	—	3 000
Lease cost	7 500	7 500	7 500	7 500	7 500	7 500	7 500
Tax reduction	2 940	2 100	2 380	2 940	2 100	2 380	2 940
Lease penalty	—	—	1 000	—	—	1 000	—

(a) Using net present value of cost, determine the best approach for the farmer.

(b) Explain why fuel cost is not included in the comparison.

21.8 Rolls-Royce provides airlines with what is called "Power By The Hour," selling its engines along with the services to maintain, repair, and overhaul them over many years.

(a) Discuss how this business model fits in the lease or buy framework discussed in this chapter.

(b) Highlight the LCC implications of this business model.

21.9 Identify other engineered objects where the business model mentioned in Exercise 21.8 may be applied. Discuss the advantages of such an arrangement.

21.10 Develop a model for optimal decision making for the arrangement described in Exercise 21.8.

21.11 Interest rates fluctuate in an uncertain manner. How should a business model this fluctuation in order to build models for capital acquisition decisions?

References

Enparantza, R., Revilla, O., Azkarate, A., and Zendoia, J. (2006) A life cycle cost calculation and management system for machine tools. Proceedings of the 13th CIRP International Conference on Life Cycle Engineering, pp. 717–723.

Hofmann, E., Maucher, D., Hornstein, J., and den Ouden, R. (2012) *Capital Equipment Purchasing*, Springer-Verlag, Berlin/Heidelberg.

Hydraulic Institute (2001) Pump Life Cycle Costs: A Guide to LCC Analysis for Pumping Systems, DOE/ GO-102001-1190.

Needles, B.E., Powers, M., and Crosson, S.V. (2010) *Principles of Accounting*, South Western Cengage Learning, Mason, OH.

22

Computerized Maintenance Management Systems and e-Maintenance

---

Learning Outcomes

After reading this chapter, you should be able to:

- Describe the evolution of management systems for maintenance;
- Define a computerized maintenance management system (CMMS);
- List the benefits of a CMMS;
- Describe the two phases of implementation of a CMMS;
- Identify the issues that need to be addressed in deciding on a CMMS;
- Describe the tasks involved in CMMS implementation;
- List the reasons for CMMS failure;
- Describe the main elements of a CMMS;
- Define e-maintenance;
- Identify the main elements of an e-maintenance management system (EMMS);
- Describe the seven layers of e-based CBM;
- List the key features of an EMMS;
- Describe the key aspects that need to be considered in deciding on the appropriate technology for an EMMS.

Introduction to Maintenance Engineering: Modeling, Optimization, and Management, First Edition.
Mohammed Ben-Daya, Uday Kumar, and D.N. Prabhakar Murthy.
© 2016 John Wiley & Sons, Ltd. Published 2016 by John Wiley & Sons, Ltd.

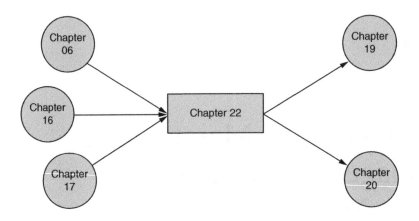

22.1   Introduction

Since 1970 technology has played an ever-increasing role in homes, businesses, and government organizations. This is also true for the maintenance of objects. A computerized maintenance management system (CMMS) is used by almost all organizations to manage the maintenance process discussed in Chapter 17. Advances in information and communication technologies (ICT) and in computers have resulted in the evolution of e-maintenance, where several maintenance activities (almost all in some cases) are carried out electronically. This trend will continue to expand with e-maintenance becoming more dominant. This chapter deals with CMMSs and e-maintenance. The outline of the chapter is as follows. In Section 22.2 we discuss briefly the evolving role of technology in maintenance management. Sections 22.3 and 22.4 deal with CMMSs and e-maintenance, respectively. In Section 22.5 we discuss the use of e-maintenance in three different industry sectors and we conclude with a summary in Section 22.6.

22.2   Role of Technology in Maintenance Management

One can define three stages (Stages A–C) in the evolution of support systems for maintenance management, and these are shown in Figure 22.1.

Stage A Manual system: The data and information were collected manually and stored on cards or log books. Often the data were not collected or stored properly, leading to a loss of useful information.

Stage B CMMS: The two important technologies responsible for the change from the manual system were (i) computers and (ii) sensors. Sensor technologies are discussed in Chapter 6 and the key feature of modern sensors is that data (relating to the states of different components of an object) are generated as digital signals. Computers enabled the storage of lots of data for analysis to help in the decision-making processes. Experts (external) provided the tools (in the form of software packages) and knowledge to help maintenance managers in their decision-making processes.

   CMMSs have been evolving since 1970. Pintelon and Parodi-Herz (2008) define three generations, and the salient characteristics of each are given in Table 22.1.

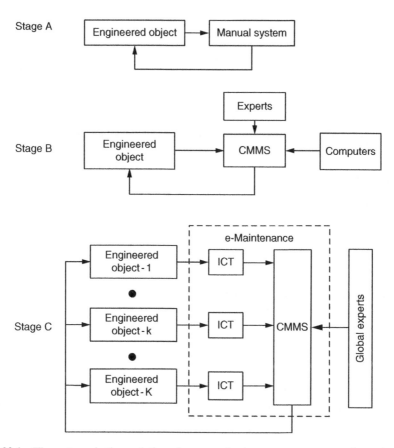

Figure 22.1 Three stages in the evolution of computerized management systems for maintenance.

Table 22.1 Evolution of CMMSs.

Decade	Generation	Salient characteristics
1970s	First	Mainly registration and data administration
		Limited scheduling of maintenance activities
		Mainly standalone system
1980s	Second	Work order management
		Link with other systems (financial, human resources, etc.)
		Connected to material management system in real time
1990s	Third	Asset monitoring and utilization
		Enhanced analytical capabilities
		Decision making and knowledge management

Stage C e-Maintenance: Two important technologies responsible for the change were (i) information and (ii) communication and the term "Information and Communication Technology" (commonly referred to as ICT) evolved. It allowed data from similar objects distributed geographically to be processed centrally with the assistance of global experts in real time for effective decision making.

The emphasis changed to knowledge management programs to capture the implicit knowledge and expertise of maintenance workers. Other knowledge-oriented approaches and techniques that evolved include expert systems to assist in the diagnosis of complex equipment failures and data mining of maintenance history records to learn about failure causes.

22.3   Computerized Maintenance Management Systems (CMMSs)

Definition 22.1

A computerized maintenance management system is a set of integrated computer software programs designed and developed to assist in the management of maintenance activities. These include planning, scheduling, monitoring, controlling, reporting, and other related administrative functions, for effective and efficient management of maintenance.

A CMMS offers core maintenance functionalities and can be adapted to any industrial or commercial organization with minor modifications to standard modules bought from external suppliers (vendors) to suit the specific maintenance needs of different types of organizations.

The key benefits of a CMMS are effective

- Life cycle management of engineered objects;
- Preventive maintenance (PM) planning and scheduling;
- Inventory planning;
- Tracking of key performance indicators;
- Use of resources.

22.3.1   Implementation of a CMMS

The implementation of a CMMS involves two phases:

- *Phase 1:* Assessment and selection.
- *Phase 2:* Installation and implementation.

We discuss these briefly as well as the reasons for CMMS failure.

22.3.1.1   Assessment and Selection

Businesses planning to implement a CMMS first need to decide among three options:

1. In-house development of CMMS.
2. Purchasing a complete CMMS from an external supplier.
3. A combination of in-house development as well as seeking input from external suppliers.

There are several CMMS packages on the market.[1] A business needs to decide which is the most appropriate for its needs. Some issues that need to be addressed in deciding on a particular CMMS include the following:

- Identifying the types of maintenance management tasks that need to be executed;
- Understanding the cost and time implications for collecting and inputting data;
- Updating the library of maintenance procedures on an ongoing basis;
- Training the personnel to use the system properly;
- Commitment to implementing the system properly.

Depending on the modules in a CMMS, one can undertake one or more of the following activities:

- Control the company's list of maintainable assets through an asset register;
- Control accounting of assets, purchase price, depreciation rates, and so on;
- Management of warranty, insurance, and so on;
- Provide documented evidence of conformance to ISO Standards (such as ISO 9000, 14000, 50000, 26000, and others);
- Schedule preventive maintenance routines;
- Control preventive maintenance procedures and documentation;
- Control the work order and documentation of planned and unplanned maintenance work;
- Organize the maintenance personnel database including shift work schedules;
- Control maintenance inventory (store's management, requisition, and purchasing);
- Provide maintenance budgeting and costing statistics;
- Process condition-monitoring inputs;
- Provide tools for benchmarking maintenance performance;
- Assist in maintenance project management;
- Contract management when some of the maintenance is outsourced;
- Integrate with other business systems.

In order to carry out these activities a CMMS requires various kinds of information, and this is discussed in the next subsection.

The process of developing/selecting a CMMS involves the following steps:

1. Developing system specifications.
2. Conducting preliminary screening.
3. Compiling, comparing, and selecting.

22.3.1.2   Installation and Implementation

This is a multi-step process involving the tasks listed below.

1. Forming a team, getting management commitment, and preparing for change.
2. Defining the scope of the project.

---

[1] The website of the Plant Maintenance Resource Centre has links to over 700 maintenance software packages. These include CMMS packages (around 365), condition-monitoring packages (around 70), maintenance audit packages (around 10), and spare part analysis and optimization packages (around 45).

3. Proper implementation of the system.
4. Setting up proper data gathering.
5. Setting up mechanisms for follow-up and monitoring.
6. Evaluating and auditing the process.

Audits (carried out either internally or by external agencies) enable the organization to check how closely their achievements meet their objectives and show conformity to the standard. In the context of certification of companies in fulfilling Asset Management Standard ISO 55000, the role of a CMMS becomes important.

22.3.1.3   CMMS Failure

CMMS failure can happen for several reasons. The most commonly cited one is the improper selection of the CMMS – the selected CMMS does not fit the needs of the organization. The lack of well-trained personnel is another reason for failure. Another issue that can hinder a successful implementation is the lack of commitment of top management and staff to support the transition and the proper use of a CMMS.

22.3.2   Elements of a CMMS

In general, a CMMS consists of four interlinked elements shown in Figure 22.2, and we discuss briefly each of these, highlighting the issues and tasks.

22.3.2.1   Database

The CMMS database stores many different kinds of data and information for all levels of management of the organization. The data include:

- *Asset register:* The asset register holds comprehensive details of each asset (engineered object). Typical data to be stored include asset number, department, asset name, model, serial number, drawing numbers, location, supplier, and so on.
- *Maintenance personnel database:* This contains the information regarding the skills and competencies of the maintenance personnel.
- *Preventive maintenance procedure library:* This is a database of all the preventive maintenance procedures required for the maintainable assets in the system.

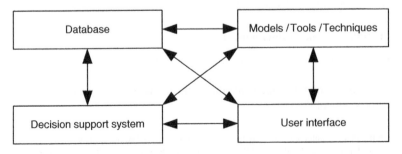

Figure 22.2   Elements of a computerized maintenance management system.

- *Condition of various engineered objects:* Based on the inspection reports.
- *Maintenance history of each asset:* Dealing with technical, economic, and other issues.

The data relating to a component are organized using a hierarchical structure:

- Asset decomposition hierarchy;
- Events;
- Maintenance plan.

The database needs to be properly maintained and managed for it to be of value to the users. This involves file maintenance, database administration, and information retrieval.

- File maintenance involves adding records to tables, updating data in tables, and deleting records from tables.
- Database administration involves the creation, deletion, and restrictions on the use of files. As time progresses, some files may become obsolete and require deletion, while additional files may be required in order to support changes to the asset configuration. It is important that only properly qualified personnel make such changes.
- Information retrieval involves the input of proper search criteria and the extraction of useful information from the data within the database.

Finally, the database needs to be linked to the corporate database. This is needed to assist in strategic-level decision making by senior management and for operational-level decision making by middle- and junior-level management. It requires the proper flow of information to various departments such as the material management and purchasing department (for spare parts), the human resource department for maintenance staff size and training, and the production department for planned shutdowns, and so on.

The database also contains the details of the various procedures for carrying out CM and PM actions, such as the creation of maintenance orders, materials and spare parts needed for maintenance, and material withdrawal. In addition, it stores the results of analysis (such as cost analysis), action logs, historical reports, and so on.

22.3.2.2 Models/Tools/Techniques

Many different kinds of models are needed to assist in decision making at the strategic, tactical, and operational levels. This element of a CMMS is a library of the different models that have been developed either internally or externally.

- The operational user should be able to select models and estimate the parameter values from the data stored in the system. Conservative default values (and probability profiles) should be offered if appropriate data are not available.
- Often, several models need to be linked to find a solution to a specific decision problem. Also, the CMMS must have the flexibility to allow for the upgrading of models and the addition of new models to the system.

A variety of tools is needed for data analysis, model building, model analysis, and determining the optimal decisions. Some of the packages are standard commercial packages whereas others are customized packages. A large number of statistical packages is available for various tasks of model building (such as model selection, parameter estimation, and model validation). Similarly, a large number of software packages is available for model analysis and optimization.

Reporting tools are needed for various tasks, and these include:

- *Viewing outstanding work:* This allows the maintenance managers, planners, and supervisors to check work orders that are outstanding with the facility to display the list in terms of work orders by trade, work type, department, and so on.
- *Reports:* This is one of the important functions of a CMMS. The output from the system is only as good as the input that is fed into the system. A good system has extensive information readily available for fault analysis, costing, work statistics, and so on.
- *Unplanned work reporting:* This deals with the reporting of CM actions. Typically this will include asset identification, the reporter's name, and brief details of the fault.

22.3.2.3 Decision Support System

Data from many different sources (maintenance, operations, suppliers, etc.) are needed for decision making at strategic, tactical, and operational levels. Some of the systems that assist in decision making are the following:

- *Preventive maintenance scheduling:* The maintenance schedule should have a flexible set-up, with each asset having a well-defined maintenance profile that includes details of maintenance frequency, maintenance personnel skills required, procedures for carrying out the maintenance, times when the asset is available for maintenance, and so on. A scheduler controls the scheduling of PM tasks.
- *Inventory management:* This deals with managing the spare part inventory and assists in making decisions regarding stock level, purchasing orders, and so on, discussed in Chapter 20.
- *Maintenance metrics and performance indices:* These are needed to assess the performance of maintenance. They are used for control and continuous improvements in maintenance, which are discussed in Chapter 17.

22.3.2.4 User Interface

Two requirements are (i) the user interface and (ii) the application interface. The user interface facilitates the flow of information from the user to the CMMS and back, while the application interface provides the link between a variety of external programs and databases that may be called upon to solve problems or upload data to, and/or download data from, the CMMS.

22.4 e-Maintenance

Definition 22.2

e-Maintenance is the integration of maintenance and ICT.

Most e-maintenance solutions seem to have evolved from the idea of providing appropriate real-time "health" information to all relevant stakeholders, 24 hours a day and 7 days a week (e-maintenance 24-7), independent of geographical location and organizational belonging. From this viewpoint, e-maintenance enables the creation of a virtual knowledge center, where users, manufacturers, and supplier organizations are participating collectively. The virtual knowledge center is the result of new and evolving ICTs that lead to improved communication between support providers and operators. e-Maintenance-enabled services include the following:

- Dynamic decision support – access to information knowledge support for maintenance personnel at any time;
- Context-driven decision support considering all the relevant factors;
- Predictive health monitoring – diagnostics and prognostics (maintenance performance assessment-benchmarking);
- Data management – data cleaning, data reduction, and creation of metadata.

Another important feature of e-maintenance is the fact that it opens up the possibility of performing remote maintenance, especially when dealing with geographically distributed systems or hazardous environments.

Definition 22.3

Remote maintenance is maintenance of a system performed without physical access of the personnel to the system (IEV 191-07-14, 2010).

In general, this allows the service provider to carry out limited maintenance remotely and to defer full maintenance to a convenient time. This ensures some level of operating capability until maintenance personnel can reach the object to fully restore it to its normal working condition. This is a common form of maintenance for space vehicles and satellites.

22.4.1 Operating Platforms and Standards

The implementation of an e-maintenance management system (EMMS) requires the integration of a variety of hardware and software components. An EMMS may be composed of a number of functional blocks or capabilities: sensing and data acquisition, data manipulation, condition monitoring, health assessment, diagnostics, prognostics, and decision reasoning. In addition, some form of human system interface (HSI) is required to provide a means of displaying vital information and provide user access to the system. Thus, there is a broad range of system-level requirements that include communication and integration with protection of proprietary data and algorithms, the need for upgradeability, modularity, and scalability, and a reduction in engineering design time and costs.

Technological advances are difficult to apply without proper and agreed standards. In this context, new standards have allowed many important advances, both in the wireless communication area, easing connectivity of many miniaturized systems, and in the logical communication and architecture of maintenance processes. These include standards for wireless data transfer protocols (WPAN, WLAN, WWAN, etc.).[2]

In particular, the following technologies and tools have played (and will continue to play) an important role in the evolution of EMMSs:

- Data acquisition (sensors, embedded sensors, sensor networks, etc.);
- Data storage (e-maintenance cloud);
- Data transmission (Internet application and wireless data transfer, networks, etc.);
- Data fusion (hybrid models);
- Data mining (pattern recognition technologies, etc.);
- Knowledge extraction (artificial intelligence (AI) tools);
- Local computers (mono station);
- Proprietary LANs (network installations with one or more servers);
- Web services and Web licencing (data hosting local or remote).

22.4.2 e-Based CBM

Definition 22.4

An e-maintenance management system is a CMMS that uses information and communication technologies to ensure that the e-maintenance services are aligned with the needs and business objectives over the object life cycle in order to deliver the planned and expected services from the assets considering the total business risks.

---

e-based condition-based maintenance (CBM) requires a proper architecture and data exchange conventions that enable interoperability of CBM components. An architecture proposed by Lebold and Thurston (2001) involves seven functional layers, as shown in Figure 22.3.

22.4.3 e-Maintenance Management System

The concept of an EMMS is that it integrates existing maintenance management principles with Web services and modern e-collaboration methods. Collaboration allows one to share and exchange not only information but also knowledge and e-intelligence.

22.4.3.1 Designing/Selecting an e-Maintenance Management System

The key features of an EMMS are the following:

- *Connectivity technologies and methodologies:* For example, MIMOSA (Maintenance Information Management Open System Alliance), Web-based, and so on;
- *Decision support systems:* AI, data mining, and so on;
- *Visualization techniques:* Augmented reality, mobile devices, and so on.

---

[2] Wireless personal area network, wireless local area network, wireless wide area network.

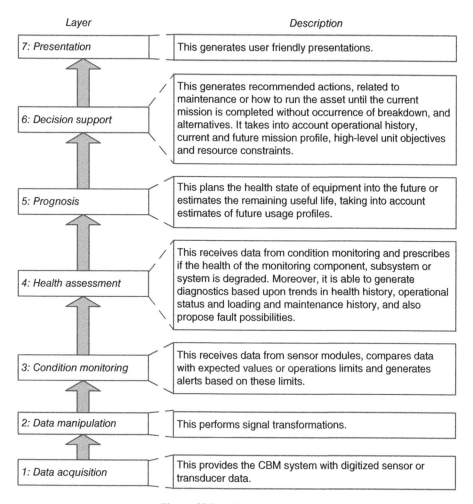

Layer — Description

7: Presentation — This generates user friendly presentations.

6: Decision support — This generates recommended actions, related to maintenance or how to run the asset until the current mission is completed without occurrence of breakdown, and alternatives. It takes into account operational history, current and future mission profile, high-level unit objectives and resource constraints.

5: Prognosis — This plans the health state of equipment into the future or estimates the remaining useful life, taking into account estimates of future usage profiles.

4: Health assessment — This receives data from condition monitoring and prescribes if the health of the monitoring component, subsystem or system is degraded. Moreover, it is able to generate diagnostics based upon trends in health history, operational status and loading and maintenance history, and also propose fault possibilities.

3: Condition monitoring — This receives data from sensor modules, compares data with expected values or operations limits and generates alerts based on these limits.

2: Data manipulation — This performs signal transformations.

1: Data acquisition — This provides the CBM system with digitized sensor or transducer data.

Figure 22.3 CBM architecture.

Technology is one of the most challenging aspects of an EMMS. Due to rapid technological changes, it is not always clear which technological solution one should choose. In this context, one needs to consider the following:

- The *dependability* aspect, addressing characteristics such as reliability, availability, maintainability, safety, and security in all components. The dependability aspect is essential for defining technology-related requirements for the solution.
- The *integration* aspect, addressing the integration models for e-maintenance services (such as diagnostics, prognostics, etc.). It provides fundamentals, conventions, rules, and guidelines for how different services can be integrated depending on the characteristics of the service (for example, internal services or external services). It also deals with other integration-related aspects such as content transfer, safety, security, authentication, and authorization.
- The *communication* aspect, referring to conventions, rules, and guidelines for service communication, such as protocols, wired/wireless communication, and synchronous/ asynchronous communication.

- The *real-time* aspect, addressing conventions, rules, and guidelines for the establishment of infrastructures for real-time solutions, and guidelines for the services requiring real-time execution. It can include aspects of robustness and physical environment.

Other important technology factors relating to knowledge include:

- *Big data and cloud computing:* "Big data" is a general term used for data sets which are so large or complex that they render traditional data-processing technologies inadequate and require new technologies to efficiently process the data within a tolerable elapsed time to deliver information in near real time or real time. Cloud computing allows sharing of resources and involves the deployment of a group of remote servers and networks that allows centralized data storage and processing capabilities. The main enabling technology for cloud computing is virtualization, facilitating the use of computing resources as a utility without any capital investment.
- *Data mining and knowledge extaction:* This is an inter-disciplinary area focusing on methodologies and technologies to extract useful information and knowledge from the big data. This involves knowledge from many areas, such as statistics, machine learning, optimization, and so on.
- *Artificial intelligence for DSS:* This is the intelligence exhibited by machines or software and is used extensively for data mining and big data analysis.

22.5 Applications of e-Maintenance

In this section we discuss briefly the application of e-maintenance in three different industry sectors. We focus on the following three topics:

1. The component (of a product, plant, or infrastructure) under consideration.
2. The variables (or parameters) characterizing the component degradation and its monitoring through sensors.
3. The data collection and transmission.

Data processing and the resulting decisions are not the focus of these examples.

22.5.1 Railway System

The two main elements of a railway system are (i) the railway infrastructure and (ii) the rolling stock. Each of these is comprised of several components. We focus on the e-maintenance of two components of the rolling stock – namely (i) bearings and (ii) wheels – and we discuss the e-maintenance being implemented by the Swedish railway sector on an iron ore track section in Sweden.[3]

The two approaches to condition monitoring of these elements are (i) wayside monitoring (where the sensors are located in a station next to the track) and (ii) on-board monitoring (where the sensor is on board the vehicle – wagon, coach, or locomotive). The type of sensor depends

---

[3] The degradation of track geometry and the maintenance of track are the focus of the second case study in Chapter 23.

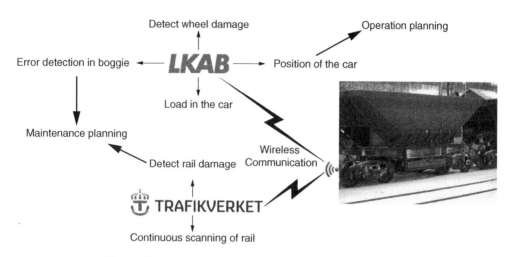

Figure 22.4   Schematic of e-Maintenance used by Trafikverket.

on what variable is being monitored. The data collected may be stored in the station or on board the vehicle and are then either transmitted using wired cables or wireless communication to a central service facility. The processing is usually done at the service facility, although in some cases it is either done at the collection station or on board and the results are communicated to the relevant people for real-time action. Figure 22.4 is a schematic representation of the e-maintenance concept under implementation by Trafikverket[4] and LKAB.[5] It shows some of the variables (for example, rail damage, wheel damage, etc.) that are monitored and the collected data are sent to the maintenance cloud for processing and analysis. Each wagon is identified using radio frequency identification (RFID) and the data collected also identify the location of the bearing and wheel in the vehicle.

22.5.1.1   Bearings

An indication of a failed bearing is a rise in its temperature. This variable is monitored with a device called a hot axle box detector. It is placed by the track at defined locations. Trafikverket has 168 such detectors spread over the country. The detector consists of an infrared array sensor to measure the heat. When a train passes by the wayside station, the triggered sensor scans every wheel bearing on predefined points for a specified number of times. All this information is sent to the traffic control center via a secured railway communication network using GPRS.[6] The measured data trigger an alarm in the traffic control center if the pre-set threshold limits are crossed.

If the wagon/coach is equipped with RFID, the traffic controller warns the train drivers with the exact ID of the wagon and wheels, and the train is stopped to check the condition of the bearings.

---

[4] Trafikverket is the Swedish transport administration.

[5] LKAB is the mining company that owns and runs the trains.

[6] GPRS (general packet radio service) is used for mobile data services and is typically charged according to volume of data transferred rather than per minute of service used.

Some of the trains are equipped with on-board hot axle bearing detectors. After measurement, a computer equipped with software calculates the state of the bearing and compares this with specified threshold limits to advise the train driver on any action to be taken.

22.5.1.2 Wheels

The degradation of a wheel can involve (i) defects on the surface of the wheel and/or (ii) changes to the wheel profile, as discussed in Chapter 16. The data for assessing the condition of a wheel are collected from two wayside measurement stations. The first data source is the automatic wheel profile measurement station. Wheel profile data measured include flange thickness, flange height, rim thickness, flange slope, tread hollow, diameter, and so on. The second measurement station collects rail force data including lateral and vertical forces, vertical load, vertical transient, angle of attack, and so on.

Defects on the Surface of a Wheel
Trafikverket uses strain gauge measurement technology (Chapter 6) with wheel-impact detectors located on a wayside measurement station. The detector measures the dynamic vertical forces induced on the rail during the whole rotation of the wheel. If there is a defect (such as a wheel flat), there can be high impact forces in the wheel/rail rolling contact. The detector also reveals out-of-round wheels.

Strain gauges are used to measure the vertical and lateral forces exerted by each wheel on the track. If the measured forces are larger than the approved axle load of the vehicle, such wheels are identified through an RFID reader and then taken up for detailed analysis. The data collected from the strain gauges are transferred to field computers for initial processing of the data. After initial processing, the data are transferred to the e-maintenance cloud using standard XML and http Internet protocols. All historical data that are stored in the e-maintenance cloud may be accessed by authorized industrial partners.

Wheel Profile Measurements
The wheel profile measurement equipment consists of four units – one on each side of the two rails. These units contain a laser, a high-speed camera, and an electronic control system. When a train passes the boxes, the first wheel triggers a sensor and the protection cover opens, the laser beam starts to shine and then the camera takes pictures of the laser beam projected onto the surface of the passing wheels. These pictures/images are saved and an algorithm transfers the pictures of the wheel profiles to an "x–y coordinate system."

Wheel profile data measured include six variables, namely flange thickness, flange height, rim thickness, flange slope, tread hollow, and rim diameter, that can be used to decide the condition of a wheel, bearing in mind its safety limits.

Data transmission is performed by a wired system from a profile measurement station to a field computer. The system converts all the measurements (images, analog signals, etc.) into real numbers and delivers them in XML format. Later, these are transferred to cloud storage in the Internet using http protocols. Figure 22.5 is a schematic diagram of the measurement and data transfer used by Trafikverket. These data are sent to the e-maintenance cloud and analyzed by the e-maintenance lab at the Lulea University of Technology. The analysis uses all the information to develop trends and identify the wheels needing immediate attention in real time. All the information is relayed to the vehicle workshop and also to the infrastructure manager.

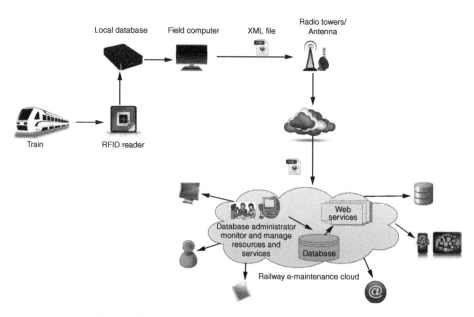

Figure 22.5 Measurement and transmission of wheel data.

22.5.2 Road Infrastructure[7]

The grip of tires and the drivability of a vehicle on a road are greatly influenced by (i) the condition of the dry road surface and (ii) weather-induced factors which result in the road surface being covered with water, ice, snow, or slush (mixture of water and ice). These result in hazards for driving and a maximum speed at which the vehicle can travel safely.

22.5.2.1 Road Eye System for Monitoring of Road Surfaces

This system has been developed to assess the state of roads due to the influence of climate factors. It involves an instrumented vehicle traveling on roads to collect data. The method used for collecting the data is an optical sensor (called MetRoad). It uses light sources of different wavelengths to illuminate the road surface and a detector to measure the reflected light from the surface. Dependent on the state (dry, wet, ice, snow, or slush) the absorption, scattering, and polarization are different, and this is used to identify the state of the road. The vehicle communicates the data in real time to special software (called the *Intelligent Road System*) as well as the position of the vehicle on the road. These data are used to divide the road into segments based on the state of the road. They are combined with Google maps (using different colors for different states) and the information is available to the public. Figure 22.6 is an illustration of the Google map with this additional information.

These maps allow the public to make their travel plans and select appropriate routes for travel. The data are also used by the road maintenance department to trigger various kinds of actions such as (i) changing speed limits on different segments of road and (ii) planning and initiating maintenance actions such as snow plowing or spreading sand on icy roads to make the roads safer for driving.

---

[7] This is based on Casselgren (2011).

Figure 22.6   Road condition (due to weather factors) displayed on Google maps.

The same technology may be used for data collection on dry roads to assess the degradation of the road over time due to age and usage. This identifies the location and size of cracks, potholes, kerb condition, and so on.

22.5.3   JAS 39 Gripen – Military Aircraft

The JAS 39 Gripen is a modern military aircraft designed to perform different mission types, such as interception, reconnaissance, and air–ground warfare. It is equipped with digital infrastructure and a fully integrated computer system that shares a common database through a standardized interface. This aircraft is equipped with state-of-the-art ICT to facilitate trouble-free operation and support effective and efficient maintenance solutions as and when needed.

The aircraft uses built-in test (BIT) systems to collect detailed information about the health of various components of the aircraft in real time and to transfer it to the base station (through highly secured data transfer protocols). It provides the base with data for health diagnostics and prognostics, leading to effective CBM programs for the aircraft and its critical components. The system provides enhanced fault forwarding, troubleshooting, and historical maintenance information to the maintenance center as well as on-board personnel. The maintenance decision support solution delivers valuable information, when and where it is needed, by converting data into useful information for reliable operation.

The BIT systems data may be transferred in real time to the base station. e-Maintenance capability is embedded in the flying platform itself. In the case of malfunction, the on-board computer controls and processes the health data of the various critical components and corrective actions are initiated (if possible). Figure 22.7 is a schematic diagram of data transfer for maintenance purposes (Candell, 2009).

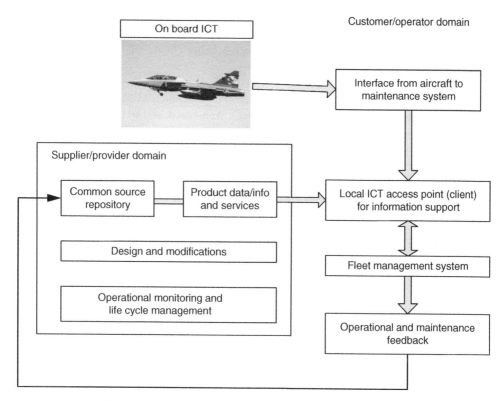

Figure 22.7   e-Maintenance concept used in the JAS 39 Gripen.

22.6 Summary

Maintenance management systems have evolved from manual systems to CMMSs and then to EMMSs. CMMSs are used by almost all organizations to manage the maintenance process including include planning, scheduling, monitoring, controlling, reporting, other related administrative functions, and so on, for effective and efficient management of maintenance.

The implementation of a CMMS involves two phases: (i) assessment and selection and (ii) installation and implementation. A business needs to decide which is the most appropriate for its needs to ensure success of the CMMS implementation. Key factors that prevent failure include management and staff commitment and proper training.

A CMMS consists of four interlinked elements: (i) a database which stores many different kinds of data and information for all levels of management of the organization; (ii) models, tools, and techniques to assist in decision making at all levels – reporting tools are of special importance for effective maintenance management; (iii) decision support systems; and (iv) a user interface.

e-maintenance involves a Web-enabled maintenance management system equipped with technologies and infrastructure to seamlessly connect all the stakeholders, diverse machines, and products built with diverse proprietary communication protocols in real time. Most e-maintenance solutions seem to have evolved around the idea of providing appropriate real-time health information to all relevant stakeholders continuously, independent of geographical location and organizational belonging.

The implementation of an EMMS requires the integration of a variety of hardware and software components for sensing and data acquisition, data manipulation, condition monitoring, health assessment/diagnostics, prognostics, and decision reasoning.

The key features of an EMMS are the following:

- *Connectivity technologies and methodologies:* For example, MIMOSA (Maintenance Information Management Open System Alliance), Web-based, and so on.
- *Decision support systems:* AI, data mining, and so on.
- *Visualization techniques:* Augmented reality, mobile devices, and so on.

e-Maintenance has been implemented in many areas and has led to more effective maintenance management. Three different applications of e-maintenance were outlined in the chapter.

Review Questions

22.1   What is a CMMS?

22.2   Describe the three stages in the evolution of support systems for maintenance management.

22.3   Describe the evolution of CMMSs.

22.4   Describe the elements of a CMMS and illustrate their functionality and inter-relationships.

22.5   What are the key advantages of implementing a CMMS ?

22.6   How does a CMMS assist in effective decision making?

22.7   What are the three main alternatives for implementation of a CMMS in an organization?

22.8   What are the main reasons for the failure of a CMMS in an organization?

22.9   What is e-maintenance? Explain the concept.

22.10 Explain the differences and similarities between e-maintenance and a CMMS.

22.11 How does remote maintenance differ from e-maintenance?

22.12 What are the key features of an e-maintenance management system?

22.13 Describe the role of new and emerging technologies in maintenance management.

22.14 There are many benefits of implementing a CMMS in an organization. Explain the role of a CMMS in effective maintenance scheduling. Illustrate your answer with examples.

22.15 Why do big data and cloud computing form the backbone of an e-maintenance system?

Exercises

22.1 How does a CMMS facilitate effective life cycle management of a railway infrastructure?

22.2 Repeat Exercise 22.1 for a civil aircraft.

22.3 Repeat Exercise 22.1 for an offshore oil platform.

22.4 As a follow-up to Exercise 6.6, discuss the benefits of e-maintenance for wind turbines.

22.5 In most developed economies, labor cost is a significant fraction of the total maintenance cost. Many organizations view a CMMS as the solution to reducing the labor cost. Is this a realistic view? Illustrate your reasoning with examples to support the argument.

22.6 List the different issues that need to be taken into account to compare the different commercially available CMMS packages.

22.7 Conduct an Internet search of the commercially available packages. How do they address the issues resulting from the solution of Exercise 22.5. What can you infer from such a study?

22.8 What kinds of training are essential to ensure the success of a CMMS implementation?

22.9 Comment on the statement "The implementation of an e-maintenance system is more likely to be successful if the organization has an effective CMMS." Give your reasons.

22.10 The effectiveness of e-maintenance is dependent on the types of information and communication technologies used for data acquisition, and the techniques for data mining and knowledge extraction in real time. Discuss the different types of technology needed for e-maintenance of rolling stock.

22.11 Repeat Exercise 22.10 for road maintenance.

22.12 Repeat Exercise 22.10 for maintenance of a manufacturing plant.

22.13 The e-maintenance capability on board an aircraft is supposed to enhance the reliability and safety of the aircraft system by providing trends and current state information about the health of critical components and, in some cases, initiating actions either automatically (self-maintaining systems) or through intervention of ground staff or pilots to compensate for the loss of functionality. Often, such monitoring systems give a false alarm, leading to a no fault found (NFF) situation. Using an Internet search, assess the impact of NFF on the applicability of e-maintenance, especially in the case of the aviation sector.

References

Candell, O. (2009) Development of information support solutions for complex technological systems using e-maintenance. Doctoral thesis. Luleå University of Technology, Sweden.

Casselgren, J. (2011) Road surface characterization using non-linear infrared spectroscopy. Doctoral thesis. Luleå University of Technology, Sweden.

IEV 191-07-14 (2010) http://www.dke.de/de/Online-Service/DKE-IEV/Seiten/IEV-Woerterbuch.aspx (accessed 5 August 2015).

Lebold, M. and Thurston, M. (2001) Open standards for condition-based maintenance and prognostic system. *Proceedings of MARCON 2001: The 5th Annual Maintenance and Reliability Conference*, Gatlinburg, TN, 2001, pp. 6–9.

Pintelon, L. and Parodi-Herz, A. (2008) Maintenance: An evolutionary perspective. In *Complex System Maintenance Handbook*, Kobbacy, K.A.H. and Murthy, D.N.P. (eds) Springer-Verlag: New York.

Part E
Case Studies

Chapter 23: Case Studies

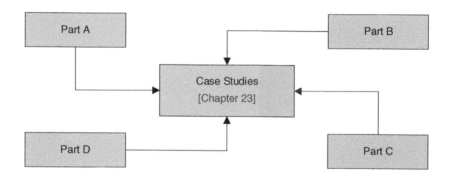

23

Case Studies

23.1 Introduction

This chapter deals with two case studies. Section 23.2 looks at the maintenance of hydraulic pumps used in an excavator. Section 23.3 looks at the longitudinal degradation of rail track and the use of tamping for preventive maintenance (PM) and corrective maintenance (CM) actions.

23.2 Case Study 1 – Hydraulic Pump Maintenance

23.2.1 Introduction

This case study deals with the maintenance of hydraulic pumps used in an excavator. Brief discussions of an excavator and of hydraulic pumps are given in Sections 23.2.2 and 23.2.3, respectively. This sets the background to discuss the case study which is the focus of Section 23.2.4. The next two subsections deal with data aspects (Section 23.2.5) and the modeling of the pump failure distribution (Section 23.2.6). Section 23.2.7 deals with modeling to decide on the optimal maintenance policy and we conclude with a brief discussion of the managerial implications in Section 23.2.8.

23.2.2 Excavators

Excavators (see Figure 23.1) are complex equipment (or machinery) used in the mining and construction sectors. In the mining sector they are used for the excavation of coal, mineral, ore, and overburden in a typical load–haul–dump (LHD) cycle. The basic operating cycle consists of a digging pass through the bank to fill an empty bucket, a loaded swing (or carry) to the dump position, a dump into a haul truck, an empty swing back to the digging face, and

Introduction to Maintenance Engineering: Modeling, Optimization, and Management, First Edition.
Mohammed Ben-Daya, Uday Kumar, and D.N. Prabhakar Murthy.
© 2016 John Wiley & Sons, Ltd. Published 2016 by John Wiley & Sons, Ltd.

Figure 23.1   A typical excavator used in the mining sector. (Courtesy Boliden Mine.)

repositioning or spotting of the bucket at the face. In the construction industry they are used for a variety of tasks and these include boring, ripping, crushing, cutting, lifting, and so on, by adding different types of attachments (such as a breaker, a grapple, or an auger) to the excavator.

23.2.2.1   Operation of an Excavator

All movement and functions of a hydraulic excavator are accomplished using a hydraulic drive system consisting of hydraulic pumps and actuators. A hydraulic pump is a mechanical device which converts mechanical energy (obtained from a diesel engine) into hydraulic energy – moving liquid (usually oil) under pressure. The high-pressure pumps supply oil (up to 5000 psi) for (i) the operation of hydraulic actuators (such as hydraulic motors and cylinders) to carry out various actions such as dig, load, and swing and (ii) the track motors for the motion of the excavator. The low-pressure pumps supply oil (up to 700 psi) for various kinds of motions and control purposes. The actuators are devices that convert the hydraulic energy into mechanical energy for linear motion (hydraulic cylinders) and rotational motion (hydraulic pumps).

23.2.3   Hydraulic Pumps

Hydraulic pumps convert mechanical power into hydraulic power by delivering different flows at different load pressures at the pump output. There are many different types of hydraulic pumps, with variable-displacement axial piston pumps being the most common as they are easy to control and also highly reliable. The main subcomponents of this kind of pump are the pistons, barrel, swash plate, bearings, drive shaft, valve plate, and the control piston. A cross-section of the pump is shown in Figure 23.2.

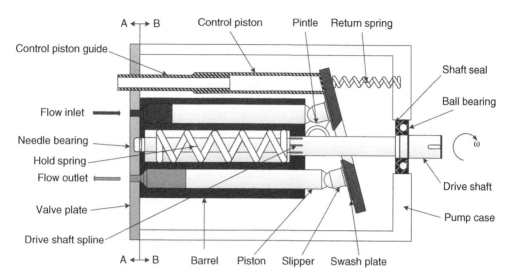

Figure 23.2   Components of a variable-displacement axial piston pump (Li, 2005).

A simple characterization of the pump operation is as follows. The barrel is attached directly to the drive shaft. As the drive shaft rotates, the barrel rotates. The pistons slide inside the barrel and the piston ends move along the swash plate on *slippers*. The pistons reciprocate inside the barrel holes (called *barrel cylinders*) due to the action of the piston ends (via the slippers) on the swash plate. This reciprocating action is responsible for the pumping capability of the unit. The pump barrel is always held against the valve plate, via a spring that is inside the barrel. There are two ports in the valve plate. One is at the suction port connected to the pump inlet. The other is the discharge port connected to the pump outlet. The barrel also contains cylinder ports on the end that is in contact with the valve plate. The movement of the cylinder ports against the valve plate directs the fluid into or out of the cylinders. In this manner, fluid is drawn in through the suction port and discharged through the discharge port repeatedly as the barrel rotates.

23.2.3.1   Pump Failures

A pump is considered to have failed if it cannot provide the required flow rate at the required pressure. Pump failure is detected by sensors and relayed to the operator. The failure is due to failure of one or more components of the pump. There can be one or more failure modes for each component, and these are listed in Table 23.1 along with the causes leading to the failure.

23.2.4   Description of the Case

A mining company (referred to as the "owner" from now onwards) used three identical excavators[1] in its mining operation. Each excavator had the following sets of pumps:

- Eight variable-flow axial piston pumps (for attachment and travel drive) arranged in two groups of four (with each group labeled as a separate engine).
- Four reversible swash plate pumps (for the swing drive).

---

[1] For reasons of confidentiality, neither the mining company nor the brand name of the excavators may be divulged.

Table 23.1 Failure modes of hydraulic pumps.

Components	Failure/faults	Causes
Pump inlet	Leakage	Fatigue, misaligned output pipe
Needle bearing	Excessive wear	Aging, spalling
		Contamination
		Improper hydraulic fluid (viscosity too low, operating temperature too high, not a hydraulic fluid, fluid breaking down)
Ball bearing	Wear, spalling	Aging
		Contamination
		Improper hydraulic fluid (viscosity too low, operating temperature too high, not a hydraulic fluid, fluid breaking down)
Barrel spline/drive shaft spline	Wear, fret wear	Contamination
		Improper hydraulic fluid (viscosity too low, operating temperature too high, not a hydraulic fluid, fluid breaking down)
Drive shaft	Broken	Over pressurization
	Misalignment	Bad assembly
Barrel face contact with valve plate	Wear	Contamination
		Improper hydraulic fluid (viscosity too low, operating temperature too high, not a hydraulic fluid, fluid breaking down)
Valve plate (inner face)	Wear	Over pressurization
Shaft seal	Wear	Improper hydraulic fluid (viscosity too low, operating temperature too high, not a hydraulic fluid, fluid breaking down)
Piston	Wear	Internal leakage increased
	Broken at neck	Over pressurization
	Seized in barrel bore	Contamination in fluid
		Improper hydraulic fluid (viscosity too low, operating temperature too high, not a hydraulic fluid, fluid breaking down)
	End-ball worn through slipper retainer	Improper hydraulic fluid (viscosity too low, operating temperature too high, not a hydraulic fluid, fluid breaking down)
Return spring	Spring constant reduce	Aging
Control piston	Piston wear	Aging
		Improper hydraulic fluid
	Piston broken	Aging
		Overload
Barrel face	Crack between cylinder ports	Over pressurization
Slipper	Pad face wear and upper face wear	Contamination in fluid
	Ball worn through shoe retainer	Improper hydraulic fluid (viscosity too low, operating temperature too high, not a hydraulic fluid, fluid breaking down)
		Misalignment
Swash plate	Face wear (friction increase)	Contamination
		Over pressurization

Table 23.2   Types of maintenance actions carried out.

Maintenance type	Policy	Planning	Duration (h)
Preventive	500-h service	Scheduled	18
	1000-h service	Scheduled	24
	2000-h service	Scheduled	24
Corrective/breakdown	Repairs	Unscheduled	Varies
Condition-based	SOS	Scheduled	Negligible
	Visual checks	Scheduled	Negligible

The excavators were operated continuously unless they were down for corrective or preventive maintenance. In a 24-h period, three different operators operated the excavator in 8-hour shifts. The usage was recorded by an on-board service meter unit (SMU) clock.

23.2.4.1   Maintenance of Pumps

The maintenance was outsourced to an independent service agent (referred to as SA from now onwards) and involved three types of maintenance actions, as indicated in Table 23.2.

Preventive Maintenance (PM)
This is done every 500 hours based on the usage clock and involves one of the three following types: 500-, 1000-, or 2000-h services. The PM follows a cycle (duration 2000 hours of usage) and is comprised of a 500-h service followed by a 1000-h service followed by a 500-h service and finally a 2000-h service. This pattern is repeated until either a pump fails or is removed for reconditioning after having operated for 12 000 hours. This figure is the pump manufacturer's recommended PM policy. The reconditioning process involves total strip down and all worn out components are replaced by new ones, so that a reconditioned pump is as good as a new pump.

Corrective Maintenance (CM)
As mentioned earlier, a pump is considered to have failed if it cannot provide the required flow rate at the required pressure. Whenever a pump fails it is sent for repair. The repair process is identical to the reconditioning process discussed earlier, so that the pump is as good as new after repair.

Condition-Based Maintenance (CBM)
The scheduled oil sampling (SOS) is performed every 500 hours based on the usage clock. The oil sample is collected by the SA after a PM service is completed and is then sent to a specialist group for analysis, the results of which are obtained after a few days. If the oil sample indicates a high level of a particular metal or mineral (indicative of probable oncoming failure), the pump is removed for CM at the subsequent PM.

Visual checks are made by the operator before the start of each shift.

The general accepted notion is that a reconditioned pump is as good as a new pump.

Resources, Spares, and Downtimes
The SA is also responsible for the resources (such as a crane, jig, and so on, since each pump weighs roughly 500 kg) needed to remove a pump needing either CM or reconditioning, and installing either a new or reconditioned pump.

Each pump sent to the SA, either under CM or PM action, is first examined to decide whether it should be reconditioned or not. If not, the unit is scrapped. The SA carries a stock of new and reconditioned pumps, so that the replacement is carried out with the least possible delay.

The downtimes to carry out 500-, 1000-, and 2000-h services are indicated in Table 23.1. The time needed to replace a failed pump by a new or a reconditioned pump is uncertain and will be discussed in a later subsection.

23.2.4.2 Decision Problem

As mentioned earlier, a pump that operates for 12 000 hours without failure is removed for reconditioning. (This limit has not been strictly followed, as the data indicate.) The owner's interest is in finding answers to the following two questions:

1. Is the current figure of 12 000 hours for the reconditioning of a non-failed unit (as recommended by the pump manufacturer) optimal (in terms of the cost of maintaining pumps)?
2. If not, what should be the optimal usage limit for reconditioning?

23.2.5 Data

The mine operates three identical excavators on site with two engines per excavator and four hydraulic pumps (variable-displacement axial piston pumps) per engine. The mine has a small maintenance department which carries out the PM and manages the outsourcing of pumps for CM actions (when a pump failure occurs) and PM actions involving the overhaul of pumps. The pumps are as good as new after each PM or CM action. In other words, reconditioned pumps are statistically similar to new pumps.

The available data consist of the failure times (for units that have failed and required CM action – these are the failure data) and service times (for units that have not failed and are sent for PM action – these are the censored data) for 102 units sent to the SA, and these are given in Table E.9. The column labeled "Type" indicates whether the data are failure data (denoted by 1) or censored data (denoted by 0). As can be seen, the data consist of 45 pieces of failure data and 57 pieces of censored data. The information regarding whether an item is new or reconditioned is missing and there is no information regarding the number of times a pump is reconditioned.

Engine number was recorded but not the location of the pump in relation to the engine. The maintenance department did not have any information regarding the failure mode. This might be due to the maintenance service contract not dealing with this issue. The SA refused to provide these data and it is not sure whether this kind of data was collected or not.

23.2.6 Modeling

Since the data consist of failure and service times, the modeling involves the selection of a distribution function $F(t; \theta)$ to model the time to failure. The starting point is the transformed plots. The plots from Minitab (for the 11 distributions) indicated that the data were not scattered along a straight line, implying that none of the distributions was appropriate to model the data set. The empirical Weibull probability plot (WPP) obtained from Minitab is shown in Figure 23.3. As can be seen, it is not scattered along a straight line.

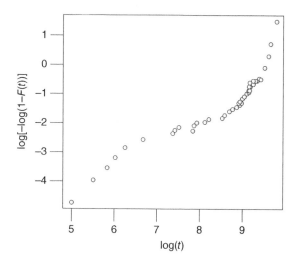

Figure 23.3 WPP of empirical distribution function (EDF).

The shape of the empirical WPP suggests that a mixture distribution involving two Weibull distributions might be appropriate for modeling the data. The twofold Weibull mixture is given by:

$$F(t) = pF_1(t) + (1-p)F_2(t) \tag{23.1}$$

with $0 \leq p \leq 1$. $F_i(t), i = 1, 2$, are two-parameter Weibull distributions given by:

$$F_i(t; \alpha_i; \beta_i) = 1 - e^{-(t/\alpha_i)^{\beta_i}}, \quad t \geq 0 \tag{23.2}$$

23.2.6.1 Parameter Estimation

The parameters were estimated using the method of maximum likelihood for the two cases: (i) shape parameter the same for both distributions and (ii) different shape parameters. The parameter estimates are given below.

- *Case 1:* $\{\hat{\beta}_1, \hat{\alpha}_1, \hat{\beta}_2, \hat{\alpha}_2, \hat{p}\} = \{2.22, 465, 2.22, 14800, 0.075\}$
- *Case 2:* $\{\hat{\beta}_1, \hat{\alpha}_1, \hat{\beta}_2, \hat{\alpha}_2, \hat{p}\} = \{1.83, 566, 2.48, 14400, 0.085\}$

The WPPs for the two cases and the empirical WPPs are shown in Figure 23.4. In each case, the model fits the data reasonably well. The sum of error squares (SESs) for the two models are 0.650 and 0.515 respectively. Although for Case 1 the SES is slightly larger, the advantage of this model is that it has one less parameter. We will use the model of Case 1 for the analysis and optimization.

Comment: The time to failure of a small fraction (around 7.5%) of the pumps has a Weibull distribution with a small scale parameter (465 hours) as compared to the rest of

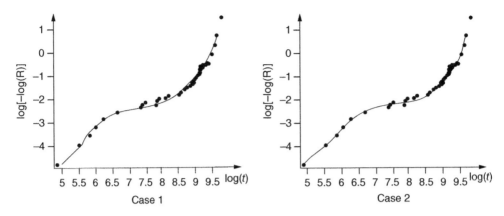

Figure 23.4   The model WPP (the smooth curve) and empirical WPP (the dots) for the two cases.

the pumps whose time to failure is Weibull with a large scale parameter (14 800 hours). We will discuss this further later in the chapter.

23.2.7   Modeling for Maintenance Optimization

23.2.7.1   Maintenance Policy and Objective Function

The maintenance policy used is Policy 4.1 (the age policy) based on a usage clock. The pump is replaced when the usage reaches T (PM action) or on failure (CM action), should it occur earlier. We assume the following:

- The time to replace is small in relation to the mean time to failure and hence can be ignored.
- The time horizon for optimization is large (around 30–40 years for a mine), so that we can approximate it as being infinite.

The pump failure distribution is given by Equations (23.1) and (23.2). Let C_f and C_p denote the average cost of a failure replacement (CM action) and a non-failure replacement (PM action), respectively, and let $\eta = C_f / C_p$. From the results of Section 14.3.2, the asymptotic expected total cost per unit time (based on expected cycle cost and expected cycle length) is given by:

$$J(T) = \frac{C_f F(T) + C_p \bar{F}(T)}{\int_0^T \bar{F}(t)\, dt} \tag{23.3}$$

As can be seen, the optimal T depends only on the cost ratio η. As a result, we will use:

$$\tilde{J}(T) = \frac{J(T)}{C_p} = \frac{1 + (\eta - 1) F(T)}{\int_0^T \bar{F}(t)\, dt} \tag{23.4}$$

as the objective function for the selection of the optimal T.

23.2.7.2 Estimating Replacement Costs under PM and CM

These costs can be divided into two categories – direct and indirect costs – and each is comprised of several elements.

Direct Costs

- *Material cost:* This the cost of the item used in the service exchange. If a new item is used, the cost is C_n (~\$50 000) and if a reconditioned item is used, the cost is C_r (~\$35 000). A new replacement is used with probability q (the probability of scrapping a unit under maintenance action), so the average material cost is $qC_n+(1-q)C_r$.
- *Labor cost:* This is the labor cost associated with the service exchange on site. The average time for service exchange is t_p (10 hours) and a crew of four is needed with a cost/hour C_{l1} (~\$500/h).

As a result, the total direct cost to the SA for each maintenance action (PM or CM) is the sum of the above two costs and is given by $[qC_n+(1-q)C_r+t_pC_{l1}]$. The cost to the owner depends on the terms of the contract. We look at a contract where the owner pays the SA $(1+\gamma)$ of this cost for each maintenance action, where γ is the mark-up (or profit ~ 0.1–0.3), so that the direct cost of each maintenance action (PM or CM) is $[qC_n+(1-q)C_r+t_pC_{l1}](1+\gamma)$.

Indirect Costs

These are the costs incurred by the owner under CM action and involve the operations being affected – labor being idle and loss of revenue.

- *Labor idle:* Each excavator operates in conjunction with three to four trucks. When waiting for CM, 4–5 (truck driver + excavator operators) people are idle. The downtime is the time for the repair crew to arrive + the service time to carry out the exchange. The average downtime is given by $t_f=(1+\delta)t_p$, with δ in the range 0–0.5. One can view δt_p as the average travel time for the crew. The average labor cost/hour (due to labor being idle) is C_{l2} (~\$1000). As a result, the average labor idle cost to the owner for each CM action is $C_{l2}t_f=C_{l2}(1+\delta)t_p$.
- *Loss in revenue:* A typical mine has three or four excavators operating 24 hours a day (~7000–8000 working hours per year or ~20 hours/day). The revenue generated per day ~\$2–4 million, so that the average loss in revenue/hour with an excavator down undergoing CM action is ~\$50 000 ($3 \times 10^6/20 \times 3$). If one assumes a profit ~20%, then the loss in profit per hour with an excavator down for CM is L_p (~\$10 000).

The average indirect cost to the owner with each CM action is $t_f[C_{l2}+L_p]$.

Average Cost of Each PM and CM Action

The average cost of a PM action is given by:

$$C_p=\left(1+\gamma\right)\left[qC_n+\left(1-q\right)C_r+t_pC_{l1}\right] \tag{23.5}$$

The average cost of a CM action is given by:

$$C_f=C_p+\left(1+\delta\right)t_p\left[C_{l2}+L_p\right] \tag{23.6}$$

From Equations (23.5) and (23.6), we have:

$$\eta=\frac{C_f}{C_p}=1+\frac{\left(1+\delta\right)t_p\left[C_{l2}+L_p\right]}{\left(1+\gamma\right)\left[qC_n+\left(1-q\right)C_r+t_pC_{l1}\right]} \tag{23.7}$$

Table 23.3 Effect of δ on η.

δ	0	1	2	3	4	5
η	2.863	4.726	6.588	8.451	10.314	12.177

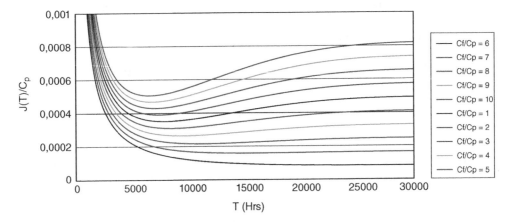

Figure 23.5 $J(T)$ versus T.

23.2.7.3 Optimization[2]

We assume the following nominal values for the relevant parameters:

$$q = 0.6, C_n = \$50\,000, C_r = \$35\,000, \gamma = 0.2$$
$$C_{l1} = \$500, t_p = 10\,(\text{h}), C_{l2} = \$1000, L_p = 10\,000$$

In general, the uncertainty in δ can be high due to the availability of the crew and the time needed for them to reach the mine site. As such, one needs to look at a range of values for this parameter. Table 23.3 shows the variations in η for a range of values for δ with the other parameters held at their nominal values. As can be seen, η is a nonlinearly increasing function of δ.

Figure 23.5 shows the plots of $J(T)$ versus T for a range of η, and the optimal T^* versus η is given in Table 23.4.

23.2.8 Managerial Implications

The optimal T^* depends critically on the cost ratio η. If the ratio is around three, then the manufacturer's recommended figure (12 000 hours) for PM action is optimal. If it is less [greater] than three, then the mine operator needs to increase [decrease] the age for PM action, as indicated in Table 23.4.

Since the failure distribution is a mixture of two Weibulls, the sub-population with distribution function $F_1(t; \alpha_1, \beta_1)$ might be the result of incorrect installation. If this is the case, then

---

[2] The optimization and the plots were done by Dr Alireza Ahmadi.

Table 23.4 T^* and $J(T^*)$ versus η.

η	1	2	3	4	5	6	7	8	9
T^*	30 791	15 467	11 294	9575	8592	7942	7476	7123	6845
$J(T^*)$	∞	0.000158	0.000215	0.000264	0.000309	0.000350	0.000391	0.000430	0.000468

through proper training of the maintenance crew, the fraction p could be reduced (possibly to zero). This will result in the optimal T^* increasing and $J(T^*)$ decreasing. This will require the owner to address proper data collection (such as the root cause) in the service contract and offer incentives to the SA to ensure that incorrect installation is avoided.

23.3   Case Study 2 – Maintenance of Rail Track[3]

23.3.1   Introduction

This case study deals with track degradation due to changes in track geometry and the associated maintenance actions. We start with a brief discussion of track geometry in Section 23.3.2, as this is needed to understand track degradation, which is the focus of Section 23.3.3. There are several factors responsible for track degradation, and these are discussed in Section 23.3.4. Tamping is the maintenance action used for rectifying the changes to track geometry, and this is discussed in Section 23.3.5. Periodic inspections of track are carried out to assess track condition, and decisions on tamping are based on the level of degradation. The decision variables of the track maintenance policy are (i) the frequency of inspections and (ii) the degradation levels for the initiation of tamping. This requires building models for track degradation and maintenance actions. A black-box approach is used to build the model for track degradation. Data collection (obtained through inspections) is discussed in Section 23.3.6 and the analysis to assess track condition is the focus of Section 23.3.7.[4] Section 23.3.8 looks at the modeling of degradation and maintenance, and Section 23.3.9 deals with model analysis and optimization. We conclude with some extensions to the model in Section 23.3.10.

23.3.2   Track Geometry

Rail tracks are designed and built to ensure that trains (goods or passenger) can travel at their maximal speed in a safe manner and with the desired ride comfort in the case of passenger trains. A brief discussion of the structure of rail track is given in Section 3.6.1. The condition of the track is characterized by track geometry.

Track geometry may be defined as the deviation of track position from its design geometry in the horizontal, vertical, and longitudinal directions. It is an important aspect of railway

---

[3] This case study was written by Dr Iman Arasteh Khouy with the assistance of the authors of the book.
[4] The data used in the case study are from a 2 km track of an iron ore line in the north of Sweden, and they were collected by Trafikverket (the infrastructure owner).

construction, which is often used as the main criterion for maintenance and renewal decision making. Not only do safety and ride quality depend on the track geometry, but it also affects the degradation of many other track components.

Track consists of straight sections and curves. Different geometric parameters have been defined to estimate geometric quality for the two sections of track. The geometric parameters used to assess track quality are longitudinal level, alignment, cant (along curved sections), twist, and gauge.

Alignment is the horizontal displacement of the left or the right rail from the mean horizontal position. *Gauge* is the minimal distance between the adjacent rails (left and right) from the top of the rail to 14 mm below. *Cant* is the height difference of the adjacent rails. *Twist* is the algebraic difference of two cant values at a specified distance. In this case study, we focus on straight-line track, where longitudinal level is the main geometric parameter of interest (EN 13848-1, 2008).

Definition 23.1

Longitudinal level is defined as the vertical deviation (Z_p in the z-direction) of consecutive running table levels (top of the rail position) on any rail (left or right) from the mean vertical position (reference line) (EN 13848-1, 2008) (see Figure 23.6).

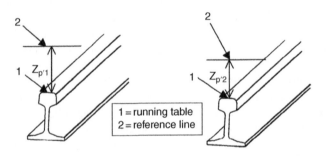

Figure 23.6   Longitudinal level (EN 13848-1, 2008).

Let x denote the position along the track from the starting point and t denote time (or the tonnage moved). Let $Z^l(x,t)$ [$Z^r(x,t)$] denote the deviations in the vertical position of the head of the left [right] rail at position x and time t relative to the position when the track was laid or upgraded. When the track is installed at $t = 0$, $Z^l(x,0)$ [$Z^r(x,0)$] are known for $0 \leq x \leq L$ where the track length is L and $x = 0$ is the reference point.

23.3.3   Track Degradation

As t increases, $Z^l(x,t)$ [$Z^r(x,t)$] change from their initial values, resulting in track degradation. These are measured at discrete time instants $t_i, i = 0,1,2,...,$ where $t_0 = 0$ is the time instant of commissioning the track. The degradation of track geometry is a complex phenomenon which occurs from vertical plastic strains due to the influence of dynamic loads. It is normally calculated as a function of traffic in mm/MGT (million gross tons), or time in mm/year.[5] The two are linearly related and, as such, we will use time as the independent variable, with degradation occurring over time.

When dealing with a straight section of track, one can use a single variable to characterize the vertical position, as indicated below:[6]

$$Z(x,t) = 0.5\{(Z^r(x,t) + Z^l(x,t)\}$$ (23.8)

Uneven track settlement results in waveform deterioration along a track. The waves can be characterized in terms of three different wavelengths (ϕ) (EN 13848-1, 2008):

- *Short wavelength:* $3\,\mathrm{m} < \phi \le 25\,\mathrm{m}$.
- *Medium wavelength:* $25\,\mathrm{m} < \phi \le 70\,\mathrm{m}$.
- *Long wavelength:* $70\,\mathrm{m} < \phi \le 150\,\mathrm{m}$.

For the short wavelengths, the track geometry degradation is characterized in terms of the standard deviation over a specified track length. Tamping is used to rectify short wavelength degradation as well as isolated defects.[7] These are geometric faults that occur over a short length of track (for example, 1–3 m) and can cause severe dynamic interaction between wheel and rail and accelerate the degradation of both.

Figure 23.7 shows a plot of deviation in the longitudinal level over a 200 m straight track at a specific time instant. In other words, it is a plot of $Z(x, t)$ at a specified time, t. As can be seen,

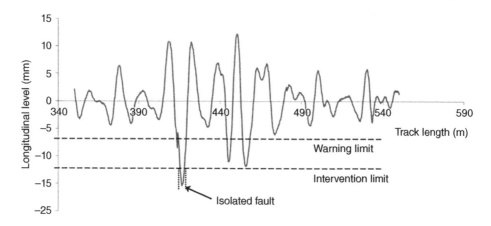

Figure 23.7 Deviations in the longitudinal level over 200 m of track.

---

[5] The track operators decide on the frequency of inspection based on MGT. As mentioned earlier, since MGT is a linear function of time, the t values are based on using this relationship.
[6] On curved sections, the two tracks are at different levels and in this case one needs to measure the vertical displacement of the left and right rails separately.
[7] No maintenance action is carried out for medium- or long-wavelength geometric faults since they do not have a major effect on safety and passenger comfort.

there is a wave pattern. Also shown in the figure are two fault limit lines. The warning limit line (referred to as the *B-fault limit*) is to alert one to the possibility of an isolated defect occurring, and an isolated defect occurs when the longitudinal level exceeds the intervention limit (referred to as the *C-fault limit*). As can be seen in the figure, an isolated fault has occurred around 415 m along the track.

The standard deviation of short-wavelength longitudinal level at a specified time instant t is computed using data over a specified track length (typically 100 or 200 m). It is the square root of the variance, which is given by $\sigma_i^2(t) = \int_{x_i}^{x_{i+1}} \{Z(x,t) - \bar{Z}_i(t)\}^2 \, dx$ with $\bar{Z}_i(t) = \int_{x_i}^{x_{i+1}} Z(x,t) dx$ and the track segment is $[x_i, x_{i+1})$.

23.3.3.1 Quantitative Measures of Track Degradation

One needs proper quantitative measures of track degradation for effective maintenance decision making. This is done by dividing the track of length L into n non-overlapping segments (usually of equal length), with segment i being the interval $[x_{i-1}, x_i)$, $i = 1, 2, \ldots, n$, with $x_0 = 0$ and $x_n = L$. One can characterize the degradation of track in segment i at a specific time instant t through the mean and standard deviation of $Z(x,t)$ over the segment. If $Q_i(t)$ denotes the degradation, then one form of characterization is the following:

$$Q_i(t) = v_1 \bar{Z}_i(t) + v_2 \sigma_i(t) \qquad (23.9)$$

where $0 \leq v_1, v_2 \leq 1$.

These degradation measures are computed at the end of an inspection, so that the time instants are discrete time points. Let t_j, $j = 1, 2, \ldots$, denote the time instant of the jth inspection. In this case, the characterization of track quality is given by the variable Q_{ij}, $1 \leq i \leq n$, $j = 1, 2, \ldots$, with:

$$Q_{ij} = v_1 \bar{Z}_i(t_j) + v_2 \sigma_i(t_j) \qquad (23.10)$$

v_1 and v_2 are non-negative parameters that need to be specified.

23.3.3.2 Causes for Degradation

Some factors which can affect the track geometry degradation are shown in the Ishikawa diagram (cause and effect diagram) in Figure 23.8 (from Arasteh Khouy *et al.*, 2014). These factors can be categorized broadly into four groups: design, construction, operation, and maintenance, as shown in the figure.

23.3.4 Maintenance of Track

Geometric faults (created by uneven settlements of track) in the wavelength range 1–25 m are rectified through tamping. Tamping involves the following five steps which are executed sequentially:

1. The tamping machine positions itself over the sleeper to be tamped.
2. The lifting rollers raise the sleeper to be tamped to the target level and thereby create a space under the sleeper.

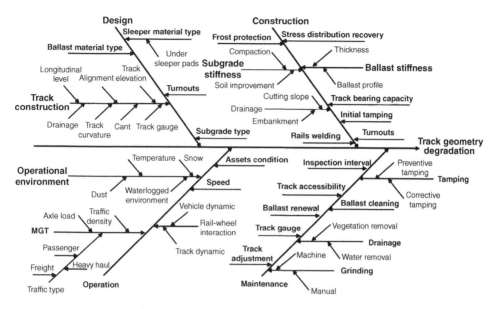

Figure 23.8 Factors influencing track geometry degradation.

3. The tamping tines are inserted into the ballast on either side of the sleeper.
4. The tamping tines squeeze the ballast into the empty space beneath the sleeper, thereby retaining the sleeper in this raised position.
5. The tamping tines are withdrawn from the ballast; the lifting rollers lower the track, and the tamper moves forward to the next sleeper.

Tamping can be performed as either a PM action or a CM action. Different infrastructure owners use different strategies for performing maintenance actions, and these are based on limits for the longitudinal level (as in the case of isolated defects) or for the track quality variable given by Equation (23.10). Three limits commonly used for maintenance actions are the following:

- *Alert limit:* This is a PM threshold. If the track geometric fault value exceeds this limit, preventive action is planned in a time horizon before the fault value exceeds the intervention limit.
- *Intervention limit:* This is a CM threshold, and if the fault value exceeds this limit, corrective action will be required.
- *Immediate action limit:* This is a safety threshold, and if the fault value exceeds this limit, due to the potential derailment risk, a speed reduction or line closure should be imposed until CM is performed.

In this case study, we have both PM and CM tamping and they are defined as follows:

- *PM tamping:* This occurs whenever the track quality variable exceeds some specified alert limit. Because of the degradation being uncertain, the time between such PM actions is a random variable.
- *CM tamping:* This occurs when an isolated defect is detected and is based on the longitudinal level exceeding some specified limit.

For tamping, the track is divided into non-overlapping segments (typically 100 m). The general assumption is that a section of track subjected to either PM or CM tamping results in the track geometry being restored to as good as new, so that the maintenance time instants are renewal points for the segment.

Inspections to assess the longitudinal levels are carried out periodically (the time between inspections is denoted by T). This inspection period and the alert limit for PM tamping (denoted by v) are the decision variables to optimize maintenance.

23.3.5 Data Collection

To obtain data to assess track condition (and for planning maintenance actions), the infra-structure owners use an inspection car with various types of sensors to measure different variables, such as optical sensors (to measure distance) and accelerometers (to measure acceleration). In track geometry monitoring, the longitudinal level, alignment, track gauge, rail elevation, twist (over 3 or 6 m), and curvature are measured for consecutive 5-cm intervals of track. The inspection cars use a tachometer to identify the position of data points along the track. The speed of the cars is also stored continuously for every point along the track. The speeds of inspection cars during measurement taking may vary due to factors such as traffic. During an inspection, the geometric data are stored on-board the car. When the whole track has been inspected, the collected data are downloaded into a database.

The output of sensors is affected by noise. To reduce the effect of this noise, Trafikverket records the data at every 5 cm of the track and then averages over five consecutive measure-ments. In other words, the geometric data consist of observations averaged over 25 cm of track and these are stored in the infrastructure owner's database.

The longitudinal levels of the right and left rails are measured by position sensors and accelerometers (shown schematically in Figure 23.9). The position sensor measures the distance between the wagon and the wheel axle (denoted by $H^l(x, t)$ for the left wheel and $H^r(x, t)$ for the right wheel). The accelerometer measures the vertical acceleration of the wagon (denoted by $A^l(x, t)$ for the left wheel and $A^r(x, t)$ for the right wheel). The measured

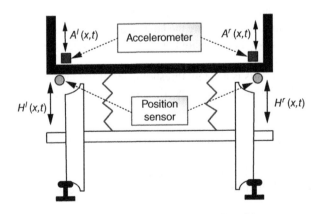

Figure 23.9 Measuring vertical rail position.

accelerations need to be integrated twice over time to obtain the position of the wagon above the left and right wheels (relative to the axle).

As a result, the measured values of the longitudinal levels of the left and right rails (relative to the reference line) are given by:

$$z^l\left(x,t\right)= H^l\left(x,t\right)+ \iint A^l\left(x,t\right)dt^2 \text{ and } z^r\left(x,t\right)= H^r\left(x,t\right)+ \iint A^r\left(x,t\right)dt^2 \quad (23.11)$$

respectively. Since we are considering the degradation over straight tracks, we have a single variable $z(x,t)$ (the observed value of $Z(x,t)$ given by Equation (23.8)) using $z^l(x,t)$ and $z^r(x,t)$ given by Equation (23.11).

The data used for the case study were obtained for a track, 2 km in length, of an iron ore line in the north of Sweden. The selected section was divided into 20 segments, each 100 m in length.

23.3.5.1 Missing Data

Sometimes, due to reasons such as ongoing maintenance actions or traffic, the inspection car needs to stop during the inspection. When the inspection car has a speed of less than 40 km/h, the collected geometric values are not acceptable. As a result of these stoppages, the speed of the car goes under 40 km/h during the break and during acceleration, and this will result in the unavailability of geometric data for a part of the track.

23.3.6 Data Processing

This involves three stages: (i) filtering, (ii) alignment of data, and (iii) estimating the track quality measure.

23.3.6.1 Filtering

The recorded signals from the measuring car comprise long, medium, and short wavelengths. Since only short-wavelength signals are used for maintenance decision making, the long- and medium-wavelength signals must be filtered. This can be done by selecting only signals in the range of 3–25 m.

23.3.6.2 Alignment of Data

Some factors, such as not being able to start the inspections from the exact position every time and also performance deficiencies of the tachometer due to sliding, may introduce errors like position shifting and scaling in the recorded data. Figure 23.10 illustrates this for the collected data from two inspections (performed during 2008 and 2010) over the same track segments. Figure 23.10a shows that the registered track positions in these inspections do not represent the same positions. The inspection data need to be aligned by shifting one measurement to the right or left along the x axis. In this example, the measurement data from 2008 were shifted to

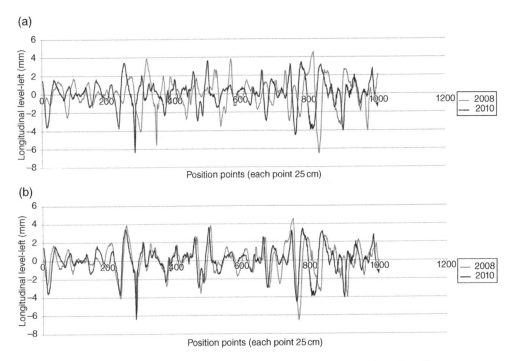

Figure 23.10   Alignment of data. (a) Data before alignment and (b) data after alignment.

the left (Figure 23.10b). Also, the scaling error may be observed by looking at the starting and ending data points in the figure. Considering these points, it can be seen that the measurement data from 2010 (dark color line) are located inside the data points from measurements from 2008 (light color line).

23.3.6.3   Estimate of Track Quality Measure

We use the quality measure given by Equation (23.10) with $v_1 = 0$ and $v_2 = 1$. This implies that the measure used is the standard deviation over 100 m track segments. The estimate is given by:

$$Q_{ij} = \sigma_i\left(t_j\right) \tag{23.12}$$

where $\sigma_i(t_j)$ is the standard deviation in the longitudinal level for segment i after the inspection at time t_j.

Estimates of the standard deviations of longitudinal level for the segments were obtained using the data from 2007 to 2014 after filtering and alignment. Table 23.5 shows the results for two segments.

Figure 23.11 shows the plot of the estimates of standard deviation of longitudinal level over 100 m of track at different time instants of inspection during the period April 2007 to

Table 23.5 Standard deviations of longitudinal level for two track segments.

Measurement date	Longitudinal level		Measurement date	Longitudinal level	
	Segment 1	Segment 2		Segment 1	Segment 2
2007-04-26	1.53	0.97	2011-11-02	0.77	0.83
2007-10-08	1.59	1.13	2012-03-27	0.90	0.89
2008-04-15	1.60	1.17	2012-04-24	1.22	1.76
2008-06-10	1.77	1.18	2012-06-05	0.95	0.92
2008-08-05	1.80	0.67	2012-08-03	1.21	2.01
2008-09-24	1.83	0.71	2012-09-18	1.01	0.95
2009-04-16	1.70	0.76	2012-10-30	1.23	1.88
2009-08-05	0.50	0.60	2013-02-05	1.18	1.12
2009-09-30	0.50	0.56	2013-06-05	1.13	1.08
2010-05-12	0.60	0.69	2013-08-03	1.21	1.11
2010-06-23	0.68	0.71	2013-09-18	1.21	1.04
2010-09-01	0.62	0.70	2013-11-05	1.22	1.08
2011-03-04	0.81	0.95	2014-01-30	1.20	1.06
2011-06-09	0.78	0.86	2014-04-09	1.24	1.11
2011-08-07	0.90	1.06	2014-05-28	1.28	1.28
2011-09-21	0.75	0.81			

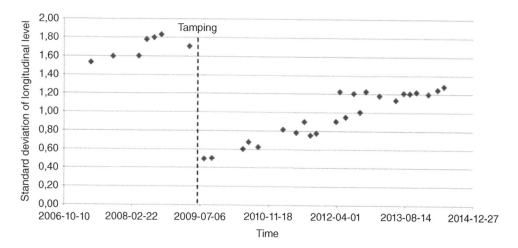

Figure 23.11 Degradation in track quality.

June 2014 for Segment 1 of the above table. As can be seen, the standard deviation increased over time (accumulated load) until around July 2009. At this time, tamping was performed over the segment, which resulted in a lowering of the standard deviation. However, after the maintenance execution, it shows an increasing trend until June 2014. The plots for the remaining 19 segments show a similar pattern.

23.3.7 Modeling

23.3.7.1 Track Quality Degradation

As can be seen from Figure 23.11, one can model the degradation of track quality $Q(t)$ with time (using a local clock which is reset to zero after tamping) by the following linear relationship:

$$Q(t) = \sigma(t) = A + Bt \tag{23.13}$$

where the unit for time is a month. For each segment the parameters A and B depend on the tamping effectiveness. For segment i, A_i is the track quality soon after tamping and B_i is the rate of degradation with higher [smaller] values indicating faster [slower] rates of degradation. We model these as random variables unknown prior to the tamping. Estimates (denoted by $\hat{A}_i$ and $\hat{B}_i$) may be obtained from the data collected from inspections after tamping using linear regression. The estimates for the 20 segments are shown in Table 23.6.

Figure 23.12 is a scatter plot and, as can be seen, there is no correlation between the parameter estimates for A and B. Figure 23.13 contains histograms of the estimates for A and B, again based on the data from Table 23.6. Also shown are the fitted normal distribution density curves together with their associated means and standard deviations.

We assume that the parameters A and B are both normally distributed random variables. If $g_1(a)$ and $g_2(b)$ denote the density functions for these parameters then, using the sample data for the parameter estimates, we have:

$$g_1(a) = \frac{e^{-\{(a-0.891)/0.234\}^2/2}}{\sqrt{2\pi}\,(0.234)} \quad \text{and} \quad g_2(b) = \frac{e^{-\{(b-0.093)/0.009\}^2/2}}{\sqrt{2\pi}\,(0.004)} \tag{23.14}$$

The mean values for A and B are given by:

$$\bar{A} = 0.892 \quad \text{and} \quad \bar{B} = 0.093 \tag{23.15}$$

Table 23.6 Estimated values of linear regression parameters for each track segment.

Segment i	$\hat{A}_i$	$\hat{B}_i$	Segment i	$\hat{A}_i$	$\hat{B}_i$
1	0.7032	0.0135	11	0.9428	0.0102
2	1.1881	0.0128	12	0.919	0.0092
3	0.9163	0.0054	13	1.0071	0.0045
4	1.0314	0.0079	14	0.8376	0.0107
5	0.8736	0.0058	15	0.8352	0.0104
6	0.9006	0.011	16	0.6742	0.0099
7	1.3474	0.0057	17	0.5734	0.007
8	1.2669	0.006	18	0.654	0.018
9	0.8962	0.006	19	0.4476	0.0126
10	1.1622	0.0063	20	0.6601	0.0145

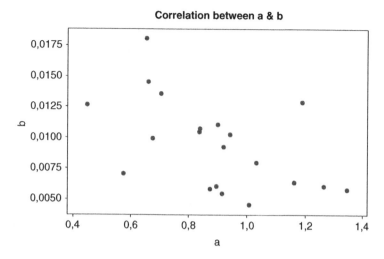

Figure 23.12 Scatter plot of *A* and *B*.

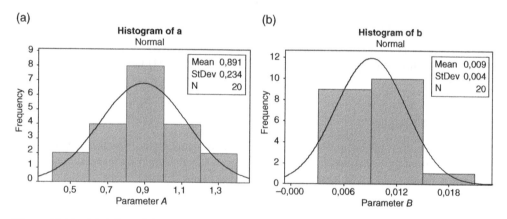

Figure 23.13 Modeling of degradation parameters. (a) Histogram for parameter *A* and (b) histogram for parameter *B*.

23.3.7.2 Occurrence of Isolated Faults

The occurrence of an isolated fault (requiring CM tamping) may be viewed as a random event (a pure chance mechanism). The probability that an isolated fault occurs in an interval T (the time between inspections) is ρ_1 and the probability no isolated fault occurs is $\rho_0 = 1 - \rho_1$. To capture the impact of the length of the interval, we assume that $\rho_1 = 1 - e^{-\lambda T}$, so that as T increases, the probability of an isolated fault increases.

Note that the time between tamping actions (CM or PM) may be viewed as a cycle, as the track degradations after each tamping are statistically similar. Let n_0 [n_1] denote the number of

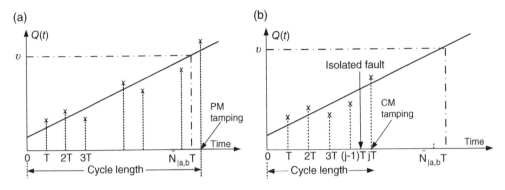

Figure 23.14 (a) PM and (b) CM tamping.

cycles with no isolated fault [an isolated fault]. These counts may be obtained from the data available for modeling. The estimates of ρ_0 and λ are then given by:

$$\hat{\rho}_0 = \frac{n_0}{n_0 + n_1} \quad \text{and} \quad \hat{\lambda} = \frac{1}{T}\ln\left\{(n_0 + n_1)/n_0\right\}. \tag{23.16}$$

23.3.7.3 Inspection and PM Tamping

The track geometry is inspected periodically with the time between inspections given by T. The first inspection after $Q(t)$ crosses the limit v results in a (PM) tamping, as indicated in Figure 23.14a.

23.3.7.4 CM Tamping

As mentioned earlier, the occurrence of isolated faults is a pure chance mechanism. Should a fault occur before $Q(t)$ crosses the limit v, then it is detected at the first inspection after the occurrence, resulting in a CM tamping, as indicated in Figure 23.14b.

23.3.7.5 Costs

The various costs are as follows:

- C_i: Cost of each inspection/segment;
- C_p: Preventive tamping cost/segment;
- C_f: Corrective tamping cost/segment ($C_f > C_p$).

23.3.7.6 Decision Variables and Objective Function

The two decision variables are the following:

- v: Limit for initiating PM tamping (alert limit);
- T: Interval between inspections.

The objective function is $J(v, T)$ – the asymptotic expected cost per unit time per segment. Since every tamping (PM or CM) may be viewed as a renewal point, $J(v, T)$ is obtained as the ratio of the expected cycle cost (*ECC*) to the expected cycle length (*ECL*), where the cycle is the interval between two tampings (see Figure 23.14).

23.3.8 Model Analysis and Optimization

We use the conditional approach by conditioning on $A = a$ and $B = b$. Define $\tilde{N}_{|a,b} = \lfloor (v - a) / bT \rfloor$.[8] This is the number of inspections before the degradation crosses the threshold limit. Let $\tilde{L}_{|\hat{a},b}[\tilde{C}_{|\hat{a}b}]$ denote the cycle length [cycle cost] conditional on $A = a$ and $B = b$. These are random variables, and the different values they may assume and the associated probabilities are given in Table 23.7.

From this we have:

$$
E\left[\tilde{L}_{|a,b}\right] = T \sum_{j=1}^{\tilde{N}_{|a,b}} j \rho_0^{j-1} \rho_1 + T\left[\tilde{N}_{|a,b} + 1\right]\left[1 - \frac{\rho_1\left(1 - \rho_0^{\tilde{N}_{|a,b}}\right)}{\left(1 - \rho_0\right)}\right] \tag{23.17}
$$

and

$$
E\left[\tilde{C}_{|\hat{a},b}\right] = C_f \rho_1 \left[\frac{1 - \rho_0^{\tilde{N}_{a,b}}}{1 - \rho_0}\right] + C_i \sum_{j=1}^{\tilde{N}_{|a,b}} j \rho_0^{j-1} \rho_1 + C_p \left[1 - \frac{1 - \rho_0^{\tilde{N}_{|a,b}}}{1 - \rho_0}\right] \tag{23.18}
$$

Table 23.7 $J(v, T)$ (SEK) for a range of v and T values.

Cycle length $\left(\tilde{L}_{|a,b}\right)$	Cycle cost $\left(\tilde{C}_{|a,b}\right)$	Probability
T	$C_i + C_f$	ρ_1
$\cdot$	$\cdot$	$\cdot$
$\cdot$	$\cdot$	$\cdot$
jT	$jC_i + C_f$	$\rho_0^{j-1}\rho_1$
$\cdot$	$\cdot$	$\cdot$
$\cdot$	$\cdot$	$\cdot$
$\tilde{N}_{|a,b}T$	$\tilde{N}_{|a,b}C_i + C_f$	$\rho_0^{\tilde{N}_{|a,b}-1}\rho_1$
$\left(\tilde{N}_{|a,b}+1\right)T$	$\left(\tilde{N}_{|a,b}+1\right)C_i + C_p$	$1 - \rho_1\left(1 - \rho_0^{\tilde{N}_{|a,b}}\right)\big/\left(1 - \rho_0\right)$

---

[8] $\lfloor y \rfloor$ is the largest integer.

On removing the conditioning, we have:

$$ECC = \underset{a,b}{E}E\left[\tilde{C}_{|a,b}\right] = \int_{-\infty}^{\infty}\int_{-\infty}^{\infty}E\left[\tilde{C}_{|a,b}\right]g_1(a)g_2(b)\,da\,db \qquad (23.19)$$

$$ECL = \underset{a,b}{E}E\left[\tilde{L}_{|a,b}\right] = \int_{-\infty}^{\infty}\int_{-\infty}^{\infty}E\left[\tilde{L}_{|a,b}\right]g_1(a)g_2(b)\,da\,db \qquad (23.20)$$

From the Renewal Reward Theorem, the asymptotic expected cost/unit time/segment is given by:

$$J(\upsilon,T) = \frac{ECC}{ECL} \qquad (23.21)$$

It is not possible to derive an analytical expression for this objective function and the integrations in Equations (23.19) and (23.20) need to be done numerically.

23.3.8.1 Special Case

If the variability in A and B is ignored, we may use their mean values as approximations. In this case, there is no need for unconditioning and we have:

$$J(\upsilon,T) = \frac{E\left[\tilde{C}_{|\bar{A},\bar{B}}\right]}{E\left[\tilde{L}_{|\bar{A},\bar{B}}\right]} \qquad (23.22)$$

Using Equations (23.17) and (23.18) with a and b replaced by $\bar{A}$ and $\bar{B}$, we have, from Equation (23.22):

$$J(\upsilon,T) = \frac{C_f\rho_1\left[\dfrac{1-\rho_0^{\tilde{N}_{|\bar{A},\bar{B}}}}{1-\rho_0}\right] + C_i\sum_{j=1}^{\tilde{N}_{|\bar{A},\bar{B}}}j\rho_0^{j-1}\rho_1 + C_p\left[1-\dfrac{1-\rho_0^{\tilde{N}_{|\bar{A},\bar{B}}}}{1-\rho_0}\right]}{T\sum_{j=1}^{\tilde{N}_{|\bar{A},\bar{B}}}j\rho_0^{j-1}\rho_1 + T\left[\tilde{N}_{|\bar{A},\bar{B}}+1\right]\left[1-\dfrac{\rho_1\left(1-\rho_0^{\tilde{N}_{|\bar{A},\bar{B}}}\right)}{(1-\rho_0)}\right]} \qquad (23.23)$$

with $\tilde{N}_{|\bar{A},\bar{B}} = \left\lceil(\upsilon - \bar{A})/\bar{B}T\right\rceil$. A computational scheme is still needed to obtain the optimal values for the two decision variables.

The cost parameters per 100-meter track segment (based on data provided by the managers of rail maintenance) are as follows:

$$C_i = 120\,\text{SEK}, C_p = 2000\text{ SEK, and } C_f = 5000\text{ SEK}$$

Table 23.8 $J(v, T)$ (SEK) for a range of v and T values.

v	T				
	12	6	4	3	2
1.5	51.036	66.953	83.804	105.871	**154.682**
1.6	45.179	61.609	80.664	104.030	154.740
1.7	40.942	58.154	79.404	103.419	154.995
1.8	37.778	55.870	78.265	103.121	155.259
1.9	35.360	54.340	77.681	**103.114**	155.516
2.0	35.360	53.776	77.477	103.190	155.671
2.1	33.477	52.936	77.341	103.330	155.781
2.2	31.990	52.378	**77.325**	103.438	155.868
2.3	30.804	52.015	77.351	103.569	155.914
2.4	29.850	51.787	77.417	103.677	155.948
2.5	29.077	51.710	77.498	103.743	155.966
2.6	28.448	51.609	77.552	103.813	155.977
2.7	27.935	51.561	77.630	103.854	155.986
2.8	27.514	**51.547**	77.699	103.896	155.990
2.9	27.515	51.550	77.739	103.919	155.992
3.0	27.170	51.567	77.792	103.9425	155.994
3.1	26.888	51.595	77.836	103.9592	155.995
3.2	26.657	51.629	77.871	103.9684	155.995
3.3	26.468	51.665	77.890	103.9774	155.996
3.4	26.314	51.683	77.915	103.9823	155.996
3.5	**26.189**	51.719	77.934	103.9871	155.996

The analysis of data from 16.4 km (164 segments of 100 m) of track yielded $n_0 = 147$ and $n_1 = 164 - 147 = 17$. From this we have an estimate of $\rho_0 = 0.896$ and $\bar{A}$ and $\bar{B}$ are given by Equation (23.15). Using these in Equation (23.23), $J(v, T)$ (in SEK) was computed numerically for a discrete set of v values in the interval (1.5–3.5 mm) and $T = 2, 3, 4, 6,$ and 12 months. The results are given in Table 23.8.

To find the optimal strategy (within the range of parameter values considered) we use a two-stage approach. In the first stage, for a given T the optimal PM limit ($v^*(T)$) that yields the minimal asymptotic cost rate is obtained. The minima are shown in Table 23.8 in bold font. As can be seen, $v^*(T)$ increases as T increases. In the second stage, the optimal T^* is obtained by the shaded entry that yields a minimum for $J(v^*(t), T)$. As can be seen, the optimal $T^* = 12$ months and $v^* = 3.5$.

According to UIC (International Union of Railways) (2008), for this speed, the track quality is classified as poor when the standard deviation of the longitudinal level reaches 1.9 mm. If it exceeds this value, the risk of train derailment increases. To reduce the risk, the maximal train speed needs to be reduced. The maximum allowable train speed on the studied line is 140 km/h and when the standard deviation of the longitudinal level reaches 1.9 mm, the passenger trains need to reduce their speed from 140 to 70 km/h to ensure the desired level of safety as well as a comfortable ride. This also results in a reduction in track capacity, which, in turn, impacts on the revenue of customers using the track and penalty costs for the track owners.

For the track used for transporting iron ore, any delay over 2 hours results in the cancelation of one iron ore train, which costs the customer (owner of the mine) 5 00 000 SEK. These costs need to be taken into account along with the safety issues in determining the optimal limit as well as the optimal frequency of inspection. If one includes these costs in the model, this will result in optimal values for T and v much smaller than the ones obtained and closer to the current value of $T = 3$ or 4 used by the Swedish track operators. Also, this implies that $v^* = 1.9$, which also ensures that the desired safety and ride comfort levels are met.

23.3.9   Improvements to the Model

As can be inferred from the above paragraph, one needs to include various other costs in the objective function to decide on the optimal alert limit and frequency of inspection. We discuss briefly how the model needs to be modified to include these costs.

Note that when an isolated fault occurs in a segment it is detected at the first inspection after the event occurs. However, an isolated fault may lead to derailment of the train before the next inspection, in which case the CM action occurs soon after the derailment. One needs to model the probability of a derailment occurring, and this would be a linear function of T (implying a higher probability of failure as T increases). Also, the cost resulting from a derailment will be very high and needs to be included in the model. This requires additional data for both modeling and model validation.

The model is based on age- (usage-) based inspection. One can use CBM to decide on optimal inspection strategies. This would require stochastic formulations (such as Markov chains or continuous-time, continuous-state stochastic processes) to model the degradation and then dynamic optimization (such as dynamic programming or close-loop stochastic control theory).

References

Arasteh Khouy, I., Larsson-Kråik, P.-O., Nissen, A., Juntti, U., and Schunnesson, H. (2014) Optimization of track geometry inspection interval, *Proceedings of the Institute of Mechanical Engineers, Part F: Journal of Rail and Rapid Transit*, **228**: 546–556.

EN 13848-1 (2008) *European Standard: Railway Applications – Track Geometry Quality – Part 1: Characterization of Track Geometry*. Swedish Standards Institute.

Li, Z. (2005) Condition monitoring of axial piston pump. Masters thesis. University of Saskatchewan, Saskatoon.

UIC (2008) *Best Practice Guide for Optimum Track Geometry Durability*. Editions Techniques Ferroviaires (ETF), Paris.

Part F

Appendices

Appendix A: Introduction to Probability Theory

A.1 Basics of Probability

A.1.1 Sample Space and Events

Consider an *experiment* whose outcome (denoted by the symbol ω) is not known in advance but is such that the set of all outcomes (called the *sample space* and denoted by the symbol $\mathfrak{I} \equiv \{\omega_1, \omega_2, \ldots \omega_n\}$) is known.

An *event* is any subset of the sample space $\mathfrak{I}$. Given a subset $E \in \mathfrak{I}$, the complement event $\bar{E}$ is the set of all elements in $\mathfrak{I}$ such that $E \cup \bar{E} = \mathfrak{I}$ and $E \cap \bar{E} = \varnothing$ (the *null* set).

A.1.2 Probability

The probability of an outcome can be interpreted as the limiting value of the proportion of times the outcome occurs in K repetitions of the random experiment as K increases beyond all bounds (Montgomery, Runger, and Huebele, 2001).

The probability $P(E)$, is a number assigned to each event E in $\mathfrak{I}$ that satisfies the following conditions:

1. $0 \leq P(E) \leq 1$;
2. $P(\mathfrak{I}) = 1$; and
3. If $E_1, E_2, \ldots, E_n$, are mutually exclusive events, then

$$P\left(E_1 \cup E_2 \cup \cdots \cup E_n\right) = P\left(E_1\right) + P\left(E_2\right) + \cdots + P\left(E_n\right)$$

Introduction to Maintenance Engineering: Modeling, Optimization, and Management, First Edition.
Mohammed Ben-Daya, Uday Kumar, and D.N. Prabhakar Murthy.
© 2016 John Wiley & Sons, Ltd. Published 2016 by John Wiley & Sons, Ltd.

A.1.3   Probability Rules

1. $P(\mathfrak{I}) = P(E \cup \bar{E}) = P(E) + P(\bar{E}) = 1.$
    *   Thus, $P(\bar{E}) = 1 - P(E)$. This rule says that the probability that an event does not occur is equal to one minus the probability that it does occur.
2. $P(E \cup F) = P(F) + P(F) - P(E \cap F)$ for any two events $E, F \in \mathfrak{I}$.
    *   This gives the probability for the event $E \cup F$. Note that $P(E \cap F)$ is subtracted because any point in $E \cap F$ is counted twice in $P(E) + P(F)$. If the events E and F are mutually exclusive, then $P(E \cap F) = \emptyset$, the null set. In this case, $P(E \cup F) = P(E) + P(F)$.

A.1.4   Conditional Probability

Consider two events, A and B, that are not mutually exclusive. The probability of the event B occurring when it is known that the event A has occurred is called the *conditional probability* of B given A. It is denoted by $P(B|A)$ and is given by:

$$P\left(B \mid A\right) = \frac{P\left(A \cap B\right)}{P\left(A\right)} \tag{A.1}$$

provided $P(A) > 0$. The rationale behind this formula is that if we know that A occurs, then for B to occur it is necessary for the outcome to be a point in both A and B; that is, the outcome belongs to $A \cap B$. Knowing that A has occurred reduces the sample space of the experiment to S. Hence, the probability of $A \cap B$ occurring is equal to the probability of $A \cap B$ relative to the probability of A.

A.2   Random Variables

A *random variable* (often abbreviated as *RV*) is a function which maps outcomes from the sample space $\mathfrak{I}$ to $\mathfrak{R}$, the space of real numbers. In other words, for every outcome ω in the sample space $\mathfrak{I}$, $X(\omega)$ assigns a real number to ω. There is no unique way of defining the mapping and the most appropriate one is based on the focus of the analysis.

A.2.1   Classification of Random Variables

Random variables can be classified into (i) discrete and (ii) continuous random variables. A discrete random variable, $X(\omega)$, takes on at most a countable number of values (for example, the set of non-negative integers). In contrast, a continuous random variable can take on values from a set of possible values which is uncountable (for example, values in the interval $(-\infty, \infty)$).

In the context of reliability and maintenance modeling, one encounters both discrete and continuous RVs, as illustrated by the following sections.

A.2.1.1   Discrete RVs

*   Number of machines waiting to be repaired each day;
*   Number of failures per week;
*   Number of PM (preventive maintenance) actions carried out per day.

A.2.1.2   Continuous RVs

- Time to first failure (for a new item);
- Time between two failures (for a repaired item);
- Reduction in strength due to a shock.

A.2.2   Notation

Because the outcomes are uncertain, the value assumed by $X(\omega)$ is uncertain before the event occurs. Once the event occurs, $X(\omega)$ assumes a certain value. The standard convention used is as follows: $X(\omega)$ (often the argument ω is dropped for notational convenience and $X(\omega)$ is simply written as X (upper case)) represents the random variable before the event, and the value it assumes after the event is represented by x (lower case).

A.3   Characterization of a Univariate Random Variable

The value assumed by a random variable is uncertain before the outcome is known. The uncertainty can be characterized either in terms of a distribution (or density) function or in terms of moments.

A.3.1   Distribution, Density, and Hazard Functions

The *distribution function* $F(x;\theta)$ is defined as the probability that the RV $X \leq x$, and is given by:

$$F\left(x;\theta\right) = P\{X \leq x\} = \sum_{i \in I} P(E_i) \qquad (A.2)$$

where I is the set $\{i : X(\omega_i) \leq x\}$. θ is the set of parameters of the distribution function. For notational ease, often θ is suppressed and one uses $F(x)$ instead of $F(x;\theta)$. The domain of $F(x)$ is $(-\infty, \infty)$ and the range is $[0, 1]$. $F(x)$ has the following properties:

1. $F(X)$ is a non-decreasing function of x.
2. $F(-\infty) = 0$ and $F(\infty) = 1$.
3. If $x_1 < x_2$, then $P(x_1 < X \leq x_2) = F(x_2) - F(x_i)$.

The distribution function is a staircase function for a discrete RV and a continuous function for a continuous RV.

- When $F(x)$ is differentiable, the *density function* $f(x)$ is given by:

$$f\left(x\right) = \frac{dF\left(x\right)}{dx} \qquad (A.3)$$

$f(x)$ can be interpreted as:

$$P\left(x < X \leq x+\delta x\right) = f\left(x\right)\delta x + O\left(\delta x^2\right). \qquad (A.4)$$

- The *hazard function* is given by:

$$h(x) = \frac{f(x)}{1 - F(x)} \tag{A.5}$$

This plays a very important role in modeling first failure and is discussed in more detail in Chapter 10.

A.3.2   Moments of a Random Variable

The *jth moment* of the random variable X, $M_j(\theta)$, is given by:

$$M_j(\theta) = E\left[X^j\right] = \begin{cases} \int_{-\infty}^{\infty} x^j f(x; \theta) dx, & \text{if } X \text{ is continuous} \\ \sum_x x^j P(X = x; \theta), & \text{if } X \text{ is discrete} \end{cases} \tag{A.6}$$

The *jth central moment* of the random variable X, $\mu_j(\theta)$, is given by:

$$\mu_j(\theta) = E\left[X^j\right] = \begin{cases} \int_{-\infty}^{\infty} (x - M_1)^j f(x; \theta) dx, & \text{if } X \text{ is continuous} \\ \sum_x (x - M_1)^j P(X = x; \theta), & \text{if } X \text{ is discrete} \end{cases} \tag{A.7}$$

The first few moments are useful in the analysis of models involving RVs and are listed below.

1. The first moment $\mu = M_1$ is called the *mean* (of the random variable or of the probability distribution). It is a measure of central tendency.
2. The second central moment ($\mu_2(\theta)$) is called the *variance* and is also denoted σ^2. The square root of the variance (σ) is called the *standard deviation*. Both are measures of dispersion or spread of a probability distribution.
3. σ/μ is called the *coefficient of variation*. It is a standardized (unit-free) measure of dispersion.

A.4   Some Basic Univariate Discrete Distribution Functions

We shall confine our attention to the case where X assumes integer values from the set $\{0, 1, 2, \ldots, n\}$, with n being either finite or infinite. Let p_i, $0 \le i \le n$, denote the probability that X assumes the value i; that is, $p_i = P\{X = i\}$.

A.4.1 Binomial Distribution

Here, X can assume values from 0 to n and p_i, $i = 0, 1, \ldots, n$, is given by:

$$p_i = \frac{n!}{i!(n-i)!} p^i (1-p)^{(n-i)}. \tag{A.8}$$

The parameter set is $\theta = \{n, p\}$ with $0 < p < 1$ and $0 \le n < \infty$.
The mean and variance are given by:

$$\mu = np \quad \text{and} \quad \sigma^2 = np(1-p). \tag{A.9}$$

A.4.2 Geometric Distribution

Here, X can assume values from 0 to ∞ and p_i, $0 \le i \le \infty$, is given by:

$$p_i = p^i (1-p) \tag{A.10}$$

The parameter set is $\theta = \{p\}$ with $0 < p < 1$.
The mean and variance are:

$$\mu = \frac{p}{(1-p)} \quad \text{and} \quad \sigma^2 = \frac{p}{(1-p)^2}. \tag{A.11}$$

A.4.3 Poisson Distribution

Here, X can assume values from 0 to ∞ and p_i, $0 \le i \le \infty$, is given by:

$$p_i = \frac{e^{-\lambda} \lambda^i}{i!}. \tag{A.12}$$

The parameter set is $\theta = \{\lambda\}$ with $\lambda > 0$. The mean and variance are given by:

$$\mu = \lambda \quad \text{and} \quad \sigma^2 = \lambda. \tag{A.13}$$

A.5 Some Basic Univariate Continuous Distribution Functions

We will look at distributions where the random variable X can assume values in the interval (a, b), with $a < b$ and $-\infty < a, b < \infty$.

A.5.1 Uniform Distribution

The distribution function is given by:

$$F(x; \theta) = \frac{(x-a)}{(b-a)}, \quad a \le x \le b. \tag{A.14}$$

The parameter set is $\theta = \{a, b\}$. The density function is given by:

$$f(x;\theta) = \frac{1}{(b-a)}, \quad a \le x \le b. \tag{A.15}$$

The hazard function is given by:

$$h(x) = \frac{1}{b-x}. \tag{A.16}$$

The first two moments are given by:

$$\mu = \frac{(b+a)}{2} \quad \text{and} \quad \sigma^2 = \frac{(b-a)^2}{12}. \tag{A.17}$$

A.5.2   Two-Parameter Exponential Distribution

The distribution function is given by:

$$F(x;\theta) = 1 - e^{-\lambda(x-\gamma)}, \quad \gamma \le x < \infty \tag{A.18}$$

The parameter set is $\theta = \{\lambda, \gamma\}$, with $\lambda > 0$ and $\gamma \ge 0$. The density function is given by:

$$f(x;\theta) = \lambda e^{-\lambda(x-\gamma)}, \quad \gamma \le x < \infty. \tag{A.19}$$

The hazard function is given by:

$$h(x) = \lambda. \tag{A.20}$$

The first two moments are given by:

$$\mu = \gamma + \frac{1}{\lambda} \quad \text{and} \quad \sigma^2 = \frac{1}{\lambda^2}. \tag{A.21}$$

The exponential distribution has been used extensively to model the failure times of electronic components.

A.5.3   Gamma Distribution

The gamma density function is given by:

$$f(x;\theta) = \frac{x^{\alpha-1}e^{-x/\beta}}{\beta^\alpha \Gamma(\alpha)}, \quad 0 \le x < \infty \tag{A.22}$$

The parameter set is $\theta = \{\alpha, \beta\}$ with $\alpha > 0$ and $\beta > 0$. Here, $\Gamma(\cdot)$ is the gamma function. It is not possible to give an analytical expression for the hazard function.

The mean and variance are given by:

$$\mu = \alpha\beta \quad \text{and} \quad \sigma^2 = \alpha\beta^2. \tag{A.23}$$

A.5.4 Three-Parameter Weibull Distribution

The density function is given by:

$$f(x) = \frac{\beta(x-\gamma)^{(\beta-1)} e^{-(\{x-\gamma\}/\alpha)^\beta}}{\alpha^\beta}, \quad \gamma \le x < \infty. \tag{A.24}$$

The parameter set is $\theta = \{\alpha, \beta, \gamma\}$, with $\alpha > 0, \beta > 0$, and $\gamma \ge 0$. The hazard function is given by:

$$h(x) = \frac{\beta}{\alpha}\left(\frac{x-\gamma}{\alpha}\right)^{\beta-1}, \quad \gamma \le x < \infty. \tag{A.25}$$

The mean and variance are given by:

$$\mu = \gamma + \Gamma\left(1 + \frac{1}{\beta}\right)\alpha \quad \text{and} \quad \sigma^2 = \left[\Gamma\left(1 + \frac{2}{\beta}\right) - \left\{\Gamma\left(1 + \frac{1}{\beta}\right)\right\}^2\right]\alpha^2 \tag{A.26}$$

where $\Gamma(\cdot)$ is the gamma function (see Abramowitz and Stegun, 1968 for more details).

The Weibull distribution has been used to model first failure times for many components.

A.5.5 Normal (Gaussian) Distribution

The density function for the normal distribution is given by Equation (10.5). It is not possible to give an analytical expression for the hazard function. The mean and variance are the location and scale parameters of the distribution, respectively.

A.5.6 Three-Parameter Lognormal Distribution

The density function, $f(x;\theta)$, is given by:

$$f(x;\theta) = \frac{e^{-\{(\log(x-\gamma)-\mu)^2/2\sigma^2\}}}{\sigma\{x-\gamma\}\sqrt{2\pi}}, \quad \gamma \le x < \infty. \tag{A.27}$$

The parameter set is $\theta = \{\mu, \sigma, \gamma\}$ with $\sigma > 0$, $-\infty < \mu < \infty$ and $\gamma \ge 0$. It is not possible to give an analytical expression for the hazard function. The mean and variance are:

$$M_1 = \gamma + e^{(\mu+\sigma^2/2)} \quad \text{and} \quad V(X) = \omega(\omega-1)e^{2\mu} \tag{A.28}$$

where $\omega = e^{\sigma^2}$.

Comment: When $\gamma = 0$, this reduces to the lognormal distribution given by Equation (10.7).

A.5.7   Exponentiated Weibull Distribution

The distribution function is given by Equation (10.8). The density function is given by:

$$f(x) = v\left\{1 - e^{-(x/\alpha)^\beta}\right\}^{v-1}\left[\frac{\beta x^{(\beta-1)}e^{-(x/\alpha)^\beta}}{\alpha^\beta}\right], \quad 0 \le x < \infty. \tag{A.29}$$

The hazard function is given by:

$$h(x) = \frac{\beta v(x/\alpha)^{\beta-1}\left[1 - \exp\left\{-(x/\alpha)^\beta\right\}\right]^{v-1}\exp\left\{-(x/\alpha)^\beta\right\}}{1 - \left[1 - \exp\left\{-(x/\alpha)^\beta\right\}\right]} \tag{A.30}$$

In general, the moments for this model are intractable. If v is a positive integer, then:

$$\mu_k = \alpha^k v \Gamma\left(1 + k/\beta\right)\sum_{j=0}^{v-1}(-1)^j\binom{v-1}{j}\frac{1}{(j+1)^{1+k/\beta}} \tag{A.31}$$

A.5.8   Logistic Distribution

The distribution function is given by:

$$F(x; \theta) = \frac{1}{1 + \exp\left\{-(x-\mu)/\sigma\right\}}, \quad 0 \le x < \infty. \tag{A.32}$$

The parameter set is $\theta = \{\mu, \sigma\}$ with $-\infty < \mu < \infty$ and $\sigma > 0$. μ is the location parameter and σ is the scale parameter.

The density function is given by:

$$f(x; \theta) = \frac{\exp\left\{-(x-\mu)/\sigma\right\}}{\sigma\left[1 + \exp\left\{-(x-\mu)/\sigma\right\}\right]^2}, \quad 0 \le x < \infty, \tag{A.33}$$

and the hazard function by:

$$h(x; \theta) = \frac{1}{\sigma\left[1 + \exp\left\{-(x-\mu)/\sigma\right\}\right]}, \quad 0 \le x < \infty. \tag{A.34}$$

The first moment is the parameter μ and the second central moment is $\pi^2/\{3\sigma^2\}$.

A.5.9 Three-Parameter Log-Logistic Distribution

The density function is given by:

$$f(x; \mu, \sigma, \lambda) = \frac{\exp\left(\dfrac{\ln(x-\lambda)-\mu}{\sigma}\right)}{\sigma\left[1 + \exp\left(\dfrac{\ln(x-\lambda)-\mu}{\sigma}\right)\right]^2}, \quad x \geq \lambda \tag{A.35}$$

The parameter set is $\theta = \{\mu, \sigma, \lambda\}$ with $-\infty < \mu < \infty$, $\lambda \geq 0$, and $\sigma > 0$. μ is the location parameter, σ is the scale parameter, and λ is the threshold parameter. It is not possible to give an analytical expression for the hazard function.

A.5.10 Log-Logistic Distribution Function

The density function is given by Equation (A.35), with $\gamma = 0$.

A.5.11 Smallest Extreme Value Distribution

The distribution function is given by:

$$F(x; \mu; \sigma) = 1 - \exp\left[-\exp\{(x-\mu)/\sigma\}\right], \quad -\infty < x < \infty. \tag{A.36}$$

The parameter set is $\theta = \{\mu, \sigma\}$, where μ $(-\infty < \mu < \infty)$ is the location parameter and $\sigma > 0$ is the scale parameter. The density function is given by:

$$f(x; \mu, \sigma) = \frac{1}{\sigma} \exp\left[(x-\mu)/\sigma - \exp\{(x-\mu)/\sigma\}\right], \quad -\infty < x < \infty. \tag{A.37}$$

The mean and variance are:

$$E(X) = \mu - \sigma\gamma \quad \text{and} \quad V(X) = \frac{\sigma^2 \pi^2}{6} \tag{A.38}$$

where $\gamma = 0.5772$ is Euler's constant.

It can be shown that the smallest extreme value distribution (SEV) reduces to the Weibull distribution Equation (10.4) under the transformation:

$$\mu = \ln(\alpha) \quad \text{and} \quad \sigma = \frac{1}{\beta} \tag{A.39}$$

A.5.12   Competing Risk Weibull (k = 2)

The distribution function is given by:

$$F(x) = 1 - \{1 - F_1(x)\}\{1 - F_2(x)\} = F_1(x) + F_2(x) - F_1(x)F_2(x) \tag{A.40}$$

where

$$F_i(x) = 1 - e^{-(x/\alpha_i)^{\beta_i}}, \quad 0 \le x < \infty, i = 1, 2. \tag{A.41}$$

The parameter set is $\theta = \{\alpha_1, \beta_1, \alpha_2, \beta_2\}$, with all parameters being greater than zero. The density function is given by:

$$f(x) = f_1(x)\left[1 - F_2(x)\right] + f_2(x)\left[1 - F_1(x)\right] \tag{A.42}$$

where

$$f_i(x) = \frac{\beta_i x^{(\beta_i - 1)} e^{-(x/\alpha_i)^{\beta_i}}}{\alpha_i^{\beta_i}}, \quad i = 1, 2. \tag{A.43}$$

The hazard function is given by:

$$h(x) = \frac{F(x)}{1 - F(x)} \sum_{i=1}^{2} f_i(x) / F_i(x) \tag{A.44}$$

The moments for this distribution are intractable.

A.5.13   Multiplicative Weibull (k = 2)

The distribution function is given by:

$$F(x) = F_1(x) F_2(x) \tag{A.45}$$

where $F_i(x; \theta)$, $i = 1, 2$, are given by Equation (A.36). The parameter set is $\theta = \{\alpha_1, \beta_1, \alpha_2, \beta_2\}$, with all being greater than zero. The density function is given by:

$$f(x) = f_1(x) F_2(x) + f_2(x) F_1(x) \tag{A.46}$$

where $f_i(x; \theta)$, $i = 1, 2$, are given by Equation (A.41). The hazard function is given by:

$$h(x) = \frac{f_1(x) F_2(x) + f_2(x) F_1(x)}{1 - F_1(x) F_2(x)} \tag{A.47}$$

The moments for this distribution are intractable.

A.5.14 Weibull Mixture (k = 2)

The distribution function is given by:

$$F(x) = pF_1(x) + (1-p)F_2(x) \tag{A.48}$$

where $F_i(x; \theta)$, $i = 1, 2$, are given by Equation (A.41). The parameter set is $\theta = \{\alpha_1, \beta_1, \alpha_2, \beta_2, p\}$, with all being greater than zero. The density function is given by:

$$f(x) = pf_1(x) + (1-p)f_2(x) \tag{A.49}$$

where $f_i(x)$, $i = 1, 2$, are given by Equation (A.43)
  The hazard function is given by:

$$h(x) = \frac{pR_1(x)}{pR_1(x) + (1-p)R_2(x)} h_1(x) + \frac{(1-p)R_2(x)}{pR_1(x) + (1-p)R_2(x)} h_2(x) \tag{A.50}$$

where $h_i(x)$ is the hazard function associated with the distribution function $F_i(x)$ and $R_i(x) = 1 - F_i(x)$, for $i = 1, 2$. The first moment is given by:

$$\mu = p\Gamma\left(1 + \frac{1}{\beta_i}\right)\alpha_1 + (1-p)\Gamma\left(1 + \frac{1}{\beta_2}\right)\alpha_2 \tag{A.51}$$

A.6 Bivariate Random Variables

For many items the degradation leading to failure depends on both age and usage. For example, in the case of an automobile, the usage would be the distance traveled in kilometers, and in the case of a photocopier, it would be the number of copies made. As a result, the age and usage at failure are random variables which are related in a statistical sense. The modeling of such failures requires a bivariate distribution with two random variables, denoted X and Y.

A.6.1 Joint, Conditional, and Marginal Distributions, and Density Functions

The *joint distribution function* $F(x,y)$ is given by:

$$F(x, y) = P\{X \leq x, Y \leq y\} \tag{A.52}$$

  The random variables are said to be jointly continuous if there exists a function $f(x,y)$, called the *joint probability density function*, such that:

$$F(x, y) = \int_{-\infty}^{x}\left[\int_{-\infty}^{y} f(u,v)\, dv\right] du \tag{A.53}$$

The *marginal distribution functions* $F_X(x)$ and $F_Y(y)$ are given by:

$$F_X(x) = F(x,\infty) \quad \text{and} \quad F_Y(y) = F(\infty,y) \tag{A.54}$$

The two *marginal density functions* are given by:

$$f_X(x) = \frac{dF_X(x)}{dx} \quad \text{and} \quad f_Y(y) = \frac{dF_Y(y)}{dy}. \tag{A.55}$$

The *conditional distribution* of X given $Y = y$ is denoted $F(x \mid y)$ and is given by:

$$F(x \mid y) = P\{X \le x \mid Y = y\} \tag{A.56}$$

The conditional distribution of Y given $X = x$, $F(y \mid x)$, is defined similarly.

For jointly continuous random variables with a joint density function, $f(x,y)$, the *conditional probability density function* of X, given $Y = y$, is given by:

$$f(x \mid y) = \frac{f(x,y)}{f_Y(y)} \tag{A.57}$$

Similarly,

$$f(y \mid x) = \frac{f(x,y)}{f_X(x)} \tag{A.58}$$

From Equations (A.57) and (A.58), we have:

$$f(x \mid y) = \frac{f(y \mid x) f_X(x)}{f_Y(y)} \tag{A.59}$$

and this is referred to as Bayes' theorem.

The random variables X and Y are said to be *independent* (or *statistically independent*) if and only if:

$$F(x,y) = F_X(x) \, F_Y(y) \tag{A.60}$$

for all x and y.

The results are similar for discrete random variables, with summation replacing integration.

A.6.2   *Moments of Two Random Variables*

The covariance of X and Y is defined by:

$$Cov(X,Y) = E\big[\{X - E[X]\}\{Y - E[Y]\}\big] = E[XY] - E[X]\,E[Y] \tag{A.61}$$

The correlation ρ_{XY} is defined as:

$$\rho_{XY} = \frac{Cov(X,Y)}{\sigma_X \sigma_Y} \tag{A.62}$$

where σ_X and σ_Y are the standard deviations of X and Y, respectively. The random variables X and Y are said to be *uncorrelated* if $\rho_{XY} = 0$. Note that independent random variables are uncorrelated but that the converse is not necessarily true.

A.6.3   Conditional Expectation

This concept is very useful in the analysis of maintenance models and is used frequently in the chapters of Part C. The reason for its significance is that often it is difficult to evaluate the expectation of a univariate random variable $E[X]$ given by Equation (A.6) with $j = 1$. This can often be overcome by finding a random variable, Y, which is statistically related to X in some manner so that the joint density function is given by $f(x,y)$. One first computes the conditional expectation $E\big[X|Y = y\big]$ of X given that $Y = y$. The unconditional expectation of X is given by:

$$E\big[X|Y = y\big] = \int_{-\infty}^{\infty} x\, f(x|y)\, dx \tag{A.63}$$

where one is using the conditional density function. The unconditional expectation is obtained by taking the expectation over Y and is given by the relation:

$$E[X] = \int_{-\infty}^{\infty} E[X|Y = y]\, f_Y(y)\, dy \tag{A.64}$$

This is easily proved by using Equation (A.63) in Equation (A.64), which yields:

$$E[X] = \int_{-\infty}^{\infty}\left[\int_{-\infty}^{\infty} x f(x|y) dx\right] f_Y(y) dy = \int_{-\infty}^{\infty} x\left[\int_{-\infty}^{\infty} f(x|y) f_Y(y) dy\right] dx$$
$$= \int_{-\infty}^{\infty} x\left[\int_{-\infty}^{\infty} f(x,y) dy\right] dx = \int_{-\infty}^{\infty} x\big[f_X(x)\big] dx \tag{A.65}$$

Often, Equation (A.64) is written symbolically as:

$$E[X] = E\big[E\big[X|Y\big]\big] \tag{A.66}$$

Similar results exist for discrete random variables.

A.7   Sums of Independent Random Variables

A.7.1   Sum of Two Random Variables

Consider the case where a failed item is replaced by a new one and the replacement time is a random variable. Let X and Y denote the time to failure and the time to replace by a new item. These are statistically independent random variables with density functions $f_X(x)$ and $f_Y(y)$, respectively. Let $Z = X + Y$. This defines a cycle and the cycle is repeated over time.[1] Here, we focus on deriving the density function of Z, which we denote by $f_Z(z)$ using the conditional approach.

Conditional on $Y = y$, $z \leq Z < z + \delta z$ if and only if $z - y \leq X < z - y + \delta z$. In other words, $f_Z(z \mid y = y)\, \delta z = f_X(z - y)\, dz$. Since Y can take any value in the interval $(-\infty, \infty)$, the unconditional probability is given by:

$$f_Z(z)\,dz = \int_{-\alpha}^{\infty} f_X(z - y)\,dz f_Y(y)\,dy = \left[\int_{-\infty}^{\infty} f_Y(y)\, f_X(z - y)\, dy \right] dz$$

and from this we have:

$$f_Z(z) = \int_{-\alpha}^{\infty} f_X(z - y) f_Y(y)\,dy \tag{A.67}$$

This operation is called the *convolution operation* and is denoted by the symbol "*". Thus:

$$f_Z(z) = f_X(z) * f_Y(z) = f_Y(z) * f_X(z) \tag{A.68}$$

The last part of the equation follows by reversing X and Y in the derivation.

A.7.2   Sum of n Random Variables

Let Z be the sum of n independent variables, X_i $(i = 1, 2, \ldots, n)$, with respective density functions $f_i(x)$ so that $Z = X_1 + X_2 + \cdots + X_n$. It could represent the total life of n identical items (with $f_i(x)$ the same for all i) or the damage caused by n shocks where the distributions for the shocks are different. The density function for Z is obtained by recursive use of the results of Section A.7.1, so that one first obtains the density function for $\tilde{Z}_2 = X_1 + X_2$, then for $\tilde{Z}_3 = \tilde{Z}_2 + X_3$, and so on, leading to $\tilde{Z}_n = \tilde{Z}_{n-1} + X_n$. This yields repeated convolutions (of multiple integrals) and is symbolically represented as:

$$f_Z(z) = f_1(z) * f_2(z) * \ldots \ldots * f_n(z) \tag{A.69}$$

Often, the distribution function for Z is denoted by $F^{(n)}(z)$, so that it indicates the number of variables in the summation and $f^{(n)}(z)$ is its derivative. In this notation, $f_Z(z) \equiv f^{(n)}(z)$.

---

[1] The concept of a cycle is useful in the analysis of many maintenance policies in Chapters 14 and 15.

References

Abramowitz, M. and Stegun, I.A. (1968) *Handbook of Mathematical Functions*, Applied Mathematics Series Vol. 55, National Bureau of Standards, Washington, DC.

Montgomery, D.C., Runger, G.C., and Huebele, N.F. (2001) *Engineering Statistics*, John Wiley & Sons, Inc., New York.

Appendix B: Introduction to Stochastic Processes

B.1 Basic Concept[1]

In Appendix A we defined a random variable, $X(\omega)$, as a function that maps outcomes from the sample space to real numbers. A stochastic process $X(t,\omega)$, $t \in T$, where T is a set of non-negative numbers, can be viewed as an extension of $X(\omega)$ in the following sense: t represents a time instant in the set T, which may be either finite or infinite. Let $X(t,\omega)$ denote the state of the process at time t. For a fixed $t \in T$, $X(t,\omega)$ is a random variable in the usual sense. For a fixed ω (outcome), $X(t,\omega)$ can be viewed as a function of t (also called the *sample path*).

B.2 Characterization of a Stochastic Process

The characterization of general stochastic processes requires multi-dimensional distributions of the form $P\{X(t_1) \le x_1;\ X(t_2) \le x_2; \ldots \ldots; X(t_n) \le x_n\}$ with $n \to \infty$. This becomes unmanageable as n increases and so is unsuitable for modeling and analysis. A class of stochastic processes which is manageable consists of those processes with the Markov property.

B.2.1 Markov Property

A stochastic process $X(t)$ is said to have the *Markov property* if the conditional distribution

$$P\{X(t+\tau) \le x \,|\, X(u) = x(u), \ -\infty < u \le t\} = P\{X(t+\tau) \le x \,|\, X(t) = x(t)\} \qquad (B.1)$$

---

[1] Our presentation is very informal. A proper, rigorous discussion requires advanced mathematics beyond the scope of this book.

Introduction to Maintenance Engineering: Modeling, Optimization, and Management, First Edition.
Mohammed Ben-Daya, Uday Kumar, and D.N. Prabhakar Murthy.
© 2016 John Wiley & Sons, Ltd. Published 2016 by John Wiley & Sons, Ltd.

In other words, the probabilistic characterization of $X(t + \tau)$ (a future uncertain event) given $\{X(u) = x(u), \ -\infty < u \leq t\}$ (the past history and present value of the process) depends only on $x(t)$ (the present value of $X(t)$) and not its past values. This simplifies the characterization considerably. Using conditional probability, we have, for an increasing sequence in t_i,

$$
\begin{aligned}
P\{X(t_1) \leq x_1; \ X(t_2) \leq x_2; \ldots\ldots; X(t_n) = x_n\} &= P\{X(t_n) = x_n \mid X(t_{n-1}) = x_{n-1}\}\cdots \\
\cdots P\{X(t_2) = x_2 \mid X(t_1) = x_1\} P\{X(t_1) = x_1\}
\end{aligned}
\tag{B.2}
$$

Thus, the joint probability distribution for $X(t)$ at n different points along the time axis can be obtained in terms of the conditional distribution of $X(t)$ involving two different values of t. In other words, the probabilistic characterization of the process can be done as a function of sets of four variables $F(t_i, x_i; t_j, x_j)$, with:

$$
F(t_i, x_i; t_j, x_j) = P\{X(t_i) \leq x_i \quad \text{and} \quad X(t_j) \leq x_j\}
\tag{B.3}
$$

for all t_i and t_j over the interval T and all x_i and x_j over the real line.

Stochastic processes which have the Markov property are called *Markov processes* and they have been used extensively in reliability and maintenance modeling; Part B of the book discusses some of the models. Processes which do not have this property are called *non-Markov processes*.[2]

B.3   Classification of Markov Processes

Markov processes can be divided into four categories depending on (i) whether the values assumed by the process $X(t)$ are discrete or continuous and (ii) whether the values assumed by the time variable t are discrete or continuous, as indicated in Table B.1. We discuss each of these four categories briefly.

B.3.1   Discrete Time Markov Chains

Here, both $X(t)$ and t assume only discrete values. Let the values assumed by $X(t)$ be denoted by s_i, $i = 1, 2, \ldots, r$ with r either finite or infinite. We use the notation $X_k \equiv X(t_k)$, $k = 0, 1, 2, \ldots$. The values assumed by t_k, $k = 1, 2, \ldots$, form an increasing sequence. The characterization of the process involves a one-step transition matrix P (of dimension $r \times r$).

Table B.1   Classification of Markov processes.

		State	
		Discrete	Continuous
Time	Discrete	A	C
	Continuous	B	D

---

[2] Sometimes, a *non-Markov process* can be transformed into a Markov process by enlarging the state space.

$$P = \begin{bmatrix} P_{1r} & \cdots & P_{1j} & \cdots & P_{1r} \\ \vdots & & \vdots & & \vdots \\ P_{i1} & \cdots & P_{ij} & \cdots & P_{ir} \\ \vdots & & \vdots & & \\ P_{r1} & \cdots & P_{rj} & \cdots & P_{rr} \end{bmatrix} \qquad (B.4)$$

where P_{ij} is the conditional probability given by:

$$P_{ij} = P\{X_{k+1} = j | X_k = i\}, \ 1 \le i, j \le r, \quad \text{and} \quad k = 0,1,2,\ldots. \qquad (B.5)$$

The elements of the matrix P need to satisfy the following conditions:

- $0 \le P_{ij} \le 1, 1 \le i, j \le r$ (since they are probabilities).
- $\sum_{j=1}^{r} P_{ij} = 1$, for $1 \le i \le r$, since once the process enters state i it either stays in the same state or transits to another state at the next time instant.

B.3.2 Continuous Time Markov Chains

Here, $X(t)$ assumes only discrete values, with r either finite or infinite as before and t assumes a continuous range of values in the interval $(-\infty, \infty)$. Once the process enters state i, the duration it stays in the state is exponentially distributed, with the parameter a function of i and state j, the next state in the transition. The characterization of the process involves $r(r-1)$ exponential distribution functions and from this a transition matrix P (of dimension $r \times r$) that characterizes the transitions over a small interval.

B.3.3 Discrete Time Markov Processes

In this case, $X(t)$ assumes a continuous range of values and t assumes discrete values. We use the notation $X_k \equiv X(t_k)$, $k = 0,1,2,\ldots$. One such case is the time series formulation given by:

$$X_{k+1} = g(X_k; W_k; \theta) \qquad (B.6)$$

where $W_k = k = 0,1,2,\ldots$, is a sequence of independent random variables and $X_0 = x_0$. If

$$X_{k+1} = g(X_k, X_{k-1},\ldots,X_{k-l}, W_k, \theta) \qquad (B.7)$$

then $X(t)$ is not a Markov process.[3]

These types of processes are useful in condition-based maintenance where the condition of a system is monitored at discrete time instants (often periodic), the model parameter (θ) is updated based on the data and then either Equation (B.6) or Equation (B.7) is used to predict the future condition and for planning maintenance actions.

---

[3] In this case, by expanding the state space so that it is a vector given by $Y_k^T = (X_k, X_{k-1},\ldots,X_{k-l})$, Y_k has the Markov property.

B.3.4   Continuous Time Markov Processes

In these processes, both $X(t)$ and t assume continuous ranges of values. They may be characterized through stochastic differential equations and are beyond the scope of the book.

B.4   Point Processes

A point process is a continuous time stochastic process characterized by events that occur randomly along the time continuum. Events can be failures, PM (preventive maintenance) or CM (corrective maintenance) actions, shocks that cause damage, and so on, occurring in an uncertain manner over time. The theory of point processes is very rich, and a variety of such processes have been formulated and studied; we discuss a few of these below.

B.4.1   Counting Processes

A point process $\{N(t),\ t \geq 0\}$ is a *counting process* if it represents the number of events (for example, the total number of failures, shocks, etc, over a time interval) that have occurred until time t. It must satisfy:

1. $N(t) \geq 0$.
2. $N(t)$ is integer valued.
3. If $s < t$, then $N(s) \leq N(t)$.
4. For $s < t$, $\{N(t) - N(s)\}$ is the number of events in the interval $(s, t]$.

We shall confine ourselves to $t \geq 0$. The behavior of $N(t)$, for $t \geq 0$, depends on whether or not $t = 0$ corresponds to the occurrence of an event. The analysis of the case with $t = 0$ corresponding to the occurrence of an event is simpler than the alternate case. Also, we assume that $N(0) = 0$.

A counting process $\{N(t),\ t \geq 0\}$ is said to have *independent increments* if, for all choices $0 \leq t_1 < t_2 < \ldots < t_n$, the $(n - 1)$ random variables

$$\left\{ N(t_2) - N(t_1) \right\}, \left\{ N(t_3) - N(t_2) \right\}, \ldots, \left\{ N(t_n) - N(t_{n-1}) \right\}$$

are independent. A counting process $\{N(t),\ t \geq 0\}$ is said to have *stationary independent increments* if, for each $s > 0$, $\{N(t_2 + s) - N(t_2)\}$ and $\{N(t_1 + s) - N(t_1)\}$ have the same distribution function, that is, if the distribution function of $\{N(t + s) - N(t)\}$ does not depend on t.

In a point process, the time between events, the number of events in an interval, and the probability of an event occurring in a short interval are all random variables. As a result, the characterization of a point process may be done in three different (equivalent) ways. For certain processes, one form of characterization is a lot simpler than the others.

Two special counting processes of particular importance to reliability modeling are the following: (i) the *Poisson process* and (ii) the *renewal process*. Both of these have been used extensively in building models to solve a variety of decision problems in reliability and maintenance.

B.5   Poisson Processes

B.5.1   *Homogeneous Poisson Process (HPP)*[4]

We give three ways of characterizing the process.[5]

Definition B.1

A counting process, $\{N(t),\ t \geq 0\}$, is a homogeneous Poisson process if:

1. $N(0) = 0$.
2. The process has independent increments.
3. The number of events in any interval of length t is distributed according to a Poisson distribution with parameter λt, that is,

$$P\{N(s+t) - N(s)\} = \frac{e^{-\lambda t}(\lambda t)^n}{n!} \tag{B.8}$$

$n = 0,1,2,\ldots$, and for all s and $t \geq 0$.

Definition B.2

Consider a counting process. Let X_1 denote the time instant of the first event occurrence, and for $j \geq 2$, let X_j denote the time interval between the $(j-1)$th and jth events. The counting process is a homogeneous Poisson process with parameter λ if the sequence $X_j,\ j \geq 1$, is a sequence of independent and identically distributed exponential random variables with mean $(1/\lambda)$.

Definition B.3

A counting process $\{N(t),\ t \geq 0\}$ is a homogeneous Poisson process if:

1. The probability of an event occurring in $[t,\ t + \delta t)$ is $\lambda \delta t + o(\delta t)$.
2. The probability of two or more events occurring in $[t,\ t + \delta t)$ is $o(\delta t)$, and
3. The occurrence of an event in $[t,\ t + \delta t)$ is independent of the number of events in $[0,\ t)$.

λ is called the *intensity* of the process.

B.5.1.1   Expected Number of Events in $[0, t)$

Let $M(t)$ denote the expected number of events in $[0, t)$. From we have:

---

[4] A homogenous Poisson process is also often called a *stationary Poisson process*.
[5] For a proof of their equivalence, see Ross (1970).

$$M(t) = E\left[N(t)\right] = \sum_{n=0}^{\infty} n \frac{e^{-\lambda t} (\lambda t)^n}{n!} = \lambda t \tag{B.9}$$

B.5.1.2 Variance of Number of Events in $[0,t)$

The variance is given by:

$$Var\left[N(t)\right] = E\left[N^2(t)\right] - \left\{E\left[N(t)\right]\right\}^2 = \lambda t. \tag{B.10}$$

B.5.2 Non-Homogeneous Poisson Process (NHPP)[6]

In a homogeneous Poisson process (HPP), the probability of an event occurring in $[t, t+\delta t)$ is $\lambda \delta t + o(\delta t)$, with λ a constant. A non-stationary, or non-homogeneous, Poisson process (NHPP) is a natural extension in which λ changes with time

---

Definition B.4

A counting process $\{N(t),\ t \geq 0\}$ is a non-homogeneous Poisson process if:

1. $N(0) = 0$.
2. $\{N(t),\ t \geq 0\}$ has independent increments.
3. $P\{N(t+\delta t) - N(t) = 1\} = \lambda(t)\delta t + o(\delta t)$.
4. $P\{N(t+\delta t) - N(t) \geq 2\} = o(\delta t)$.

$\lambda(t)$ is called the *intensity function*. Define the cumulative intensity function as:

$$\Lambda(t) = \int_0^t \lambda(x)\ dx \tag{B.11}$$

Definition B.5

A counting process $\{N(t),\ t \geq 0\}$ is a non-homogeneous Poisson process if:

1. $N(0) = 0$.
2. The process has independent increments.
3. The probability of j events occurring in an interval of length t is given by:

$$p_n(s, s+t) = P\{N(s+t) - N(s) = j\} = \frac{e^{-\{\Lambda(s+t) - \Lambda(s)\}} \{\Lambda(s+t) - \Lambda(s)\}^j}{j!} \tag{B.12}$$

for $j \geq 0$ and for all s and $t \geq 0$. When $s = 0$, define $p_n(t) = p_n(t, 0)$.

---

[6] A non-homogenous Poisson process is also often called a *non-stationary Poisson process*.

Comment: The characterization of an NHPP in terms of the time between events is not easy.

B.5.2.1   Expected Number of Events in $[0, t)$

The expected number of events in $[0, t)$, $M(t)$, is given by:

$$M(t) = E\left[N(t)\right] = \sum_{n=0}^{\infty} n p_n(0, t) = \sum_{n=0}^{\infty} n \frac{e^{-\Lambda(t)}\left\{\Lambda(t)\right\}^n}{n!} = \Lambda(t) \tag{B.13}$$

B.5.2.2   Variance of Number of Events in $[0, t)$

The variance is given by:

$$Var\left[N(t)\right] = E\left[N^2(t)\right] - \left\{E\left[N(t)\right]\right\}^2 = \Lambda(t). \tag{B.14}$$

B.6   Renewal Processes

B.6.1   *Ordinary Renewal Processes*

This is a generalization of the homogeneous Poisson process in the sense that the inter-event times are independent and identically distributed with an arbitrary distribution function as opposed to having an exponential distribution.

Definition B.6

A counting process $\{N(t), \ t \geq 0\}$ is an ordinary renewal process if:

1. $N(0) = 0$.
2. X_1, the time to occurrence of the first event (from $t = 0$) and $X_j, j \geq 2$, the time between the $(j - 1)$th and the jth events, are a sequence of independent and identically distributed random variables with distribution function $F(x)$.
3. $N(t) = \text{Max}\{n : S_n \leq t\}$, where:

$$S_0 = 0, \qquad S_n = \sum_{i=1}^{n} X_i, \ n \geq 1 \tag{B.15}$$

Note: The homogeneous Poisson process is a special case of the ordinary renewal process with $F(x)$ an exponential distribution function.

Events for a renewal process are often referred to as *renewals*.

Distribution of $N(t)$

Note that S_n is the time instant for the nth renewal (or event) and is the sum of n independent and identically distributed random variables. Since the $X_i's$ are distributed with distribution function $F(x)$, from the result in Appendix A, the distribution of S_n is given by the n-fold convolution of $F(x)$ with itself - i.e.,

$$P(S_n \le x\} = F^{(n)}(x) = F(x) * F(x) * \ldots\ldots * F(x) \tag{B.16}$$

Note that $N(t) \ge n$ if and only if $S_n \le t$. As a result, $p_n(t) \equiv P\{N(t) = n\}$, is given by

$$p_n(t) = P\{N(t) \ge n\} - P\{N(t) \ge (n+1)\} = P\{S_n \le t\} - P(S_{n+1} \le t\} \tag{B.17}$$

for $n = 0, 1, \ldots$, where $S_0 \equiv 0$. Since

$$P\{S_n \le t\} = F^{(n)}(t) \tag{B.18}$$

where $F^{(0)} \equiv 1$, we have:

$$p_n(t) = F^{(n)}(t) - F^{(n+1)}(t) \tag{B.19}$$

From this, expressions for the moments of $N(t)$ can be obtained. Of particular interest are the expected number and variance of the number of renewals in $[0,t)$.

B.6.1.1   Expected Number of Renewals in $[0,t)$

The expected number of renewals in $[0,t)$, $M(t)$, is given by:

$$M(t) = E[N(t)] = \sum_{n=0}^{\infty} n \, p_n(t) \tag{B.20}$$

Using Equation (B.17), this can be written as:

$$M(t) = \sum_{n=0}^{\infty} n\{F^{(n)}(t) - F^{(n+1)}(t)\} = \sum_{n=1}^{\infty} F^{(n)}(t) \tag{B.21}$$

Using Laplace transforms,[7] it can be shown that:

$$M(t) = F(t) + \int_0^t M(t-x) f(x) dx \tag{B.22}$$

An alternative derivation for $M(t)$ based on conditional expectation, is as follows.[8] Conditioned on X_1, the time to first failure, $M(t)$ can be written as:

$$M(t) = \int_0^{\infty} E[N(t)|X_1 = x] f(x) dx \tag{B.23}$$

But,

---

[7] The Laplace transform of a function $g(t)$ is given by $\hat{g}(s) = \int_0^{\infty} g(t) e^{-st} dt$. The inverse transform is given by $g(t) = \int \hat{g}(s) e^{st} ds$. The Laplace transform of an n-fold convolution is given by the product of the Laplace transforms of the variables inside the integral sign. As a result, $\hat{f}^{(n)}(s) = [\hat{f}(s)]^n$. Details of this can be found in any good book on mathematical analysis.

[8] Conditional expectation is discussed in Section A.6.3.

$$E\left[N(t)|X_1 = x\right] = \begin{cases} 0, & \text{if } x > t \\ 1 + M(t-x), & \text{if } x \leq t \end{cases} \tag{B.24}$$

Note that one is using the "renewal property" in deriving the above expression. If the first failure occurs at $x \leq t$, then the renewals over $(t - x)$ occur according to an identical renewal process and hence the expected number of renewals over this period is $M(t-x)$. Using Equation (B.24) in Equation (B.23) yields Equation (B.22).

Equation (B.22) is called the *renewal integral equation* and $M(t)$ is called the *renewal function* associated with the distribution function $F(t)$. $M(t)$ plays an important role in reliability analysis. In general, it is difficult to obtain $M(t)$ analytically and one needs to evaluate the function computationally.

The *renewal density function, m(t)*, is given by:

$$m(t) = \frac{dM(t)}{dt} \tag{B.25}$$

and it is easily shown that:

$$m(t) = f(t) + \int_0^t m(t-x) f(x) dx \tag{B.26}$$

where $f(t)$ is the density function associated with $F(t)$.

B.6.1.2   Variance of the Number of Renewals in $[0,t)$

The variance of $N(t)$ is given by:

$$Var\left[N(t)\right] = \sum_{n=1}^{\infty} \left[n - M(t)\right]^2 p_n(t) \tag{B.27}$$

Using Equation (B.19) in Equation (B.27) and after some analysis, we have:

$$Var\left[N(t)\right] = \sum_{n=1}^{\infty} (2n-1) F^{(n)}(t) - \left[M(t)\right]^2 \tag{B.28}$$

Even for simple distribution functions it is not possible to derive analytical expressions for $Var[N(t)]$ and one needs to use computational methods.

B.6.2   Delayed Renewal Process

A delayed renewal process is a counting process similar to the ordinary renewal process with the following important differences:

1. X_1, the time to the first event, is a non-negative random variable with distribution function $F(x)$.
2. $X_j, j \geq 2$, the time intervals between the jth and $(j - 1)$th events, are independent and identically distributed random variables with a distribution function $G(x)$ different from $F(x)$.

When $G(x)$ and $F(x)$ are the same, the delayed renewal process reduces to an ordinary renewal process.

B.6.2.1 Expected Number of Renewals in $[0,t)$

Let $M_d(t)$ denote the expected number of renewals over $[0,t)$ for the delayed renewal process. Using the conditional approach discussed earlier, we have:

$$M_d(t) = E\big[N(t)\big] = \int_0^\infty E[N(t)|X_1 = x]\, f(x)dx \tag{B.29}$$

with

$$E\big[N(t)|X_1 = x\big] = \begin{cases} 0, & \text{if } x > t \\ 1 + M_g(t-x), & \text{if } x \le t \end{cases} \tag{B.30}$$

where $M_g(t)$ is the renewal function associated with the distribution function $G(t)$. This follows from the fact that, if the first event occurs at $x \le t$, then, over the interval (x,t), the events occur according to a renewal process with distribution $G(\cdot)$. As a result, $M_d(t)$ is given by:

$$M_d(t) = F(t) + \int_0^t M_g(t-x)\, f(x)\, dx \tag{B.31}$$

B.6.3 Alternating Renewal Process

In an ordinary renewal process, the inter-event times are independent and identically distributed. In an alternating renewal process, the inter-event times are all independent but not identically distributed. More specifically, the odd-numbered inter-event times $X_1, X_3, X_5, \ldots$ have a common distribution function $F(x)$ and the even-numbered ones $X_2, X_4, X_6, \ldots$ have a common distribution function $G(x)$.

B.6.3.1 State of an Item at a Given Time

At any given time, an item can be either in its working state or in a failed state and undergoing repair. A variable of interest is the probability $P(t)$ that the item is in its working state at time t. $P(t)$ is given by:[9]

$$P(t) = 1 - F(t) + \int_0^t P(t-x)\, h(x)dx \tag{B.32}$$

---

[9] For a proof, see Cox and Miller (1965).

where $H(x)$ is the convolution of $F(x)$ and $G(x)$; that is:

$$H(x) = F(x) * G(x) \qquad \text{(B.33)}$$

and $h(x) = dH(x) / dx$.

B.6.4 Renewal Reward Theorem

Consider an ordinary renewal process with time between events (renewals) $X_1, X_2, X_3, \ldots$. Suppose that a reward (or cost) of Z_i is earned (incurred) at the time of the ith renewal. Then, the total reward (cost) earned by time t is given by:

$$Z(t) = \sum_{i=1}^{N} Z_i \qquad \text{(B.34)}$$

where $N(t)$ is the number of renewals in $[0, t)$.[10] If $E[|Z_i|]$ and $E[X_i]$ are finite, then:

$$\lim_{t \to \infty} E\left[\frac{Z(t)}{t}\right] \to \frac{E[Z_i]}{E[X_i]}. \qquad \text{(B.35)}$$

This result is very useful in the asymptotic cost analysis of various maintenance policies and is discussed in Chapters 14 and 15.[11]

B.7 Marked Point Processes

A marked point process is a point process with an auxiliary variable, called a *mark*, associated with each event.[12] Let Y_i, $i \geq 1$, denote the mark attached to the ith event.

B.7.1 Simple Marked Point Process

A simple marked point process is characterized by:

1. $\{N(t), t \geq 0\}$, a stationary Poisson process with intensity λ;
2. A sequence of independent and identically distributed random variables $\{Y_i\}$, called marks, which are independent of the Poisson process.

This point process is also called a *compound Poisson process*. Various extensions (for example, a non-stationary point process, and marks constituting a dependent sequence, to name a few) yield more complex marked point processes.

---

[10] $Z(t)$ is a cumulative process (see Section B.7.2).

[11] For a proof, see Ross (1970).

[12] An example of this is where an item gets damaged due to the impact of shocks that occur randomly and the mark denotes the severity of shock.

B.7.2 Cumulative Process

A cumulative process, $X(t)$, is given by:

$$X(t) = \sum_{i=1}^{N(t)} X_i \tag{B.36}$$

with $\{N(t),\ t \geq 0\}$ a counting process and X_i representing a mark attached to event i. The mark is a random variable with distribution function $G(\cdot)$. Note that here, $X(t)$ experiences random jumps at discrete points in time, corresponding to the event times for the point process, and hence is discontinuous at these points.

References

Cox, D.R. and Miller, H.D. (1965) *The Theory of Stochastic Processes*, John Wiley & Sons, Inc., New York.

Ross, S.M. (1970) *Applied Probability Models*, Prentice-Hall, New York.

Appendix C: Introduction to the Theory of Statistics

C.1 Introduction

In this appendix we review some basic material from the theory of statistics and focus on two topics: (i) descriptive statistics and (ii) inferential statistics. The former deals with concepts and techniques and is used in extracting information from a sample data set. The latter uses this information to draw inferences concerning the population from which the sample was drawn. Both of them are very important in the context of data-based model building, discussed in Part D of the book. Our discussion of the topic is brief and details can be found in most introductory books on statistics, see, for example, Dalgaard (2008) and Hogg, McKean, and Craig (2012).

C.2 Descriptive Statistics

Descriptive statistics are statistical methods that are used in the summarization and presentation of data. Summary values provide concise measures of the basic information content; graphical presentations give the overall picture. Note that the entire collection of data is called the *sample*. In this context, each individual measurement is called an *observation* and the number of observations in the sample is called the *sample size* and denoted n.

One begins with a sample, that is, a set of observations (measurements, responses, etc.), and performs various calculations and operations in order to focus and understand the information content of the sample data. The word "statistic" is used to refer to any quantity calculated from the data – averages, ranges, percentiles, and so forth. In this section, we will look at a number of statistics that are intended to describe the sample and summarize the sample information. These statistics also provide a foundation for building models for degradation, failure, and maintenance.

Introduction to Maintenance Engineering: Modeling, Optimization, and Management, First Edition.
Mohammed Ben-Daya, Uday Kumar, and D.N. Prabhakar Murthy.
© 2016 John Wiley & Sons, Ltd. Published 2016 by John Wiley & Sons, Ltd.

Let $y = \{y_1, y_2, \ldots, y_n\}$ denote the values of the random variables $Y = \{Y_1, Y_2, \ldots, Y_n\}$ after the events have all taken place. If the data available are $y = \{y_1, y_2, \ldots, y_n\}$ then they are called *complete data*. Often, the data available are not complete and are called *incomplete data*. This arises when one or more of the pieces of data are censored. There are different kinds of censoring and the two that are relevant in the context of maintenance and reliability data are the following:

- *Right-censored data:* The data for Y_i are said to be right censored if the data available are $Y_i > a_i$, with a_i known.
- *Interval-censored data:* The data for Y_i are said to be interval censored if the data available are $a_i < Y_i \le b_i$, with a_i and b_i known.

In the context of maintenance and reliability data, right-censored data are often referred to as *service times*. These represent the ages of items that are still in operation whereas *censored data* (given by $Y_i = y_i$) refers to all items that have failed.

The discussion in this section is for complete data and $Y = \{Y_1, Y_2, \ldots, Y_n\}$ are n statistically independent variables with distribution function $F(y; \theta)$.

C.2.1 Fractiles

The *p-fractile* (related terms are *percentile* and *quantile*) of a continuous probability distribution is any value y_p such that $F(y_p) = p$, where $0 \le p \le 1$. For a continuous cumulative distribution function, y_p is almost always uniquely determined. In cases where it is not, the *p-fractile* can be taken to be any value in an interval, and there are several commonly used definitions for the term.

The *p-fractile* of a sample is defined as that value y_p such that at least a proportion p of the sample lies at or below y_p and at least a proportion $1 - p$ lies at or above y_p. This value may also not be unique and there are a several alternative definitions that may be used. We define the *p-fractile* of a sample of observed values as follows: Let $k = [p(n + 1)]$ and $d = p(n + 1) - k$, where [x] denotes the integer part of x. If $k = 0$ or n (corresponding to very small or very large values of p), the fractile is not defined. If $k = 1, \ldots, n - 1$, then y_p is given by:

$$y_p = y_{(k)} + d\left(y_{(k+1)} - y_{(k)}\right) \tag{C.1}$$

Of particular interest in descriptive statistics are the 0.25-, 0.50-, and 0.75-fractiles, called the *quartiles*, and denoted Q_1, Q_2, and Q_3. Fractiles also have important applications in reliability, where the interest is in fractiles for small values of p.

C.2.2 Measures of Center

The most common measures of the center of a sample (also called measures of location, or simply averages) are the *sample mean* and *median*. The sample median is the 0.50-fractile or Q_2 and is a natural measure of center, since at least one-half of the data lie at or above this value and at least one-half lie at or below Q_2. The sample mean of Y, denoted by $\bar{y}$, is the simple arithmetic average given by:

$$\bar{y} = \frac{1}{n}\sum_{i=1}^{n} y_i \tag{C.2}$$

For many statistical purposes, the sample mean is the preferred measure. It is the basis for many statistical inference procedures and is a "best" measure for these purposes for many types of populations. A problem is that the mean is sensitive to extreme values (*outliers*) in the data and may provide a somewhat distorted indication of center as a result of the outliers. In such cases, the median, which is not affected by extreme values, provides a more meaningful measure of the location of the center of the data.

Although both are measures of center, $\bar{y}$ and Q_2 measure this differently, and a comparison of the two values provides additional information about the sample (and, by inference, about the population from which it was drawn). If the sample is perfectly symmetrical about its center, the mean and median are identical. If the two differ, this is an indication of *skewness*. If $Q_2 < \bar{y}$, the data are skewed to the right; if $Q_2 > \bar{y}$, the data are skewed to the left.[1]

Another approach to dealing with the distortion caused by outliers is to calculate a trimmed mean – remove a fixed proportion of both the smallest and largest values from the data and calculate the average of the remaining values.[2]

A few other measures are sometimes used. These include the mode, which is not of much use in statistical inference, and various measures that may be defined as functions of fractiles, for example, $(Q_3 - Q_1)/2$, $(y_{0.90} - y_{0.10})/2$, and so forth. In analyzing data, we will use the mean and median as measures of center, and occasionally look at the trimmed mean as well.

C.2.3   Measures of Dispersion

A second descriptive measure commonly used in statistical analysis is a measure of dispersion (or spread) of the data. These measures reflect the variability in the data and are important in understanding the data and in properly interpreting many statistical results.

The most important measures of dispersion for most purposes are the *sample variance* and *standard deviation*. The sample variance, s^2, is given by:

$$s^2 = \frac{1}{n-1}\sum_{i=1}^{n}(y_i - \bar{y})^2 = \frac{1}{n-1}\left\{\sum_{i-1}^{n}y_i^2 - \frac{1}{n}\left(\sum_{i=1}^{n}y_i\right)^2\right\} \tag{C.3}$$

The *sample standard deviation* is $s = \sqrt{s^2}$, and is the preferred measure for most purposes since it is in the units of the original data.

---

[1] Failure data and the distributions used to model these data are often skewed to the right, which results from (usually) small numbers of items with exceptionally long lives.

[2] Minitab removes the smallest and largest 5% (using the nearest integer to $0.05n$). This usually removes the values causing the distortion and provides a more meaningful measure.

Another measure of variability sometimes used is the *interquartile range, I,* given by $I = Q_3 - Q_1$. An advantage of the interquartile range is that it is not affected by extreme values. A disadvantage is that it is not readily interpretable as is the standard deviation.

Finally, a useful measure of dispersion in certain applications is the *coefficient of variation,* defined as c.v. $= s / \bar{y}$.

C.2.4   Measures of Relationship

When the data include two or more variables, measures regarding the relationship between the variables are of interest. One such measure of strength of relationship for two variables is the *sample correlation coefficient, r.* Consider a sample of bivariate data (x_i, y_i), $i = 1, 2, \ldots, n$. The sample correlation coefficient is given by:

$$r = \frac{\dfrac{1}{n-1} \sum_{i=1}^{n} (x_i - \bar{x})(y_i - \bar{y})}{s_x s_y}, \tag{C.4}$$

where s_x and s_y are the standard deviations of the two variables, X and Y, respectively. The numerator of this expression, called the *sample covariance*, is itself used as a measure of relationship in certain applications.

r lies in the interval $[-1, 1]$, with the values -1 and $+1$ indicating that the variables are co-linear, with lines sloping downward and upward, respectively. The general interpretation is that values close to either extreme indicate a strong relationship and values close to zero indicate very little relationship between the variables.

C.2.5   Histograms

The most commonly used graphical presentation of data is the *histogram.* To form a histogram, the observations are grouped into intervals (usually contiguous intervals of equal length) and counts are made of the number of observations falling into each interval. These counts are called *frequencies*, and the set of intervals and associated counts is a *frequency distribution.* A *histogram* is simply a plot of the frequency distribution. A second type of frequency distribution is a table of counts of occurrence of events, for example, failures of components or modes of failure of a system.

C.2.6   Empirical Distribution Function (EDF)

One of the key tools for investigating the distribution underlying the data is the sample equivalent of $F(y)$, denoted $\hat{F}(y)$ and called the *empirical distribution function* (EDF). The EDF plots as a "step-function" with steps at data points. The form of the function depends on the type of population from which the sample was drawn. On the other hand, the procedure is non-parametric, in the sense that no specific form is assumed in calculating the EDF.

The EDF and its calculation require the reordering of the data. The *ordered data* (for the case of complete data) are given by $y = \{y_{(1)}, y_{(2)}, \ldots, y_{(n)}\}$ with $y_{(i)} \leq y_{(i+1)}$, $i = 1, \ldots, n$. The EDF is given by:

$$\hat{F}(y) = \begin{cases} 0 & y < y_{(1)} \\ \dfrac{1}{n+1} & y_{(i)} \le y < y_{(i+1)}, i = 1, \ldots, n-1 \\ 1 & y \ge y_{(n)} \end{cases} \tag{C.5}$$

For incomplete (right-censored) data, the procedure is slightly more involved and is as follows. To calculate $\hat{F}(t)$, the observations are ordered, including both censored and uncensored values in the ordered array. Suppose that m observations in the ordered array are uncensored. Denote these $t_1, t_2, \ldots, t_m$. These are the locations of the steps in the plot of the EDF. To determine the heights of the steps, for $i = 1, \ldots, m$, form the counts $n_i =$ "number at risk" = number of observations greater than or equal to t_i in the original ordered array, and $d_i =$ number of values tied at t_i (=1 if the value is unique), then calculate the "survival probabilities":

$$\hat{S}_1 = 1 - \frac{d_1}{n_1} \quad \text{and} \quad \hat{S}_i = \left(1 - \frac{d_i}{n_i}\right)\hat{S}_{i-1}, \quad i = 2, \ldots, m \tag{C.6}$$

and

$$\hat{F}(t_i) = 1 - \hat{S}(t_i), \quad 1 \le i \le m \tag{C.7}$$

C.3   Inferential Statistics

Inferential statistics are statistical methods that use the information from a sample to draw inferences concerning the population from which the sample was drawn. Inferences about populations typically are expressed in terms of population *parameters* (for example, the shape and scale parameters of the Weibull distribution) or related population characteristics such as the mean, median, and so forth. This approach is called *parametric analysis* and depends crucially on, amongst other things, specific model assumptions (such as the form of distribution function) regarding the source of the data. For some inference problems, there exist alternative techniques that do not involve population parameters or require that the form of the model be known. These are variously called *distribution-free* or *non-parametric* procedures.

The following three topics from inferential statistics are relevant in the context of stochastic or probabilistic models.

C.3.1   *Parameter Estimation*

Parameter estimation deals with methods that use the data to assign numerical values to any parameters whose values are unknown. The data consist of a random sample of size n from the population in question and we wish to estimate a parameter θ (which may be a vector). An estimation procedure is a formula or equation (explicit or implicit) for estimating θ. The result is denoted $\hat{\theta}$. To express $\hat{\theta}$ as a function of the sample, we write $\hat{\theta}(Y_1, Y_2, \ldots, Y_n)$ or

$\hat{\theta}(y_1, y_2, ..., y_n)$, depending on whether we are considering $\hat{\theta}$ to be a random variable or a numerical value. To distinguish these, the first is called an *estimator* (or *point estimator*) and the second an *estimate* (or *point estimate*).

It is important to recognize that there is uncertainty in this process – different samples from the same population will lead to different numerical results for $\hat{\theta}$. This must be taken into account in selecting an appropriate methodology and in interpreting the results. One approach to this problem is *confidence interval estimation*. Confidence intervals are intervals of numbers determined in such a way that the probability that the interval contains the true value of θ is a specified quantity, γ. Typically, γ is taken to be 0.95 or 0.99 and the results are called 95% and 99% confidence intervals, respectively.

Various methods have been proposed for parameter estimation. The two well-known ones are: (i) the method of moments and (ii) the maximum likelihood method.

C.3.1.1   Method of Moments

The method of moments is based on expressing the moments of a distribution (also called *population moments*) in terms of the parameters of the distribution, equating sample moments to population moments, and solving the resulting equations for the unknown parameters. The population moments M_i (also called *moments about zero*) are given by Equation (A.6) and the corresponding sample moments $\hat{m}_i$ are given by:

$$\hat{m}_i = \frac{1}{n}\sum_{j=1}^{n} y_j^i, \quad i = 1, 2, ... \tag{C.8}$$

If more convenient, the population and sample *central moments* or *moments about the mean* given by Equation (A.7) may be used instead. In this case, the sample central moments $\hat{\mu}_i$ are given by:

$$\hat{\mu}_i = \frac{1}{n}\sum_{j=1}^{n}(y_j - \bar{y})^i, \quad i = 1, 2, ... \tag{C.9}$$

The conventional approach to estimating k parameters is to use the first k moments and apply the method as indicated.

C.3.1.2   Method of Maximum Likelihood

The method of maximum likelihood (ML) is a broadly applicable estimation procedure. In applying the method, we formulate the *likelihood function,* which is basically the joint distribution of the data, expressed as a function of the parameters of the distribution and the data $Y_1, Y_2, ..., Y_n$. The *maximum likelihood estimator* (MLE) of the parameters is the set of values that maximizes this function. The idea is that the likelihood function, in a sense, represents the joint probability of the data and we are choosing parameter values that maximize the probability of obtaining the sample actually observed. (This is often called the *likelihood principle*.)

Parameters of a Failure Distribution Function

If the failure times of n items are statistically independent, the likelihood function is given by:

$$L(\theta) = \prod_{i=1}^{n} f(y_i; \theta) \qquad (C.10)$$

where $f(y; \theta)$ is the failure density function. In the case of incomplete data (failures statistically independent), reorder the data set so that for the first r data, $Y_i = y_i$ and for the remaining data, we have $Y_i > a_i$, $i = r+1, \ldots, n$, and the likelihood function is given by:

$$L(\theta) = \prod_{i=1}^{r} f(y_i; \theta) \prod_{i=r+1}^{n} \left[1 - F(a_i; \theta) \right] \qquad (C.11)$$

where $F(y; \theta)$ is the failure distribution function.

Maximization of the likelihood function is straightforward if the parameter space is unbounded and the distribution is differentiable with respect to θ (or the components of θ in the multi-parameter case). The likelihood equations are obtained by equating derivatives to zero and solving the equations (numerically, if necessary) to obtain estimates of the parameters. In other cases, numerical methods of maximization are used.

C.3.2   Hypothesis Testing

In hypothesis testing, various hypotheses (usually two) are formulated concerning a population, and one of these is selected on the basis of sample information. The hypotheses are usually expressed in terms of relevant population parameters, but may be in terms of other population characteristics as well (for example, a population fractile or a function of the parameters such as a coefficient of variation).[3]

In the classical approach to hypothesis testing, two hypotheses, the *null hypothesis,* H_0, and the *alternative hypothesis,* H_1, are formulated. By definition, H_0 is the hypothesis that is tested; H_1 is the hypothesis that is accepted if H_0 is rejected. In essence, H_0 is tested by determining statistically whether or not the data support the hypothesis. If the data are found to be unlikely to occur under H_0, the hypothesis is rejected and H_1 is accepted. The definition of "unlikely" requires specification of the probability of occurrence of an event and, in parametric analysis, of a probability model underlying the data. If the data do not lead to rejection of H_0, the statistical conclusion is that we fail to reject H_0 rather than that we accept it.

Because of the way in which the null and alternative hypotheses are structured, one or the other must be true. Correspondingly, there are two possible errors that may occur:

- *Type I error:* Reject H_0 when it is true;
- *Type II error:* Fail to reject II_0 when it is false.

The probability of a Type I error, called the *level of significance* of the test and denoted α, is under the control of the data analyst. Typical values of $\alpha = P\{\text{Reject } H_0 | H_0 \text{ true}\}$ used in data analysis

---

[3] Fractiles and coefficients of variation are discussed in Sections C.2.1 and C.2.3.

are 0.10, 0.05, and 0.01 (called *testing at the 10% level*, etc.). The probability of a Type II error, denoted β, depends on α, the underlying probability distribution, and the distance between the hypothesized value and the true value. $(1 - \beta)$ is called the *power* of the test. The "best" tests are those with the highest power (i.e., the highest probability of rejecting H_0 when it is false).

C.3.3 Goodness-of-Fit Tests

The primary objective of a goodness-of-fit test is to investigate how well a specified model (for example, a probability distribution) "fits" a given set of data. This type of test is used for determining the adequacy of a model. One uses statistical tests to judge the adequacy of the fit. This provides a rigorous framework for the analysis. It also provides a method of comparing fits of a data set to many different model formulations (for example, probability distributions) – an important issue in model building.

The two well-known goodness-of-tests for models involving distribution functions are (i) the Kolmogorov–Smirnov and (ii) the Anderson–Darling tests for fitting continuous distributions.

C.3.3.1 Kolmogorov–Smirnov Test

We assume that the null-hypothesized distribution $F_0(\cdot)$ is completely specified. In this case, the Kolmogorov–Smirnov (K–S) test is non-parametric, in that the distribution of the test statistic does not depend on $F_0(\cdot)$. The test statistic, denoted D_n, is the maximal difference between the EDF and $F_0(\cdot)$. This is calculated as $D_n = \max\left\{D_n^-, D_n^+\right\}$, with:

$$D_n^- = \max_{i=1,\ldots,n}\left[\frac{i}{n} - F_0\left(y_{(i)}\right)\right], \; D_n^+ = \max_{i=1,\ldots,n}\left[F_0\left(y_{(i)}\right) - \frac{i-1}{n}\right] \tag{C.12}$$

where the $y_{(i)}$ are the order statistics. H_0 is rejected at level α if D_n exceeds the critical value $d_\alpha(n^{1/2} + 0.11n^{-1/2} + 0.12)^{-1}$, where $d_\alpha = 1.224$, 1.358, and 1.628, for $\alpha = 0.10$, 0.05, and 0.01, respectively.

C.3.3.2 Anderson–Darling Test

The Anderson–Darling (A–D) test is also non-parametric if $F_0(\cdot)$ is completely specified. The test statistic is:

$$A^2 = \frac{-1}{n}\sum_{i=1}^{n}(2i-1)\log\{(F_0\left(y_{(i)}\right)\left[1 - F_0\left(y_{(n-i+1)}\right)\right]\} - n \tag{C.13}$$

The critical values of A^2 do not depend on n. For $\alpha = 0.10$, 0.05, and 0.01, they are $a_\alpha = 1.933$, 2.492, and 3.857.

References

Dalgaard, P. (2008) *Introductory Statistics with R*, 2nd edition. Springer-Verlag.
Hogg, R.V., McKean, J., and Craig, A.T. (2012) *Introduction to Mathematical Statistics*, 7th edition. Macmillan Publishing Co., Inc., New York.

Appendix D: Introduction to Optimization

D.1 Introduction

Optimization problems involve one or more decision makers, each with a set of decision variables and an objective function. Decision makers select their optimal decisions to maximize or minimize their objective functions. Optimization problems can be grouped broadly into two categories: (i) deterministic optimization (appropriate if the uncertainty is very insignificant, so that it can be ignored) and (ii) stochastic optimization (when uncertainty is significant and needs to be taken into account). Each of these can be subdivided, based on the number of decision makers involved, into (i) single decision maker and (ii) two or more decision makers (also referred to as *game theory problems*). A further subdivision of each, based on the objective function, is: (i) scalar objective function and (ii) vector objective function (also referred to as *multi-criteria decision problems*). The next subdivision, based on the decision variables, is: (i) static (the variables do not change with time, also referred to as *static optimization*) and (ii) dynamic (where the variables change with time, also referred to as *dynamic optimization*). Finally, each of these may be subdivided into three categories: (i) discrete optimization (decision variables are given by sets of discrete values), (ii) continuous optimization (decision variables assume a continuous range of values), and (iii) mixed optimization (some variables are discrete-valued and others continuous-valued). Figure D.1 shows the hierarchical decomposition of optimization problems for the deterministic case and a single decision maker.

In this appendix we discuss the following cases:

- *Case A:* Deterministic and static optimization with one decision maker and a scalar objective function, discussed in Section D.2.
- *Case B:* Deterministic and static optimization with one decision maker and a vector objective function, discussed in Section D.3.

Introduction to Maintenance Engineering: Modeling, Optimization, and Management, First Edition.
Mohammed Ben-Daya, Uday Kumar, and D.N. Prabhakar Murthy.
© 2016 John Wiley & Sons, Ltd. Published 2016 by John Wiley & Sons, Ltd.

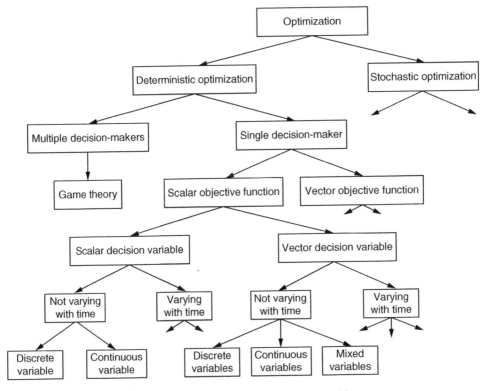

Figure D.1 Classification of optimization problems.

- *Case C:* Deterministic game theory with two decision makers, discussed in Section D.4.
- *Case D:* Stochastic optimization, discussed in Section D.5.

These cases are used in Part C of the book to derive the optimal parameters for a variety of maintenance decision policies. We use the following notation:

K	Number of decision makers given by the set $\{k : 1 \leq k \leq K\}$
$J^k(\cdot)$	Objective function for decision maker $k : 1 \leq k \leq K$ with range $\mathbb{R}^m$ if $J^k(\cdot)$ is an m-dimensional vector
$x, y, \ldots$	Decision variables of different decision makers (scalar or vector)
S_k	The set of decision variables for decision maker $k : 1 \leq k \leq K$ (also the domain for $J^k(\cdot)$)[1]

---
[1] $J^k(\cdot)$ can be viewed as a mapping from S_k to $\mathbb{R}^m$.

D.2   Case A

$K = 1$ and the problem is to select the decision variable(s) x to maximize a scalar objective function $J(x)$ $[\equiv J^1(x)]$.[2] Note that $m = 1$ (as the objective function is scalar) and S $[\equiv S_1] \in \mathbb{R}^n$. $n = 1$ if x is scalar and $n > 1$ if x is a vector.

D.2.1   Preliminaries

A set of points S is said to be *convex* if, for any points $x_1, x_2 \in S$, all points on the line segment joining x_1, x_2 also belong to S.

A function $J(x)$ is said to be *convex* if, for any two points x_1, x_2, the function is below the line segment joining the two points $(x_1, J(x_1))$ and $(x_2, J(x_2))$.[3]

When $n = 1$, the first and second derivatives of $J(x)$ are given by $dJ(x)/dx$ and $d^2J(x)/dx^2$ respectively. When $n > 1$, the first derivative is given by the column vector of the partial derivatives of J (also called the *gradient* of $J(x)$), given by:

$$\nabla J(x) = \begin{bmatrix} \dfrac{\partial J(x)}{\partial x_1} \\ \vdots \\ \dfrac{\partial J(x)}{\partial x_n} \end{bmatrix},$$

and the second derivative is a matrix (also called the *Hessian* of $J(x)$) and is given by:

$$H(J(x)) = \begin{bmatrix} \dfrac{\partial^2 J(x)}{\partial x_1^2} & \cdots & \dfrac{\partial^2 J(x)}{\partial x_1 \partial x_n} \\ \vdots & & \vdots \\ \dfrac{\partial^2 J(x)}{\partial x_n \partial x_1} & \cdots & \dfrac{\partial^2 J(x)}{\partial x_n^2} \end{bmatrix}.$$

The gradient and the Hessian play an important role in the characterization of the optimal solutions and in designing algorithms to find the optimal solution using computational methods.

The following result from matrix theory is useful in the context of optimization. A matrix A is said to be positive [negative] semi-definite if $x^T A x \geq 0$ $[x^T A x \leq 0]$ for all $x \in \mathbb{R}^n$. The matrix A is said to be positive [negative] definite if $x^T A x > 0$ $[x^T A x < 0]$ for all $x \in \mathbb{R}^n$. If $J(x)$ is a convex [concave] function, then its Hessian $H(J(x))$ is a positive [negative] semi-definite matrix for all $x \in S$.

---

[2] If the problem is to find a local minimum for a function $J(x)$, then it is equivalent to finding the local maximum for the function $\{-J(x)\}$. Hence, without loss of generality, we will confine our attention to finding the local maximum for a function.

[3] An interesting property of a convex function is that it has a unique global minimum.

There are several methods to check whether a matrix A is positive semi-definite or not, two of these are listed below.

1. All eigenvalues of A are non-negative.[4]
2. *Sylvester's Rule:* The determinants of A and all its principal submatrices are non-negative.

D.2.2 Optimality Conditions for Scalar x

D.2.2.1 Local versus Global Optima

A *local maximum* [minimum] is a point x^* such that $J(x^*) \geq J(x)$ $[J(x^*) \leq J(x)]$ for all x in some open region (or interval if $n = 1$) containing x^*. A *global maximum* [minimum] is a point x^* such that $J(x^*) \geq J(x)$ $[J(x^*) \leq J(x)]$ for all $x \in S$.

D.2.2.2 Unconstrained Optimization

We have two cases depending on whether x is continuous- or discrete-valued. When x is continuous-valued ($x \in \mathbb{R}^1$) then x^* must satisfy the first-order necessary condition given by:

$$J'\left(x^*\right) = \frac{dJ\left(x\right)}{dx}\bigg|_{x=x^*} = 0 \tag{D.1}$$

as well as the condition:

$$J''\left(x^*\right) = \frac{d^2J\left(x\right)}{dx^2}\bigg|_{x=x^*} \leq 0 \tag{D.2}$$

A sufficient condition for x^* to yield a local maximum is given by Equation (D.1) and:

$$J''\left(x^*\right) = \frac{d^2J\left(x\right)}{dx^2}\bigg|_{x=x^*} < 0 \tag{D.3}$$

In general, it is necessary to use a computational method to obtain the solution(s) to Equation (D.1) and then check Equation (D.2) to determine which yields a local (or global) maximum. Many commercial software packages are readily available for this purpose.

When x takes on only discrete values x^i, $i = 1, 2, \ldots$ (for example, the set of positive integers), then $x^* = x^m$ is a local maximum if:

$$J\left(x^m\right) > J\left(x^{m\pm1}\right) \tag{D.4}$$

---

[4] The eigenvalues are obtained by finding the roots of the equation $|A - I\lambda| = 0$ where $|\ |$ is the determinant and the equation is a polynomial equation of order n giving the n − eigenvalues $\{\lambda_1, \lambda_2, \ldots, \lambda_n\}$.

D.2.2.3 Constrained Optimization

Often, x is constrained (for example, $a \leq x \leq b$). In this case, the optimal x that maximizes the objective function must not violate the constraint. As a result, in some instances, it might be one of the end points of the interval and the first-order condition does not hold.

D.2.3 Optimality Conditions for Vector x

D.2.3.1 Unconstrained Optimization

As with the scalar case discussed earlier, we have two cases depending on whether x is continuous- or discrete-valued. When x is continuous-valued ($x \in \mathbb{R}^n$), then x^* must satisfy the first-order necessary condition given by:

$$\nabla J\left(x^*\right) = \nabla J\left(x\right)\big|_{x=x^*} = 0 \tag{D.5}$$

as well as the condition that the Hessian $H(J(x^*)) = H(J(x))\big|_{x=x^*}$ is a negative semi-definite matrix. A sufficient condition for x^* to yield a local maximum is given by Equation (D.5) and the condition that the Hessian matrix $H(J(x^*))$ is negative definite.

We conclude this section with the following remarks that highlight the importance of concave functions.

- If the objective function is (strictly[5]) concave, then a stationary point is a (unique) global maximum.
- At a local maximum, there is no local information that helps to determine whether it is global. This makes global optimization difficult, in general.

D.2.3.2 Constrained Optimization

Optimality conditions for constrained optimization problems are more involved, since they also have to take into account constraint information. The conditions may be summarized briefly as follows: A feasible point ($x \in S$) is a local maximum if it satisfies some constraint qualifications and, at this point, there is no direction which is both feasible and improving. The reader is referred to Bazaraa *et al.* (2006) for details.

D.3 Case B

D.3.1 Pareto Optimality

We confine our attention to the case where the objective function is a two-dimensional vector given by $J^T(x) = \left[J_1(x), J_2(x)\right]$ and x is a scalar continuous variable. We assume that the objective is to maximize both elements. Note that in this case, the value of x that maximizes $J_1(x)$ will, in general, not be the same as that which maximizes $J_2(x)$. This is best illustrated by plotting $J_1(x)$ and $J_2(x)$ on a two-dimensional plane with x varying. Figure D.2 is an illustrative plot where the Pareto optimal solution is given by the interval $a \leq x \leq b$, that is a set of values

---

[5] A function is strictly concave if the condition in Equation (D.2) holds as a strict inequality.

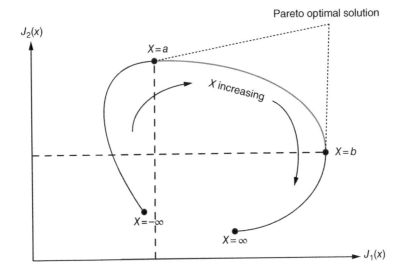

Figure D.2   Pareto optimality.

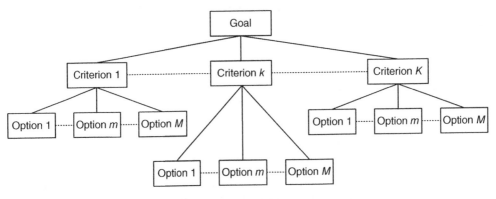

Figure D.3   The AHP process.

of x. All values of x belonging to this set are superior to other values of x, in the sense that one can obtain a higher value for one or both elements of $J(x)$ by moving to the Pareto optimal set. Within the Pareto optimal set, there is a trade-off, in the sense that an increase in one element is obtained at the expense of a reduction in the other.

D.3.2   *Analytic Hierarchy Process (AHP)*[6]

The AHP (analytic hierarchy process) is a structured method of modeling the decision at hand. It consists of an overall *goal*, a group of options or *alternatives* for reaching the goal, and a group of factors or *criteria* that relate the alternatives to the goal, as indicated in Figure D.3 where there

---

[6] The AHP methodology for decision making was conceived and developed by Thomas Saaty, and his book on the topic (Saaty, 1980) is a classic. The literature on the AHP and its application is vast and overviews of the applications can be found in Vargas (1990) and Vaidya and Kumar (2006).

are M options and K different criteria. The criteria can be further broken down into subcriteria, subsubcriteria, and so on, to produce as many levels as the decision problem requires.

The first step is to decompose the decision problem into a hierarchy of subproblems, each of which can be analyzed independently. The elements of the hierarchy can relate to any aspect of the decision problem – tangible or intangible, carefully measured or roughly estimated, well- or poorly understood – anything at all that applies to the decision at hand.

Once the hierarchy is built, one evaluates the various elements by comparing them to one another two at a time, with respect to their impact on an element above them in the hierarchy. In making the comparisons, one uses data about the elements as well as subjective judgments about the elements' relative meaning and importance. The essence of the AHP is that it combines human judgments with the underlying information in performing the evaluations.

The AHP converts these evaluations to numerical values that can be processed and compared over the entire range of the problem. A numerical weight (or priority) is derived for each element of the hierarchy, allowing diverse and often incommensurable elements to be compared with one another in a rational and consistent way. This capability distinguishes the AHP from other decision-making techniques.

In the final step of the process, numerical priorities are calculated for each of the decision alternatives. These numbers represent the alternatives' relative ability to achieve the decision goal, so they allow a straightforward consideration of the various courses of action.

D.4 Case C

A game consists of several elements: The *players* (the decision makers who participate in the game), their decision choices, and the objective functions (which are functions of the outcome of interactions) affect the outcome of the game. The optimal decision is one which achieves the goals of both players. Figure D.4 shows this for the case of two players.

In any game, an action is the decision that a player makes at a particular point in the game, whereas a strategy specifies what actions the player will take at each point in the game. A solution concept is a technique that is used to predict the outcome (equilibrium) of the game. It identifies the strategies that the players are actually likely to play in the game.

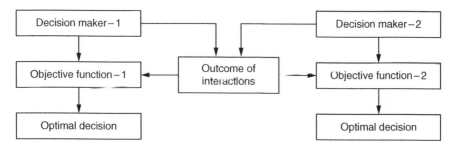

Figure D.4   Decision problem structure for two decision makers.

D.4.1   Key Issues

There are several issues that play a critical role in the decision-making process. The most important one involves the power structure, as outlined in the following subsection.

D.4.1.1   Power Structure

The two possible types of power structure between the two players (denoted by P_1 and P_2) are: (i) *dominance,* which is indicated by $P_1 \rightarrow P_2$ and (ii) *no dominance* (equal power), indicated by $P_1 \leftrightarrow P_2$. In the former case, the dominant player P_1's decisions are known to, and influence the decisions made by, the dominated player P_2. P_1 is known as the *leader* and P_2 the *follower* in this type of power structure. In the latter case, the two players are assumed to make their decisions simultaneously or at least be unaware of each other's decisions.

D.4.2   Classification of Games

Game theory problems may be classified in a number of different ways. The timing of actions by the players and also the number of periods during which games are played lead to different solution approaches. In some games, the players may choose their actions simultaneously, so that no player knows exactly what the others have done when they make a decision. Alternatively, in games with sequential timing, the players choose their actions in predetermined order. These two situations are termed *Nash games* and *Stackelberg games*, respectively and are discussed later in the section.

Some games take place during a single time period whereas others occur over multiple time periods, and the actions taken by the players in each period affect the actions and rewards of the players in subsequent periods. These two situations are termed *static games* and *dynamic games*, respectively.

To describe a game it is also important to specify the information available to each player. In a game with complete information, all elements of the structure of the game are known to all players, whereas in games with incomplete information, some players may have private information. In a game with perfect information, all the players know exactly what has happened in the game prior to choosing an action. Imperfect information implies that at least one of the players is unaware of the full history of the game. *Another important assumption of GT is that the players will always act rationally (choose their best decision).*

Finally, games may be either *cooperative* or *non-cooperative.* In a cooperative game, the players communicate with each other to coordinate their strategies and, most importantly, make binding agreements. This type of game can be formulated as a multi-objective optimization problem. In a non-cooperative game, the players may communicate, but binding agreements are not made.

D.4.2.1   Two-Person Games

The two-person non-cooperative game is the simplest game formulation appropriate for deciding on optimal decisions in the context of maintenance outsourcing. We denote the two players P_1 and P_2. The actions available to player P_1 are denoted $x \in X$ and for player P_2 they

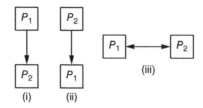

Figure D.5 Different power structures in a two-person game.

are $y \in Y$. Let $J_1(x, y)$ and $J_2(x, y)$ be the objective functions for players P_1 and P_2, respectively.

The different power structures in a two-person game are shown in Figure D.5.

D.4.2.2 Nash Game

Here, each player selects a single action without knowing the particular action chosen by their rival. This effectively means that the two players, P_1 and P_2, choose their actions simultaneously and so have equal decision-making power. This power configuration corresponds to case (iii) in Figure D.5. In such a game, the players' strategies are just the single actions they choose, so the terms "actions" and "strategies" will be used interchangeably. The most well-known and widely-used solution concept for this static game is called *Nash equilibrium* (NE). An NE is a set of strategies (strategy profile) for the two players such that no player has an incentive to change their strategy unilaterally, given the strategy chosen by the other player. More formally, the strategy profile $(x*, y*)$ is an NE if:

$$
\begin{aligned}
J_1\left(x*, y*\right) &\geq J_1\left(x, y*\right), \text{ for all } x \in X, \text{ and} \\
J_2\left(x*, y*\right) &\geq J_1\left(x*, y\right), \text{ for all } y \in Y.
\end{aligned}
\tag{D.6}
$$

D.4.2.3 Stackelberg Game

We assume that P_1 is the leader who chooses an action $x \in X$ and then P_2 (follower) observes x and chooses an action $y \in Y$. This corresponds to case (i) of Figure D.5. (In case (ii), P_2 is the leader and P_1 is the follower.) The *backward induction* method of solution for the two-stage *Stackelberg* game in case (i) is as follows:

- *Stage 2:* Given the action x previously chosen by P_1, P_2's problem is to find the value of y that maximizes $J_2(x, y)$. The solution to this problem is the best response function:

$$
BR_2\left(x\right) = \arg\max_{y \in Y} J_2\left(x, y\right)
\tag{D.7}
$$

Thus, P_2 responds optimally to P_1's action.

- **Stage 1:** P_1 anticipates what P_2 will do in stage 2, so P_1's problem in this part of the game is to:

$$\max_{x \in X} J_1 \left(x, BR_2 \left(x \right) \right) \qquad (D.8)$$

If x^* is the optimal solution to the above optimization problem, then the outcome of the game is that P_1 chooses x^* and P_2 chooses $BR_2(x^*)$.

D.5   Case D

When uncertainty is significant it affects the outcome of any decision made. Since the exact outcome is unknown before the decision is made, the optimal decision must take into account the different possible outcomes resulting from the uncertainty. We illustrate how this has been tackled by taking a simple scenario to introduce some new concepts that are important in the context of stochastic optimization.

Consider a single decision (player) maker with an initial wealth S_0 that he/she can invest in a gamble where the uncertain outcome (denoted by S) is determined by the toss of a coin. If the player decides to gamble, the wealth increases to S_m if the toss results in a Head, or decreases to S_l if a Tail is obtained, with $0 \le S_l < S_0 < S_m$. The probability that the toss results in a Head [Tail] is p [$1 - p$]. For a fair coin, $p = 0.5$. Note that the decision variable x is discrete-valued (1 to denote gamble and 0 not gamble).

Utility theory provides the basis for the player's optimal decision (gamble or not). The utility that the player derives from a wealth S is given by a function $u(S)$ and this captures the attitude of the player to the risk associated with the uncertain outcome. Utility theory postulates that a rational player will choose the optimal decision to maximize his/her expected utility. The expected utility if the player decides to gamble is given by:

$$E\left[u\left(\tilde{S} \right) \right] = p\left[u\left(S_m \right) \right] + \left(1 - p \right)\left[u\left(S_l \right) \right] \qquad (D.9)$$

If the player does not gamble, the utility derived is $u(S_0)$. Hence, the optimal decision is to gamble if $E\left[u(\tilde{S}) \right] > u(S_0)$ and not gamble if $E\left[u(\tilde{S}) \right] < u(S_0)$.

The form of the utility function plays a critical role. Figure D.6 shows three different utility functions (A–C). Based on this, we have the following results:

- If the utility function is A, then the optimal decision is to gamble (in other words, the player is *risk-seeking*).
- If the utility function is C, then the player is indifferent (in other words, *risk neutral*).
- If the utility function is B, then the optimal decision is not to gamble (in other words, the player is *risk averse*).

Most senior decision makers in a business are either risk neutral or risk averse. A form of utility function that is used extensively is given by:

$$u\left(S \right) = \frac{1 - e^{-\gamma S}}{\gamma}, \ \gamma \ge 0. \qquad (D.10)$$

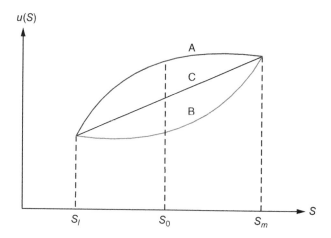

Figure D.6 Risk-seeking, risk neutral, and risk-averse utility functions.

γ is called the *risk parameter*. $\gamma = 0$ corresponds to the decision maker being risk neutral and, in this case, $u(s) = S$ (a linear function). $\gamma > 0$ implies risk aversion, with higher values indicating higher risk aversion. Two special cases of this are:

1. The objective function is the expected utility given by $J(x) = E\left[u(\tilde{S})\right]$ (implying that the decision maker is risk neutral).
2. The objective function is given by $J(x) = E\left[u(\tilde{S})\right] - \left(\gamma/2\right)E\left[u(\tilde{S})^2\right]$ (implying that the decision maker is risk averse).

References

Bazaraa, M.S., Sherali, H.D., and Shetty, C.M. (2006) *Nonlinear Programming: Theory and Algorithms*, 3rd edition. John Wiley & Sons, Inc., New York.

Saaty, T.L. (1980) *The Analytic Hierarchy Process*, McGraw-Hill, New York.

Vaidya, O.S. and Kumar, S. (2006) Analytic hierarchy process: An overview of applications, *European Journal of Operational Research* **169**: 1–29.

Vargas, L.G. (1990) An overview of the analytic hierarchy process and its applications. *European Journal of Operational Research* **48**: 2–8.

Appendix E: Data Sets

In this appendix we present some real data sets that will be used in various chapters of the book.

Data Set E.1 Battery (Component of a Bus)

Incomplete failure data on a sample of 54 batteries used in buses are given in Table E.1. The data include failure times for 39 items that failed under warranty and service times for 15 items that had not failed at the time of observation.

Data Set E.2 Automobile (Repair Costs)

A large car rental agency has its own workshop for maintenance. Table E.2 shows the costs of fixing 32 engine problems. The data include the odometer readings at failure and the costs of repair.

Data Set E.3 Photocopier

Data were collected on failure times (expressed in days and in counts of numbers of copies) and type of failure of an office copier. The data were recorded on 98 service calls over a period of about 4½ years. Failures of the copier were due to failure of 17 different components. The original data are given in Table E.3.

Introduction to Maintenance Engineering: Modeling, Optimization, and Management, First Edition.
Mohammed Ben-Daya, Uday Kumar, and D.N. Prabhakar Murthy.
© 2016 John Wiley & Sons, Ltd. Published 2016 by John Wiley & Sons, Ltd.

Table E.1 Battery failure and censored data.

Time to failure			Service time
64	599	852	131
66	619	929	162
164	631	948	163
178	639	973	202
185	645	977	232
299	656	1084	245
319	681	1100	286
383	722	1100	302
385	727	1350	315
405	738		337
482	761		845
492	765		983
506	788		1259
548	801		1384
589	848		1421

Table E.2 Automobile engine failure data.

Engine #	km at failure (10^3)	Cost of repair ($)
1	13.1	24.60
2	29.2	5037.28
3	13.2	253.50
4	10.0	26.35
5	21.4	1712.47
6	14.5	127.20
7	12.6	918.53
8	27.4	34.68
9	35.5	1007.27
10	15.1	658.36
11	17.0	42.96
12	27.8	77.22
13	2.4	77.57
14	38.6	831.61
15	17.5	432.89
16	14.0	60.35
17	15.3	48.05
18	19.2	388.30
19	4.4	149.36
20	19.0	7.75
21	32.4	29.91
22	23.7	27.58
23	16.8	1101.90
24	2.3	27.78
25	26.7	1638.73
26	5.3	11.70
27	29.0	98.90
28	10.1	77.24
29	18.0	42.71

(*Continued*)

Table E.2 (*Continued*)

Engine #	km at failure (10^3)	Cost of repair ($)
30	4.5	1546.75
31	18.7	556.93
32	31.6	78.42
32	31.6	78.42

Table E.3 Photocopier failure data.

Count	Day	Component	Count	Day	Component
60152	29	Cleaning Web	769384	1165	Feed Rollers
60152	29	Toner Filter	769384	1165	Upper Fuser Roller
60152	29	Feed Rollers	769384	1165	Optics PS Felt
132079	128	Cleaning Web	787106	1217	Cleaning Blade
132079	128	Drum Cleaning Blade	787106	1217	Drum Claws
132079	128	Toner Guide	787106	1217	Toner Guide
220832	227	Toner Filter	840494	1266	Feed Rollers
220832	227	Cleaning Blade	840494	1266	Ozone Filter
220832	227	Dust Filter	851657	1281	Cleaning Blade
220832	227	Drum Claws	851657	1281	Toner Guide
252491	276	Drum Cleaning Blade	872523	1312	Drum Claws
252491	276	Cleaning Blade	872523	1312	Drum
252491	276	Drum	900362	1356	Cleaning Web
252491	276	Toner Guide	900362	1356	Upper Fuser Roller
365075	397	Cleaning Web	900362	1356	Upper Roller Claws
365075	397	Toner Filter	933637	1410	Feed Rollers
365075	397	Drum Claws	933637	1410	Dust Filter
365075	397	Ozone Filter	933637	1410	Ozone Filter
370070	468	Feed Rollers	933785	1412	Cleaning Web
378223	492	Drum	936597	1436	Drive Gear D
390459	516	Upper Fuser Roller	938100	1448	Cleaning Web
427056	563	Cleaning Web	944235	1460	Dust Filter
427056	563	Upper Fuser Roller	944235	1460	Ozone Filter
449928	609	Toner Filter	984244	1493	Feed Rollers
449928	609	Feed Rollers	984244	1493	Charging Wire
449928	609	Upper Roller Claws	994597	1514	Cleaning Web
472320	677	Feed Rollers	994597	1514	Ozone Filter
472320	677	Cleaning Blade	994597	1514	Optics PS Felt
472320	677	Upper Roller Claws	1005842	1551	Upper Fuser Roller
501550	722	Cleaning Web	1005842	1551	Upper Roller Claws
501550	722	Dust Filter	1005842	1551	Lower Roller
501550	722	Drum	1014550	1560	Feed Rollers
501550	722	Toner Guide	1014550	1560	Drive Gear D
533634	810	TS Block Front	1045893	1583	Cleaning Web
533634	810	Charging Wire	1045893	1583	Toner Guide
583981	853	Cleaning Blade	1057844	1597	Cleaning Blade
597739	916	Cleaning Web	1057844	1597	Drum
597739	916	Drum Claws	1057844	1597	Charging Wire
597739	916	Drum	1068124	1609	Cleaning Web

(*Continued*)

Table E.3   (*Continued*)

Count	Day	Component	Count	Day	Component
597 739	916	Toner Guide	1 068 124	1609	Toner Filter
624 578	956	Charging Wire	1 068 124	1609	Ozone Filter
660 958	996	Lower Roller	1 072 760	1625	Feed Rollers
675 841	1016	Cleaning Web	1 072 760	1625	Dust Filter
675 841	1016	Feed Rollers	1 072 760	1625	Ozone Filter
684 186	1074	Toner Filter	1 077 537	1640	Cleaning Web
684 186	1074	Ozone Filter	1 077 537	1640	Optics PS Felt
716 636	1111	Cleaning Web	1 077 537	1640	Charging Wire
716 636	1111	Dust Filter	1 099 369	1650	TS Block Front
716 636	1111	Upper Roller Claws	1 099 369	1650	Charging Wire

From Bulmer and Eccleston, 2003.

Table E.4   Throttle failure and censored (denoted by +) data.

0.478	0.959	1.847+	3.904	6.711+
0.484+	1.071+	2.400	4.443+	6.835+
0.583	1.318+	2.550+	4.829	6.947+
0.626+	1.377	2.568+	5.328	7.878+
0.753	1.472+	2.639	5.562	7.884+
0.753	1.534	2.944	5.900+	10.263+
0.801	1.579+	2.981	6.122	11.019
0.834	1.610+	3.392	6.226+	12.986
0.850+	1.729+	3.393	6.331	13.103+
0.944	1.792+	3.791+	6.531	23.245+

Data Set E.4 Throttle Valve (Automobile Component)

The "throttle" is a component of an automobile. The data presented in Table E.4 are from Carter (1986) and give the distance traveled (in thousands of kilometers) before failure, or the item being suspended before failure, for a pre-production, general-purpose, load-carrying vehicle. Thus, the independent variable is the distance traveled, with 1000 km being the unit of measurement.

Data Set E.5 Valve Seat Replacement for Diesel Engines

Table E.5 shows engine age (in days) at the time of a valve seat replacement for a fleet of 41 diesel engines. These data on a sample of systems appeared in Nelson and Doganaksoy (1989) and Nelson (1995) and also in Meeker and Escobar (1998, p. 635).

Data Set E.6 Heavy Vehicle

Table E.6 shows the time to repair and downtime for 69 failures of a heavy vehicle.

Data Set E.7 Buses

Table E.7 shows the weekly repair costs for a fleet of 22 buses in a city in China for 175 weeks.

Table E.5 Diesel engine age at time of replacement of valve seats.

System ID	Days observed	Engine age at replacement time			
251	761	—	—	—	—
252	759	—	—	—	—
327	667	98	—	—	—
328	667	326	653	653	—
329	665	—	—	—	—
330	667	84	—	—	—
331	663	87	—	—	—
389	653	646	—	—	—
390	653	92	—	—	—
391	651	—	—	—	—
392	650	258	328	377	621
393	648	61	539	—	—
394	644	254	276	298	640
395	642	76	538	—	—
396	641	635	—	—	—
397	649	349	404	561	—
398	631	—	—	—	—
399	596	—	—	—	—
400	614	120	479	—	—
401	582	323	449	—	—
402	589	139	139	—	—
403	593	—	—	—	—
404	589	573	—	—	—
405	606	165	408	604	—
406	594	249	—	—	—
407	613	344	497	—	—
408	595	265	586	—	—
409	389	166	206	348	—
410	601	—	—	—	—
411	601	410	581	—	—
412	611	—	—	—	—
413	608	—	—	—	—
414	587	—	—	—	—
415	603	367	—	—	—
416	585	202	563	570	—
417	587	—	—	—	—
418	578	—	—	—	—
419	578	—	—	—	—
420	586	—	—	—	—
421	585	—	—	—	—
422	582	—	—	—	—

Data Set E.8 Buses

The data in Table E.8 come from the maintenance records of a fleet of 22 buses, all of the same model, that started to operate at the same time (24 August 2005) on the same route. The records were available from 1 September 2006 to 15 October 2009. Repair time is in minutes and cost is in Yuan (RMB).

Table E.6 Repair and downtimes for a heavy vehicle.

Time to repair	Downtime	Time to repair	Downtime	Time to repair	Downtime	Time to repair	Downtime
3.5	4.5	1	1.5	3	8	1	1
5.5	6	1.5	2	1.5	2	3.5	4
2	2	1	1	4	5	1	1
4	5	3	3	2	2	0.5	0.5
1.5	1.5	1.5	1.5	1.5	2	0.5	0.5
2	—	0.5	0.5	10	96	2	3
0.5	0.5	1.5	1.5	2.5	2.5	1	1
4	5	2	2.5	1	2.5	1	1
4	5	0.5	0.5	1.5	1.5	1	1
6	6	10	16	1.5	1.5	0.5	0.5
1	1	2.5	2.5	1	2	7	8
2	2.5	3.5	3.5	6.5	7	1.5	1.5
1.5	2.5	1.5	1.5	2	2	2	2
1	1	2	5.5	2	2	0.5	0.5
1.5	1.5	1	1	8.5	8.5	5	6
1	1	2	2	10	11	—	—
6	6	1	1.5	1	1	—	—
0.5	0.5	2	3.5	8	9	—	—

Table E.7 Weekly repair costs for a fleet of 22 buses.

Week	Repair cost	Week	Repair cost	Week	Repair cost
1	150.58	60	160.33	119	109.43
2	17.08	61	236.46	120	276.86
3	268.52	62	96.46	121	253.02
4	182.89	63	25.65	122	430.61
5	215.23	64	35.95	123	229.92
6	56.16	65	30.59	124	192.3
7	277.8	66	286.65	125	48.01
8	16.21	67	292.52	126	287.24
9	260.89	68	72.43	127	34.78
10	470.52	69	333.02	128	210.64
11	396.81	70	252.53	129	70.25
12	250.27	71	101.77	130	83.79
13	200.29	72	318.1	131	235.75
14	86.38	73	375.86	132	176.88
15	139.13	74	476.92	133	168.36
16	161.13	75	191.07	134	131.2
17	19.47	76	31.89	135	195.37
18	269.36	77	141.35	136	110.8
19	24.13	78	198.07	137	245.42
20	237.85	79	441.1	138	121.8
21	109.53	80	221.85	139	415.59

Table E.7   (*Continued*)

Week	Repair cost	Week	Repair cost	Week	Repair cost
22	568.25	81	391.17	140	205.42
23	105.18	82	205.63	141	257.65
24	130.33	83	63.94	142	247.54
25	100.52	84	119.19	143	185.04
26	21.45	85	399.65	144	183.73
27	59.03	86	226.48	145	113.56
28	49.9	87	208.32	146	214.27
29	79.89	88	40.9	147	112.68
30	16.6	89	80.15	148	219.26
31	201.8	90	141.3	149	411.32
32	76.26	91	329.27	150	357.17
33	1119.37	92	290.73	151	271.44
34	533.67	93	235.37	152	185.98
35	448.57	94	111.76	153	219.4
36	588.99	95	138.69	154	151.51
37	558.18	96	181.11	155	48.34
38	186.39	97	224.07	156	238.67
39	1070.28	98	191.81	157	325.93
40	879.65	99	538.62	158	127
41	383.2	100	182.46	159	599.81
42	294.02	101	306.77	160	147.15
43	482.85	102	129.66	161	261.71
44	191.3	103	255.48	162	151.32
45	80.44	104	103.39	163	370.56
46	499.02	105	354.24	164	120.97
47	169.44	106	98.6	165	382.43
48	270.59	107	205.7	166	150.31
49	160.22	108	325.04	167	123.93
50	123.81	109	138.52	168	322.24
51	786.57	110	200.16	169	294.69
52	141.5	111	222.7	170	332.16
53	220.49	112	665.78	171	105.93
54	86.8	113	276.84	172	102.37
55	99.68	114	181.32	173	120.98
56	129.38	115	71.14	174	313.57
57	304.76	116	80.13	175	164.7
58	38.79	117	124	—	—
59	481.41	118	250.36	—	—

Data Set E.9 Hydraulic Pumps

Table E.9 shows the data for hydraulic pumps (102 units) recorded by the maintenance department of a mining company and consists of failure times (Type 1), and service times (Type 0).

Table E.8 Time and cost to fix bus generator failures.

Time	Cost	Time	Cost	Time	Cost	Time	Cost	Time	Cost	Time	Cost
No. 01		No. 02		No. 03		No. 04		No. 05		No. 06	
33	153.4	119	185.13	140	483.08	39	552.28	33	185.13	42	202.08
32	160.08	69	185.13			144	80.13	32	185.13	53	160.08
49	160.08	53	160.08			79	280.08	62	202.08	40	160.08
41	160.08	78	160.08			49	940.08	39	160.08		
50	160.08	48	160.08					70	160.08		
86	160.08	75	49.58					95	236.08		
32	160.08	64	160.08					71	160.08		
61	160.08	32	160.08					63	133.36		
117	160.08	5	160.08								
53	146.72	162	871.36								
105	884.72										
No. 07		No. 08		No. 09		No. 10		No. 11		No. 12	
60	185.13	54	185.13	58	185.13	29	185.13	50	185.13	35	185.13
59	89.13	52	81.8	84	68.13	70	64.08	74	227.13	23	185.13
129	67.42	111	310.08	80	255.08	41	430.08	47	185.13	48	140.04
62	214.08	67	160.08	89	44.48	62	201.18	44	185.13	46	43.08
75	160.08	39	160.08	36	56.08	28	121.72	46	160.08	46	160.08
146	205.88	114	160.08	67	140.04			50	160.08	47	160.08
		80	74.08	32	145.05			30	483.08	33	160.08
		34	160.08	35	160.08			49	160.08	60	148.62
		96	100.6	49	62.08			49	160.08	42	148.62
		66	160.08					36	194.08		
		50	202.08					149	165.09		
								63	160.08		
								226	120.1		
								100	701.72		
No. 13		No. 14		No. 15		No. 16		No. 17		No. 18	
38	82.08	48	68.13	50	878.04	50	185.13	43	41.28	117	185.48
32	160.08	165	185.13	33	70.08	61	94.08	32	160.08	48	185.13
81	125.01			59	74.08	32	53.48	83	160.08	48	119.13
110	160.08			31	74.08	38	160.08	28	45.28	60	160.38
				35	878.04	20	161.98	28	423.11	113	160.08
								65	160.08	55	160.08
								37	160.08	18	82.08
								25	498.72	25	143.38
								60	158.38		
								25	60.72		
								26	60.72		
No. 19		No. 20		No. 21		No. 22					
61	185.13	27	57.4	16	315.49	72	168.13				
155	188.83	44	160.08	35	185.13	83	893.07				
40	74.63	115	156.08	63	160.08	60	73.53				
98	185.13			34	160.08	71	185.13				
25	160.08			82	80.08	65	160.08				

(Continued)

Table E.8 (*Continued*)

Time	Cost	Time	Cost	Time	Cost	Time	Cost	Time	Cost	Time	Cost
49	74.08			35	163.08	45	61.08				
30	160.08			34	160.08	41	878.04				
23	160.08			49	190.08	48	160.08				
169	285.54			83	202.08	40	29.54				
189	965.78			149	96.72	37	22.64				
49	74.08			62	146.72	286	186.08				
136	160.08					58	160.08				
						118	160.08				
						50	140.04				
						64	140.04				

Table E.9 Pump replacement data.

t_i (h)	Type	t_i (h)	Type	t_i (h)	Type
81	1	5 923	0	11 923	1
149	0	6 333	0	12 005	1
245	0	6 717	0	12 082	1
340	0	7 207	0	12 090	1
407	0	7 265	0	12 136	1
461	0	7 624	0	12 141	1
629	0	7 625	1	12 143	1
856	1	7 973	0	12 163	1
947	1	8 183	0	12 198	1
1 460	0	8 217	0	12 198	1
1 513	0	8 390	0	12 198	1
1 670	0	8 462	0	12 198	1
1 688	1	8 728	0	12 198	1
2 093	1	8 817	0	12 198	1
2 242	1	8 870	0	12 236	1
2 242	1	8 884	1	12 236	1
2 242	1	9 055	0	12 236	1
2 242	1	9 182	0	12 236	1
2 242	1	9 334	0	12 236	1
2 607	0	9 368	0	12 236	1
2 668	0	9 729	0	12 394	1
2 806	0	9 751	1	12 459	1
3 132	1	10 299	0	13 097	1
3 132	1	10 389	1	13 497	1
3 132	1	10 413	1	13 497	1
3 132	1	10 557	0	13 497	1
3 333	0	10 944	0	13 497	1
3 569	0	10 970	0	13 497	1
3 837	1	11 647	1	13 497	1
3 837	1	11 678	0	13 497	1
4 150	1	11 686	0	14 407	0
5 123	0	11 798	1	15 536	0
5 258	0	11 869	1	16 289	0
5 662	1	11 869	1	17 517	0

Table E.10 Shock absorber data.

	km	Failure mode	FC indicator		km	Failure mode	FC indicator
1	6700	M1	1	20	17 520	M1	1
2	6950	Censored	0	21	17 540	Censored	0
3	7820	Censored	0	22	17 890	Censored	0
4	8790	Censored	0	23	18 450	Censored	0
5	9120	M2	1	24	18 960	Censored	0
6	9660	Censored	0	25	18 980	Censored	0
7	9820	Censored	0	26	19 410	Censored	0
8	11 310	Censored	0	27	20 100	M2	1
9	11 690	Censored	0	28	20 100	Censored	0
10	11 850	Censored	0	29	20 150	Censored	0
11	11 880	Censored	0	30	20 320	Censored	0
12	12 140	Censored	0	31	20 900	M2	1
13	12 200	M1	1	32	22 700	M1	1
14	12 870	Censored	0	33	23 490	Censored	0
15	13 150	M2	1	34	26 510	M1	1
16	13 330	Censored	0	35	27 410	Censored	0
17	13 470	Censored	0	36	27 490	M1	1
18	14 040	Censored	0	37	27 890	Censored	0
19	14 300	M1	1	38	28 100	Censored	0

Data Set E.10 Shock Absorber

Table E.10 shows the data for a shock absorber. The data consist of failure data (Type 1) and censored data (Type 0).

References

Bulmer, M. and Eccleston, J. (2003) Photocopier reliability modelling using evolutionary algorithms. In *Case Studies in Reliability and Maintenance*, pp. 399–422, W.R. Blischke and D.N.P. Murthy (eds), Wiley-Interscience, New York.

Carter, A.D.S (1986) *Mechanical Reliability and Design*, John Wiley & Sons, Inc., New York.

Meeker, W.Q. and Escobar, L.A. (1998) *Statistical Methods for Reliability Data*, John Wiley & Sons, Inc., New York.

Nelson, W. (1995) Confidence limits for recurrence data – applied to cost or number of repairs, *Technometrics*, **37**: 147–157.

Nelson, W. and Doganaksoy, N. (1989) A Computer Program for an Estimate and Confidence Limits for the Mean Cumulative Function for Cost or Number of Repairs of Repairable Products. TIS report 89CRD239, General Electric Company Research and Development, Schenectady, New York.

Index

Introduction to Maintenance Engineering: Modeling, Optimization, and Management, First Edition.
Mohammed Ben-Daya, Uday Kumar, and D.N. Prabhakar Murthy.
© 2016 John Wiley & Sons, Ltd. Published 2016 by John Wiley & Sons, Ltd.

Printed and bound by CPI Group (UK) Ltd, Croydon, CR0 4YY

27/10/2024

14580307-0001